Mechanisms and Machine Science

Volume 25

Series editor

Marco Ceccarelli, Cassino, Italy

More information about this series at http://www.springer.com/series/8779

Marco Ceccarelli
Eusebio Eduardo Hernández Martinez
Editors

Multibody Mechatronic Systems

Proceedings of the MUSME Conference held in Huatulco, Mexico, October 21–24, 2014

Springer

Editors
Marco Ceccarelli
Laboratory of Robotics and Mechatronics LARM
University of Cassino and South Latium
Cassino
Italy

Eusebio Eduardo Hernández Martinez
Section of Graduate Studies
National Polytechnic Institute
Mexico
Distrito Federal
Mexico

ISSN 2211-0984 ISSN 2211-0992 (electronic)
ISBN 978-3-319-09857-9 ISBN 978-3-319-09858-6 (eBook)
DOI 10.1007/978-3-319-09858-6

Library of Congress Control Number: 2014946398

Springer Cham Heidelberg New York Dordrecht London

Printed on acid-free paper

Springer is part of Springer Science+Business Media (www.springer.com)

Preface

The MUSME 2014, IFToMM-FeIbIM–International Symposium on Multibody Systems and Mechatronics is the fifth event of a series that was started in 2002 as a conference activity mainly for promoting these topics in South American community. The first event was held at Universidad Panamericana de la Ciudad de México, Mexico in May 2002, the second was held at Federal University of Uberlandia, Brazil in March 2005, the third was hosted at Universidad Nacional de San Juan, Argentina, in April 2008, and the fourth was celebrated at Universidad Politecnica deValencia, Spain, in October 2011. This year the MUSME event has come to SUNEO (Sistema de Universidades Estatales de Oaxaca) at UMAR Campus Huatulco, Oaxaca, under the auspices of Instituto Politecnico Nacional (IPN) and Universidad Autonoma de Queretaro, Mexico.

The MUSME aim was decided at the funding meeting in 2002 as: a conference stimulating integration between Mechatronics and Multibody Systems Dynamics disciplines; a forum for facilitating contacts among research people and students; and a match conference for communities from IFToMM (International Federation for the Promotion of Mechanism and Machine Science) and FeIbIM (Federación Iberoamericana de Ingeniería Mecánica). In addition, since the beginning it has been considered preferable to have the MUSME Symposium in a location within South America, but without neglecting the possibility to bring MUSME in other parts of the world, since it is supported both by IFToMM and FeIbIM.

The aim of the MUSME Symposium is to bring together researchers, industry professionals, and students from a broad ranges of disciplines referring to Mechatronics and Multibody Systems, in an intimate, collegial, and stimulating environment. Again, in the 2014 MUSME event we received a significant attention to the initiative, as can be seen by the fact that this Proceedings volume contains contributions by authors from all over the world.

The Proceedings volume of the MUSME Symposium is published within the Springer series on MMS (Mechanism and Machine Science) and contains 53 papers that have been selected from 63 submitted papers after peer review for oral presentation. The accepted papers cover several aspects of the wide field of Multibody Systems and Mechatronics. Special attention has been given to organizing student sessions with good works from young researchers, who are still in the formation process.

This is the first time that the Proceedings is published by Springer whereas the previous proceedings were published as CD proceedings that nevertheless are available at the hosting institutions.

We would like to express grateful thanks to the members of the International Scientific Committee for MUSME Symposium for cooperating enthusiastically for the success of the MUSME 2014 event:

Prof. Marco Ceccarelli (Chair), Italy
Prof. Mario Acevedo, Mexico
Prof. Jorge A.C. Ambrósio, Portugal
Prof. Alberto Cardona, Argentina
Prof. Osvaldo H. Penisi, Argentina
Prof. João Carlos M. Carvalho, Brazil
Prof. Javier Cuadrado, Spain
Mario Fernandez Fernandez, Chile
Prof. Manfred Husty, Austria
Prof. Tatu Leinonen, Finland
Prof. Vicente Mata, Spain
Prof. Carlos Munares, Perù
Prof. Pietro Fanghella, Italy

We thank the authors who have contributed with interesting papers in several subjects, covering many fields of Multibody Systems and Mechatronics and, additionally, for their cooperation in revising papers in a short time in agreement with the reviewers' comments. We are grateful to the reviewers for the time and effort they spent in evaluating the papers with a very tight schedule that has permitted the publication of this Proceedings volume in time for the symposium event.

We thank Rector Modesto Seara Vazquez for accepting to host the MUSME 2014 event at SUNEO. We thank our colleagues for their help at the LARM Laboratory of Robotics and Mechatronics of University of Cassino and at the ESIME Ticoman del Instituto Politecnico Nacional. We thank the Director of ESIME Ticoman, Javier Roch Soto. We thank COFAA-IPN for its economic support for some activities of the MUSME event.

We also thank the auspices of IFToMM (International Federation for the Promotion of Mechanism and Machine Science) and FeIbIM (Federación Iberoamericana de Ingeniería Mecánica).

We thank the publisher and Editorial staff of Springer for accepting and helping the publication of this Proceedings volume, since the early step in 2012.

We are grateful to our families since without their patience and comprehension it would not have been possible for us to organize MUSME 2014, IFToMM-FeIbIM–International Symposium on Multibody Systems and Mechatronics and this Proceedings volume.

June 2014

Marco Ceccarelli
Eusebio Eduardo Hernandez Martinez

Contents

Kinematic and Workspace-Based Synthesis of a 2-DOF Mechanism for Haptic Applications

R. Roberts and E. Rodriguez-Leal

Abstract This paper presents the development of a mechanism aimed to haptic applications. The basic design proposed in this work is intended to interact with a finger without the use of a fixture attached to the body. This work investigates the theoretical workspace of a human index finger and proposes a two degree-of-freedom 7-bar linkage mechanism that is synthesized based on such workspace. The paper determines the closed-form solutions to the forward and inverse position, and presents a prototype that is built and tested as a proof of concept of the novel device. The workspace of the constructed mechanism is compared with theoretical models in order to assess their similarity and the viability of accelerometers as position sensing instruments is also tested.

Keywords Haptics · Workspace analysis · Kinematics · Human–machine interaction · Prototype

1 Introduction

The enhancement of the user-experience in virtual environments has been a highly studied topic in recent years [1, 2]. Audiovisual devices are capable to provide engaging interactive experiences to users, and can be classified as one-way or two-way communication systems, e.g. watching TV or playing videogames respectively. One of the most challenging issues in two-way communications is the saturation of the communication channels that result in the loss of information. To overcome this issue, a haptic device could be used as an alternative communication channel. For several decades, haptic devices have been commercially available for

R. Roberts (✉) · E. Rodriguez-Leal
Tecnológico de Monterrey, Monterrey, Mexico
e-mail: re.roberts.phd.mty@itesm.mx

E. Rodriguez-Leal
e-mail: ernesto.rodriguez@itesm.mx

M. Ceccarelli and E.E. Hernández Martinez (eds.), *Multibody Mechatronic Systems*, Mechanisms and Machine Science 25, DOI 10.1007/978-3-319-09858-6_1

different applications, e.g. using vibrotactile actuators in cell phones and pagers [3], or guiding tools that are designed to help soldiers to navigate in battlefields [4]. Moreover, the gaming industry is a niche for haptic devices, enhancing the multimedia experiences of players and providing a new sense of realism [5]. Furthermore, force feedback haptic devices have received a special interest from the medical community including applications such as palpation, needle insertion, laparoscopy, endoscopy, endovascular procedures or arthroscopy [6].

Several of the commercial single-point force feedback devices are designed to be manipulated using an entire hand, displaying three translational degrees-of-freedom (DOF), and can be provided with three additional rotational DOF [6, 7]. Some of the benefits of this type of devices include a workspace in which comfortable movement is allowed, and a mechanism design that is low weight while is capable to provide reliable force feedback. Some of the drawbacks that are encountered in single-point force feedback devices include accuracy limitations in multiple-object simulation, which is essential for object recognition in haptics [8, 9].

This paper describes the development and synthesis of the novel haptic device and is organized as follows: Sect. 2 presents a description of the desired characteristics and features of this mechanism. Section 3 performs a thorough mathematical analysis of the mechanism. A practical approach for conducting the synthesis of the mechanism is discussed in Sect. 4, where the workspace of a human finger is used to determine the dimensional parameters of the mechanism. Section 5 discusses the prototype and validates the theoretical workspace with experimental data. Finally, the paper presents conclusions and suggestions for further work.

2 Mechanism Description

The new haptic mechanism proposed in this paper considers the following features: (i) compact size suitable for finger movement, (ii) non-collapsibility, (iii) low mechanical impedance. A compact apparatus is desirable since a long term goal of this project is to build a multi-point haptic device. Hence, it is important to reproduce this mechanism five times within the workspace of a hand. A non-collapsible mechanism eliminates the inclusion of components that force contact with the finger. This feature is required in cases in which movement flexibility is desired. Finally, reducing inertia and friction in the device is desired to achieve a realistic haptic experience.

Figure 1a shows the proposed seven-link and six-joint mechanism, which can be thought as two four-bar mechanisms that share a common bar. The base is connected to link 2 with links 1 and 3 with the use of joints, while the platform is connected to link 2, with links 4 and 5. Note that all the axes of the joints are parallel to each other and to the Z axis. Consider that links 1, 3, 4 and 5 have a length l while link 2, the base and the platform have a length that is equal for the three elements, although the particular value of this parameter is irrelevant as the following equations will show. The abovementioned dimensional considerations

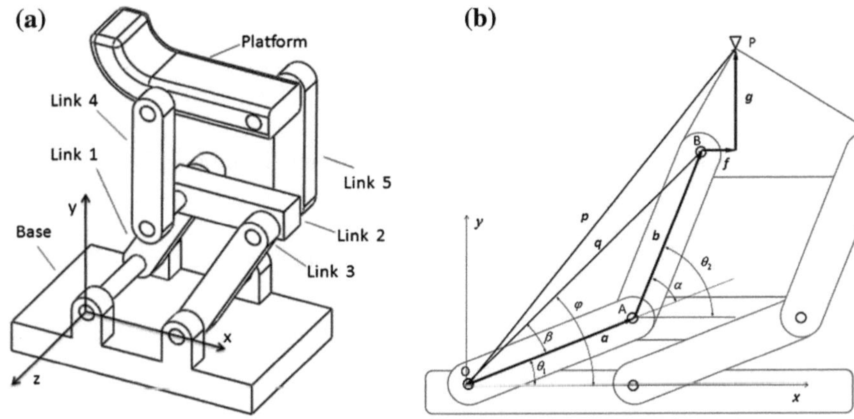

Fig. 1 **a** CAD model and **b** vector representation of the seven bar linkage

guarantee that the mechanisms comply with the Grashof condition [10]; they also assure that the base, platform, and link 2 remain parallel for all positions of the mechanism. Note that the joint connecting links 2 and 5 is active, meaning that a motor applies torque to this kinematic pair; this is also the case for the joint connecting the base and link 1. The following section includes the kinematics analysis that determines the closed-form solutions to the position, velocity and acceleration of the seven-bar mechanism.

3 Mechanism Kinematics

The forward *kinematics* consists in determining the position vector p of an arbitrary point P of the platform. See Fig. 1b for a vector representation of the mechanism. Note that for analysis convenience, all vectors lie on the XY plane and is possible to determine p as follows:

$$\boldsymbol{p} = \left[lc\theta_1 + lc\theta_2 + f, \quad ls\theta_1 + ls\theta_2 + g \right]^T \tag{1}$$

where f and g are the magnitudes of vectors f and g. θ_1 and θ_2 denote the angular position of the active joints connected to links 1, and 4, respectively. Note that links 1 and 3 are parallel, the same is true for links 4 and 5.

The *inverse kinematics* consists in finding the magnitude of the angular positions of the active joints for a given position of the platform. Consider from Fig. 1 that vector q describes the position of point B and can be expressed as:

$$\boldsymbol{q} = \left[lc\theta_1 + lc\theta_2, \quad ls\theta_1 + ls\theta_2 \right]^T \tag{2}$$

where the magnitude q of vector $\boldsymbol{q}$ results in

$$q = |\boldsymbol{q}| = \sqrt{(lc\theta_1 + lc\theta_2)^2 + (ls\theta_1 + ls\theta_2)^2} = \sqrt{(l + lc\alpha)^2 + (ls\alpha)^2} \tag{3}$$

where α is the angular position of link 5, relative to link 1. Solving for α, φ, and β results in

$$\alpha = acos\left(\frac{q^2 - 2l^2}{2l^2}\right), \beta = atan\left(\frac{ls\alpha}{l + lc\alpha}\right), \; and \; \varphi = atan\left(\frac{q_y}{q - f}\right) \tag{4}$$

where q_x and q_y are the X and Y components of q, respectively. The angular positions of the active joints θ_1 and θ_2 can be written as

$$\theta_1 = \varphi - \beta \quad and \quad \theta_2 = \alpha + \theta_1 \tag{5}$$

All the above calculations use vector q rather than p for simplicity. The mechanism design produces a constant translational relationship between these two vectors.

$$\boldsymbol{p} = [q_x + f \quad q_y + g]^T \tag{6}$$

Note that for any given point P the magnitudes f and g are constant and independent of θ_1 and θ_2. The Jacobian matrix for the mechanism results in

$$\mathbf{J_p} = \mathbf{J_q} = \begin{bmatrix} -ls\theta_1 & -ls\theta_2 \\ lc\theta_1 & lc\theta_2 \end{bmatrix} \tag{7}$$

Note that the Jacobian of vectors p and q is identical. The study of singularities is fundamental for haptic mechanisms which are defined as the configurations where the matrices relating input and output parameters (Jacobian) become rank deficient [11]. Therefore is essential to identify all possible singularity configurations since they might lead to undesirable operation of the mechanism.

The singular configurations of the mechanism occur when the determinant $|\,\mathrm{J}\,|$ in Eq. (7) equals zero, resulting in

$$-l^2 s\theta_1 c\theta_2 + l^2 c\theta_1 s\theta_2 = 0 \tag{8}$$

By inspection of Eq. (8) is possible to determine that the singularity occurs when

$$\theta_1 = \theta_2 \tag{9}$$

$$\theta_1 = \theta_2 \pm \pi \tag{10}$$

Note that the first singularity configuration in Eq. (9) occurs when links 1 and 5 are aligned and the mechanism finds itself at the edge of its reachable workspace.

By using polar coordinates it is possible to observe the loss of one DOF of the mechanism in the singularity configuration. The second singularity conditions are met when links 1 has an angle of 180° relative to link 5. Note that the mechanism depicted in this work is unable to reach this condition due to mechanical constraints.

4 Mechanism Synthesis by Workspace Analysis

This section presents the synthesis of the mechanism following an approach based on the study of the human finger workspace in order to define the link dimension parameters for this mechanism. According to the literature, it is common to model a human finger as a three bar linkage with four rotational DOF where the first joint has two rotational DOF, the second and third joints have one DOF [12]. For the following workspace analysis, it is considered that each joint provides one DOF. The dimensional parameters of a human index finger and the range of angular displacement allowed by each of its joints differ unevenly among the human population [13]. However, the values used in this work represent a median and have been used in related work [12]. The mechanism selected to interact with the finger has to comply with various characteristics. The mechanism has to be safe; therefore the forces that the apparatus can exert cannot pass determined levels. The mechanism range has to intersect with the bulk of the finger workspace, although not completely. Low mass and friction are desired to optimize the haptic experience.

A configuration that complies with the mentioned characteristics is the seven bar mechanism described in the previous sections. Consider that the length and maximum angular displacement are optimized considering its interaction with the finger model. The synthesis of the seven bar mechanism is thus performed by iterating several combinations of dimensional parameters. The approach followed is to produce different mechanism workspaces and identify the best fit in the theoretical finger workspace. To show how the angle and length parameters modify the mechanism workspace, multiple combinations were studied. The lengths of links 1, 3, 4, and 5 are defined to be more than 30 mm; similarly the joints rotation range is required to be at least 150°. Given that the links have a width of 10 mm a simple cosine rule shows that for this angle, the base, link 2 and the platform have to be of at least 40 mm in length. This length is also applied to links 1, 3, 4 and 5 since it has been concluded that an acceptable length is between 30 mm and 50 mm. The mechanism with the mentioned dimensions is simulated using a Labview program to generate its theoretical workspace, which is compared with the model of the finger. The resulting intersection is presented in Fig. 2.

For the following analysis, the area of the mechanism and finger workspaces depicted in Fig. 2 are named m_A and f_A, respectively. The intersection between both workspaces referred as i_A. Note that f_A is exogenous and cannot be modified, and while the maximization of i_A is desired, conversely minimization of m_A is also preferred as it implies a more compact mechanism. To further explore this concept,

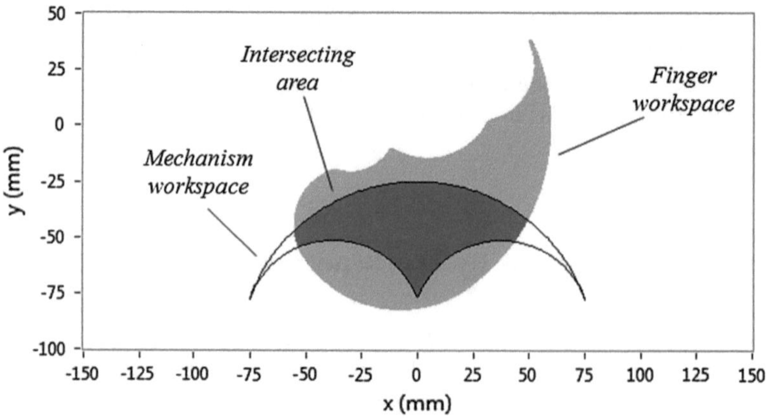

Fig. 2 Mechanism and finger workspaces and the intersecting area

two indexes are defined. The *finger utilization index* n_f is the ratio between the intersecting area i_A and the entire finger workspace f_A. Similarly, the *mechanism utilization index* n_m is the ratio between the intersecting area i_A and the entire mechanism workspace m_A. By definition both indexes can vary between 0 and 1 where the latter represents the complete use of the workspace.

$$n_f = \frac{i_A}{f_A} \quad and \quad n_m = \frac{i_A}{m_A} \tag{11}$$

In ordinary robotic applications the mechanism utilization is a parameter of interest; in haptic devices other parameters take importance such as finger utilization and device weight. These parameters are taken into account in the following equation:

$$\varepsilon = n_f + n_m - w_n \tag{12}$$

where ε is the proposed efficiency parameter and w_n is a normalized value of the apparatus mass. It is known that the device weight is proportional to the cube of the length l, given that proportions are kept. Due to the fact that n_f and n_m are non-linear, it is difficult to estimate them for complex workspaces as the ones considered in this document. Consequently, Fig. 3 presents the results of computational simulations that were performed in order to obtain n_f and ε_f for different values of l.

An inverse relation between n_f and n_m is observable in Fig. 3. Note that as l varies from 5 to 95 mm, n_f increases while n_m decreases; however, for any value of l equal or smaller than 25 mm, n_m saturates at 1, i.e. the finger workspace completely engulfs its mechanism counterpart. Extrapolation of the observed curves should yield a continuous decrease of n_m and an eventual saturation of n_f at 1; nevertheless for the purpose of this work, bar sizes above 100 mm are not suitable for a compact haptic device.

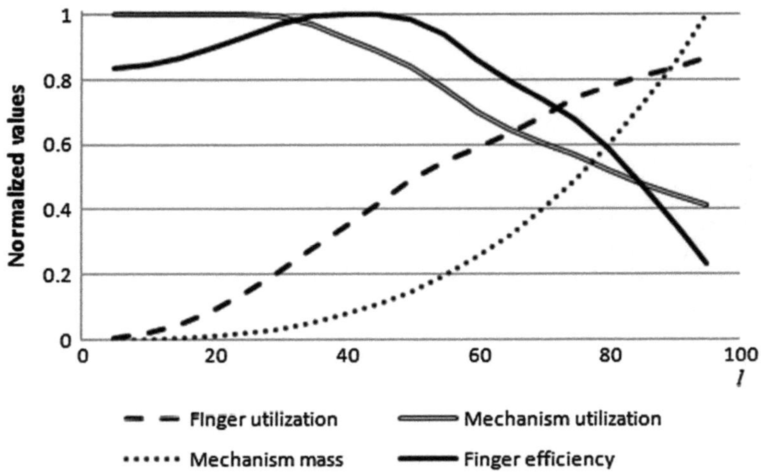

Fig. 3 Values of ε_f, n_m, and n_f for different mechanism dimension configurations

As the mechanism size increases, so does its mass. At low values this increase in weight is justified with a similar increase of finger utilization n_f; above certain point however, this relationship breaks since the mass increases at a rate higher than the utilization. Considering this relationship, the abovementioned plot suggests an optimal value of 40 mm for the links of the haptic device.

This parameter is used in the construction of the haptic device, which is presented in the next section. Workspace analysis is validated comparing the theoretical model with real data.

5 Prototype

A prototype of the haptic device was constructed using the parameters defined in the previous sections. 3D printing technology was used in the construction of structural pieces, and DC motors are connected to the mechanism without the typical geared speed reduction. This produces a mechanism with low mechanical impedance, especially when the motors are inactive, which is essential in haptic simulations. Low mass helps to this effect as well as low friction joints. When active, the motors can apply a torque of up to 200 mNm, which is relatively high for this size of motors but commercially available nonetheless. The torque of the upper portion of the mechanism is transmitted with a toothed belt. The sprockets connected to the base and link 4 have a 2:1 ratio; which produces a torque increase of the same magnitude. The reactive force that the mechanism can apply to any user depends on the mechanism configuration and the torque applied to the motors. If the mechanism is in the singularity conditions described in Eqs. (9) and (10) and the user applies a force parallel to link 1, then the theoretical reactive force is infinite.

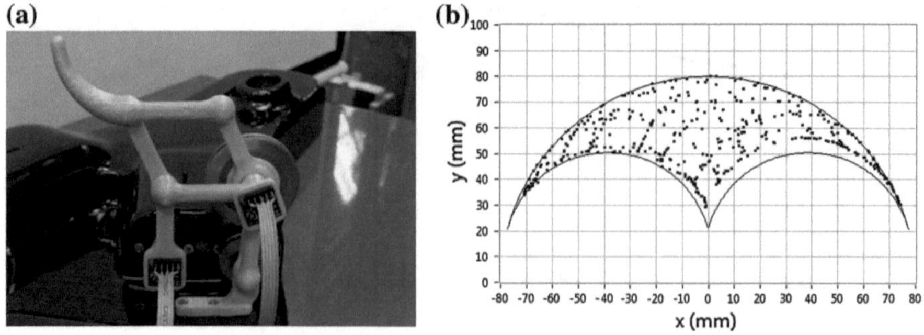

Fig. 4 **a** Prototype, and **b** comparison of the mechanism theoretical and experimental workspace

Conversely, the minimum reactive force occurs when the user applies force perpendicular to the links; in this case, the lower link will yield at 2.5 N. In case that the lower part of the mechanism is blocked (e.g. it has reached its angular limit), the upper mechanism will yield at 10 N due to the sprockets ratio and the shortening of the lever. The end-effector position is determined using Eq. (1). The values of l, f and g are constant for a given mechanism construction, while angles θ_1 and θ_2 are determined with accelerometers. The resolution of the position estimation depends on factors such as the accelerometer sensitivity and the characteristics of the analog to digital converter (ADC). Resolutions of less than 1° are possible using commercially available components such as a 10 bit ADC and a 300 mV/g accelerometer sensitivity. For the mechanism shown in Fig. 4a the accelerometers are placed on links 1 and 5.

The effectiveness of the accelerometer based positioning is tested by gathering their data as the end effector moves throughout its workspace. Five hundred points are estimated and compared with the theoretical workspace as shown in Fig. 4b. The overlap between the theoretical and experimental workspaces proves the effectiveness of the accelerometers as position measuring instruments. Note that the slight asymmetry can be attributed to the interference between link 5 and the platform. The latter has a diameter larger than the rest of the pieces in the joint that holds the sprocket. This feature reduces the angular liberty by a few degrees but does not affect the workspace shape.

6 Conclusions

This work presented the development of a seven-link mechanism force feedback device. The position analysis was performed and their equations obtained for solving the forward and inverse problems. Mechanism dimensions are suggested and tested using simulation software, performance indexes are proposed to compare various designs under an objective structure. An optimal configuration is obtained considering factors such as the device weight and the utilization of the finger and the mechanism workspaces. The resulting parameters are used to construct the haptic device; the workspace of the resulting apparatus is compared with simulations and is shown that the CAD tools effectively predicted the emergent workspaces. Unlike conventional robots that use encoders to assess joint angles, the constructed apparatus uses accelerometers which effectively track the end effector position.

Acknowledgments The authors acknowledge the support received from Consejo Nacional de Ciencia y Tecnología (CONACyT), the e-Robots Research Chair from Tecnológico de Monterrey, and the Laboratorio de Robótica del Área Noreste y Centro de México.

References

1. Coles T, Meglan D, John N (2012) Adaptive virtual environments for neuropsychological assessment in serious games. IEEE Trans Consum Electron 58(2):197–204
2. Jones JA, Swan JE, Bolas M (2013) Peripheral stimulation and its effect on perceived spatial scale in virtual environments. IEEE Trans Vis Comput Graph 19(4):701–710
3. Levesque V, MacLean K (2011) Do-it-yourself haptics a practical introduction to haptics for consumer electronics. In: International conference on consumer electronics
4. Elliott L, van Erp J, Redden E, Duistermaat M (2010) Field-based validation of a tactile navigation device. IEEE Trans Haptics 3(2):78–87
5. Ramsamy P, Haffegee A, Jamieson R, Alexandrov V (2006) Using haptics to improve immersion in virtual environments. In: Lecture notes in computer science, vol 3992. Springer, Berlin, pp 603–609
6. Coles T, Meglan D, John N (2011) The role of haptics in medical training simulators: a survey of the state of the art. IEEE Trans Haptics 4(1):51–66
7. Nakao M, Kitamura R, Minato K (2010) A model for sharing haptic interaction. IEEE Trans Haptics 3(4):155–165
8. Plaisier M, Bergmann W, Kappers A (2010) Haptic object individuation. IEEE Trans Haptics 3(4):257–265
9. Pauluk D, Kitada R, Abramowicz A, Hamilton C, Lederman S (2011) Figure ground segmentation via haptic glance attributing initial finger contacts to objects or their supporting surfaces. IEEE Trans Haptics 4(1):2–13
10. Chang W, Lin C, Wu L (2005) A note on Grashof's theorem. J Mar Sci Technol 13 (4):239–248
11. Gosselin C, Angeles J (1990) Singularity analysis of closed-loop kinematic chains. IEEE Trans Robot Autom 6(3):281–290
12. Abdel-Malek K, Yang K, Brand R, Tanbour E (2004) Towards understanding the workspace of human limbs. Ergonomics 47(13):1386–1405
13. Imrhan S, Sarder M, Mandahawi N (2009) Hand anthropometry in Bangladeshis living in America and comparisons with other populations. Ergonomics 52(8):987–998

Fuzzy Logic Control on FPGA for Solar Tracking System

Ricardo Antonio-Mendez, Jesus de la Cruz-Alejo and Ollin Peñaloza-Mejia

Abstract This paper proposes a control technique for solar tracking system through the processing of fuzzy logic implemented on FPGA. The control is designed by calculating and storing membership values in lookup tables as integer values which are addressed in all stages involved in the fuzzy logic system according to Mamdani rules and max—min implication model. A defuzzification method using alpha-levels is proposed. Results are presented to validate the theoretical analysis.

Keywords Fuzzy logic · Alfa-levels · Sun tracking · FPGA · Mamdani

1 Introduction

Currently one of the mathematical disciplines with the highest number of followers is the fuzzy logic technique, which is the logic that uses expressions that are neither completely true nor completely false, i.e., is the logic applied to concepts that can take a any truth value in a set of values ranging between two extremes, absolute truth and complete falsity. Central to those based on the theory of fuzzy logic systems is that, unlike those based on classical logic, have the ability to acceptably reproduce the usual modes of reasoning, considering the certainty of a proposition is a matter of degree [1, 2]. Thus, the most attractive features of fuzzy logic are its

R. Antonio-Mendez (✉) · J. de la Cruz-Alejo · O. Peñaloza-Mejia
Mechatronic Department, Tecnológico de Estudios Superiores de Ecatepec, Ecatepec, Estado de México, Mexico
e-mail: ric_xof@hotmail.com

J. de la Cruz-Alejo
e-mail: jesus_ch517@hotmail.com

O. Peñaloza-Mejia
e-mail: openaloza@tese.edu.mx

M. Ceccarelli and E.E. Hernández Martinez (eds.), *Multibody Mechatronic Systems*,
Mechanisms and Machine Science 25, DOI 10.1007/978-3-319-09858-6_2

flexibility, tolerance for imprecision, its ability to model non-linear problems, and its basis in natural language.

In the other hand, implementation can be using general-purpose processors and depending fully on software in the realization of the system or adapting a general-purpose processor to perform dedicated fuzzy instructions. The approach is a trade-off between speed and generality. An alternative is using an exclusive hardware to perform the fuzzy operations as a closely related approach, through dedicated fuzzy circuits or Application Specific Integrated Circuits (ASICs). The approach leads to relatively high-speed operation, but is more costly. FPGAs are hardware devices used as user-programmable ASICs. The availability of software tools to generate efficient and flexible hardware description configurations automatically also brings easiness to the reconfiguration process. Moreover, FPGA designs can already be modeled, simulated and verified.

For the case of solar-concentrating collectors, some form of tracking mechanism is usually employed to enable the collector to follow the sun. This is done by monitoring the variation in degrees. Monitoring tracking systems can be classified based on their movement. This can be a single axis or two axes. In the case of a single-axis mode, the motion can be in various ways: east-west, north-south, or parallel to the earth's axis [2—4].

The paper describes the fastest way to program the FPGA based on the corresponding application program, which in this case is the code for the sun tracking control. Section 2 characteristics of the sensors and actuators, as well as communication protocols implemented on the FPGA program for each device are described. Section 3, experimental results are presented. Section 4, a comparison between defuzzification method used in the paper against centroid method is presented.

2 Fuzzy Logic System

Figures 1, 2 and 3 shows the block diagram of the overall system. Data acquisition is carried out by four brightness sensors which convert light intensity into digital output values. The output values are stored in internal registers which are accessed via the I^2C communication protocol. It serves as interface for input values to the FLC. Outputs values of the FLC are the control actions which are communicated to the servomotors through the RS-232 protocol in digital mode, which contain the commands for positioning, according to the own configuration of each one servomotor. Hardware description, processing, setup and characteristics of the devices used in the system are presented next.

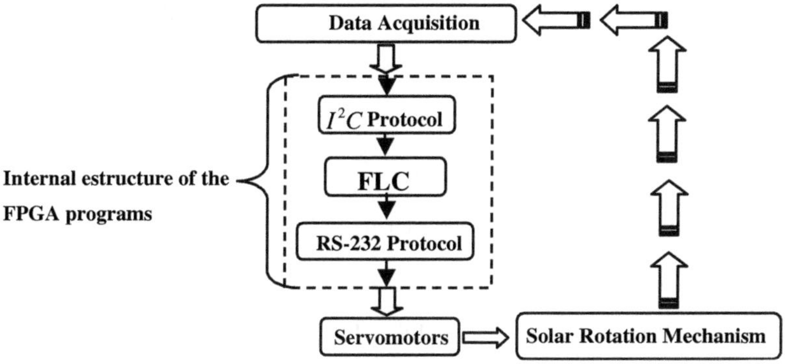

Fig. 1 Block diagram for the sun tracking system

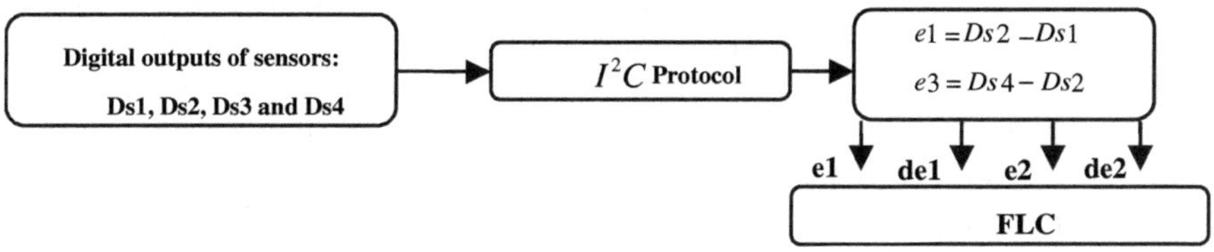

Fig. 2 Data acquisition for the FLC system

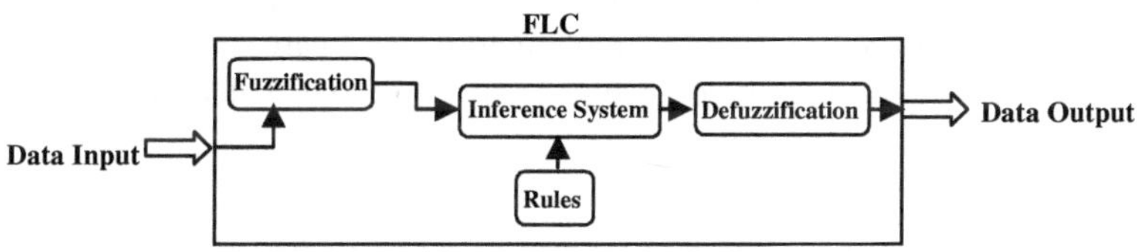

Fig. 3 Fuzzy logic control (FLC)

2.1 Fuzzification

The membership function chosen for this control was triangular, which is given by the parameters *a*, *b*, *c* as follows:

$$Triangulo(x; a, b, c) = \begin{cases} 0, & x \leq a \\ \frac{x-a}{b-a}, & a \leq x \leq b \\ \frac{c-x}{c-b}, & b \leq x \leq c \\ 0, & c \leq x \end{cases} \tag{1}$$

Using max—min operators, (1) can be represented as:

Fig. 4 Membership function for inputs and outputs

$$triangulo(x; a, b, c) = \max\left(\min\left(\frac{x-a}{b-a}, \frac{c-x}{c-b}\right), 0\right) \quad (2)$$

Using (2), the membership values for each input or output variable are obtained. For fuzzification stage, alpha-levels method was chosen. Input variables were expressed by five linguistic variables positive big (PB), positive small (PS), zero (ZO), negative small (NS) and negative big (NB). Membership functions of the fuzzy system are shown in Fig. 4. The universe of discussion for each input and output variable was expressed by five linguistic variables positive big (PB), positive small (PS), zero (ZO), negative small (NS) and negative big (NB).

The membership values are obtained through the programming of (2) in a spreadsheet on a pc. The resultants values are captured in look up tables which are stored in the internal program in order to be accessed to them in subsequent stages of the fuzzy logic system control. For example: Given an input value x, the membership value obtained is a floating-point value in a range between 0 and 1. This kind of value is complicated to implement in hardware, besides that involves the use of more resources on the device. So, the values are scaled depending on the resolution required, according to:

$$\left[\mu(x) = Eq.(2)_{floathing\ point} \times 2^n - 1\right] = \mu(x)_{integer\ value} \quad (3)$$

This manner, using (3), floating point value is scaled to a $2^n - 1$ resolution, where *n* represent the number of bits used in the system. The code to obtain all membership values for negative_big set is as follows:

```
type Negative_Big is array (a to c) of integer range 0 to 2^n − 1.
constant f_m1: Negative_Big: = (a => 0,1 => 6,2 => 12,4 => 18,....,c => 2^n − 1);
for i in a to c loop
if (x = i) then
μ(x) = f_m1(i)
end if;
end loop;
```

where *a* and *c* are the parameters of the membership function given by (1) and (2).

In this case, because the sensors have a resolution of 8 bits($2^n - 1$), the input universe of discussion is partitioned from −127 to 127 as is shown in Fig. 4. A

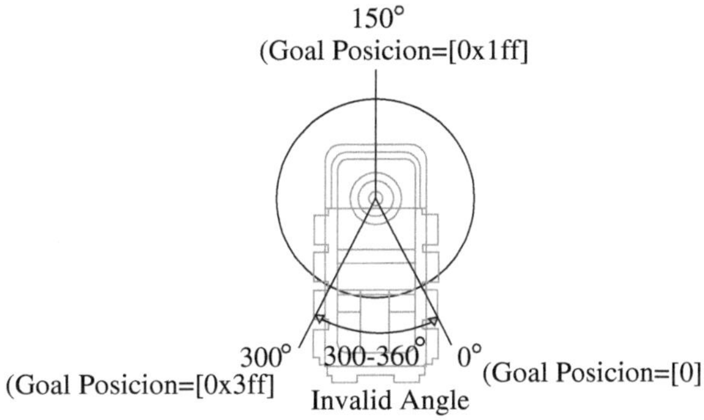

Fig. 5 Range of movement of the servomotors

value of 0 indicates the lowest value of lighting and a value of 255 highest level. This indicates that the entry for the maximum value $e1 = Ds2 - Ds1$either negative or positive is $2^n - 1$. Depending on whether *Ds2* is greater than *Ds1,* the result is a positive number or otherwise *Ds1* is greater than *Ds*2, which are negatives values. The universe from −512 to 512 indicated in Fig. 4 for the output of the system is proposed due to the movement of the servomotors have 1024 possible values. So, 1024 represents a position of 300°, 512 represents the midpoint, as shown in Fig. 5.

2.2 Rules

In this work, Mamdani's method is used. In fuzzy logic controller a descriptive verbal rules (If—Then rules) are used to describe the relation among inputs and outputs, according to:

$$R^{(l)} : IF\ x_1\ is\ F_1^l\ and\ldots and\ x_n\ is\ F_n^l\ THEN\ y\ is\ G^l \tag{4}$$

where:

$(x_1 \ldots x_n)$ *Represent the input variable*

(y) *Represent the output variable*

$(F\,y\,G)$ *Represent the membership function of fuzzy set.*

So, rules are activated according to (4). Table 1 shows the corresponding rules for the fuzzy logic system.

Table 1 Corresponding rules for the fuzzy logic system

				el			
		NG	NP		Z	PP	PG
	NG	NG	NP		NP	NP	NP
	NP	NP	Z		PP	NP	Z
del	Z	NP	PP		Z	PG	PP
	PP	Z	PP		PP	Z	PP
	PG	PP	PP		PP	PP	PG

2.3 Inference

The inference method chosen for this work is given by:

$$\mu_{A \to B}(x, y) = \min[\mu_A(x), \mu_B(y)] \tag{5}$$

It involves the comparison between two integer values, selecting the minimum value to activate rules. Now, in order to programming the fuzzy system, two arrays of rows and columns numbers which contain the maximum membership functions are programming. In these arrays are store the membership values which, corresponding to the fuzzy sets where they will be kept, i.e., if an input value falls within one or two fuzzy sets, the corresponding membership values for each one will be different in each set, therefore they will be stored in the array. Figure 6 shows the flow chart for the search and storing for all values in the array. The value *Data_inx* having *el,* is stored in this variable and *a* and *c* values corresponding to the parameters of (1) (triangular membership function). These are the ranges in which the input value is looked *Data_inx*.

Whenever *Data_inx* has a constant value that matches the value previously stored in the tables corresponding to the membership function in question, this will be stored in the array *Wnx*. Table 2 shows vectors obtained from the search processes for *Data_inx* (*el*) and *Data_iny* variable (*del*).

Taking into account the resulting vectors from the search, the inference process shown in Fig. 7, is performed. Inference process results are given in Table 3, as can be seen only the array of rules denominated *Wn_rules*(i,j), are stored. This array has the same dimension as Table 1 have. This means that the input values which activate output linguistic rules are stored in the array *Wn_rules*(i,j).

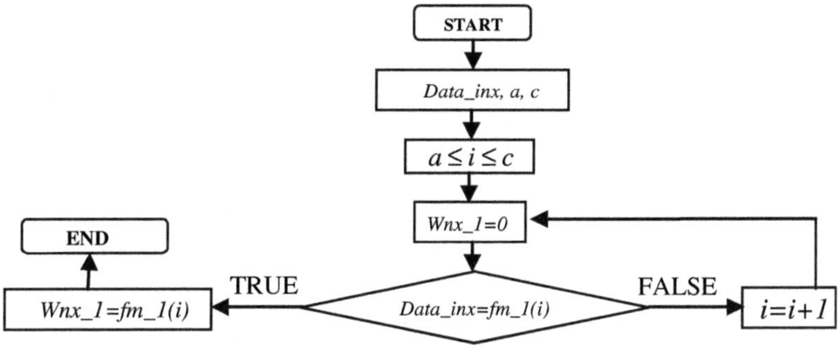

Fig. 6 Flow chart for the search process

Table 2 For data *inx* and data *iny*

Digital inputs	*Vectors*	*fm_1*	*fm_2*	*fm_3*	*fm_4*	*fm_5*
Data_inx	*Wnx=*	*Wnx_1*	*Wnx_2*	*Wnx_3*	*Wnx_4*	*Wnx_5*
Data_iny	*Wny=*	*Wny_1*	*Wny_2*	*Wny_3*	*Wny_4*	*Wny_5*

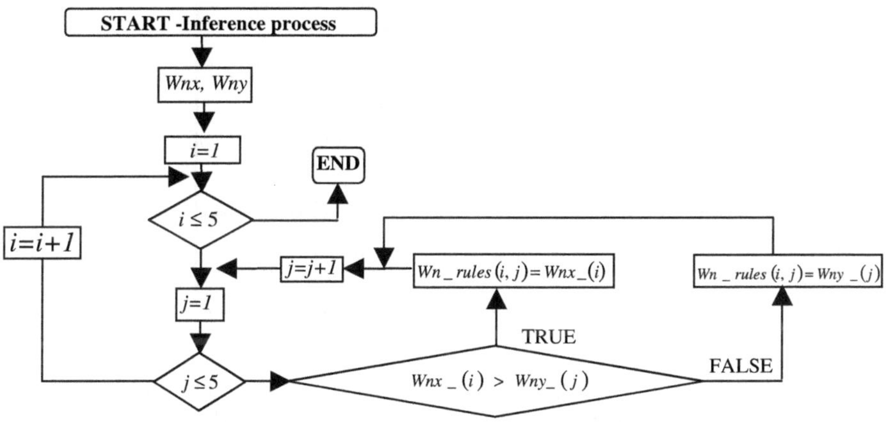

Fig. 7 Flow chart for the inference process

2.4 Aggregation

Aggregation stage is done by:

$$\mu_{B^l}(y) = \max\left[\mu_{F_1^l}(x_1), \ldots, \mu_{F_n^l}(x_n)\right] \tag{6}$$

It shows the global union of fuzzy sets and rules activated. It takes into account only those rules that have a non-zero value. The aggregation process is carried out by the method of maximum Eq. (6), for each one enabled rule whose value is

Table 3 Vector obtained by the Inference process

				el		
		NG	NP	Z	PP	PG
	NG	*Wn_rules (1,1)*	*Wn_rules (1,2)*	*Wn_rules (1,3)*	*Wn_rules (1,4)*	*Wn_rules (1,5)*
	NP	*Wn_rules (2,1)*	*Wn_rules (2,2)*	*Wn_rules (2,3)*	*Wn_rules (2,4)*	*Wn_rules (2,5)*
del	Z	*Wn_rules (3,1)*	*Wn_rules (3,2)*	*Wn_rules (3,3)*	*Wn_rules (3,4)*	*Wn_rules (3,5)*
	PP	*Wn_rules (4,1)*	*Wn_rules (4,2)*	*Wn_rules (4,3)*	*Wn_rules (4,4)*	*Wn_rules (4,5)*
	PG	*Wn_rules (5,1)*	*Wn_rules (5,2)*	*Wn_rules (5,3)*	*Wn_rules (5,4)*	*Wn_rules (5,5)*

Table 4 Process for the aggregation process

				el		
		NG	NP	Z	PP	PG
	NG	Wn_rules(1,1)	Wn_rules(1,2)	Wn_rules(1,3)	Wn_rules(1,4)	Wn_rules(1,5)
		↓	↓	↓	↓	↓
	NP	Wn_rules(2,1)	Wn_rules(2,2)	Wn_rules(2,3)	Wn_rules(2,4)	Wn_rules(2,5)
		↓	↓	↓	↓	↓
del	Z	Wn_rules(3,1)	Wn_rules(3,2)	Wn_rules(3,3)	Wn_rules(3,4)	Wn_rules(3,5)
		↓	↓	↓	↓	↓
	PP	Wn_rules(4,1)	Wn_rules(4,2)	Wn_rules(4,3)	Wn_rules(4,4)	Wn_rules(4,5)
		↓	↓	↓	↓	↓
	PG	Wn_rules(5,1)	Wn_rules(5,2)	Wn_rules(5,3)	Wn_rules(5,4)	Wn_rules(5,5)

different of 0. Then, it is compared with another to know which one is the maximum, and so on as is shown in Table 4. It shows the way and how the maximum is saved by comparing the vector $Wn\_rules(i,j)$. In this case, e.g., taking NG column, membership function values are compared following, (1,1,), (2,1), (3,1), (4,1) and (5,1). So, the maximum value between them is the result that activates the output. It is the value that was used in the alpha-levels defuzzification (αi) process, illustrated in Fig. 8, where represents the membership value as a result for the aggregation.

2.5 Defuzzification

The method chosen for defuzzification is the Center of slice Area Average (COSAA) [1], which is given by:

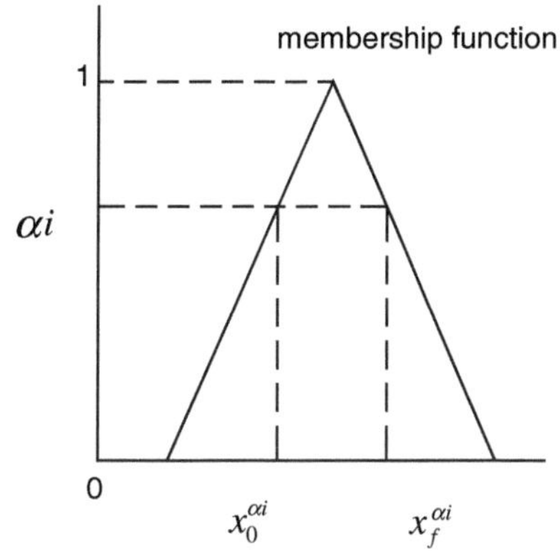

Fig. 8 Parameters of membership function

$$COSAA = \frac{\sum_{i=0}^{\alpha \max} \left(\frac{x_f^{\alpha i} - x_0^{\alpha i}}{2} \right) + x_0^{\alpha i}}{\alpha \max} \tag{7}$$

where $x_0^{\alpha i}$ and $x_f^{\alpha i}$ are parameters given according to the Fig. 8.

2.6 Experimental and Simulated Results

Figure 9a, b show the position of the sensors in the system, where we can see the position of the sensors and the direction of the movement that each motor have in the FLC system. Figure 10 shows a comparison between the implemented method proposed and the MatLab toolkit. The toolkit of Matlab as Bisector and Centroid establish a non-significant difference for the control. The inputs to the controller are:

$$e1 = Ds2(k) - Ds1(k), \ and \ de1 = e1(k) - e1(k-1) \tag{8}$$

where k, is the sampling time. So, the process adapts the characteristics of the nonlinear behavior of the solar tracking. This means that either cloudy days or sunny days, the system adapts to the minimum or maximum conditions of solar radiation or in this case lighting. The novelty of this system is the reduction of operations performed on the FPGA. Therefore it only uses basic operations such as addition, subtraction, divisions and comparisons. In Table 5, the experimental output results values for the implemented proposed method and the MatLab are given. The discrepancy between them is less than an 5 % for all parameters over a full range contained in the universe of discussion X.

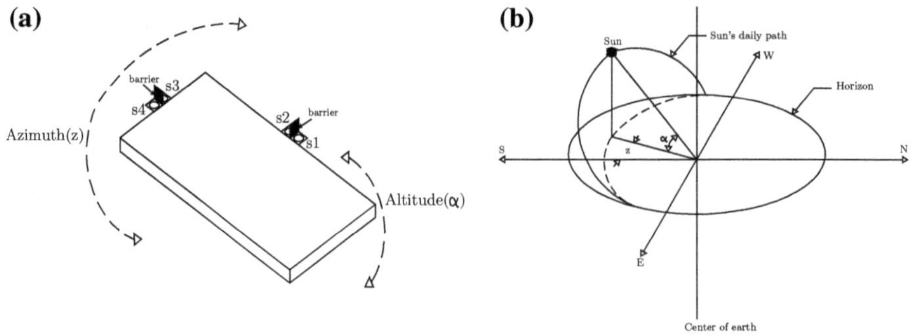

Fig. 9 **a** Motion Overviews, **b** daily path of the sun across the sky from sunrise to sunset

Fig. 10 Comparison between the system proposed and Matlab defuzzification methods

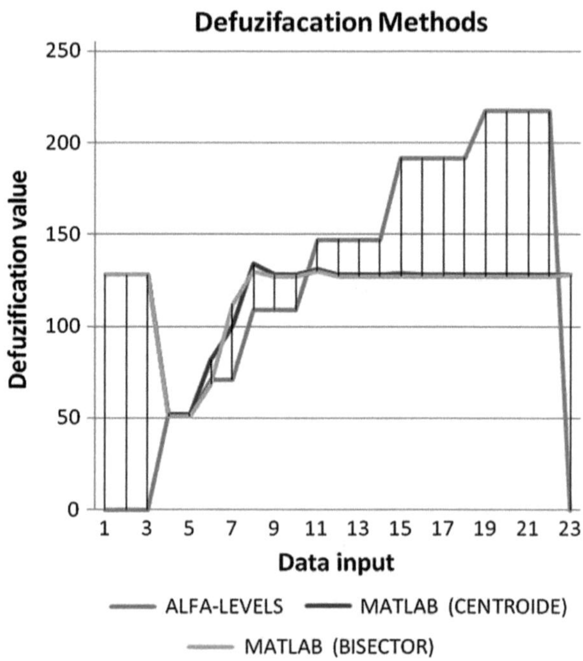

3 Conclusions

The proposed FPGA sun tracking control, based on fuzzy logic technique was implemented. It focuses on the optimization of hardware resources, conducting operations to obtain the values of membership functions for either input or output, accessing them as constant values using look-up tables. All the simulation and experimental results demonstrate the effectiveness of the FLC system. Through the comparison between them, the proposed system achieves a control performance and

Table 5 Comparative methods defuzification

Comparative table				
Inputs value		Outputs value (Z)		
X	Y	Methods of defuzification		
		Alfa-levels (proposed method)	MATLAB (centroide)	MATLAB (bisector)
0	0	0	128.0	128
14	14	0	128.0	128
14	20	0	128.0	128
25	30	52	52.0	51
35	40	52	52.0	51
50	65	71	81.7	68.8
70	80	71	99.8	112
85	95	109	134	130
100	110	109	128	127
115	120	109	128	127
245	255	0	128	128

allowing a faster and precise control of the sun tracking. Besides the proposed fuzzification method (COSAA), does not imply higher processing calculations. Moreover the devices used both data acquisition and those performing the control actions have a good behavior.

References

1. Zavala AH (2009) Arquitectura de Alto Rendimiento para sistemas difusos, Tesis, IPN, Enero
2. Chekired F, Larbes C, Mellit A (2012) Comparative study between two intelligent MPPT-controllers implemented on FPGA: application for photovoltaic systems. Int J Sustain Energ. doi:10.1080/14786451.2012.742896
3. Othman AM et al, Real world maximum power point tracking simulation of PV system based on Fuzzy Logic control. NRIAG J Astron Geophys 1(2):186–194
4. Nader Barsoum Curtin University of Technology, Sarawak, Malaysia, Implementation of a prototype for a traditional solar tracking system, UKSim 3rd European modelling symposium on computer modelling and simulation, Athens, Greece, 25 November 2009, pp 23–30

Dynamic Balancing of a Nutating Planetary Bevel Gear Train

P. Fanghella, L. Bruzzone and S. Ellero

Abstract The paper presents a dynamic model of a planetary gear train based on nutating bevel gears. The goal is to determine the conditions under which the mechanism do not transmit shaking inertial forces and moments to its frame. The paper first presents system geometry and kinematics; then an analytical model of its dynamics is developed and explicit balancing conditions are obtained. Analytical results are the verified through multibody simulations of a sample geometry. A discussion about balancing geometries is finally carried out.

Keywords Mechanism design · Multibody dynamics · Nutating bevel gears · Planetary gear train

1 Introduction

Planetary gear trains represent an important class of mechanical transmissions used in a wide variety of applications. When very compact solutions and high transmission ratios are required, special configurations based on bevel gears become interesting [1, 5, 8]. This particular type of planetary gear train has been extensively studied from a kinematic point of view, see for example [3, 6, 7, 9, 12], and in order to evaluate the efficiency and power-flows of different configurations, see for example [4, 10]. Only a limited number of results, especially for the bevel gear mechanisms, is available regarding mechanism dynamics: beyond specific results

P. Fanghella (✉) · L. Bruzzone
University of Genoa, Genoa, Italy
e-mail: pietro.fanghella@unige.it

L. Bruzzone
e-mail: bruzzone@dimec.unige.it

S. Ellero
Stam Srl, Genoa, Genoa, Italy
e-mail: s.ellero@stamtech.com

M. Ceccarelli and E.E. Hernández Martinez (eds.), *Multibody Mechatronic Systems*,
Mechanisms and Machine Science 25, DOI 10.1007/978-3-319-09858-6_3

regarding gear meshing, for example [11] discusses the internal inverse dynamics (torques and forces) of a bevel gear train used in robotic applications.

This paper considers the dynamics of planetary bevel gear trains from a completely different point of view, that, as far as the authors know, has not been previously considered: since this kind of mechanisms, even in steady state conditions, are characterized by unbalanced bodies with angular velocities varying in direction, they possibly present centrifugal and gyroscopic effects that, at high speed, can transmit high loads the system frame. Therefore, it is relevant to assess the conditions required to have, if possible, static and dynamic balancing of this kind of systems. In particular, the paper examines in detail the balancing conditions for a specific type of bevel gear train characterized by the presence of a body with a nutating motion [1, 5, 6, 8, 9], that allows a very high reduction ratio with a very compact geometry.

The paper first presents the geometry and kinematics of the mechanism, then a fully analytical model of its dynamics is developed and explicit static and dynamic balancing conditions are obtained. They are validated by numerical multibody analyses, and, finally, actual application of the balancing conditions to body geometries is discussed.

2 Geometry and Kinematics

Figure 1 presents a scheme of the mechanism:

- the input shaft, realized by the carrier *c,* is attached to the frame *f* by a revolute joint R_{fc}
- the output shaft 5 and the output gear 2 are fixed together and connected to the frame by a revolute joint R_{f5}, coaxial with R_{fc}
- the gear body 4 is fixed to the frame
- the intermediate gears 1 and 3 are fixed together to form a unique nutating body (*n*) attached by a revolute joint R_{cn} to the carrier *c*; the axis of R_{cn} intersects the common axis of R_{fc} and R_{f5} at point **O**, that is the center of the spherical motion of all bodies; the angle α between the joint axes is fixed by the shape of the carrier body and constitutes one of the key parameters in the dynamics of the mechanism.

Figure 1 also shows the two bevel gearings composing the planetary gear train:

- g_a, formed by gears 3 and 4 (traces of primitive cones in white in Fig. 1)
- g_b: formed by gears 1 and 2 (traces of primitive cones in grey in Fig. 1)

The transmission ratio τ_{5_c} between the angular velocities of input and output shafts is [5, 6]:

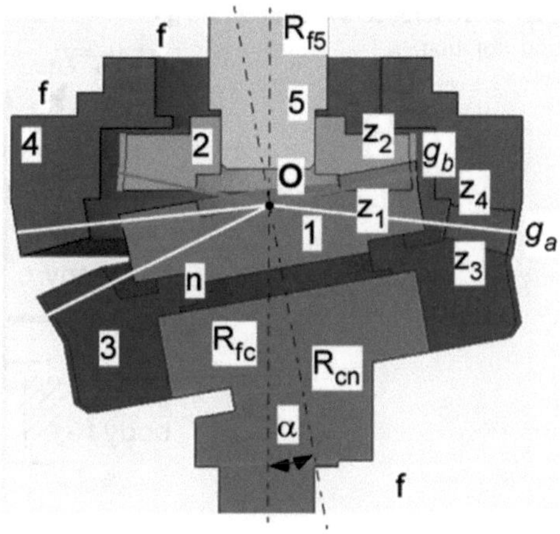

Fig. 1 Mechanism geometry

$$\tau_{5\_c} = \frac{\omega_5}{\omega_c} = 1 - \frac{z_4 z_1}{z_3 z_2} \tag{1}$$

while the transmission ratio τ_{nc_c} between the input velocity ω_c and the relative velocity ω_{nc} of bodies c and n is:

$$\tau_{nc\_c} = \frac{\omega_{nc}}{\omega_c} = -\frac{z_4}{z_3} \tag{2}$$

According to the goals of this paper, the scheme of Fig. 1 can be simplified by considering only the motion of bodies c and n; the output bodies 2 and 5 are balanced and rotating around a fixed axis, so they do not yield any static or dynamic unbalance effect.

Figure 2 shows the reduced scheme and the used reference systems and notation. To simplify the development of kinematic and dynamic equations, according to the spherical nature of the mechanism, the origins of all used reference systems coincide with the center of the motion **O**; therefore:

(a) reference systems are attached to frame f and carrier c; they share the revolute axis (y direction); the rotation angle $\vartheta_c(t)$ from x_f to x_c completes the information about their relative position
(b) a (constant) rotation around z_c of angle α yields to an intermediate reference system named a; the nutating body n is reached by a rotation around the $y_a = y_n$ axis, i.e. the second revolute joint, with angle $\vartheta_n(t)$ from x_a to x_n.

By design, the carrier body is symmetric with respect to the x_c-y_c plane, therefore in its local reference frame c its center of mass $\mathbf{G}_\mathrm{c}$ has coordinates ${}_c[G_{cx}, G_{cy}, 0]^T$; similarly, by design, the nutating body has its center of mass $\mathbf{G}_\mathrm{n}$ lying on the local y_n axis, with coordinates ${}_n[0, G_{ny}, 0]^T$.

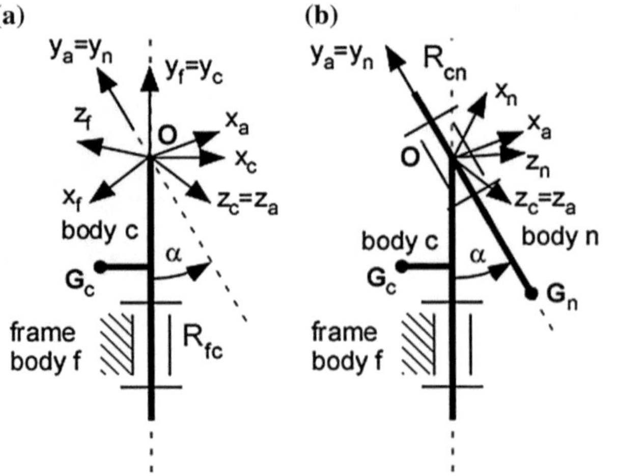

Fig. 2 Reference systems and notation

Based on previous considerations and notation, it is possible to express the rotations of the moving bodies and the positions of their center of masses, along with their time derivatives with respect to the fixed reference frame; according to the study purpose, the system is considered in steady state motion ($\ddot{\vartheta}_n = 0; \ddot{\vartheta}_c = 0;$), so we obtain:

(1) rotation matrices

$$[\mathbf{R}_{fc}] = \begin{bmatrix} \cos(\vartheta_c) & 0 & \sin(\vartheta_c) \\ 0 & 1 & 0 \\ -\sin(\vartheta_c) & 0 & \cos(\vartheta_c) \end{bmatrix} \tag{3}$$

$$[\mathbf{R}_{fn}] = \begin{bmatrix} \cos(\vartheta_c)\cos(\alpha)\cos(\vartheta_n) - \sin(\vartheta_c)\sin(\vartheta_n) & -\cos(\vartheta_c)\sin(\alpha) & \cos(\vartheta_c)\cos(\alpha)\sin(\vartheta_n) + \sin(\vartheta_c)\cos(\vartheta_n) \\ \sin(\alpha)\cos(\vartheta_n) & \cos(\alpha) & \sin(\alpha)\sin(\vartheta_n) \\ -\sin(\vartheta_c)\cos(\alpha)\cos(\vartheta_n) - \cos(\vartheta_c)\sin(\vartheta_n) & \sin(\vartheta_c)\sin(\alpha) & -\sin(\vartheta_c)\cos(\alpha)\sin(\vartheta_n) + \cos(\vartheta_c)\cos(\vartheta_n) \end{bmatrix} \tag{4}$$

(2) angular velocities

$${}_f\boldsymbol{\omega}_{fc} = \begin{bmatrix} 0 & \dot{\vartheta}_c & 0 \end{bmatrix}^T \text{ and } {}_a\boldsymbol{\omega}_{cn} = \begin{bmatrix} 0 & \dot{\vartheta}_n & 0 \end{bmatrix}^T \tag{5}$$

and w.r.t. fixed reference frame:

$$ {}_f\boldsymbol{\omega}_{cn} = \left[-\cos(\vartheta_c)\sin(\alpha)\dot{\vartheta}_n \quad \cos(\alpha)\dot{\vartheta}_n \quad \sin(\vartheta_c)\sin(\alpha)\dot{\vartheta}_n \right]^T \tag{6} $$

$$ {}_f\boldsymbol{\omega}_{fn} = \left[-\cos(\vartheta_c)\sin(\alpha)\dot{\vartheta}_n \quad \dot{\vartheta}_c + \cos(\alpha)\dot{\vartheta}_n \quad \sin(\vartheta_c)\sin(\alpha)\dot{\vartheta}_n \right]^T \tag{7} $$

(3) angular accelerations

$$ {}_f\dot{\omega}_{fc} = [0 \quad 0 \quad 0]^T \text{ and } {}_f\dot{\omega}_{fn} = \sin(\alpha)[\sin(\vartheta_c) \quad 0 \quad \cos(\vartheta_c)]^T\dot{\vartheta}_c\dot{\vartheta}_n \tag{8} $$

(4) center of masses positions and accelerations w.r.t. fixed reference frame

$$ \begin{aligned} {}_f\mathbf{G}_c &= [\cos(\vartheta_c)G_{cx} \quad G_{cy} \quad -\sin(\vartheta_c)G_{cx}]^T \\ {}_f\underset{c}{\ddot{\mathbf{G}}} &= [-\cos(\vartheta_c) \quad 0 \quad \sin(\vartheta_c)]^T G_{cx}\dot{\vartheta}_c^2 \end{aligned} \tag{9} $$

$$ \begin{aligned} {}_f\mathbf{G}_n &= [-\cos(\vartheta_c)\sin(\alpha) \quad \cos(\alpha) \quad \sin(\vartheta_c)\sin(\alpha)]^T G_{ny} \\ {}_f\underset{n}{\ddot{\mathbf{G}}} &= \sin(\alpha)[\cos(\vartheta_c) \quad 0 \quad -\sin(\vartheta_c)]^T G_{ny}\dot{\vartheta}_c^2 \end{aligned} \tag{10} $$

As a final remark concluding kinematics, we remind that the system has only 1 degree of freedom, and according to Eq. 2 it is $\dot{\vartheta}_n = \tau_{nc\_c}\dot{\vartheta}_c$.

3 Dynamics

The inverse dynamic model is obtained by applying the Newton-Euler equations to the nutating body and then to the carrier body, so that the total reaction forces ant torques transmitted to the frame are determined [2]. All vector entities in such equations are projected to the fixed reference frame, thereby obtaining:

(1) nutating body

$$ \begin{cases} {}_f\mathbf{F}_{cn} = m_n\left({}_f\underset{n}{\ddot{\mathbf{G}}}\right) \\ {}_f\mathbf{M}_{cn} = {}_f\left([\mathbf{J}_n]\dot{\omega}_{fc}\right) + {}_f\left([\tilde{\omega}_{fc}][\mathbf{J}_n]\boldsymbol{\omega}_{fc}\right) \end{cases} \tag{11} $$

in which

- ${}_f\mathbf{F}_{cn}$ and ${}_f\mathbf{M}_{cn}$ are force and torque transmitted to the carrier body due to inertial effects
- ${}_f[\mathbf{J}_n]$ is the inertial tensor of body n, expressed in the absolute reference frame through the transformation:

$$_f[\mathbf{J}_n] = [\mathbf{R}_{fn}]({}_n[\mathbf{J}_n])[\mathbf{R}_{nf}] \tag{12}$$

Due to the rotational symmetry of the nutating body, its inertia tensor is diagonal, with $J_{nx} = J_{nz}$:

$$_n[\mathbf{J}_n] = \begin{bmatrix} J_{nx} & 0 & 0 \\ 0 & J_{ny} & 0 \\ 0 & 0 & J_{nx} \end{bmatrix} \tag{13}$$

(2) carrier body

$$\begin{cases} {}_f\mathbf{F}_{fc} = m_c\left({}_f\ddot{\underset{c}{\mathbf{G}}}\right) - (-{}_f\mathbf{F}_{cn}) \\ {}_f\mathbf{M}_{fc} = {}_f\left([\tilde{\omega}_{fc}][\mathbf{J}_c]\boldsymbol{\omega}_{fc}\right) - (-{}_f\mathbf{M}_{cn}) \end{cases} \tag{14}$$

in simplified form since ${}_f\dot{\omega}_{fc} = 0$, from which reaction forces and torques transmitted to the frame due to inertia forces, i.e. the static and dynamic unbalance of the system, can be finally determined (${}_f\mathbf{F}_{fc}$ and ${}_f\mathbf{M}_{fc}$). The inertia tensor w.r.t. fixed frame ${}_f[\mathbf{J}_c]$ is obtained as previously, starting from the inertial tensor of carrier c, that, according to its symmetry, is:

$$_c[\mathbf{J}_c] = \begin{bmatrix} J_{cxx} & J_{cxy} & 0 \\ J_{cxy} & J_{cyy} & 0 \\ 0 & 0 & J_{czz} \end{bmatrix} \tag{15}$$

By substituting all previous relations into the Newton-Euler equations and simplifying, the following results are obtained:

$$_f\mathbf{F}_{cn} = m_n \sin(\alpha) G_{ny} \begin{bmatrix} \cos(\vartheta_c) \\ 0 \\ -\sin(\vartheta_c) \end{bmatrix} \dot{\vartheta}_c^2 \tag{16}$$

$$_f\mathbf{M}_{cn} = -\sin(\alpha)\left(\cos(\alpha)(J_{nx} - J_{ny})\dot{\vartheta}_c - J_{ny}\dot{\vartheta}_n\right) \begin{bmatrix} \sin(\vartheta_c) \\ 0 \\ \cos(\vartheta_c) \end{bmatrix} \dot{\vartheta}_c \tag{17}$$

and then, as a complete result

$$_f\mathbf{F}_{fc} = (m_n \sin(\alpha) G_{ny} - m_c G_{cx}) \begin{bmatrix} \cos(\vartheta_c) \\ 0 \\ -\sin(\vartheta_c) \end{bmatrix} \dot{\vartheta}_c^2 \tag{18}$$

$$
{}_f\mathbf{M}_{fc} = \left(-\sin(\alpha)\left(\cos(\alpha)\left(J_{nx} - J_{ny}\right)\dot{\vartheta}_c - J_{ny}\dot{\vartheta}_n\right) - J_{cxy}\dot{\vartheta}_c\right)\begin{bmatrix}\sin(\vartheta_c)\\ 0\\ \cos(\vartheta_c)\end{bmatrix}\dot{\vartheta}_c \quad (19)
$$

Obviously, ${}_f\mathbf{F}_{fc}$ and ${}_f\mathbf{M}_{fc}$ are two orthogonal vectors rotating with the carrier body, whose amplitudes depend quadratically on the rotation velocities; moreover all terms coming out from the nutating body vanish for design angle $\alpha = 0$. Due to the gyroscopic effect, the moment ${}_f\mathbf{M}_{fc}$ depends on both velocities $\dot{\vartheta}_c$ and $\dot{\vartheta}_n$, but, reminding that such velocities are related by the transmission ratio τ_{nc_c}, we get the following final expressions for the amplitudes of force and moment over $\dot{\vartheta}_c^2$:

$$
S_u = \frac{\left|{}_f\mathbf{F}_{fc}\right|}{\dot{\vartheta}_c^2} = m_n \sin(\alpha) G_{ny} - m_c G_{cx} \quad (20)
$$

$$
D_u = \frac{\left|{}_f\mathbf{M}_{fc}\right|}{\dot{\vartheta}_c^2} = \sin(\alpha)\left(\tau_{nc\_c} J_{ny} - \cos(\alpha)\left(J_{nx} - J_{ny}\right)\right) - J_{cxy} \quad (21)
$$

That depend only on the kinematic and inertial parameters of the system.

In order to avoid static and dynamic unbalances, the system must be designed so that such amplitudes are null or sufficiently small to yield, at nominal speed, acceptable values of force and moment. The first relation has a clear and simple physical interpretation: in order to have null static unbalance, the center of mass of the joined carrier and nutating bodies must lie on the fixed axis of rotation of the carrier. The second relation is more complex and hard to be physically discussed, as it involves a mix of kinematic, geometrical and inertial parameters.

4 Multibody Simulations

The analytical results of Sect. 3 have been validated through dynamic simulations carried out by a commercial multibody software (Creo 2.0-Mechanism Dynamics), for a geometrical configuration similar to that in Fig. 1, with the following values:

$$
\begin{aligned}
\alpha &= 10.3^\circ \quad \tau_{nc\_c} = -1.05\\
m_c &= 0.04\ \text{kg} \quad G_{cx} = -3.83\ \text{mm}\\
m_n &= 0.1\ \text{kg} \quad G_{ny} = -8.92\ \text{mm}\\
J_{cxy} &= -3.08\ \text{kgmm}^2\\
J_{nx} &= 19.7\ \text{kgmm}^2_{ny} = 20\ \text{kgmm}^2
\end{aligned}
$$

For which, from previous relations, it is: $S_u = 0.00062$ kgmm and $D_u = -0.62\ \text{kgmm}^2$. For an input velocity $\dot{\vartheta}_c = 3000$ rpm, previous analytical

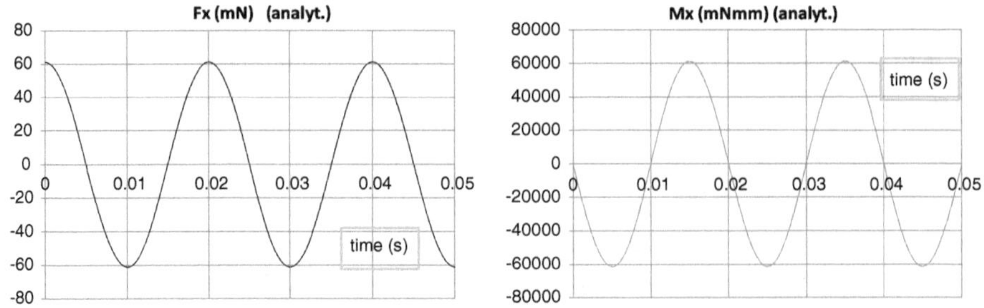

Fig. 3 *x* component of the unbalance force: analytical (*left*), numerical (*right*)

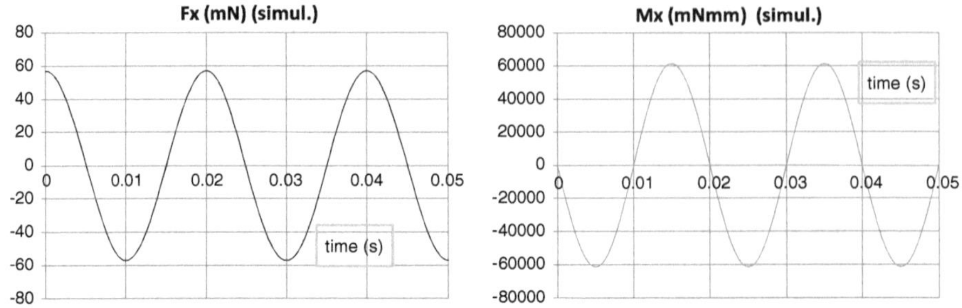

Fig. 4 *x* component of the unbalance moment: analytical (*left*), numerical (*right*)

results provide a value of force transmitted to the frame ${}_fF_{fc} = 61.2$ mN and of moment ${}_fM_{fc} = 61297$ mNmm. Figure 3 shows the *x* component of the static unbalance force ${}_f\mathbf{F}_{fc}$ as obtained by the analytical model presented in previous Sections and by the numerical multibody model. Similarly, Fig. 4 shows the x component of the dynamic unbalance moment ${}_f\mathbf{M}_{fc}$, analytical and numerical. Both figures show a perfect match of the two approaches.

5 Discussion About a Balancing Geometry

The balancing of the planetary gear train requires that both relations (20) and (21) are zero. Some design parameters, namely α and τ_{nc_c}, are related to the functional and gear design of the mechanism and cannot be used to balance it.

Regarding the static balancing, the first term of Eq. 20 ($m_n \sin(\alpha) G_{ny}$) is mostly determined by embodiment requirements of gears 1 and 3 (Fig. 1), so the only possibility is to add a counterweight to the carrier body, opposite to the position of nutating body center of mass, creating a static moment to equilibrate the nutating body unbalance (Fig. 5).

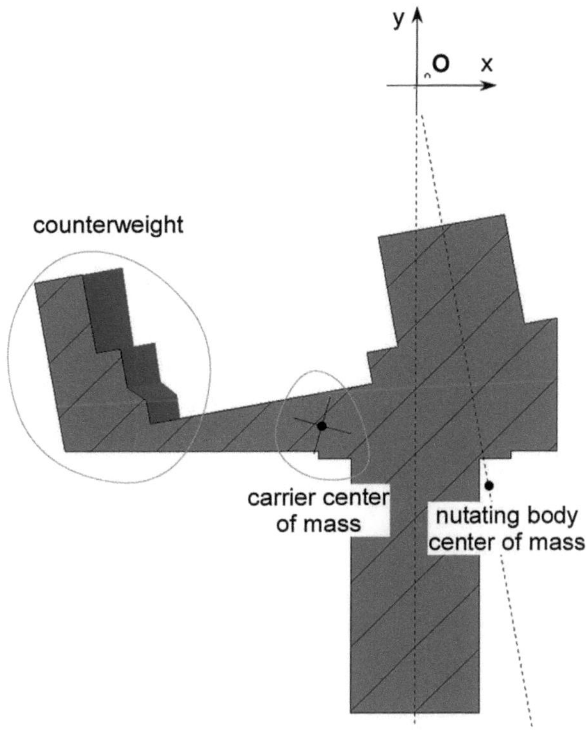

Fig. 5 Geometry of the carrier body, static balancing

Regarding dynamic balancing, the value of D_u (Eq. 21) is composed by the sum of three terms: (a) $\sin(\alpha)\tau_{nc\_c}J_{ny}$, (b) $-\sin(\alpha)\cos(\alpha)(J_{nx} - J_{ny})$ and (c) $-J_{cxy}$.

As shown by the data in the numerical example, for practical geometries, the term (b) is negligible as $J_{nx} \cong J_{ny}$, so the balancing may be mainly obtained equating (a) and (c). Again, term (a) depends on the mechanism functional design and on the inertial properties of nutating body; reducing inertia J_{ny} may help, but there are limitations due to embodiment design of nutating gears. Then the main balancing effect may be achieved by adding couple of masses to the carrier body to obtain the required J_{cxy} value, as shown schematically in Fig. 6.

As a matter of fact, the actual balancing of a planetary bevel gear train requires a complex geometric design of the real shapes of carrier and nutating bodies that can be hardly carried out manually. Nevertheless, current software technologies for the parametric manipulation and optimization of 3D solid models allow solutions of the considered problem that are feasible both from the functional and manufacturing points of view.

In any case, no matter the approach used to reduce the values of unbalance coefficients S_u and D_u, both analytical and multibody models have demonstrated a corresponding reductions of shaking force and moment, thereby validating the result presented in the paper.

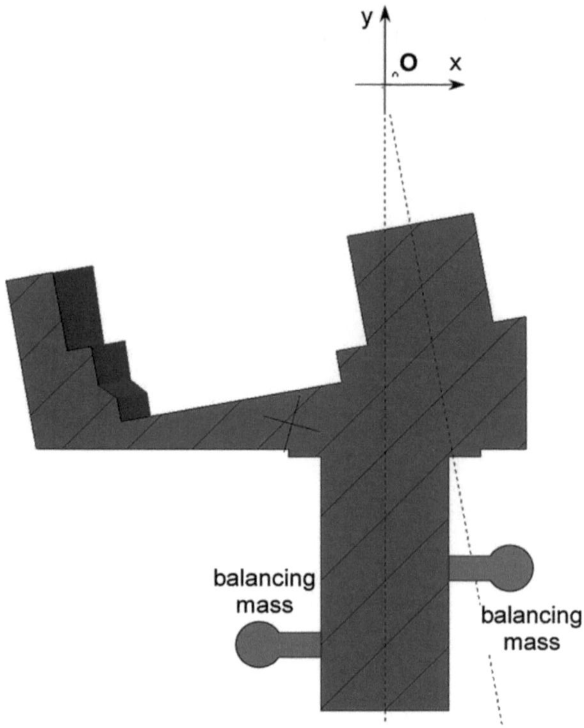

Fig. 6 Geometry of the carrier body, dynamic balancing

References

1. Barbagelata A, Fanghella P, Landò R (2003) Innovative Drive system for electric windows for cars based on a space born gearbox (Spacegear). In: International symposium CONVERGENCE '03: aeronautics, automotive and space, Paris, France
2. Callegari M, Fanghella P, Pellicano F (2013) Mechanics of machines, Cittastudi. ISBN 8825173814 (in Italian)
3. Day C, Akeel HA, Gutkowsky LJ (1983) Kinematic design and analysis of coupled planetary bevel-gear trains> J Mech Trans ASME 105:441–445
4. Del Pio G, Pennestrì E, Valentini PP (2013) Kinematic and power-flow analysis of bevel gears planetary gear trains with gyroscopic complexity. Mech Mach Theory 70:523–537
5. Landò R (2001) Analisys and design of a nutating planetary gear-train, MS thesis, University of Genoa (in Italian)
6. Ligang Y, Bing G, Shujuan H, Guowu W, Dai JS (2010) Mathematical modeling and simulation of the external and internal double circular-arc spiral bevel gears for the nutation drive. J Mech Des Trans ASME 132(2):0210081–02100810
7. Litvin FL, Zheng Y (1986) Robotic bevel-gear differential train. Int J Robot Res 5:75–81
8. Molyneux WG (1997) The internal bevel gear and its applications. Proc Inst Mech Eng 211 (Part G):39–61
9. Nelson CA, Cipra RJ (2005) Similarity and equivalence of nutating mechanisms to bevel epicyclic gear trains for modeling and analysis. J Mech Des Trans ASME 127(2):269–277

10. Nelson CA, Cipra RJ (2005) Simplified kinematic analysis of bevel epicyclic gear trains with application to power-flow and efficiency analyses. J Mech Des Trans ASME 127(2):278–286
11. Staicu S (2008) Inverse dynamics of a planetary gear train for robotics. Mech Mach Theory 43:918–927
12. Uyguroglu M, Demirel H (2005) Kinematic analysis of bevel-gear trains using graphs. Acta Mech 177:19–27

Output Stabilization of Linear Systems with Disturbances

Francisco Javier Bejarano and Jorge Dávila

Abstract In this paper a control design is proposed to stabilize a controlled output. It is assumed that only a part of the state vector is known. One part of the disturbances affecting the controlled output is decoupled from it by using feedback, the other part of the disturbances that affects the output, which cannot be decoupled by using feedback, is compensated by using its estimation. An example dealing with the output regulation of a Boeing 767 is presented to illustrate the effectiveness of the proposed methodology.

Keywords Output stabilization · Systems with disturbances · Sliding mode control

1 Introduction

The classical point of view for output stabilization with disturbances is that of using a feedback control for decoupling the disturbances from the output (see [1]). This allows one to design a feedback to stabilize the part of the state that is not affected by the disturbances and that appears in the output. Other approach is that of compensating the matched disturbances by using sliding modes (see [2, 3]). For instance, in [4] a finite-time convergent controller is designed for the linear-time invariant systems with matched perturbations, the controller is restricted to the class of systems with full relative degree. A third approach is to use the estimation of the disturbances to compensate them from the part of the state vector that affects directly the output (see e.g. [5]). In case the disturbance has a particular structure, a different method can be carried out, for instance, in [6] the exponential stabilization is addressed when the measured output is affected by sinusoidal disturbances generated by an unknown exosystem with known upper bound of the exosystem order.

F.J. Bejarano (✉) · J. Dávila (✉)
SEPI, ESIME Ticomán, IPN, Av. San José Ticomán 600, C.P. 07340 Mexico
Mexico
e-mail: fjbejarano@ipn.mx

J. Dávila
e-mail: jadavila@ipn.mx

M. Ceccarelli and E.E. Hernández Martinez (eds.), *Multibody Mechatronic Systems*,
Mechanisms and Machine Science 25, DOI 10.1007/978-3-319-09858-6_4

Main contribution: in this paper a control design is proposed to do the controlled output stabilization by using only partial state information. It is shown that under suitable assumptions the part of the unknown input that cannot be decoupled from the controlled output may be compensated by using unknown input estimation.

Paper structure: the description of the system and the problem formulation is given in Sect. 2. Main assumptions required to follow the proposed control design are also given in that section. The main results dealing with the proposed control law allowing the output stabilization are given in Sect. 3, which is split into two subsections. Section 3.1 contains some basic results allowing for to express the original assumptions (in terms of the original system) to the assumptions expressed in terms of the transformed system. In Sect. 3.2, the specific control design is given. An example of a linear system of a Boeing 767 is presented in Sect. 4.

2 Problem Formulation

We consider the class of systems whose dynamics is governed by the following equations

$$\begin{aligned} \dot{x}(t) &= Ax(t) + Bu(t) + Ew(t) \\ y(t) &= Cx(t) + Du(t) + Fw(t) \\ y_c(t) &= C_c x(t) + D_c u(t) + F_c w(t) \end{aligned} \tag{1}$$

where the state vector $x(t) \in \mathbb{R}^n$, the input vector $u(t) \in \mathbb{R}^m$, the disturbance vector $w(t) \in \mathbb{R}^q$, and the system output $y(t) \in \mathbb{R}^p$. All matrices are constant of suitable dimension.

The aim is to design a control law $u(t)$ so that $y_c(t)$ tends to zero asymptotically in spite of the disturbances $w(t)$ and assuming that $x(t)$ is not available.

We define the system matrix $\Sigma = (A, B, C_c, D_c)$ (representing the system involving x and y_c when $w = 0$) and the system matrix $\bar{\Sigma} = (A, E, C, F)$ (representing the system involving x and y when $u = 0$).

Let $\mathscr{V}(\Sigma)$ and $\mathscr{V}(\bar{\Sigma})$ be the weakly unobservable subspaces of Σ and $\bar{\Sigma}$, respectively, let $n_{\mathscr{V}}$ be the dimension of $\mathscr{V}(\Sigma)$, $\mathscr{V}(\Sigma)^{\perp}$ be the orthogonal complement of $\mathscr{V}(\Sigma)$ (see the appendix for a review $\mathscr{V}(\Sigma)$), and $\langle A|B\rangle$ be the controllability subspace of the (A, B) pair. Let M be a full row rank matrix chosen so that $M\mathscr{V}(\Sigma) = 0$. Along the paper we will assume that,

A1. $M\langle A|B\rangle = \mathbb{R}^{n-n_v}$.

A2. $\operatorname{im}\begin{bmatrix} E \\ F_c \end{bmatrix} \subset \left((\mathscr{V}(\Sigma) \times 0) + \operatorname{im}\begin{bmatrix} B \\ D_c \end{bmatrix} \right)$

A3. The system $\bar{\Sigma} = (A, E, C, F)$ is strongly detectable (i.e., the zeros of $\bar{\Sigma}$ belong to left open half-plane of the complex plane).

3 Output Stabilization

Let V be a full column rank matrix whose columns form a basis of $\mathcal{V}(\Sigma)$. In view of (9) in Appendix, there exists a matrix K_v so that $(A - BK_vV^+)V$ belongs to $\mathcal{V}(\Sigma)$ and $(C_c - D_cK_vV^+)V = 0\left(V^+ := (V^TV)^{-1}V^T\right)$. Thus, let us define the following nonsingular matrix

$$P = \begin{bmatrix} M \\ V^+ \end{bmatrix}$$

where M is a full row rank matrix so that $MV = 0$. In view of the previous considerations, system (1) is transformed, by means of $z = Px$, into the following system

$$\begin{bmatrix} \dot{z}_1(t) \\ \dot{z}_2(t) \end{bmatrix} = \begin{bmatrix} A_{11} & 0 \\ A_{21} & A_{22} \end{bmatrix}\begin{bmatrix} z_1(t) \\ z_2(t) \end{bmatrix} + \begin{bmatrix} B_1 \\ B_2 \end{bmatrix}(u(t) + K_v z_2(t)) + \begin{bmatrix} E_1 \\ E_2 \end{bmatrix} w(t)$$
$$y_c(t) = C_1 z_1(t) + D_c(u(t) + K_v z_2(t)) + F_c w(t)$$

where $z_1 \in \mathbb{R}^{n-n_v}$ and $z_2 \in \mathbb{R}^{n_v}$ (n_v is the dimension of $\mathcal{V}(\Sigma)$).

3.1 Preliminary Results

Lemma 1 Under assumption A2, there exists a matrix $\Gamma \in \mathbb{R}^{m \times q}$ so that

$$\begin{bmatrix} B_1 \\ D_c \end{bmatrix}\Gamma = \begin{bmatrix} E_1 \\ F_c \end{bmatrix} \tag{2}$$

Proof Firstly, we notice that

$$P\mathcal{V}(\Sigma) = \mathrm{im}\begin{bmatrix} 0 \\ I_{n_v} \end{bmatrix} \tag{3}$$

Thus, the inclusion of A2, in new coordinates takes the form

$$\mathrm{im}\begin{bmatrix} E_1 \\ E_2 \\ F_c \end{bmatrix} \subset \left((0 \times \mathbb{R}^{n_v} \times 0) + \mathrm{im}\begin{bmatrix} B_1 \\ B_2 \\ D_c \end{bmatrix}\right). \tag{4}$$

By inspection it is clear that (4) is equivalent to the following inclusion

$$\operatorname{im}\begin{bmatrix} E_1 \\ F_c \end{bmatrix} \subset \operatorname{im}\begin{bmatrix} B_1 \\ D_c \end{bmatrix} \tag{5}$$

Hence, (2) is a direct consequence of (5).

What Lemma 1 implies is that under assumptions A2, the perturbations are matched (to the control) when considering only the system involving z_1 and y (taking out z_2). Thus, using the estimation of the perturbation $w(t)$ we could compensate the part of it that affects z_1 and $y(t)$. □

Lemma 2 (A_{11}, B_1) is controllable provided assumption A1 is satisfied.

Proof By (3), in new coordinates $\mathscr{V}(\Sigma)^{\perp}$ is equal to $\operatorname{im}\begin{bmatrix} I_{n_v} \\ 0 \end{bmatrix}$, and in the same coordinates $\langle A - BK^* | B\rangle$ takes the form

$$\operatorname{im}\begin{bmatrix} B_1 \\ B_2 \end{bmatrix} + \operatorname{im}\begin{bmatrix} A_{11}B_1 \\ * \end{bmatrix} + \cdots + \operatorname{im}\begin{bmatrix} A_{11}^{n-1}B_1 \\ * \end{bmatrix}$$

Thence, regarding that $\langle A|B\rangle$ is identical to $\langle A - BK^*V^+|B\rangle$, the assumption $M\langle A|B\rangle = \mathbb{R}^{n-n_v}$ implies, that

$$\begin{aligned} \mathbb{R}^{n-n_v} &= MP^{-1}P\langle A - BK^*V^+|B\rangle \\ &= [I \quad 0]\operatorname{im}\begin{bmatrix} B_1 & A_{11}B_1 & \cdots & A_{11}^{n-1}B_1 \\ * & * & \cdots & * \end{bmatrix} \\ &= \operatorname{im}\begin{bmatrix} B_1 & A_{11}B_1 & \cdots & A_{11}^{n-1}B_1 \end{bmatrix} = \langle A_{11}|B_1\rangle \end{aligned}$$

which, in turn implies that (A_{11}, B_1) is controllable. □

3.2 Control Design

Since $\bar{\Sigma} = (A, E, C, F)$ is assumed to be strongly detectable (i.e., the zeros of $\bar{\Sigma}$ are of negative real part), then the state $x(t)$ and $w(t)$ can be reconstructed asymptotically ($w(t)$ is assumed to be unknown) by following a differentiation procedure. In [7–10] different ways of estimating $x(t)$ and $w(t)$ asymptotically were proposed. We refer the reader to such works for more details of the estimation procedure. Thus, the estimates of $x(t)$ and $w(t)$ is denoted by $\hat{x}(t)$ and $\hat{w}(t)$, respectively.

Thus, we propose to design the control $u(t)$ by means of the following formula,

$$u(t) = -K\hat{z}_1(t) - K_v\hat{z}_2(t) - \Gamma\hat{w}(t) \tag{6}$$

where K is to be designed so that $(A_{11} - B_1K)$ matrix be stable, which is possible due to Lemma 1.

Theorem 1 Under assumptions A1–A3, if $u(t)$ is designed according to (6), $y(t)$ tends exponentially to zero.

Proof Taking into account (2), substitution of $u(t)$ from (6) into (1) yields the following dynamic equations involving $z_1(t)$ and $y_c(t)$,

$$\begin{aligned}\dot{z}_1 &= (A_{11} - B_1K)z_1 + B_1K(z_1 - \hat{z}_1)\\ &\quad + B_1K_v(z_2 - \hat{z}_2) + E_1(w - \hat{w})\\ y_c &= (C_1 - D_cK)z_1 + D_cK(z_2 - \hat{z}_2)\\ &\quad + D_cK_v(z_2 - \hat{z}_2) + F_c(w - \hat{w})\end{aligned} \tag{7}$$

From the equations above, the proof is straightforward since $\hat{x}(t)$ and $\hat{w}(t)$ converge asymptotically to $x(t)$ and $w(t)$, respectively, (i.e., $\hat{z}(t) = P\hat{x}(t)$ tends to $z(t) = Px(t)$). Therefore, the errors $(z_1 - \hat{z}_1)$, $(z_2 - \hat{z}_2)$, and $(w - \hat{w})$ will vanish, and, since $(A_{11} - B_1K)$ is Hurwitz, z_1 will tend asymptotically to zero. Thereby, $y(t)$ will tend to zero also. □

4 Example: Control Regulation of a Boeing 767

In this example we apply the theoretical results to the regulation problem of an airplane (see, e.g., [11]). Let us consider the following linear longitudinal model of a Boeing 767,

$$\begin{aligned}\dot{\bar{x}}(t) &= \bar{A}\bar{x}(t) + \bar{B}u(t) + \bar{E}w(t)\\ \bar{y}(t) &= \bar{C}x(t) + \bar{D}u(t) + \bar{F}w(t)\end{aligned}$$

where $\bar{x}$ is the state vector whose components are the surge velocity u_l, the angle of attack α, the pitch rate q, and the pitch angle θ. The input vector u has two components: elevator deflexion δ_e and thrust angle δ_T. The flight conditions are: speed 980 ft/s, altitude 35,000 ft, mass 184,000 lbs, and Mach number 0.8. Matrices of the system take the following values:

$$\bar{A} = \begin{bmatrix} -0.0168 & 0.1121 & 0.0003 & -0.5608\\ -0.0164 & -0.7771 & 0.9945 & 0.0015\\ -0.0417 & -3.6595 & -0.9544 & 0\\ 0 & 0 & 1 & 0\end{bmatrix}$$

$$\bar{B} = \begin{bmatrix} -0.0243 & 0.0519\\ -0.0634 & -0.0005\\ -3.6942 & 0.0243\\ 0 & 0\end{bmatrix}, \bar{E} = \begin{bmatrix} 1 & 0\\ 0 & 1\\ 0 & 0\\ 0 & 0\end{bmatrix}$$

$$\bar{C} = \begin{bmatrix} 1 & 0 & 0 & 0\\ 0 & 0 & 0 & 1\end{bmatrix}, \quad \bar{D} = 0, \quad F = 0.$$

The **goal** is to maintain a constant surge velocity of 800 ft/s, and a pitch angle equal to 2°. This in spite of the disturbances affecting the performance of the airplane. Thus, by defining $r = \begin{bmatrix} u_d \\ \theta_d \end{bmatrix} = \begin{bmatrix} 800 \\ 2 \end{bmatrix}$, we can define a extended state vector $x = \begin{bmatrix} \bar{x}^T & r^T \end{bmatrix}^T$ and so we have a linear system as in (1) with

$$A = \mathrm{diag}(\bar{A}, 0),\ B = \begin{bmatrix} \bar{B} \\ 0 \end{bmatrix}, C = \mathrm{diag}(\bar{C}, I),$$
$$D = 0, E = \begin{bmatrix} \bar{E} \\ 0 \end{bmatrix}, F = 0$$

and the control output $y_c = \bar{y} - r$, therefore,

$$C_c = \begin{bmatrix} \bar{C} & -I \end{bmatrix}, D_c = \bar{D}, F_c = \bar{F}$$

Thus, $\bar{y}(t)$ tends to r, when y_c tends to zero. With this matrices the control takes the following form

$$u(t) = -K\hat{z}_1(t) - K_v\hat{z}_2(t) - \Gamma\hat{w}(t)$$

with

$$K = \begin{bmatrix} -3.46 & 1.64 & -13.24 \\ 80.64 & 179.42 & 286.83 \end{bmatrix}$$
$$K_v = \begin{bmatrix} 0.53 & -0.71 & -0.46 \\ 6.84 & -1.70 & 3.99 \end{bmatrix}$$
$$\Gamma = \begin{bmatrix} 0.12 & 0 \\ 19.32 & 0 \end{bmatrix}$$

In view of the structure of Γ, it is clear that only the first unknown input is needed to be estimated. In Fig. 1 the estimation error of the state vector $x(t)$ and the

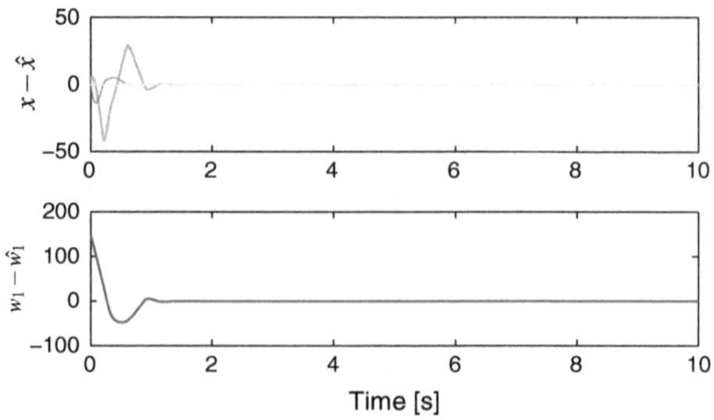

Fig. 1 Estimation error of the states (*above*) and the first unknown input (*below*)

Fig. 2 Output using partial perturbation compensation

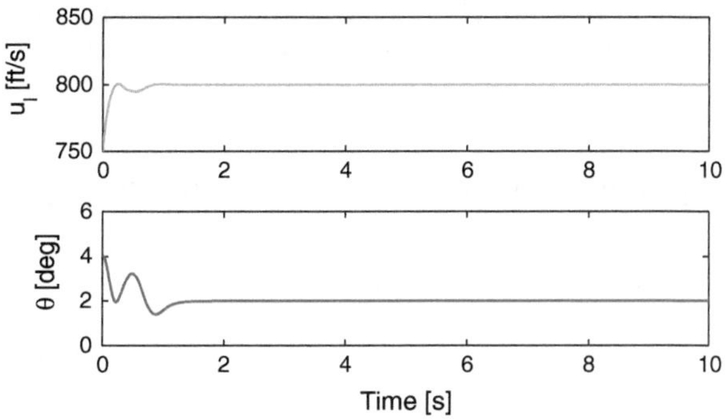

Fig. 3 Output without perturbation compensation

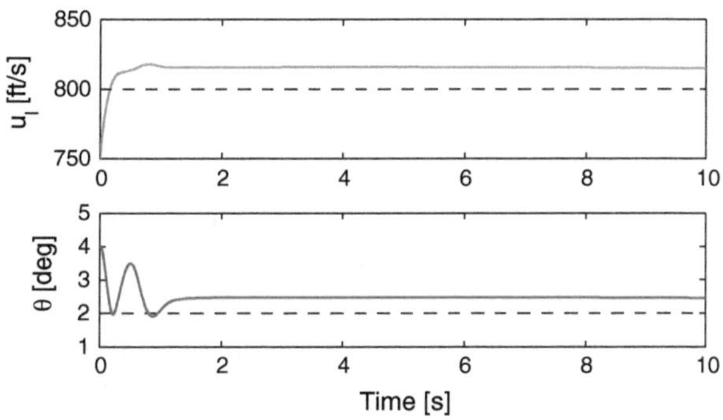

unknown input $w_1(t)$ are shown. The output variables u_l and θ are depicted in Fig. 2, there the effectiveness of the proposed control is shown since both come near to its respective reference signal. Figure 3 presents the case when the term $\Gamma\hat{w}$ is excluded from the control law. There we can see that u_l and θ do not approach to the signals references.

5 Conclusions

We have proposed a control design methodology for stabilizing a controlled output. The main idea of the control design has been the direct compensation of the part of that unknown input that cannot be decoupled (by using a feedback control) from the controlled output. To apply the control law to the system, it was needed to estimate the estate vector and the part of the unknown input that is to be compensated. We have shown by means of an example that the proposed control design may be used for output regulation also. For a future work, it is aimed to relax the assumptions needed to accomplish the output stabilization.

Acknowledgments Authors acknowledge the support of joint research project CONACyT-CNRS under grant 209247 and the IPN SIP project 20141364.

Appendix: Weakly Unobservable Subspace

Let us associate to the matrix system $\Sigma = (A, B, C, D)$ the dynamic system

$$\begin{aligned}\dot{x}(t) &= Ax(t) + Bu(t)\\ y(t) &= Cx(t) + Du(t)\end{aligned}$$

For this triple, the subspace $\mathscr{V}(\Sigma)$ related to Σ is the maximal subspace satisfying (8) (see, e.g. [1, 12])

$$\begin{bmatrix} A \\ C \end{bmatrix} \mathscr{V}(\Sigma) \subset (\mathscr{V}(\Sigma) \times 0) + \operatorname{im} \begin{bmatrix} B \\ D \end{bmatrix} \tag{8}$$

This subspace is known as the weakly unobservable subspace (for the case when $D = 0$, $\mathscr{V}(\Sigma)$ coincides with the maximal (A, B) invariant subspace (or controlled invariant subspace). $\mathscr{V}(\Sigma)$ consists of the vectors x_0 for which there exits an input $u_0(t)$ (depending of x_0) so that the corresponding output $y(t; u_0(t), x_0) = Cx(t; u_0(t), x_0) + Du_0(t)$ is identically equal to zero for all t.

Properties involving $\mathscr{V}(\Sigma)$:

1. It is known that, for any initial condition x_0 and any input $u(t)$,

   $$y(t) = 0 \text{ for all } t \text{ implies } x(t) = 0 \text{ for all } t$$

   if and only if $\mathscr{V}(\Sigma) = 0$ (in such a case Σ is called strongly observable)
2. Let V be a full column rank matrix whose columns span $\mathscr{V}(\Sigma)$, then there exists a pair (K_v, Q) so that

   $$\begin{aligned} AV &= VQ + BK_v \\ CV &= DK_v \end{aligned} \tag{9}$$

   From the first equation above, K_v may be obtained by solving the algebraic equation

   $$\begin{bmatrix} AV \\ CV \end{bmatrix} = \begin{bmatrix} V & B \\ 0 & D \end{bmatrix} \begin{bmatrix} Q \\ K_v \end{bmatrix}$$

3. By the defining the Rosenbrock matrix $R(\lambda)$ as

$$R(\lambda) = \begin{bmatrix} \lambda I - A & -B \\ C & D \end{bmatrix}, \lambda \in \mathbb{C}$$

A complex number λ_0 is called a zero of Σ if

$$\operatorname{rank} R(\lambda_0) < n + rank \begin{bmatrix} -B \\ D \end{bmatrix}.$$

It was shown in [13] that $\mathscr{V}(\Sigma) = \{0\}$ if, and only if,Σ has no zeros. There it was shown also that for any initial condition x_0 and any input $u(t)$,

$$y(t) = 0 \text{ for all } t \text{ implies } x(t) = 0 \text{ for all } t$$

if, and only if, all the zeros of Σ belong to the open left half-plane of the complex plane (in this case Σ is called strongly detectable). This property allows for estimating $x(t)$ asymptotically in spite of the lack of knowledge of $u(t)$.

References

1. Trentelman H, Stoorvogel A, Hautus MLJ (2001) Control theory for linear systems. Communications and control engineering. Springer, New York
2. Hashimoto H, Utkin V, Xu J, Suzuki H, Harashima F (1990) In: Proceeding of IECON'90, Pacific Grove CA, 1990, pp 34–39
3. Utkin V (1992) Sliding modes in control and optimization. Springer, Berlin
4. Angulo MT, Fridman LM, Levant A (2011) Int J Syst Sci 42(11):1847
5. Ferreira A, Bejarano F, Fridman L (2013) Int J Robust Nonlinear Control 23(7):754
6. Marino R, Santosuosso GL, Tomei P (2008) Eur J Control 14(2):131
7. Fridman L, Levant A, Davila J (2007) Int J Syst Sci 38(10):773
8. Bejarano F, Fridman L, Poznyak A (2009) SIAM J Control Optim 48(2):1155
9. Fridman L, Davila J, Levant A (2011) Nonlinear Anal: Hybrid Syst 5(2):189
10. Bejarano F, Fridman L (2010) Int J Control 83(9):1920
11. Bottasso CL, Leonello D, Savini B (2008) IEEE Trans Control Syst Technol 16(6):1152
12. Molinari B (1976) IEEE Trans Automat Control 21(5):761
13. Hautus M (1983) Linear Algebra Appl 50:353

Development of a Heliostat for a Solar Tower Power Plant

Jorge A. García Pitol, Alicia Hernandez Gutierrez, Manuel Toledano Ayala, Juan C.A. Jáuregui Correa, Enma V. Godoy Avendaño and Oswaldo Mendoza Herbert

Abstract Since renewable energies are gaining significance, the development of efficient devices is vital to make them competitive against the current used sources of energies. The present paper contains the development of a Heliostat for a Solar Tower Plant, the description of its elements as well as the Finite Element Analysis that was taken into account to design the different mechanisms that form the heliostat. The main innovation in the design is the modularity of the whole device, the simplification of the movement mechanisms and a high resistance under severe wind conditions. According to the Finite Element Method analysis the heliostat is able to withstand a wind speed up to 34 m/s, it accomplishes the modularity desired but still present backlash issues in the azimuth movement. The next step in the design is to reduce this resultant backlash with the modification of the azimuthal mechanism, to employ another transmission method different from gearboxes, and to compensate the inaccuracy within the dynamic control system.

Keywords Heliostat · Solar · Plant · Mechanism · Analysis

J.A. García Pitol (✉) · A.H. Gutierrez · M.T. Ayala · J.C.A. Jáuregui Correa
E.V. Godoy Avendaño · O.M. Herbert
Autonomous University of Queretaro, Queretaro, Mexico
e-mail: j.garcia.pitol@gmail.com

A.H. Gutierrez
e-mail: alicia.hernandez.gutierrez@outlook.com

M.T. Ayala
e-mail: toledano@uaq.mx

J.C.A. Jáuregui Correa
e-mail: jc.jauregui@uaq.mx

E.V. Godoy Avendaño
e-mail: enma.godoy@hotmail.com

O.M. Herbert
e-mail: oswaldo.herbert@hotmail.com

M. Ceccarelli and E.E. Hernández Martinez (eds.), *Multibody Mechatronic Systems*,
Mechanisms and Machine Science 25, DOI 10.1007/978-3-319-09858-6_5

1 Introduction

Nowadays human society depends mostly on fossil fuels, according to the International Energy Agency (IEA) in 2011 the world electricity generation was of 22,126 TWh; 41.3, 21.9 and 4.8 % came from Coal, Natural Gas and Oil, respectively. Renewable energies are produced from natural sources and have a low environmental impact, for example hydroelectric, wind energy, solar energy, etc.

Solar tower power plants consist in a field of heliostats and a central tower, the purpose of the heliostats is to follow the sun during the day and redirect the solar radiation to a receiver on top of the tower in order to obtain a heat spot to concentrate energy and then introduce it to an energy generation system using a heat transfer fluid. One of the main challenges of the heliostat design is to eliminate the backlash present in the gearboxes of the different movement mechanisms, although it is small within the axes when is translated to the complete structure it increases and becomes an important deviation of the reflection.

This paper aims to explain the development of a Heliostat for a Solar Tower Power Plant using Finite Element Analysis and graphic modeling, the design tries to eliminate the backlash issue, obtain a modular structure to facilitate its transportation and storage, and to be able to stand high speed wind conditions.

1.1 Solar Tower Plants

Concentrated solar power has been under investigation for several decades, and is based on a simple general scheme: using mirrors, sunlight can be redirected, focused and collected as heat, which can be used to power a turbine or a heat engine to generate electricity.

According to Mousazadeh et al. [3] there are two general categories of solar collectors. The first includes stationary, non-concentrating collectors, in which the same area is used for both interception and absorption of incident radiation. The second category consists of sun-tracking, concentrating solar collectors, which utilize optical elements to focus large amounts of radiation onto a small receiving area and follow the sun throughout its daily course to maintain the maximum solar flux at their focus.

Power towers (also called central receivers) are in the practice already in the commercialization stage but it can be considered a high technology because of all the infrastructure needed to operate a plant like this. Because of their higher operating temperatures, power towers have the potential to achieve higher efficiency and lower-cost Thermal Energy Storage (TES) compared with current trough technology. Power towers use heliostats—reflectors that rotate about both the azimuth and elevation axis—to reflect sunlight onto a central receiver. A large power tower plant can require from several thousand to more than one-hundred thousand heliostats (Fig. 1), each under computer control. Because they typically

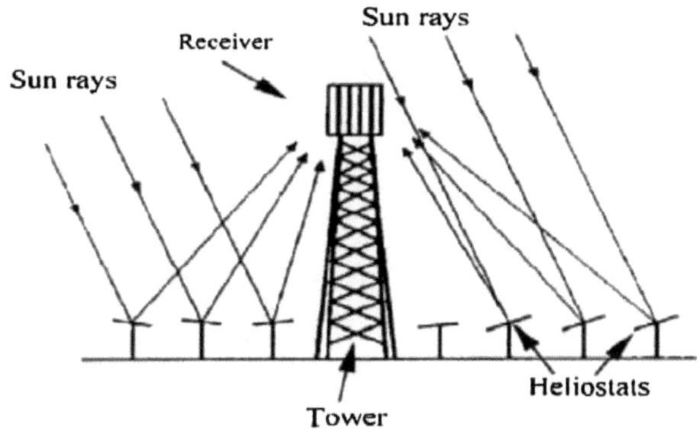

Fig. 1 Heliostat field collector—*source* [1]

constitute about 50 % of the plant cost, it is important to optimize the heliostat design. Heliostat size, weight, manufacturing volume, and performance are important design variables, and developers have selected different approaches to minimize cost. Some heliostat technology can be installed on relatively uneven land, with 5 % or more slope, thereby reducing site-preparation costs for new projects .

The two principal power tower technology concepts currently being pursued by developers are defined by the Heat Transfer Fluid (HTF) in the receiver: steam or molten salt [4]. In direct-steam power towers, heliostats reflect sunlight onto a receiver on a tower where water is evaporated and superheated to produce steam which feeds a turbine generator to generate electricity. In the molten salt power towers the HTF is the molten salt which is used as heat storage and also to transfer heat to water and then generate electricity [1].

2 Heliostat Design

In order to reach the basic requirements of an heliostat the design took into account as main priorities to create a modular device, capable of following the sun and reflecting the incident energy to a central tower; also easy to transport, to maintain and repair due to its configuration. The main challenges of the design were to be able to achieve high resistance to meteorological conditions, mainly to winds.

The aim was to develop a unit that could resist winds up to 34 m/s, which are the speed of a category 1 hurricane; the purpose of the heliostat's high resistance is to provide a tool to act also in case of a moderate impact natural disaster. This means that a solar tower power plant with this heliostat design can become a source of energy on the daily basis, but also an emergency energy supply. The heliostat is powered through a photovoltaic cell that generates enough energy to feed the two engines for each movement, azimuth and elevation.

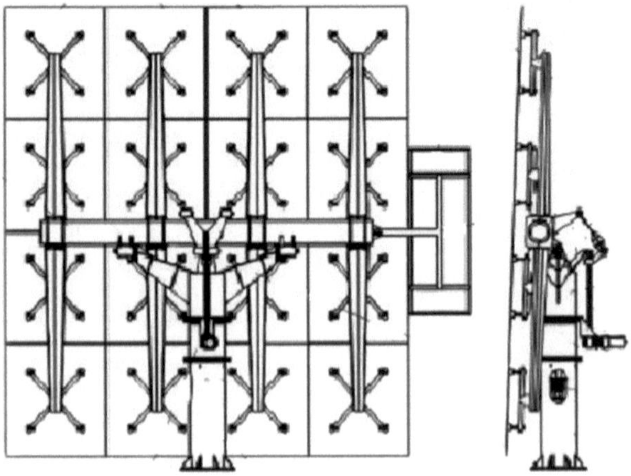

Fig. 2 Heliostat design

2.1 Modularity

One of the main obstacles in this kind of devices is the transportation of the elements that conform a heliostat, mostly due to their dimensions. Part of the design intended to overcome this difficulty without sacrificing the size of the reflecting surface.

In order to do so the heliostat in its upper section is constituted by a main support tube, then four arms on which the mirror frames are placed (Fig. 2). The total reflecting surface of this model of heliostat is 16 m^2 with individual mirrors of 1 m^2. This represents a considerably lower storage and transportation space. Also, when necessary, there is the possibility of repairing any damage mirror in small units instead of large ones.

3 Azimuth Mechanism

The azimuth mechanism has a hole-body rotor that allows the superior section of the heliostat to rotate 360°. The mechanism is built in such a way that friction between the superior and lower sections of the heliostats are avoided, because of the gap left between the main support and the mirror frame. The Fig. 3 shows a diagram of the mechanism and a detail of the lowest part of the main support.

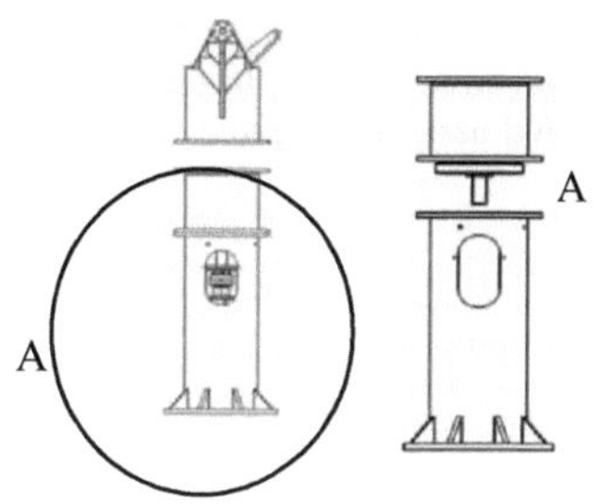

Fig. 3 Azimuth mechanism and detail

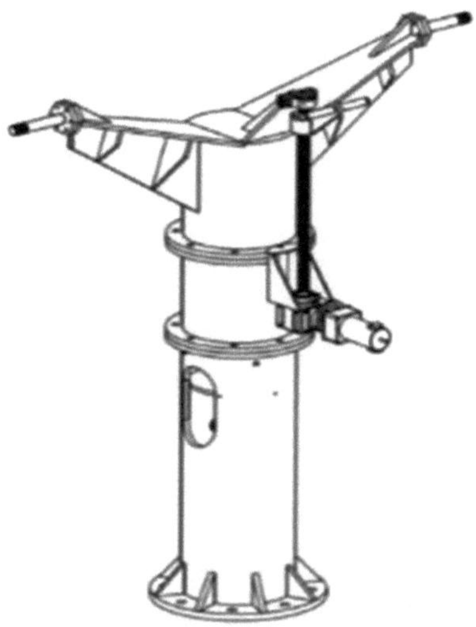

Fig. 4 Elevation mechanism

4 Elevation Mechanism

The elevation mechanism (Fig. 4) consists in a four bars mechanism working with a ball screw and actioned by an electric motor. It moves the reflective surface up and down until 90°. The best quality of this configuration is the zero backlash of the ball screw, this means that the position of the mirrors would be exact depending on the time of the day.

5 Finite Element Method Analysis

To justify the structural ideal design, the main proposals were subjected to Finite Element Method (FEM) Analysis in order to find possible weak points and allow reinforcement as a security measure. The FEM cuts a structure into several elements, and then reconnects elements at nodes as if nodes were drops that hold elements together. This process results in a set of simultaneous algebraic equations.

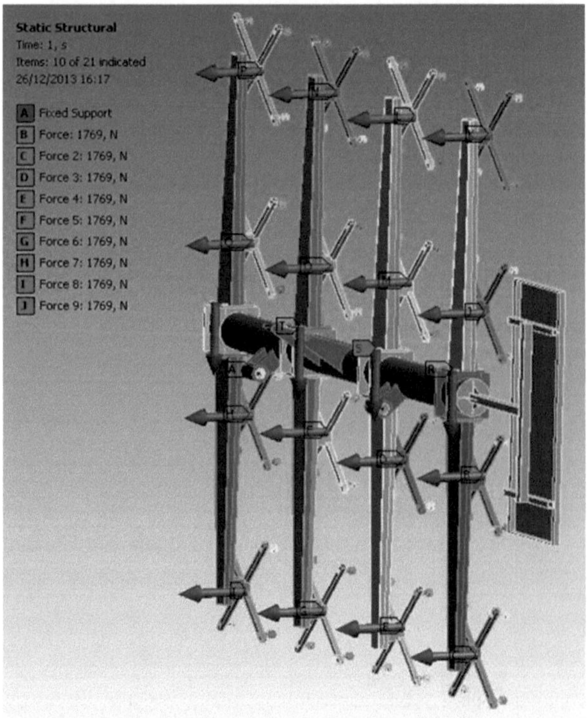

Fig. 5 Upper movable structure subjected to the wind and weight forces

[2] However the FEM has several disadvantages: only approximate solutions are obtained so it is interpretive and is subject to thc knowledge and experience of the user, an error caused by the user is common and fatally modifies all the final result.

To take this orderly and avoid user's mistakes, the system was divided into three main components to evaluate separately: upper movable structure (Fig. 5), the moving middle structure (Fig. 6) and the lower fixed structure (Fig. 7). Each of these components may have in turn interrelated modules with each other and depend on each other. Starting from different structures to a final model that was tested was reached and based on the results was modified to establish a reliable architecture and a possible construction. The movable upper structure has different modules that were analyzed as parts were considered critical to the proper functioning of the entire system.

This mechanism was simulated fully assembled, removing dynamic considerations as motion analysis could present a high risk of providing inaccurate data. As is already defined previously, the materials and thicknesses of the structural elements are being reconsidered, this time taking into account the production systems used to make the prototype.

For the analysis of the upper movable structure (Fig. 5) the wind load is perpendicularly applied to the surface of every arm, the forces are distributed in different points of the model so finally they are fixed to the support structure. As a result the free-body diagram in Fig. 5 shows the upper movable structure in a time

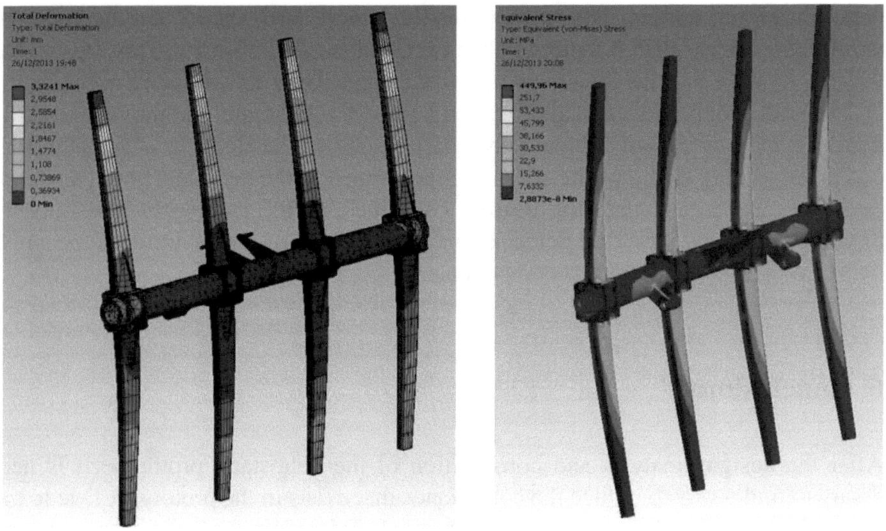

Fig. 6 Upper movable structure's stress and deformation

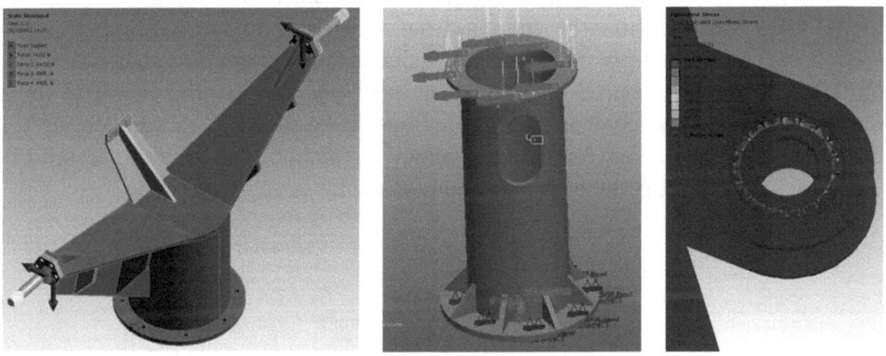

Fig. 7 Examples of some heliostat elements under the FEM analysis

of 1 s under the action of nine forces, applied to each mirror frame's base, with a value of 1,769 N. The magnitude of the forces was set from the general formula of the aerodynamic resistance Eq. (1)

$$F = 1/2 * \rho * C_v * v^2 * S \tag{1}$$

The vertical load due to the weight of the structure is also considered to lay on the central support axis of the upper movable structure, with a maximum measurement of 1,000 kg.

For the evaluation of the deformation and stress of the upper movable structure the Von Mises yield criterion was used because it represents the system's behavior better than other criteria. The result of the analysis (Fig. 6) shows that the maximal deformation is less than 3.5 mm and the maximal stress is 53 MPa. This means that the deformation and the stress are below the capacity of the structure.

The same analysis was made for each part of the structure, the movable middle structure, the lower fixed structure, the bearings and all the elements of the heliostat.

On the joining elements the stress is concentrated in the bearings, but in any case the forces are greater than the bearing capacities. In the entire structure, mechanisms, bearing and joints the deformation never exceeds 3.5 mm nor does the stress overtake the resistance of the components.

6 Conclusions

After the design, analysis and construction of the heliostat's prototype it is necessary to find a way to reduce the compliance that exists in the prototype. Due to the sum of the backlash, elasticity and thermal deformation of the structure and mechanisms within the heliostat, there is a strong compliance that decreases the accuracy of the reflection towards the heliostat objective.

The system dynamic (Fig. 8) became an important issue, due to the reduction mechanisms that transmit the motion from the motor to the medium movable section of the heliostat, the magnitude of reduction is so important that a deviation of 0.01° within the source of movement becomes 30° on the heliostat upper movable section. The control system has now to be redesign to compensate this deviation and try to achieve a better accuracy of the heliostat, also the design of a transmission without gears is the second part of the development.

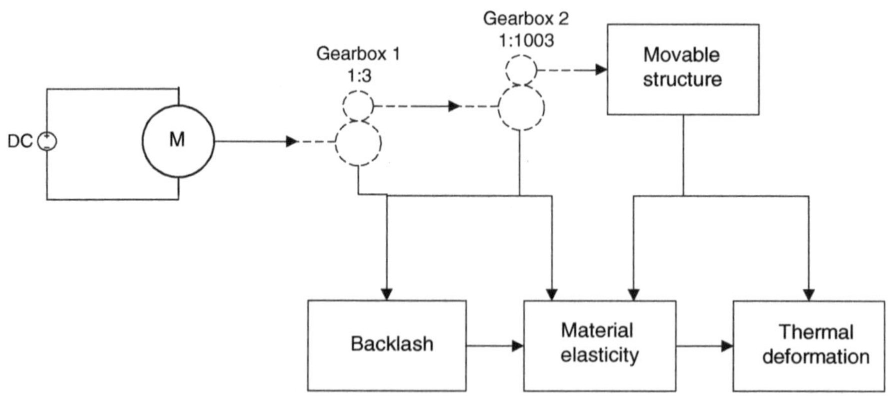

Fig. 8 Azimuthal movement compliance

As for the capability of the structure to endure strong wind conditions in any position the design approved the Finite Element Method Analysis and the prototype so far is stable and without trouble on location. Although no wind tests have been made so far.

Further experimentation is required to improve the present heliostat design, but the experience acquired through this first model shows that it is possible to develop high-technology devices with common elements. It is important to continue with the application of the FEM analysis since these tools open a new gap to improve the design from the structural analysis.

Acknowledgments The working team would like to acknowledge the Autonomous University of Queretaro for the support through the elaboration of this project, also CaruArdica for the trust invested in our institution and staff.

References

1. International Energy Agency (2013) Key world energy STATISTICS, p 24
2. de Weck O, Yong Kim I (2004) Finite element method. http://web.mit.edu/16.810/www/16.810_L4_CAE.pdf. Accessed 12 Jan 2004
3. Mousazadeh H, Keyhani A, Javadi A, Mobli H, Abrinia K, Sharifi A (2009) A review of principle and sun-tracking methods for maximizing solar systems output. Renew Sustain Energy Rev 13(18):1800–1818
4. Barlev D, Vidu R, Stroeve P (2011) Innovation in concentrated solar power. Sol Energy Mater Sol Cells 95(10):2703–2725

Mechatronic Sizing of Ball-Screw Feed Drives

R. Hecker, D. Vicente and G. Flores

Abstract This work proposes a procedure to sizing a ball-screw feed drive considering the closed-loop performance. This is achieved by additional steps to the traditional mechanical sizing, where a high frequency model is used to tune the controller in order to achieve adequate stability margins. After that, the tracking error is evaluated for different screw candidates, under nominal and non-nominal operational conditions for a high speed machine tool. Finally, the optimum screw assemble is selected based on minimum torque requirement. As a main conclusion, the screw lead is identified as a high influencing factor in the system dynamics and then on the closed-loop performance, showing the need of an integrated design.

Keywords Mechatronic sizing · Ball screw · Feed drive · High-frequency model

1 Introduction

The ball screw is a power transmission element with high degree of precision, low friction, reduced backlash, and can withstand high forces and high rotational velocities. It is used in combination with a rotational motor to generate lineal motion in precise machines [1] used to handle and manufacture parts. Applications of this type of feed drives can be found in CNC machine tools, positioning systems in electronic manufacturing, inspection equipment, and automated metrology. Also it can be found in actuators for multibody systems [2, 3] like surgical parallel robots, pan-tilt mechanism for a parallel robot, actuation module for wearable robot, and prismatic actuators for hexapods.

R. Hecker (✉) · D. Vicente · G. Flores
Universidad Nacional de La Pampa, Santa Rosa, Argentina
e-mail: hecker@ing.unlpam.edu.ar

R. Hecker
CONICET, Santa Fe, Argentina

M. Ceccarelli and E.E. Hernández Martinez (eds.), *Multibody Mechatronic Systems*,
Mechanisms and Machine Science 25, DOI 10.1007/978-3-319-09858-6_6

In general these applications require precise positioning. For a system with a ball screw and a rotary encoder, the overall accuracy is composed by the screw error and the closed-loop position error. There are different accuracy classes for each screw size; however, it is important to mention that the screw error coming from its manufacturing process has a large deterministic component, therefore it can be measured by a laser interferometer and online compensated [1]. However, the error coming from the closed-loop dynamics is the error that must be addressed by adequate controller selection and tuning. In order to prevent that mechanical dynamic limits the controller action, the mechanical design must be integrated with the controller design.

Traditionally, a ball screw is selected to achieve a desired service life under mechanical requirements given by some typical machine cycle. After that, a servomotor is selected to fulfill torques and velocities required also by the machine cycle. Finally, a controller is selected and tuned to accomplish low tracking errors for specific trajectories. Although some variations, the ball screw and motor selection processes are in general well established in commercial catalogs [1, 4] and traditional books [1]. On the other side, different control strategies can be found in the literature but always for predefined systems [5, 6]. Therefore, an optimal design solution is not accomplished because these three design stages are sequentially executed without interaction between them.

Few integrated design methodologies are found in the literature for this kind of systems. The effect of the mechanical parameters, such as the screw diameter and lead, on the closed-loop performance was study in [7, 8]. However, the approach of these works requires considerable effort to obtain expressions of closed-loop performance, for instance the band width and the tracking error, in terms of mechanical parameters. In addition some mechanical constraints are neglected. A more complete mechatronic design is presented in [9], where the motor constraints, the ball screw constraints, and the closed loop performance are represented as bounds in the lead-diameter space. This graphical approach allows a suitable range of solutions given the many conflicting requirements. However, a simplified system model is used to study the performance of the closed loop. The accurate prediction of the system natural modes is a major component to study the closed loop performance, especially for trajectories with demanding acceleration profiles.

This work presents a mechatronic design by integrating the screw and motor selection with the control performance. Particularly, an accurate model to predict the system natural modes is used to tune the controller in order to reach stable responses. The screw assemble is simulated to evaluate the tracking error for nominal and non-nominal operational conditions. Finally, the optimum screw assemble is selected based on minimum torque requirement.

2 Proposed Mechatronic Procedure

The proposed procedure is summarized as steps that can be executed iteratively until an optimal solution is found. The steps from 1 to 5 are traditional in the literature and are included here for completeness. On the other hand, steps from 6 to 8 are included to obtain a mechatronic approach. Finally, step 9 is included to take into account optimality.

1. A typical machine cycle is defined, including percentage of different types of operations with their characteristic forces and velocities.
2. The dynamic load rate is calculated based on the required service life.
3. A screw is selected that closely meets the dynamic load rate. It becomes a candidate with a given diameters and lead. In general, the type of nut, its preload, and the tolerance grade is previously selected.
4. The candidate is tested to fulfill the maximum rotational speed due to maximum velocity of ball recirculation, the critical speed to prevent the excitation of the first natural bending frequency, and the possibility of buckling using the maximum expected axial force. At this point, suitable configuration of end fixity must be selected.
5. After that, a motor candidate must be selected. The main operation points must be calculated as pairs of torque-velocity that describe nominal and extreme transient operations. Additionally, the inertia ratio criterion must be fulfilled, that is the relation between the rotor inertia and the total reflected inertia from the mechanical system. This recommendation is to prevent limit cycles in precise control of systems with backlash and nonlinearity behavior, as it is general found in the screw-nut interface.
6. A controller is selected. Many alternatives can be used at this stage. Although nonlinear strategies can be found in the literature, most modern machine controllers use linear feedback control with additional feedforward compensations.
7. A system model covering a convenient frequency range is evaluated to include the first natural modes. For completeness, other models must be included, such as the motor current response and a model for the friction.
8. The controller is tuned and simulated with the integrated models to evaluate closed-loop parameters, such as stability margins and tracking errors.
9. Finally, the screw candidate is checked for optimum lead based on minimum torque requirement.

3 Example

This example shows the described mechatronic procedure in the selection of a screw assemble for a machine tool. Based on the machine requirements, screw candidates are evaluated simultaneously instead of evaluating one by one iteratively.

Table 1 Machine cycle

Cycle	Velocity (m/min)	Axial load (N)	Percentage of share (%)
Machining	9	1,125	30
Rapid traverse	42	120	20
Acceleration	4, 5	1,837	10
Stand still with machining	0	1,125	20
Stand still without machining	0	0	20

3.1 Machine Duty Cycle

A characteristic cycle for a high speed milling machine is presented in Table 1. First, velocities and forces for the machining stage are estimated, combining different tools diameter and different material removal rates for milling of aluminum. Second, the rapid traverse is the maximum velocity without machining, where only the friction forces are considered. After that, the acceleration and deceleration requirements are calculated including the inertia forces and the machining forces as well. Also, the feed drive must be able to standstill machining forces when another machine axis is moving. Finally, a non operation time must be included to account for dwell times.

3.2 Screw Selection and Verification

The dynamic load rate is calculated using the data from the machine cycle and the expected machine life. First, the average axial force acting on the screw is calculated as the cubic mean value, where each individual load is weighed with the fraction of share and also with the fraction of velocity respect to the overall average velocity. In this case the mean force value is 832 N.

After that, the screw service life must be calculated from the total machine desired life and the percentage of utilization of the ball screw. In this case the machine must last 100,000 h where the screw executed movements only 60 % of the total cycle, as Table 1 shows. The screw service life is converted from hours to number of revolutions using the screw lead and the mean linear velocity from the cycle. Then, the average axial force and the service life are used to calculate the dynamic load rate, C_{dyn}, given 16853, 13376, 10616, and 9855 N for screw leads of 5, 10, 20 and 25 mm, respectively.

Searching directly in a manufacturer catalog, the screw with smaller diameter that fulfill C_{dyn} for each lead is selected, given the results shown in Table 2. Where the size is represented by a pair data of diameter-lead. Additionally, each screw must be checked for static load rate, C_{stat}, and maximum velocity, V_{max}. Each C_{stat} from Table 2 is well above the axial forces from Table 1; however, the screw size

Table 2 Main parameters of different screw assemblies

Diameter-lead (mm-mm)	32–5	25–10	32–20	25–25
C_{dyn} (N)	21,600	15,700	13,500	10,100
C_{stat} (N)	40,000	27,000	21,800	15,100
V_{max} (m/min)	23	60	94	150
V_{adm} (m/min)	–	58	150	144
F_b (N)	–	41,711	109,873	38,030
J_{eq} (kgm^2)	–	0,00031	0,00102	0,00091

32–5 has a V_{max} below the 42 m/min required for rapid traverse. Therefore, this screw size is discarded from the candidates.

The admissible speed, V_{adm}, to prevent the excitation of the first natural bending is calculated, considering a screw length of 750 mm and an end-fixity of fixed-floating. Also, for this end fixity, the maximum allowable force to prevent buckling, F_b, is calculated. Both results are shown in Table 2. Clearly, the V_{adm} and F_b are above the values of velocity and axial load from the proposed cycle of Table 1.

3.3 Motor Selection

The first criterion to select a motor for a feed drive is the inertia match requirements. For precision positioning, the motor manufacturer recommends 1.5 $J_m > J_{eq}$ [10], where J_m is the motor inertia and J_{eq} is the system equivalent rotational inertia. This later inertia is calculated adding the corresponding screw and coupling inertia, and the linear inertia converted to rotational inertia using the corresponding transmission ratio. According with the calculated J_{eq} shown in Table 2, a brushless servo motor with $J_m = 0.00073$ kgm^2 is selected for all the screw assemblies.

Additionally, the motor must provide the torque and speed required for all the operations. In this case, the capabilities of this motor are: peak torque of 23 Nm, maximum speed of 4550 1/min, rated torque of 6.36 Nm, rated speed 3600 1/min, and standstill torque of 12.53 Nm. The speed values are adequate for the proposed machine cycle and for every system. Furthermore, the characteristic motor torques largely overcome the required torques. This suggests that a smaller motor can be selected but at expenses of a not recommended inertia match, with consequents that must be studied. However, this matter is out of the aim of this work.

3.4 Control Strategy

Most comercial machine tools use collocated control and high precision ball screws for accurate table positioning. In this case, collocated control means that the control signal, the motor toque τ_m, is applied at the feedback signal location, the angular position of the motor θ_m. Indirectly, the carriage position is controlled by the relation $x_c = l\theta_m$, where x_c is carriage position and l is the screw lead.

State-space feedback in combination with a ZPETC (Zero Phase Error Tracking Controller) prefilter is used as the control strategy. Where the feedback gains are selected based on desired closed-loop pole locations considering a rigid body model for the mechanical system. This approach gives a simple controller implementation with two state-space variables: the motor angular position and its time derivative.

3.5 System Dynamics with Natural Modes of Vibrations

This kind of feed drives can be considered as a multibody system with mechanical links given a high frequency model with multiple vibration modes. A comprehensive model presented in [11] is used, where the screw axial and angular deformations are represented by continuous field functions. After that, the energy and work equations were used to obtain the complete dynamics by Lagrange formulation. Finally, Ritz series, with suitable basis functions for this case, were used to approximate the continuous field functions and then to obtain a finite model. This model requires additional screw parameters that are found in catalogs, such as the stiffness for the coupling k_a, for the end fixity bearing k_b, and for the nut-screw k_n.

3.6 Controller Tuning and Stability Analysis

A stability analysis is used to study the performance of the position loop for each system. It is important to note that the system parameters can change during the machining operations, mainly the load mass and the ball-nut position. The nominal condition (NC) considered for the previous calculations were a load mass of 30 kg and a ball-nut centered in the ball-screw length, $x_c = 0.5\ L$, where L is the total screw length. This gives a bode plot represented by the blue line in Fig. 1, in this case for the screw size of 20–10.

However, the non-nominal condition (NNC) given by the most adverse combination of both factors, that is a maximum load mass of 145 kg and an extreme nut position of $x_c = 0.9\ L$, is represented by the green line in Fig. 1.

Clearly, the NNC represents the worst stability case due to: the movement of the first natural mode to lower frequency values and the increment of its magnitude. Therefore, the controller gains were iteratively selected until a gain margin (GM) close to 2 and a phase margin (PM) close to 22° were achieved for the NNC, ensuring in this way stability for all the operational conditions.

This procedure is repeated for the other two screw sizes with the results shown in Figs. 2 and 3. Table 3 summarizes the main bode parameter for both NC and NCC to compare the tendency among the three screw sizes. In each case, the overall stability improves for the NC respect to the NNC, as it is expected. On the other

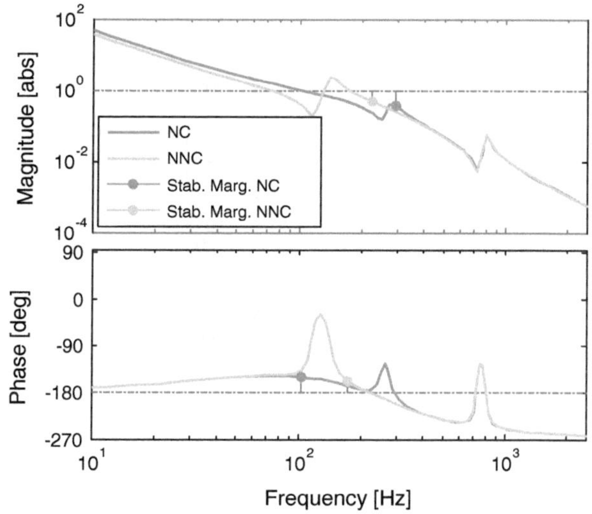

Fig. 1 Bode plot for 20–10

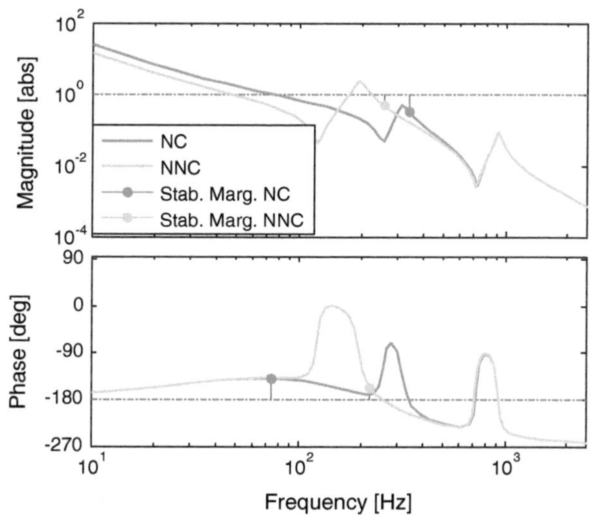

Fig. 2 Bode plot for 32–20

hand, the cross over frequency (CF) increases considerably as the screw lead decreases, for both NC and NNC.

3.7 Trajectory Tracking

In the previous section, the controller was tuned to obtain stable responses for each system, resulting with different crossover frequencies. As the crossover frequency increases, the closed-loop bandwidth also increases with the consequent reduction

Fig. 3 Bode plot for 25–25

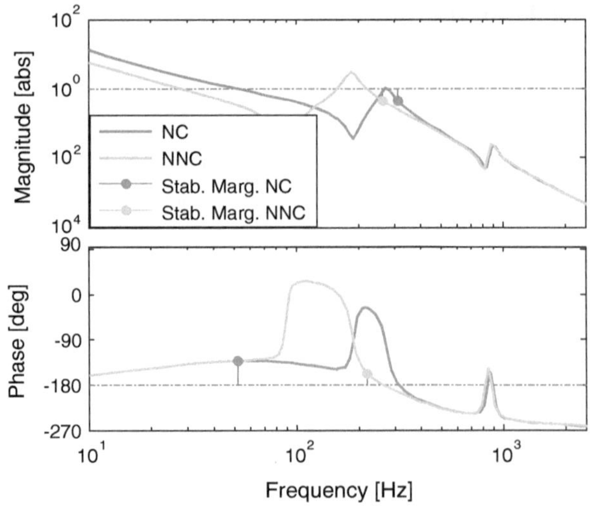

Table 3 Summary of bode parameters

Size	Parameter	NC	NNC
25–10	PM	29.9°	22.6°
	GM	2.69	2.06
	CF	102.7	72.8
32–20	PM	40.4°	22.4°
	GM	2.98	2.08
	CF	74.13	46.7
25–25	PM	47.9°	22.4°
	GM	2.16	2.18
	CF	51.9	27.5

in the tracking error. Therefore, the tracking error of a forward and backward movement with sine shape is analyzed as a direct implication of changes in the closed-loop bandwidth. The trajectory has amplitude of 68 mm, maximum speed of 170 mm/s, maximum acceleration of 2 m/s^2 and jerk of 50 m/s^3, which is used for high speed positioning. Additionally, a friction model is included for a more realistic simulation, especially for high precision positioning. The model includes the nonviscous friction component and the Stribeck effect, similar to the one presented in [12].

Figure 4 shows the tracking errors for NC of each screw assemble. The maximum errors values are observed at the motion reversal, which are given mainly by the nonlinear friction effect. The maximum error values are 6.6, 14, and 38.7 μm for the screw leads of 10, 20, and 25 mm, respectively. This results correspond

Fig. 4 Tracking error for NC

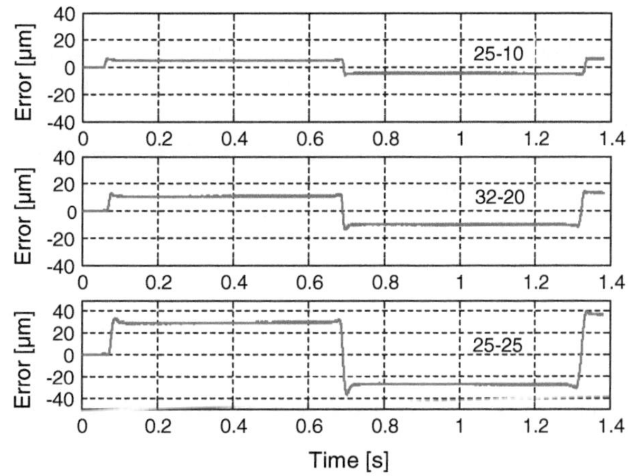

qualitatively with the observed tendency in the bode parameters for the NC and their effects on closed loop bandwidth.

On the other side, the maximum tracking errors for NNC are 7.4, 18.8, 57.7 μm, for the screw leads of 10, 20, and 25 mm respectively, as is shown in Fig. 5. The screw assemble 25–25 is eliminated as a candidate because it presents an unacceptable error for a CNC high speed milling machine.

It is important to mention that in all the cases the required torque was below the rated motor torque.

3.8 Optimal Selection

The different aspect studied up to this stage can be considered as constraints toward the selection of screw-motor candidates. These constraints are: dynamic load rate, static load rate, maximum linear velocity, admissible linear velocity, motor-screw

Fig. 5 Tracking error for NNC

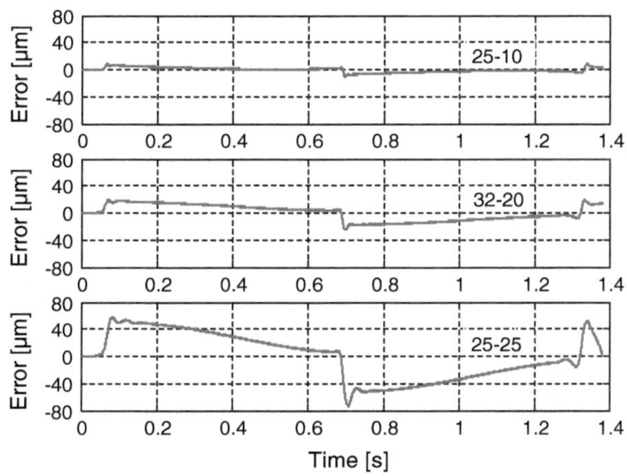

inertia ratio, maximum motor torque-velocity, and rated motor torque-velocity. Furthermore, the cross over frequency from each controlled system with suitable phase and gain margins, can be used as an extra constraint, based on the admissible tracking error.

Now, all the candidates that fulfill the constraints must be checked for optimality. A common approach is to calculate the optimal screw lead, l_{opt}, to minimize the root mean square (RMS) motor torque [9] as:

$$l_{opt} = 2\pi \sqrt{\frac{J_t \ddot{x}_{RMS}}{(F_{ext} + M\,\ddot{x})_{RMS}}} \tag{1}$$

where J_t is the rotational inertia composed by the screw, the coupling and the rotor motor inertia and M is the moving mass composed by the carriage mass and the load mass. The parameter $\ddot{x}$ is the linear acceleration and F_{ext} is the overall external force. These last two parameters are related to the machining cycle presented in Table 1.

The optimal screw lead for the two candidates left are 8, 1 mm for the 25–10 and 8, 7 mm for the 32–20 assemble. Clearly, the first assemble has a near optimal screw lead and it is selected as the best candidate.

4 Conclusions

The mechatronic procedure presented here allows to achieve a system design based on the closed loop behavior, for which a high frequency model becomes an important factor.

The screw lead highly influences the system dynamics, which affects the closed-loop bandwidth and then the tracking error. A low value of screw lead is preferable for low tracking error, as long as the maximum velocity can be reached, the required motor power does not become too high and the screw lead is near optimum.

This mechatronic methodology can be used with different control strategies. Furthermore, the motor selection can be optimized by relaxing the inertia match criterion and studying its effect on the closed loop behavior.

References

1. Slocum A (1992) Precision machine design, society of manufacturing engineers
2. Lenarčič J, Stanišić M (2010) Advances in robot kinematics: motion in man and machine. Springer, New York
3. Taghirad HD (2013) Parallel robots, mechanics and control. CRC Press, Boca Raton

4. Precision Ball Screw Assemblies, RE 83 301/2002-09. http://www.boschrexroth.com. Accessed 21 May 2014
5. Altintas Y, Erkorkmaz K, Zhu W-H (2000) Sliding mode controller design for high speed feed drives. Annals of the CIRP 49(1):265–270
6. Erkorkmaz K, Altintas Y (2001) High speed CNC system design, Part III: High speed tracking and contouring control of feed drives. Int J Mach Tools Manuf 41:1637–1658
7. Kim MS, Chung SC (2006) Integrated design methodology of a ball-screw driven servomechanisms with discrete controllers. Part II: Formulation and synthesis of the integrated design. Mechatronics 16:503–512
8. Varanasi K (2002) On the design of a precision machine for closed-loop performance. MS thesis, Massachusetts Institute of Technology, Cambridge
9. Caracciolo R, Richiedei D (in press) Optimal design of ball-screw driven servomechanisms through an integrated mechatronic approach. Mechatronics. http://dx.doi.org/10.1016/j.mechatronics.2014.01.004
10. Ball Rail Tables TKK, RE 82 501/200-04. http://www.boschrexroth.com
11. Vicente D, Hecker R, Flores M (2012) Modeling and vibration modes analysis of a ball screw drive. Int J Adv Manuf Technol 58(1):257–265
12. Erkorkmaz K, Altintas Y (2001) High speed CNC system design, Part II: modeling and identification of feed drives. Int J Mach Tools Manuf 41:1487–1509

Behavior of the Robot with Vibratory Excitation

T. Majewski and D. Szwedowicz

Abstract The paper presents a mini robot which moves as a result of vibratory forces. The robot has elastic bristles which allow the robot to vibrate. The vibration of the robot is generated by one or two unbalanced motors. With one motor the robot moves forward and with two motors the trajectory of the motor can be controlled. The paper presents the mathematical models of the robots, simulation of their behavior, prototypes, and some parameters which characterize their properties.

Keywords Robot · Locomotion · Vibration · Friction · Simulation

1 Introduction

Many robots exist with different methods of locomotion. Larger robots use legs, caterpillars, or wheels for changing their position. Small robots may use wheels or they may move in a similar way to small animals, e.g. snakes, fishes, birds. New materials such as carbon fiber, intelligent material and small batteries with greater stored energy allow the construction of a mobile robot for terrestrial locomotion, swimming or even for flying. The Festo's mobil bird is able to fly in controlled ways using an electric motor for moving the wings and some actuators to control the direction. The propulsion system is made of carbon fibers to make the bird very light. Some robots are able to take energy from the environment. The wave glider uses the waviness of the ocean to generate a pushing force what allows its locomotion. The propulsion force changes in time but its average magnitude during one

T. Majewski (✉)
Universidad de Las Americas-Puebla, Cholula, Mexico
e-mail: tadeusz.majewski@udlap.mx

D. Szwedowicz
Centro Nacional de Investigacion Y Desarrollo Tecnologico-Cuernavaca, Cuernavaca, Mexico
e-mail: d.sz@cenidet.edu.mx

M. Ceccarelli and E.E. Hernández Martinez (eds.), *Multibody Mechatronic Systems*, Mechanisms and Machine Science 25, DOI 10.1007/978-3-319-09858-6_7

period of waviness is constant and depends on the amplitude of waviness of the ocean. The wave glider gains energy from the environment and the solar panels supply the energy to the control system. Some robots have actuators built of intelligent materials and their periodic motions generate a vibratory force which pushes the robot forward.

For very small robots the principle of working should be as simple as possible. Therefore the principle of locomotion of small animals is an inspiration for building very small or even micro robots [1, 2]. Usually a robot is designed for performing one task or a few related tasks [3–6]. A collection of mobile robots is able to self-assemble and self-organize in order to perform complex tasks—swarm robots. Some of vibratory robots are used as a toy—there is no any different of principle of locomotion of a toy and real robot and vice versa (Fig. 1).

To understand the principle of locomotion, its theory should be established, e.g. physical and mathematical models. One of the simplest actuators is an electric motor with unbalance which generates vibration and such a robot can be mobile because the motor needs a little power.

The paper presents the model of the robot for rectilinear and curvilinear locomotion, the forces acting on the robot, their change in time, the form of the robot vibrations, and the velocity of locomotion.

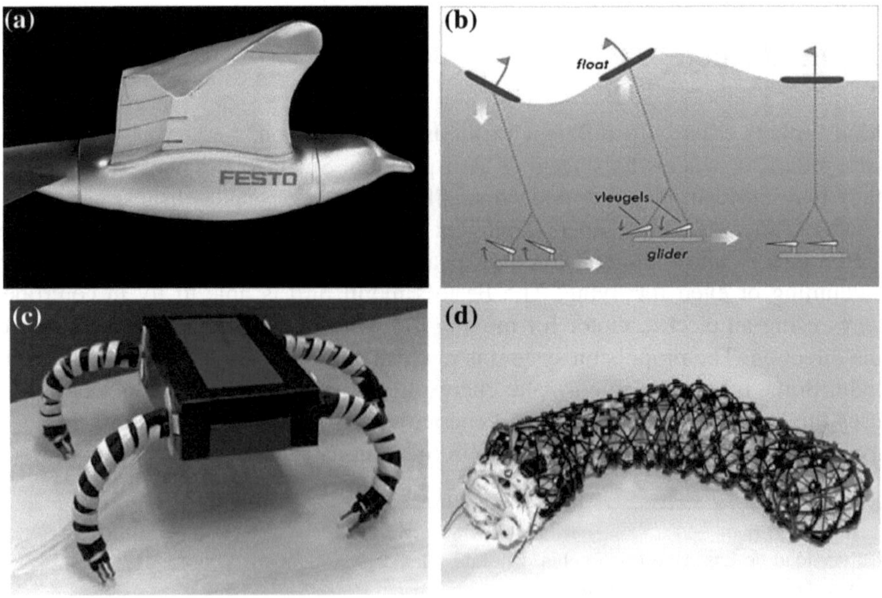

Fig. 1 Example of robots. **a** Festo's bird, **b** wave glaider, **c** continuum limbs [3], **d** peristaltic locomotion [2, 5]

2 Description of the Vibratory Robot

The base of the robot 1 has some bristles 2 which are inclined towards the supporting plane 3—Fig. 2. The robot has one or two motors 4 and each of them has an unbalance which generates rotating force in the plane *Cxy* which is perpendicular to the supporting surface. The bristles are inclined in the direction of locomotion by an angle γ—Fig. 3.

The centrifugal force of the motor is $P = mR\omega^2$and it can be presented as two components

$$P_x = mR\omega^2 \cos(\omega t), \quad P_y = mR\omega^2 \sin(\omega t), \tag{1}$$

where ω is the angular velocity of the motor, the angle ωt defines the position of the unbalance with respect to the axis x, m is the mass of the unbalance and R is the distance from its axis of rotation (mR is a static unbalance of the motor).

The stiffness of all bristles in the direction perpendicular to them is k and it depends on the material of the bristles (Young modulus), their length and diameter.

A small deformation of bristles and contact forces are shown in Fig. 3.

The force P_y deforms the bristles and the robot vibrates in a vertical direction $y(t)$ which generates the normal reaction N and the friction force F. A deflection of the bristles Δ depends on the vertical displacement of the robot y and with a small deformation it can be written as

$$y = \Delta \sin \gamma.$$

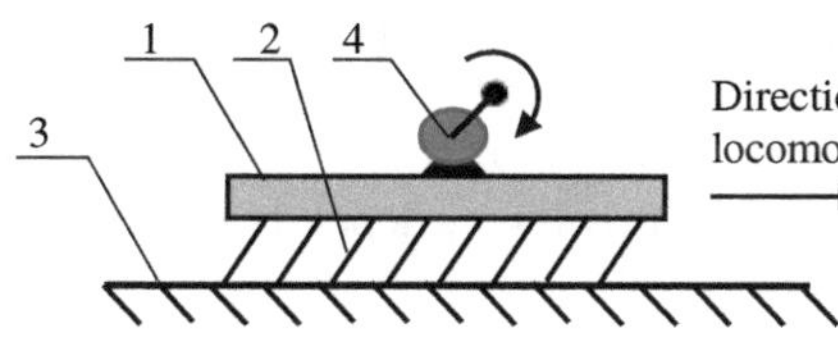

Fig. 2 Sketch of the robot. *1* base of the robot, *2* bristles, *3* supporting plane, *4* motor with unbalance

Fig. 3 Deflection of the bristle

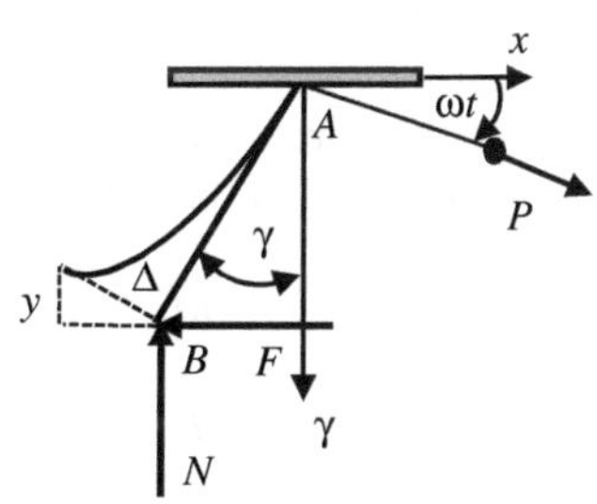

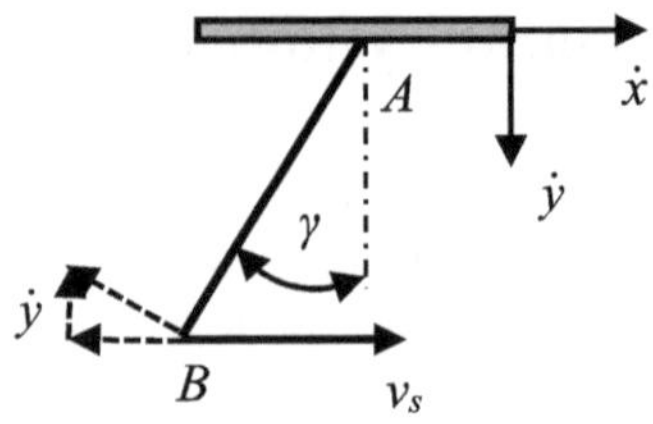

Fig. 4 Velocity of bristle's slipping

If the slipping velocity of point B is not zero then the friction force $F = \mu N$ is opposite the velocity, where μ is the coefficient of friction. The deformation Δ depends on normal and friction forces.

$$\Delta = (N \sin \gamma + F \cos \gamma)/k. \tag{2}$$

Therefore the normal and friction forces are the function of the vertical displacement of the robot y and the direction of the bristle's slipping

$$N = \frac{k}{\sin \gamma(\sin \gamma + \mu \cos \gamma sign(v_s))} y, \quad F = \mu N \cdot sign(v_s) \text{ for } v_s \neq 0. \tag{3}$$

Slipping velocity of the bristles has a significant influence on the normal and friction forces. For positive slipping these forces are smaller and for negative slipping of the bristles these forces are bigger. A difference ΔF is a vibratory force which generates the robot locomotion

$$\Delta F = F(v_s < 0) - F(v_s > 0) > 0$$

If there is no friction then $N = \frac{k}{\sin^2 \gamma} y$.

The slipping velocity of the bristles v_s depends on the horizontal and vertical velocities of the robot (Fig. 4)

$$v_s = \dot{x} - \dot{y}/\tan \gamma. \tag{4}$$

3 First Model and Its Behavior

When the forces acting on the robot are known then the equations defining the behavior of the robot are also determined

$$M\ddot{x} = P_x - F, \qquad M\ddot{y} = P_y + Mg - N. \tag{5}$$

with forces N and F defined by Eq. (3).

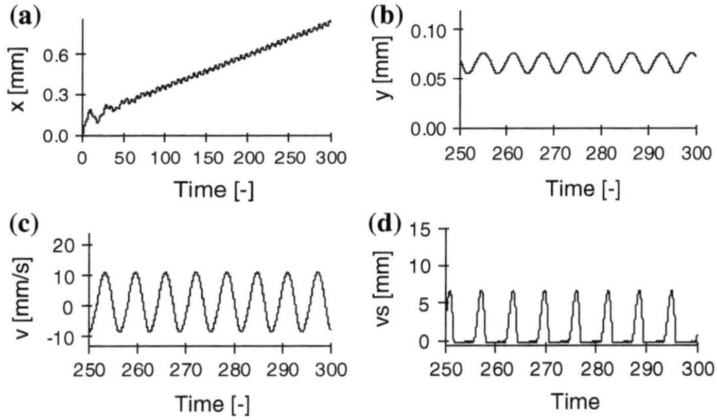

Fig. 5 Robot with two degrees of freedom and one motor. **a** Horizontal displacement, **b** vertical vibration, **c** instant velocity, **d** slipping velocity of bristles

The forces N and F change in complex ways with the robot motion so one possibility to establish the behavior of the robot is to analyze the Eq. (5) with a help of a numerical method for differential equations in the library of Matlab or other software.

The behavior of a robot which moves along a straight line is presented in Fig. 5 (Time = ωt). The parameters of the system are as follows: $f = 100$ Hz, $mR = 2$ gmm, $\gamma = 0.5$, $k = 4{,}000$ N/m, $M = 0.12$ kg, $\mu = 0.4$. The motion of the robot occurs with vibrations in x and y directions and the average velocity of locomotion for these parameters is 0.3 mm/s.

If the inertial force increases the robot loses periodically the contact with the ground and the robot moves faster.

4 Second Model and Its Behavior

If the plane of the unbalance rotation is not exactly in the plane of the bristles then the robot has a tendency to slowly change the direction into one direction. Usually a user of the robot wants to control its direction of motion. Such a robot should have two rotating unbalances and also two rows of bristles at some distance each other—Fig. 6.

As the bristles are located in two rows, the robot has more degrees of freedom. Its motion is more complicated; position of the center of the robot is defined by the coordinates x, y, z in the reference frame XYZ and the rotation about the axes x and y. When the excitations of each motor are different then the friction forces of the bristles are also different which results in a motion of the robot along a curvilinear trajectory. So the direction of motion may be controlled by changing the spin velocities of the motors.

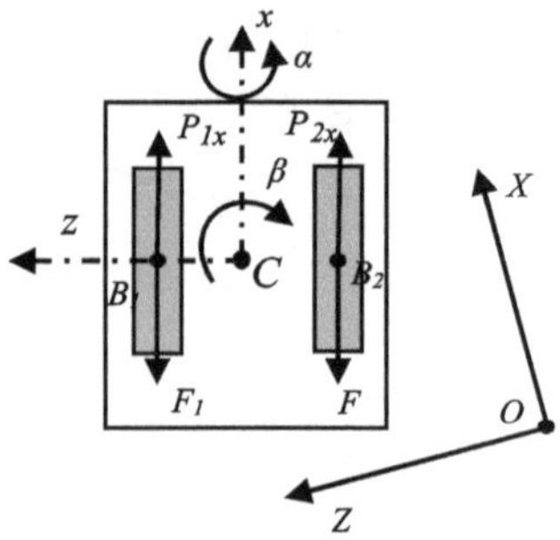

Fig. 6 Robot with two motors

$$
\begin{aligned}
M\ddot{x} &= (P_1\cos(\omega_1 t) + P_2 os(\omega_2 t))\cos\beta - (F_1 + F_2)\cos\beta, \\
M\ddot{y} &= P_1\sin(\omega_1 t) + P_2\sin(\omega_2 t) + Mg - (N_{B1} + N_{B2}), \\
M\ddot{z} &= -(P_1\sin(\omega_1 t) + P_2\sin(\omega_2 t))\sin\beta + (F_{B1} + F_{B2})\sin\beta \\
I_x\ddot{\alpha} &= -0.5a[(P_1\sin(\omega_1 t) - P_2\sin(\omega_2 t)) + (N_{B1} - N_{B2})] \\
I_y\ddot{\beta} &= 0.5a[(P_1\cos(\omega_1 t) - P_2\cos(\omega_2 t)) - (F_{B1} - N_{B2})].
\end{aligned}
\tag{6}
$$

When the motors spin with the same velocity the behavior of the robot is show in Fig. 7. The trajectory is a rectilinear line. With the excitation of each motor given by an unbalance $m_1e_1 = m_2e_2 = 1$ gmm the robot does not lose a contact with the plane (Fig. 7b).

The diagrams in Fig. 8 show the position of the robot on the trajectory, the shape of the trajectory and friction force of the left bristles for the frequency of excitations $f_1 = 100$ Hz and $f_2 = 110$ Hz ($f_2/f_1 = 1.1$) and the unbalance $m_1e_1 = m_2e_2 = 1$ gmm. The friction force for *Time* = 0–20 is different than for *Time* = 280–300. Greater friction force at the beginning accelerates the robot. For $m_1e_1 = m_2e_2 = 2$ gmm the bristles periodically lose contact with the plane.

A prototypes were built using tooth brushes as is shown in Fig. 9. The robot has one or two motors DC powered from a 9 V battery.

The spin velocities of the motors are controlled in a continuous way by an infrared remote method. So it is possible to obtain different velocity of locomotion along a straight line or change the direction of locomotion at any moment (robot with two motors). Figure 10a presents the measured vibration of the prototype $y(t)$ which was measured by the accelerometer B&K type 4507, and Fig. 10b a frequency spectrum of vibration in a vertical direction. Measured vertical vibration shows that it is not harmonic and has a wide spectrum of frequency.

The experiments confirmed the applied models of the robot for rectilinear and curvilinear motion. With a battery of 9 V it is possible to obtain fast motion of the robot and also rapid change of the direction of locomotion.

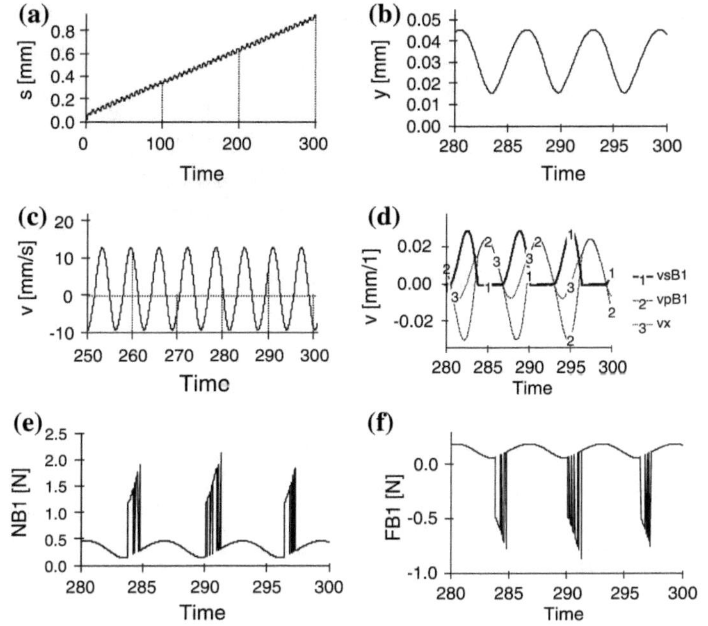

Fig. 7 Behavior of the robot for the frequency of excitations $f_1 = f_2 = 100$ Hz. **a** Displacement, **b** vertical vibration, **c** instant velocity of the robot and slipping velocity of bristle, **d** normal reaction of the *left* bristle, **e** normal force of the *left* bristle, **f** friction force of the *left* bristle

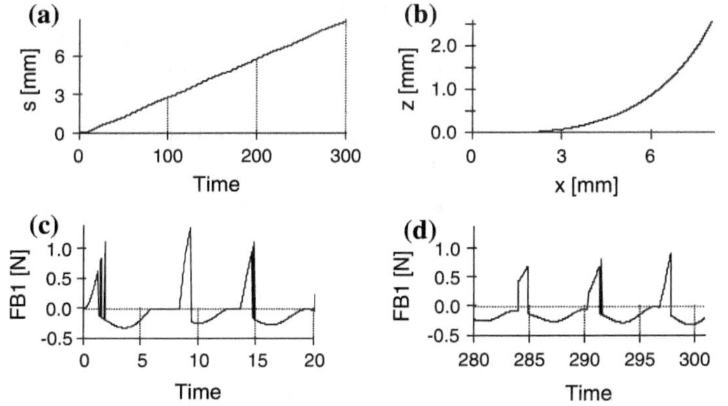

Fig. 8 Behavior of the robot for $f_2/f_1 = 1.1$. **a** position on trajectory, **b** trajectory, **c** friction force at beginning of motion, **d** friction force for *Time* = ωt = 280–300

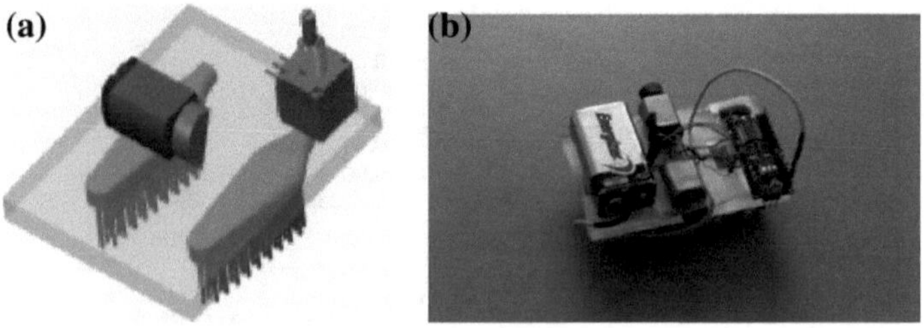

Fig. 9 Prototypes of robots. a a CAD design, b a prototype

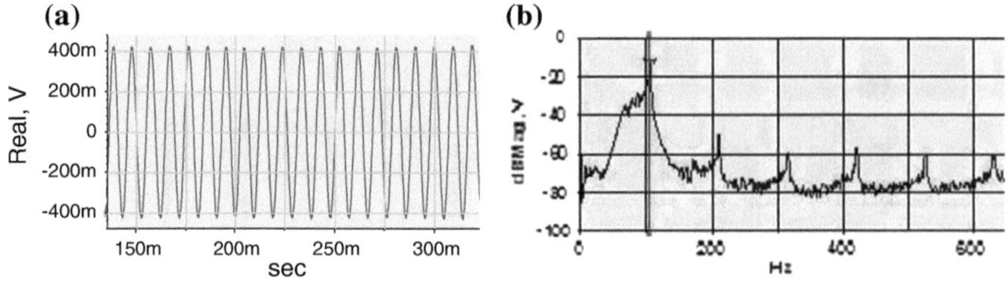

Fig. 10 a Vibration and, b its frequency spectrum

5 Conclusions

The robot with inertial excitation is a simple one, the motor does not use too much energy which allows the building of a simple mobile robot with very small dimensions which can work for a relatively long time. If the robot is equipped with two motors then its direction of locomotion can be controlled by changing the motors angular velocities. The mean velocity of the robot increases by increasing the unbalance or spin velocity of the motor. Remote control allows the operator to control its velocity and direction at some distance from the robot.

References

1. Alexander RM (1983) Animal mechanics. Wiley-Blackwell, Oxford
2. Guillot A (2007) Handbooks of robotics, chapter 61, biologically-inspired robots. Springer, New York
3. Jatsun S, Bolotnik N, Zimmerman K, Zeidis I (2007) Modelling of motion of vibratory robots. In: 12th IFToMM world congress (France)
4. Boxerbaum AS, Kendrick M, Shaw KM et al (2012) Continuous wave peristaltic motion in a robot. Int J Robot Res 31(3):302–318
5. Godage IS, Nanayakkara T, Caldwell DG (2012) Locomotion with continuum limbs. In: IEEE/RSJ international conference on intelligent robots and Systems, Portugal
6. Chernousko FL (2013) Snake-like locomotion of multilink mechanisms. J Sound Vib 9:237–256

Attitude Determination System Based on Vector Observations for Satellites Experiencing Sun-Eclipse Phases

J. Rodrigo Cordova-Alarcon, Mario A. Mendoza-Barcenas and Arturo Solis-Santome

Abstract Due to mission requirements, fault detection and isolation protocols or budget restrictions, a satellite is required to use most reliable attitude determination hardware, such as magnetometers and sun sensors, in order to keep 3-axis attitude information available during its complete orbit. However, satellites experiencing sun-eclipse phases, sun sensors become no operational. In this paper, we propose an attitude determination system which provide 3-axis attitude information in both sun and eclipse phases, considering vector observations acquired from sun and magnetic measurements. To compensate the unavailability of sun sensors during eclipse phase, two variations of innovation processes merged into the Extended Kalman Filters are proposed. In order to keep the accuracy of attitude estimation process during eclipse mode, angular rates must be accurately estimated during sun phase. To solve this issue, rough angular rate information is calculated based on previous attitude information calculated by Gauss-Newton method, which fuse magnetic and sun sensor data. Numerical simulation results show the performance of the proposed attitude determination system, considering the use of vector measurement hardware with different precision degree.

Keywords Satellite attitude determination · Kalman filter · Gauss-Newton method

J. Rodrigo Cordova-Alarcon (✉) · M.A. Mendoza-Barcenas · A. Solis-Santome
Centro de Desarrollo Aeroespacial Del Instituto Politecnico Nacional,
Belisario Dominguez 22, Mexico 06010, Mexico
e-mail: jcordovaa@ipn.mx

M.A. Mendoza-Barcenas
e-mail: mmendozab@ipn.mx

A. Solis-Santome
e-mail: sol_mecanicohotmail.com

M. Ceccarelli and E.E. Hernández Martinez (eds.), *Multibody Mechatronic Systems*,
Mechanisms and Machine Science 25, DOI 10.1007/978-3-319-09858-6_8

1 Introduction

Several methods of attitude estimation algorithms for guidance, navigation and control systems have been developed for spacecraft with precision pointing requirements. The use of star sensors and gyroscopes in attitude determination systems allows the availability of precise 3-axis attitude information during the satellite orbit [1, 2]. However, satellite missions under budget restriction, fault detection and isolation routines or safe operation modes, require the use of reliable or inexpensive 3-axis attitude determination hardware, such as coarse sun sensors and magnetometers.

Magnetic measurements are available during the satellite orbit and is used to estimate its attitude [3–5]. The use of linear and nonlinear filters can be useful to improve attitude information. Kalman Filter (KF) deals with noisy measurements, non-modeled dynamics and nonlinear measurement models. The tuning process of covariance values is a challenging task that KF programmers face and solve mostly via successive approximations [6] or taking into account a calculated covariance values from sensor noise.

When two vector observations are available (e.g. data provided by sun sensors and magnetometers data) attitude information can be calculated by deterministic methods such as TRIAD method [7] or Gauss-Newton method [8]. When sun information is not available in eclipse phase, deterministic methods become ineffective. In addition, those methods can not compensate the presence of sensor noise and bias.

Several approaches have been studied using interlaced deterministic methods and KF in order to compensate sensor noise [9, 10], as well as bias features [11, 12]. Concerning small satellite missions, KF-based ADS using sun sensors and magnetometers were designed and successfully implemented in recent projects [13], also adding rate gyroscopes [14, 15].

During eclipse phase, KF-based ADS propagate the last attitude information. It can be performed by switching the state update step of KF considering just magnetic measurements [16]. Another alternative is the implementation of local filters, which output is fused afterwards, according to the combination of available sun sensors and magnetometers information [17]. Gyroscopic measurements are also used to provide better performance of ADS during attitude propagation. According to experimental results, the presence of gyroscope drift reduces ADS accuracy [14, 15].

This work proposes an attitude determination system based on EKF to fulfill attitude availability requirements. Due to the absence of sun information during eclipse phase, two innovation processes merged into an EKF are proposed to maintain attitude estimation even in transition of phases. The use of Gauss-Newton method in sun phase, allows the calculation of rough attitude and angular rate information. Rough attitude information contributes to the convergence of attitude estimation process during sun phase. Rough angular rate information contributes to the convergence of angular rate estimation process during sun phase and accuracy maintenance of the EKF during eclipse phase. The performance of this attitude determination system is exhaustively analyzed by numerical simulations, considering the use of sun sensors and magnetometers with diverse precision degree.

The present work is organized as follows. In Sect. 2, rigid body dynamics, kinematics and measurement models are explained. Section 3 is devoted to the Gauss-Newton method and the EKF equations. Section 4 presents the proposed attitude determination system. Results from numerical simulations are analyzed in Sect. 5. Finally, some conclusion remarks are presented in Sect. 6.

2 System Model

To describe the satellite dynamic model, it is considered as a rigid body, which angular velocity depends on external torques and mass distribution in a certain geometry:

$$I\dot{\omega}_b^i = -S(\omega_b^i)I\omega_b^i + \tau \tag{1}$$

where I represents the inertia tensor, ω_b^i the angular velocity of the satellite b in terms of a inertial reference frame i, τ the external and control torques and $S(\omega_b^i)$ the cross-product operator matrix build as $x \times y = S(x)y$:

The satellite kinematic model is expressed in terms of a unit quaternion $\mathbf{q} = [\, q \; q_4 \,]^\top$, where $q = [\, q_1 \; q_2 \; q_3 \,]^\top$, which represents the satellite attitude in terms of a inertial reference frame i. The kinematics equation can be expressed as a quaternion product, where $V_\omega = [\, \omega_b^i \; 0 \,]^\top$:

$$\dot{\mathbf{q}}_b^i = \frac{1}{2}\mathbf{q}_b^i \otimes V_\omega = \begin{bmatrix} \frac{1}{2}[q_{b_4}^i I + S(q_b^i)]\omega_b^i \\ -\frac{1}{2}q_b^{i\top}\omega_b^i \end{bmatrix} \tag{2}$$

The measurement model given by (3), consider the assumption that the $[3 \times 1]$ $n-$measurement vector is aligned to the satellite body axes.

$$y_m^b = R(\mathbf{q}_b^i)y_r^i + v(t) \tag{3}$$

where $R(\mathbf{q}_b^i) = I + 2q_{b_4}^i S(q_b^i) + 2S(q_b^i)^2$ is the $[3 \times 3]$ rotation matrix which relates the body and inertial coordinate system, expressed in terms of the unit quaternion $\mathbf{q}_b^i$ and $v(t)$ is an independent noise process.

3 Gauss-Newton and EKF Equations

Concerning attitude determination methods, they can be classified in deterministic and analytical methods. The first requires reference and measurement vectors, providing rough a priori attitude estimate. The second provides an optimal attitude estimation solution, as well as other attitude parameters, such as sensor biases and orbit parameters [18]. In this work, the Gauss-Newton method [9] and EKF were used to design the proposed Attitude Determination System.

3.1 Gauss-Newton Method

This method is based on a minimum squared error criterion S^k, defined as:

$$S^k = \sigma^\top \sigma = \left(y_r^i - My_m^b\right)^\top \left(y_r^i - My_m^b\right) \tag{4}$$

where $y_r^i = [y_{r_1}^{i\top}\ y_{r_2}^{i\top}\ \cdots\ y_{r_n}^{i\top}]^\top, y_{r_n}^{i\top}$ is the $[3 \times 1]$ n–reference vector in a inertial reference frame, $y_m^b = \left[y_{m_1}^{b\top}\ y_{m_2}^{b\top}\ \cdots\ y_{m_n}^{b\top}\right]^\top, y_{m_n}^{b\top}$ is the $[3 \times 1]$ n–measurement vector expressed in body frame, and $M = \text{diag}\left(R_1(\mathbf{q}_b^i), \ldots, R_n(\mathbf{q}_b^i)\right)$.

To find the quaternion $\mathbf{q}_b^i$, an iterative algorithm is used in order to minimize the error function (4). This is given as:

$$\hat{\mathbf{q}}_{k+1} = \hat{\mathbf{q}}_k - [J^\top(\hat{\mathbf{q}}_k)J(\hat{\mathbf{q}}_k)]^{-1}J^\top(\hat{\mathbf{q}}_k)\sigma(\hat{\mathbf{q}}_k) \tag{5}$$

where $\hat{\mathbf{q}}_k$ is the estimated quaternion according to an iteration k and $J(\hat{\mathbf{q}}_k)$ is the Jacobian of the product My_m^b given by:

$$J(\hat{\mathbf{q}}_k) = \left[\left.\frac{\partial My_m^b}{\partial \mathbf{q}}\right|_{\mathbf{q}=\hat{\mathbf{q}}_k}\right] \tag{6}$$

3.2 Extended Kalman Filter

Considering the nonlinear system:

$$\dot{x} = f(x, u, t) + w(t), \quad y = h(x, t) + v(t) \tag{7}$$

where it is assumed that x is the state vector, u the control command vector, z is the measurement process, $f(x, u, t)$ and $h(x, t)$ are nonlinear functions and $v(t)$ and $w(t)$ is the independent noise processes, which covariance matrices are given by:

$$E[w_j w_i^\top] = \begin{cases} Q, & i = j \\ 0, & i \neq j \end{cases} \quad E[v_j v_i^\top] = \begin{cases} R, & i = j \\ 0, & i \neq j \end{cases} \tag{8}$$

$$E[w_j v_i^{\top}] = 0, \text{ for all } j \text{ and } i$$

The first step of EKF is the calculation of Kalman gain K_k:

$$K_k = P_k^- H_k^{\top}(\hat{x}_k^-)[R + H_k(\hat{x}_k^-)P_k^- H_k^{\top}(\hat{x}_k^-)]^{-1} \quad (9)$$

The next step is to update $\hat{x}_k$ with the measurement z_k and the Kalman gain K_k:

$$\hat{x}_k = \hat{x}_k^- + K_k v_k \quad (10)$$

where $v_k = z_k - H_k(\hat{x}_k^-)\hat{x}_k^-$ is the innovation process. Once obtaining $\hat{x}_k$, a propagation step for the next $k+1$ iteration is performed as indicated as follows:

$$\begin{aligned} \hat{x}_{k+1}^- &= \hat{x}_k + \int_k^{k+1} f(\hat{x}_k, u, t)dt \\ P_{k+1}^- &= F_k(\hat{x}_k)P_k F_k(\hat{x}_k)^{\top} + Q \end{aligned} \quad (11)$$

where $P_k = P_k^- - K_k H_k(\hat{x}_k)P_k^-$. After this step, the iteration $k+1$ turns into k, closing the loop by repeating estimation process with (9). The main characteristic of the EKF yields in the continuous update of the linearized functions f and h from (7) along the trajectory of the estimated variables $\hat{x}$:

$$F_k = \left.\frac{\partial f(x,u,t)}{\partial x}\right|_{x=\hat{x}_k} \qquad H_k = \left.\frac{\partial h(x,t)}{\partial x}\right|_{x=\hat{x}_k} \quad (12)$$

The main challenge during the EKF design is the tuning of the EKF and to choose an adequate innovation process to keep filter convergence to obtain a good estimation.

4 Gauss-Newton and EKF Based Attitude Determination System

As mentioned in Sect. 3.1, rough attitude information can be computed when two or more vector observations with its respective references are available. In this case, both magnetic and sun vector measurements are available in the orbit sun phase and rough attitude computation is possible. In eclipse mode, only magnetic vector observation is available and the satellite attitude information cannot be computed directly by using Gauss-Newton method.

To guarantee the availability of satellite attitude information during its whole orbit, the EKF becomes an attractive method to filter and estimate both attitude and angular rates information. In Fig. 1, the ADS flow diagram is illustrated, where two

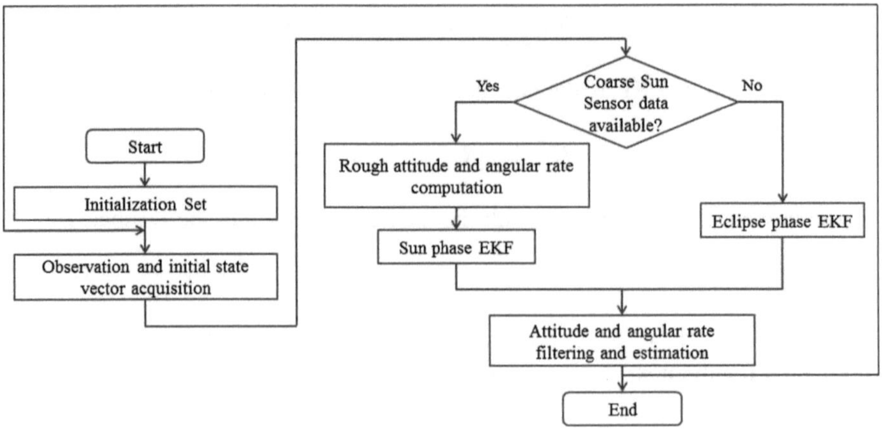

Fig. 1 ADS flow diagram

attitude determination schemes were designed and implemented according to coarse sun sensors information availability.

To compute the Kalman gain with the Eq. (9), the matrix R is set by the selection of the measurement covariance values cov_m.

The update process of the state vector $\hat{x}_k$ is performed as indicated in (10). The quaternion update suggested in [3] is performed by a quaternion product as indicated in Eq. 13.

$$\hat{x}_k = \begin{bmatrix} \hat{\mathbf{q}}_k \\ \hat{\omega}_k \end{bmatrix} = \begin{bmatrix} \hat{q}_{k_4}^- \Delta q_{ud} + \hat{q}_k^- \Delta q_{ud_4} - S(\hat{q}_k^-)\Delta q_{ud} \\ \hat{q}_{k_4}^- \Delta q_{ud_4} - \hat{q}_k^{-\mathrm{T}} \Delta q_{ud} \\ \omega_k^- + K_\omega v \end{bmatrix} \tag{13}$$

where $\Delta q_{ud} = K_q v$ and $\Delta q_{ud_4} = \sqrt{1 - \Delta q_{ud}^{\mathrm{T}} \Delta q_{ud}}$.

Note that the update process is performed by dividing the Kalman gain as shown below, which is calculated as indicated in Eq. (9):

$$K_k = \begin{bmatrix} K_\omega \\ K_q \end{bmatrix} \tag{14}$$

To compute P_{k+1}^- from (11), a matrix Q is set by the tuning of process covariance values cov_{p_q} (related to kinematic model) and cov_{p_ω} (related to dynamic model). A linearization of the system 1 and 2 along the trajectory $\hat{x}_k$ is needed to compute F_k.

To compute $\hat{x}_{k+1}^-$ from (11), the system described by Eqs. 1 and 2 is evaluated in terms of the estimated variables $\hat{x}$, resulting in the propagation step for the estimated variables:

$$\hat{x}_{k+1}^{-} = \hat{x}_k + \int_k^{k+1} f(\hat{x}_k, u_k, t)dt \tag{15}$$

where the numerical integration of Eq. (15) was proposed to be performed via trapezoid rule. Therefore, the computation of $\hat{x}_{k+1}^{-}$ is performed as follows:

$$\hat{x}_{k+1}^{-} = \hat{x}_k + \frac{1}{2}\left(f(\hat{x}_k, u_k, t) + f(\hat{x}_k^{-}, u_k^{-}, t)\right)\Delta t \tag{16}$$

Remark: At the beginning of each KF iteration, it has to be ensured a positive value of the fourth element of the quaternion, also it is important to keep the unity of the quaternion, in other words $q^{\top}q = 1$, as well as each measured and calculated magnetic and sun vectors.

4.1 EKF Innovation Process in Sun Phase

During this mode, both sun and magnetic field vector in body frame are available, in consequence, the attitude computation is performed via deterministic methods. Due to the possibility of obtaining noisy results with this algorithm, an EKF is used to get smoothed results.

In this work, it is proposed a rough angular rate calculation from the attitude computation step in sun phase, to improve the EKF performance, particularly in the eclipse phase. This technique is based on the quaternion product from Eq. 2, in such a way that:

$$V_{\omega} = 2\dot{\hat{q}}_c \otimes \hat{q}_c^* = \begin{bmatrix} \hat{\omega}_c \\ 0 \end{bmatrix} \tag{17}$$

where $\hat{q}_c^* = [-\hat{q}_c \; \hat{q}_{c_4}]^{\top}$ is the conjugate of the quaternion $\hat{q}_c$, which represents a previously calculated quaternion via Gauss-Newton method and $\hat{\omega}_c$ is the rough angular rate calculated based on quaternion calculation. Though $\hat{\omega}_c$ is quite noisy, EKF softs it and provides a better estimation of the satellite attitude in both sun and eclipse phases.

After obtaining the quaternion that fuse both magnetic and sun sensors, as well as the rough angular rate, the measurement process yields:

$$\hat{y}_k = h(\hat{x}_k, t) + v_k(t) = \begin{bmatrix} \operatorname{diag}(\hat{\omega}_c) & 0^{[3\times3]} \\ 0^{[3\times3]} & \operatorname{diag}(\hat{q}_c) \end{bmatrix} \tag{18}$$

Once defined $h(x, t)$, the linearization process according to the Eq. (12) results in the matrix H described as:

$$H = \begin{bmatrix} I^{[6\times 6]} & 0^{[6\times 1]} \end{bmatrix} \tag{19}$$

the last row of the matrix H was omitted considering the fact that the fourth term of the quaternion q_4 depends on the vector part, being a redundancy due to quaternion unicity, additionally it was taken into account the existence of singularity when the covariance matrix is calculated using the 4 elements of the quaternion.

Finally, the innovation process is proposed as follows:

$$v = \begin{bmatrix} \hat{q}_c - \hat{q}_k^- \\ \hat{\omega}_c - \hat{\omega}_k^- \end{bmatrix} \tag{20}$$

4.2 EKF Innovation Process in Eclipse Phase EKF

During this mode, just magnetic information is available. To estimate the satellite attitude, previous information is computed by the Sun mode EKF or initial conditions programmed after ADS initialization.

The measurement process yields:

$$\hat{y}_k = h(\hat{x}, t) + v_k(t) = R(\hat{\mathbf{q}}_k^-) b_i \tag{21}$$

where b_i is the Earth magnetic field in inertial reference frame and $R(\mathbf{q}_b^i)$ is the rotation matrix which related the body and inertial reference field.

The linearization process to compute H is described by the Eq. (12) and due to the nonlinear structure of the function $h(x,t)$ shown in Eq. (21), the matrix H yields:

$$H = \begin{bmatrix} 0^{3\times 3} & \left.\frac{\partial (R(\mathbf{q}) b_i)}{\partial \mathbf{q}}\right|_{\mathbf{q}=\hat{\mathbf{q}}_k^-} \end{bmatrix} \tag{22}$$

where it will be used to compute the Kalman gain as indicated in Eq. (9).

Finally, the innovation process is defined as:

$$v = b_B - R(\mathbf{q}_k^-) b_i \tag{23}$$

5 Numerical Simulations

As a numerical validation platform, a virtual satellite flight simulator was programmed, considering satellite dynamics, orbit parameters (LEO and Sun-

Table 1 EKF initial parameters

EKF Tuned covariances	Values
Gauss-Newton quaternion covariance	0.8
Rough angular rate covariance	0.9
Magnetometer covariance	0.6
Kinematics process covariances	$5\,e^{-4}$
Dynamics process covariances	$8\,e^{-5}$
Initial P_o	$0.01[I^{7\times7}]$

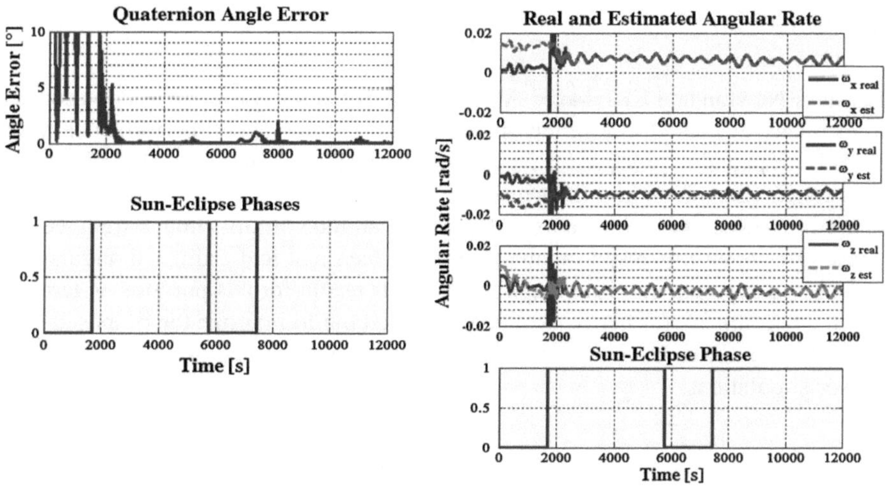

Fig. 2 Attitude and angular rate estimation performance

Synchronous), environmental models and simulated sensors and actuators. Concerning reference vectors, the local magnetic field vector was computed from the IGRF model (*International Geomagnetic Reference Field*). A Sun model for radiation applications was implemented in this work, which was adapted for this particular application [19].

In Table 1 shows the EKF tuned covariances used in the simulations performed in this work, according with noise and bias properties from sensors and the calculated rough attitude and angular rate data.

The results presented below are based on the fact that the satellite performed a detumble maneuver before initializing the ADS. It means that an angular rate reduction maneuver was performed after satellite deployment. This condition was considered to establish a small angular rate as an initial parameter in the simulation.

The Fig. 2 shows the angle error behavior between the real and estimated attitude quaternion during the satellite orbit including sun-eclipse phases. The Fig. 2 also shows a comparison between real and estimated stellite angular rates. The Fig. 2 shows that the ADS cannot estimate precisely the satellite attitude and angular rate during its deployment in eclipse phase. After entering sun phase, the ADS estimate the attitude estimation and angular rates with a convergence time of 200 s. Covariance matrices must be tuned appropriate according to sensor noise properties to achieve good results.

When phase change occurs, previous satellite attitude and angular rate information are used as initial condition for the eclipse phase EKF. The small error between real and estimate angular rates allow the propagation of satellite attitude with an angle error below 2°.

6 Conclusion

A Gauss-Newton and EKF based Attitude Determination System was designed and numerically evaluated, where the obtained results show a good performance of the attitude propagation process during eclipse phase. The angular rate estimation based on a rough calculation and filter process, as well as the reduction of its bias in sun phases, allowed the propagation of satellite attitude information during eclipse phase. To obtain this result, a small error between real and estimated angular rate must be guaranteed, in this concern EKF results useful for this purpose. In terms of space environment, further work must be carried out to improve the proposed ADS by including Earth albedo models and magnetic perturbance in order to improve sensor simulations.

References

1. Xiong K, Liang T, Yongjun L (2011) Multiple model Kalman filter for attitude determination of precision pointing spacecraft. Acta Astronaut 68:843–852
2. Chaurais JR, Ferreira HC, Ishihara JY, Borges RA, Kulabukhov AM, Larin VA, Belikov VV (2013) A high precision attitude determination and control system for the UYS-1 nanosatellite, 2013 IEEE aerospace conference, USA, pp 1–12
3. Mark LP (1990) Three-axis attitude determination via Kalman filtering of magnetometer data. J Guidance Control Dynam 13:506–514
4. Ahn H-S, Lee S-H (2005) Gyroless attitude estimation of sun-pointing satellites using magnetometers. IEEE Geosci Remote Sens Lett 2(1):8–12
5. Khosravian A, Namvar M (2012) Rigid body attitude control using a single vector measurement and Gyro. IEEE Trans Autom Control 57(5):1273–1279
6. Cordova Alarcon JR, Rodriguez Cortes H, Vicente Vivas E (2009) Extended Kalman filter tuning in attitude estimation from inertial and magnetic field measurements. In: 6th international conference on electrical engineering, computing science and automatic control CCE, pp 518–523
7. Lerner GM (1990) Three-axis attitude determination, spacecraft attitude determination and control, Kluwer Academic Publishers, USA, pp 420–426
8. Martin TH, Howard BD, Beale M (2002) Neural network design. PWS Publishing Company, Boston

9. João LM, Xiaoping Y, Eric RB, Robert BM, Michael JZ (2001) An extended Kalman filter for quaternion-based orientation estimation using MARG sensors. In: Proceedings of the 2001 IEEE/RSJ, pp 2003–2011
10. Kutlu A, Haciyev Ch, Tekinalp O (2007) Attitude determination and rotational motion parameters identification of a leo satellite through magnetometer and sun sensor data. In: 3rd international conference on recent advances in space technologies, pp 458–461, Turkey
11. Tang X, Liu Z, Zhang J (2012) Square-root quaternion cubature Kalman filtering for spacecraft attitude estimation. Acta Astronaut 76:84–94
12. Wei Q, Liang H, Huijuan Z, Jianchen F (2013) Interlaced optimal-REQUEST and unscented Kalman filtering for attitude determination. Chin J Aeronaut 26(2):449–455
13. Ovchinnikov M, Ivanov D (2013) Approach to study satellite attitude determination algorithms. Acta Astronaut 98:133–137
14. Springmann JC, Sloboda AJ, Klesh AT, Bennet MW, Cutler JW (2012) The attitude determination system of the RAX satellite. Acta Astronaut 75:120–135
15. Xiang T, Meng T, Wang H, Han K, Jin Z-H (2012) Design and on-orbit performance of the attitude determination and control system for the ZDPS-1A pico-satellite. Acta Astronaut 77:182–196
16. Clements R, Tavares P Lima (2000) Small satellite attitude control based on a Kalman filter. In: Proceedings of the 2000 IEEE international symposium on intelligent control, pp 79–84
17. Lu C, Weiwei Y, Xiaoqian C, Yiyong H (2011) Application of multi-sensors data fusion based on improved federal filtering in micro-satellite attitude determination. In: 2011 international workshop on multi-platform/multi-sensor remote sensing and mapping, pp 1–5, China
18. James RW (1990) Spacecraft attitude determination and control. Kluwer Academic Publishers, USA, pp 436–438
19. Reda I, Andreas A (2008) Solar position algorithm for solar radiation applications, technical report. National Renewable Energy Laboratory, USA

Nonlinear Identification of Inverted Pendulum System Using Volterra Polynomials

G. Ronquillo, G.J. Ríos Moreno, E. Hernández Martínez and M. Trejo Perea

Abstract The inverted pendulum problem is one of the most important problems in control theory and has been studied excessively in control literatures. When control systems have strong requirements, the adjustment of the controller is a complex problem. The nonlinear model is useful for control design. In the present work, Volterra polynomial basis function (VPBF) networks have been used to identify a single inverted pendulum on a moving cart (SIPC) system. The inverted pendulum is a benchmark problem of nonlinear multivariable systems with inherent instability. The multivariable system has been considered with a force produced by a DC motor as the input, and four states variables as the outputs. A Fuzzy Logic controller has been used to stabilize the system for closed-loop identification. Here, the nonlinear model of the inverted pendulum has been implemented. The offline structure selection through orthogonal least square algorithm is used for the nonlinear system identification via the basis function selection of Volterra polynomial networks. The neural network is trained using the error between the model's outputs and the plant's actual outputs. The results show good match between predicted and actual outputs.

Keywords Volterra polynomials · Basis function · Inverted pendulum · Nonlinear systems · Identification

G. Ronquillo (✉)
Centro de Ingeniería Y Desarrollo Industrial, Querétaro, Mexico
e-mail: gronquillo@cidesi.mx

G.J. Ríos Moreno · M. Trejo Perea
Universidad Autónoma de Querétaro, Querétaro, Mexico
e-mail: riosg@uaq.mx

M. Trejo Perea
e-mail: mtp@uaq.mx

E. Hernández Martínez
Instituto Politécnico Nacional, Mexico, Mexico
e-mail: euhernandezm@ipn.mx

M. Ceccarelli and E.E. Hernández Martinez (eds.), *Multibody Mechatronic Systems*,
Mechanisms and Machine Science 25, DOI 10.1007/978-3-319-09858-6_9

1 Introduction

The study of nonlinear systems has not been part of many engineering curricula for some time. This is partly because nonlinear systems have been perceived as difficult. A good reason for this was that there were not many good analytical tools like the ones that have been developed for linear, time-invariant systems over the years. Linear systems are well understood and can be easily analyzed. It is well-known that the inverted pendulum can describe a variety of inherently unstable systems and has wide applications in many areas, such as, human body self-balancing explanation [1–4] and robot design technology [5, 6]. Lately, due to the wide range of applications of the inverted pendulum, the development of its control strategy has become more attractive, but in order to make a good controller it is needed a good model of the system.

Although there are many techniques available for identification of linear dynamical systems, in general, dynamic systems are complex and nonlinear [7]. The main problem of using linearization techniques is that the resulting model is valid only in a certain operating range; consequently, the use of nonlinear identification methods has become a need, but nonlinear system identification is much more difficult than linear system identification [8].

Looking backward, is easy to realize that the artificial neural networks have been used for nonlinear identification for many years due to their capabilities, being the most important ones (1) their ability to learn and (2) their good performance for approximation of nonlinear functions [9]. The use of Neural Networks (NN) for identification of nonlinear models has already been explored [10–12]. Suthradar et al. [13] demonstrates both identification and control of nonlinear dynamical systems using static and dynamic back-propagation methods. Riedmiller [14] applied supervised learning techniques to learn control of unstable systems. More recently, Wang [15] studied the problem in two axis movement with a computer aided simulation and PID controllers. Some other works have been devoted to friction modelling and parameter identification, further a feedback linearization controller which takes into account the viscous friction was presented in [16]. Finally, [17] made physical modelling of various simulated systems using computer software for educational purposes.

However, the above results mentioned mainly concern control problems of inverted pendulums. Few results have been obtained as a solution to the problem of modelling structure for such systems, especially expansion-like solution, which usually plays an important role in the engineering calculus.

The present work is to model the inverted pendulum system driven by a DC motor within the chosen framework through data acquired directly from the real plant. The acquired data are generally classified as input and output, measured at discrete instants of time t and collected as an array of finite duration data. A VPBF NN has been constructed for the inverted pendulum nonlinear system driven by a DC motor to fit the data instead of mathematical equations. These models were trained and then, for each of the VPBF NN architecture, the approximation error was calculated in order to select which structure is better for the actual plant identification.

2 Methodology

2.1 Inverted Pendulum System: Mathematical Model

The system modelling is done in two parts: First, the nonlinear dynamics of the SIPC is considered having as the system input the translational force; second, the lineal system existing between the actuator, the DC motor and the generated force, is analyzed.

2.1.1 Dynamics of Inverted Pendulum System

Consider the SIPC system, Fig. 1, given by

$$\frac{d}{dt}\left(\frac{\partial \mathscr{L}}{\partial \dot{q}}\right) - \left(\frac{\partial \mathscr{L}}{\partial q}\right) = Bu + D \qquad (1)$$

where $q = (q_a, q_u)^T \in \mathbb{R} \times \mathbb{R}$ and $\dot{q} = (\dot{q}_a, \dot{q}_u)^T \in \mathbb{R} \times \mathbb{R}$ are vectors of generalized coordinates and velocities, respectively; q_a and q_u are actuated and unactuated variables, respectively; $B = (1, 0)^T$ and $u \in \mathbb{R}$ is a control input variable; the function

$$\mathscr{L}(q, \dot{q}) = \frac{1}{2}\dot{q}^T M(q)\dot{q} - V(q) \qquad (2)$$

is the Lagrangian; $M(q)$ is a symmetric positive-definite inertia matrix; and $V(q)$ is the potential energy of the system. By introducing a lumped disturbance vector $D \in \mathbb{R}^2$, the system Eq. (1) can also be rewritten in a matrix form as:

$$M(q)\ddot{q} + C(q, \dot{q})\dot{q} + G(q) = Bu + D \qquad (3)$$

or, equivalently

$$\begin{bmatrix} m_{11}(q) & m_{12}(q) \\ m_{21}(q) & m_{22}(q) \end{bmatrix} \begin{bmatrix} \ddot{q}_a \\ \ddot{q}_u \end{bmatrix} + \begin{bmatrix} n_1(q, \dot{q}) \\ n_2(q, \dot{q}) \end{bmatrix} = \begin{bmatrix} u + d_1 \\ 0 \end{bmatrix} \qquad (4)$$

where $C(q, \dot{q})\dot{q}$ is the Coriolis and centrifugal loading vector, $G(q)$ is the gravitational loading vector, and $D = (d_1, 0)^T$ is assumed to satisfy the classical matching condition. d_1 includes parameter uncertainties, external disturbance, and unmodelled dynamics, such as viscous and Coulomb's friction forces exerted on the actuated joint, and it is assumed that $|d_1| \leq \bar{d}_1$ for a known constant bound $\bar{d}_1$. In the

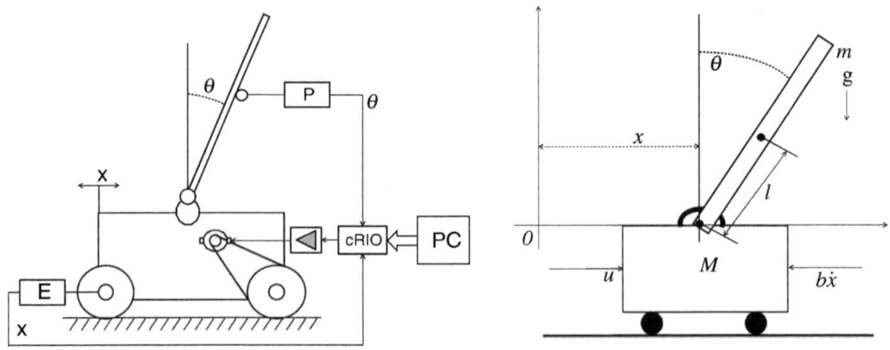

Fig. 1 Block diagram of the system, and free body diagram of the SIPC

case of the SIPC in Fig. 1, according to [18], the equations of motion in Eq. (4) can be rewritten as:

$$\begin{bmatrix} \gamma & \beta cos(\theta) \\ \beta cos(\theta) & \alpha \end{bmatrix} \begin{bmatrix} \ddot{x} \\ \ddot{\theta} \end{bmatrix} + \begin{bmatrix} -\beta sin(\theta)\dot{\theta}^2 \\ -\eta sin(\theta) \end{bmatrix} = \begin{bmatrix} u + d_1 \\ 0 \end{bmatrix} \tag{5}$$

where $\alpha = ml^2, \beta = ml, \gamma = M + m$, and $\eta = mgl$; M and m are masses of the cart and pendulum, respectively; l, θ, x are length of pendulum, angle of pendulum with respect to vertical line, and moving distance of cart from initial position, respectively; g is a gravitational acceleration; and u is a control force applied to the cart.

2.1.2 Modelling of the Mechanical System for the DC Motor

The DC motor is coupled via a mechanical system band-pulley-wheel which pushes the cart generating the force u. The motor is driven through the armature voltage $v_{in}(t)$. The torque generated by the DC motor is proportional to the armature current like is indicated in the next equation

$$t_m(t) = K_t i_a(t) \tag{6}$$

where $t_m(t)$ is the torque developed by the motor, K_t is the torque constant of the motor, $i_a(t)$ is the armature current of the motor. The electrical circuit equations for the motor are:

$$v_{in}(t) - v_{emf} = L_a i_a(t) + R_a i_a(t) \tag{7}$$

$$v_{emf} = K_v \dot{\theta}_m(t) = K_v \omega_m(t) \tag{8}$$

where $v_{in}(t)$ is the motor voltage, v_{emf} is the electromotive voltage, R_a armature resistance, K_v back electromotive force's constant, $\theta_m(t)$ motor's shaft angular position, and $\omega_m(t)$ motor's shaft angular velocity.

The torque equation for motor output shaft is:

$$t_m(t) - t_L(t) = J_m \ddot{\theta}_m(t) + B_m \dot{\theta}_m(t) \tag{9}$$

where J_m is inertia moment of the rotor with load, B_m is the damping coefficient.

The motor is coupled to the cart with a radius wheel r_w and a coupling gear ratio n, $t_L(t)$ is the required torque to generate the linear force u. Converting the rotational variables into translational variables using:

$$t_L(t) = r_w n u \quad \text{and} \quad \theta_m(t) = \frac{x}{r_w n} \tag{10}$$

The equilibrium torque is obtained from Eqs. (9) and (10),
Assuming zero inductance, and using Eqs. (6), (7) and (8),

$$t_m(t) = J_m \frac{\ddot{x}}{n r_w} + B_m \frac{\ddot{x}}{n r_w} + n r_w u \tag{11}$$

$$t_m(t) = K_i \left[\frac{V_{in} - K_v \dot{\theta}_m}{R_a} \right] = K_i \left[\frac{V_{in} - K_v \left(\frac{\dot{x}}{n r_w} \right)}{R_a} \right] \tag{12}$$

The Eqs. (5), (11) and (12) describe the nonlinear model of the SIPC impulsed by a DC motor.

2.2 Nonlinear Modelling by VPBF Networks

Consider the nonlinear discrete system described by:

$$X_{t+1} = G(X_t, u_t) \tag{13}$$

$$y_t = h(X_t, u_t) \tag{14}$$

where $G(\cdot)$ is a nonlinear function vector, $h(\cdot)$ a nonlinear function, X_t the state vector, y_t the output and u_t the input.

Based on the input and output relation of a system, the above nonlinear discrete system can also be expressed by a nonlinear auto-regressive moving average (NARMA) model [19], that is,

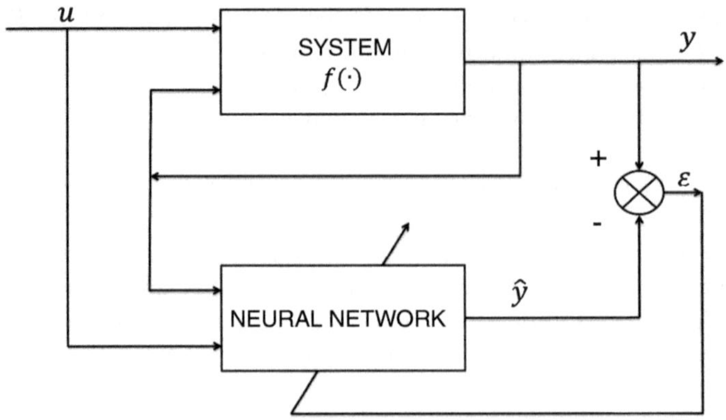

Fig. 2 Neural network based identification

$$y_t = f(y_{t-1}, y_{t-2}, \ldots, y_{t-n_y}, u_{t-1}, u_{t-2}, \ldots, u_{t-n_u}) \quad (15)$$

where $f(\cdot)$ is a nonlinear function, n_y and n_u are the corresponding maximum delays.

It is well known that NN provide a good nonlinear function approximation techniques. A nonlinear identification structure by neural networks is shown in Fig. 2. Here it assumes that the nonlinear function $f(\cdot)$ in the NARMA model is approximated by a single layer neural network, which consists of a linear combination of basis functions.

$$\widehat{f}(x_t) = \sum_{k=1}^{N_f} w_k \varphi_k(x_t) \quad (16)$$

where $\varphi_k(x_t)$ is the basis function, $x_t = [y_{t-1}, y_{t-2}, \cdots, y_{t-n_y}, u_{t-1}, u_{t-2}, \cdots, u_{t-n_u}]$ and w_k the weight.

The representation of the nonlinear function $f(x_t)$ is given by

$$\begin{aligned}\widehat{f}(x_t) =& w_1 + w_2 y_{t-1} + \cdots + w_{n_y+1} y_{t-n_y} + w_{n_y+2} u_{t-1} + \cdots \\ &+ w_{n_y+n_u+1} u_{t-n_u} + w_{n_y+n_u+2} y_{t-1}^2 + + w_{n_y+n_u+3} y_{t-1} y_{t-2} + \cdots + w_{N_f} u_{t-n_u}^o\end{aligned} \quad (17)$$

where o is the system order.

The number of polynomial basis function is given by:

$$N_f = \frac{(n_u + n_y + o)!}{o!(n_u + n_y)!} \quad (18)$$

Using the VPBF network, the nonlinear function $f(\cdot)$ can be obtained by:

$$f(x_t) = \widehat{f}(x_t) - \varepsilon(x_t^o) \tag{19}$$

where $\varepsilon(x_t^o)$ is the approximation error.

It is assumed that a set of input-output data ($y_t, u_t, t = 1, 2, \cdots, M_t$) of the system is given. Based on Eq. (17) the input-output relation may compactly be written in the following vector form:

$$Y = \Phi(x)W + E(x^o) \tag{20}$$

where the output vector is $Y \in \mathbb{R}^{M_t \times 1}$, the weight vector is $W \in \mathbb{R}^{N_f \times 1}$, the approximation error vector is $E(x^o) \in \mathbb{R}^{M_t \times 1}$ and the basis function matrix is $\Phi(x) \in \mathbb{R}^{M_t \times N_f}$.

The weight vector W is usually found minimizing the Euclidean norm, i.e.,

$$\hat{W} = \arg\min_W \|Y - \Phi(x)W\|_2 \tag{21}$$

which is a least squares solution.

The classical Gram Schmidt method can be used to transform from the set of basis vectors $\{\Phi_i\}$, into a set of orthogonal basis vectors, and thus makes it possible to calculate the individual contribution to the desired output from each basis vector. An orthogonal decomposition of the matrix $\Phi(x)$ gives:

$$\Phi(x) = PQ \tag{22}$$

where $P = [P_1, P_2, \cdots, P_{N_f}]$ is an $M_t \times N_f$ matrix with orthogonal columns and Q is an $N_f \times N_f$ unit upper triangular matrix with 1 on the diagonal and 0 below the diagonal.

The corresponding optimal weight vector is:

$$\hat{W} = Q^{-1}\hat{V} \tag{23}$$

where $\hat{V} = [\hat{v}_1, \hat{v}_2, \cdots, \hat{v}_{N_f}]^T \in \mathbb{R}^{N_f \times 1}$

$$\hat{v}_i = \frac{Y^T P_i}{P_i^T P_i}, \quad for \quad i = 1, 2, \cdots, N_f \tag{24}$$

$\hat{v}_i^2 P_i^T P_i$ is the increment to the desired output variance introduced by P_i, the error reduction ratio due to P_i may be defined by:

$$r_i = \frac{\hat{v}_i^2 P_i^T P_i}{Y^T Y} \tag{25}$$

The ratio offers a simple and effective means of seeking a subset of significant basis function. This implementation based on the classical Gram Schmidt is given by [20, 21].

Changing the order of the VPBFs will lead to a change in the error reduction ratio r_i. The normalized residual sum of squares (NRSS), is given by:

$$NRSS = 1 - \sum_{j=1}^{L_s} r_j \tag{26}$$

The procedure is terminated when $NSSR < e_0$, where e_0 is a chosen tolerance. This gives a subset model containing L_s significant terms. For more information on offline structure selection see [8, 22].

2.3 Experiment Description

The experimental results from SIPC are presented in this section. SIPC parameters are shown in Table 1. The SIPC is inherently unstable, which means that in open loop the pendulum will fall. One of the requirements for the systems identification is that the data must satisfy the persistent excitation condition and have been obtained inside of an operation interval, for this is necessary stabilize the inverted pendulum using a controller.

Table 1 Parameters from the real SIPC system

Parameter	Description	Value
M	Mass of the cart	2.00 kg
l	Length of the pendulum	1.00 m
m	Mass of the pendulum rod	0.10 kg
r	Radius of the wheel	0.0335 m
g	Gravity	9.81 m/s^2
B_m	Damping coefficient	1.568×10^{-4} Nm/rad/s
R_a	Armature resistance	1.8 Ω
K_t	Torque constant	0.0168 Nm/A
K_v	Back EMF constant	0.0168 V/rad/s
J_m	Moment of inertia	0.000011 Kgm^2
n	Gear ratio	0.2
P_s	Potentiometer for angle position	5 kΩ
E	Encoder for linear position	50 pulses per revolution

To stabilize the inverted pendulum a Fuzzy controller was used, it was implemented on a SIPC didactic system (RT-124) from G.U.N.T. which has an encoder, for measuring the linear position and linear velocity, and a potentiometer, for measuring the position and velocity angular.

The controlled system was continuously disturbed in order to accomplish the persistent excitation condition. A cRIO system from National Instruments with encoder and analog input cards was used to acquire the inverted pendulum experimental data with a sampling time of $T_s = 10$ ms which meet Shannon theorem taking mechanical time constant. Experimental data was acquired during 75 s which are 7,500 samples, the first 60 s were used to train the network, the last 15 s were used for test and validation of the network. The experimental data were obtained using LabView real time software. All signals were saved to a file in cRIO. The experimental data are shown in Fig. 3.

Several VPBF NN algorithms were built, using MATLAB with training data, using different VPBF architectures, changing maximum delays n_u, n_y and system's order o, in order to determine which is the architecture model that best fits the experimental data.

3 Results

We assume that the model structure has the same delay in the input and output. They were built several models with different structures, which are combinations between order (o) and delays of the inputs/outputs ($n_u = n_y$) being varied between 2 and 4. This was done for each output variable of the SIPC, i.e. x, $\dot{x}$, θ and $\dot{\theta}$ using training experimental data (0–60 s).

Table 2 shows the results of all built models with classical Gram Schmidt method to get orthogonal decomposition. The model's structure that best fits the experimental training data are marked with bold.

The VPBF NNs were built with the first 20 VPBFs and their respective weights, the basis functions were reordered from highest to lowest contribution to the outputs using r_i from Eq. (25).

These models were tested with experimental data different to the training data, experimental test data (60–75 s) and its respective calculated NRSSs, which were different but consistent with the training NRSSs. The model test results are shown in Table 3, it is clear that NRSSs from model's test are not equal to NRSSs from model's training.

The difference between NRSSs is due to that model information from experimental data test is not incorporated to model. Hence, the NRSSs from model with training data and NRSSs from the model with test data are different but very close. These mean that the model has enough information to emulate the real system in states near to experimental training data.

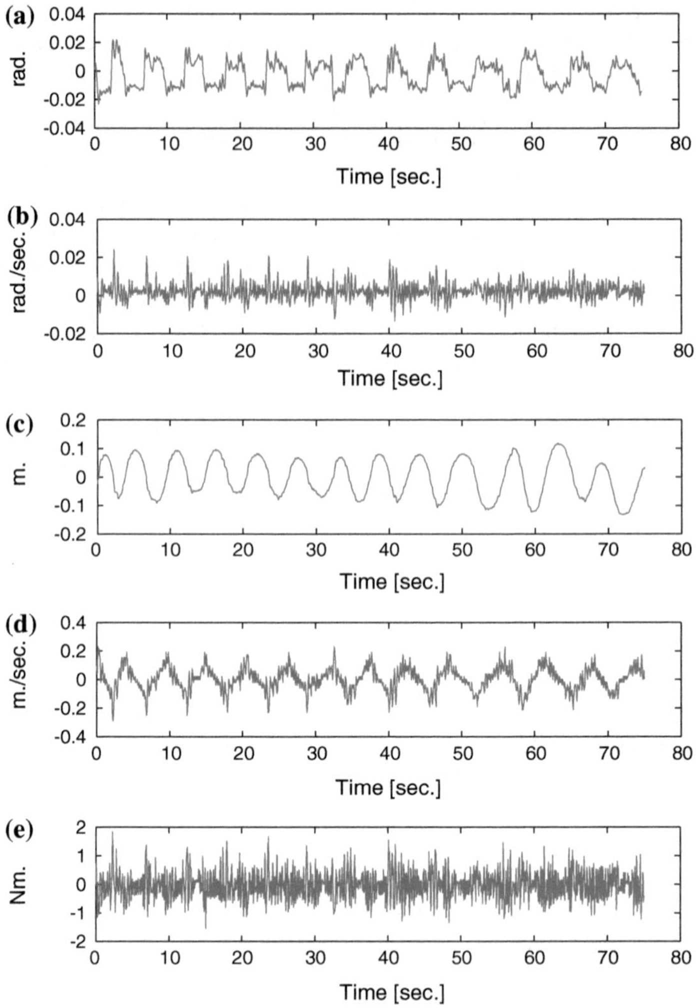

Fig. 3 SIPC acquired experimental data

Table 2 Normalized residual sum of squares (NRSS) result train

$n_{u,y}$	o	x	$\dot{x}$	θ	$\dot{\theta}$
2	2	0.0092	0.1856	3.06×10^{-4}	0.0492
3	2	0.0081	0.1759	1.87×10^{-4}	0.0444
4	2	0.0076	0.1657	1.78×10^{-4}	0.0419
2	3	0.0082	0.1704	1.86×10^{-4}	0.047
3	3	0.0069	0.1489	1.62×10^{-4}	0.0396
4	3	**0.0059**	0.1293	$\mathbf{8.18 \times 10^{-5}}$	**0.0333**
2	4	0.0078	0.161	1.80×10^{-4}	0.0457
3	4	0.0062	**0.1289**	1.47×10^{-4}	0.0359

Table 3 Normalized residual sum of squares (NRSS) result test

$n_{u,y}$	o	x	$\dot{x}$	θ	$\dot{\theta}$
2	2	0.0121	0.3463	63×10^{-4}	0.0484
3	2	0.011	0.3167	62×10^{-4}	0.0421
4	2	0.0095	0.2668	58×10^{-4}	0.0372
2	3	0.0113	0.305	62×10^{-4}	0.0424
3	3	0.0089	0.2476	61×10^{-4}	0.0329
4	3	**0.0052**	0.1251	$\mathbf{15 \times 10^{-4}}$	**0.016**
2	4	0.0107	0.2839	62×10^{-4}	0.0414
3	4	0.0066	**0.1629**	49×10^{-4}	0.0,236

The graphs in Fig. 4 show the experimental test and estimated output signals by selected models of VPBF NNs. In this figure it can be seen that the selected models fit well to the experimental data test but these have not been used for training the network, which means that the model is well trained and has the actual system information.

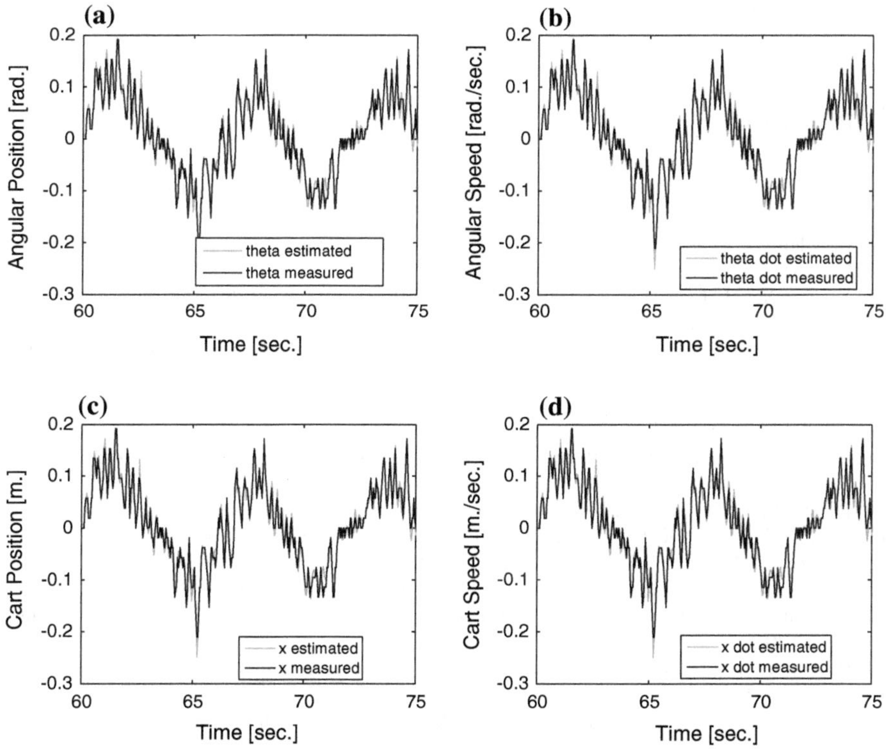

Fig. 4 Validation curves for: **a** Pendulum angular position, **b** Pendulum angular velocity, **c** Cart displacement, **d** Cart velocity

4 Conclusions

The dynamic systems containing nonlinear relations are difficult to model with conventional techniques. The Inverted Pendulum is a nonlinear system object with non-typical uncertainty. The uncertainty is not dealt with. However it affects the performance. For identification of the present problem, the VPBF NN has the ability to approximate nonlinear functions with a single structure. In this work, the VPBF NN has been developed using data from experiment with a real pendulum system. The orthogonal least squares algorithm has been used.

The advantages of the VPBF neural networks are its reduced computational complexity, improved rate of convergence and identification accuracy. The validation test results indicate numerical stability and high precision. The performance of the VPBF neural networks is superior to that conventional neural networks. The VPBF neural network may be used for off-line plant identification due to its performance.

Acknowledgements This work has been supported by the Programa de Estímulos a la Investigación, Desarrollo Tecnológico e Innovación del CONACYT—C003 V-2014-01, contract number 217217 and Fondo Sectorial de Investigación para la Educación (SEP-CONACyT contract number 133353-2009-México).

References

1. Milton J, Cabrera JL, Ohira T, Tajima S, Tonosaki Y, Eurich C, Campbell S (2009) Chaos: An Interdisc J Nonlinear Sci 19(2):0261101
2. Bowden C, Holderbaum W, Becerra VM (2012) IEEE Trans Autom Control 57(11):2891
3. Kuo AD (2007) Hum Mov Sci 26(1):617
4. Young-Dae H, Jong-Hwan K (2013) IEEE/ASME Trans Mech 18(2):657
5. Erbatur K, Kurt O (2009) IEEE Trans Ind Electron 56(3):835
6. Vanderborght B (2010) Dynamic stabilisation of the biped Lucy powered by actuators with controllable stiffness. Springer, Heidelberg
7. Todorovic N, Klan P (2006) In: Neural network applications in electrical engineering. NEUREL 2006. 8th Seminar
8. Liu GP (2001) Nonlinear identification and control: a neural network approach. Springer Verlag, London
9. Purwar S, Kar I, Jha A (2007) Appl Soft Comput 7(1):364
10. Ruan XG, Ding MX, Gong DX, Qiao JF (2007) Neurocomputing 70(1):770
11. Rankovic VM, Nikolic IZ (2008) FME Trans 36(2):87
12. Xing Z, Pang X, Ji H, Qin Y, Jia L (2011) Int J Control Autom Syst 9(4):759
13. Sutradhar A, Sengupta A, Challa V (2010) in India Conference (INDICON), 2010 Annual IEEE (2010)
14. Riedmiller M (2005) In: IEEE International conference on systems, man and cybernetics, 2005
15. Wang JJ (2011) Simul Model Pract Theory 19(1):440
16. Carrillo-Serrano RV, Hernandez-Guzman VM (2010) Int J Innov Comput Inf Control 6 (12):5553
17. Zupančič B, Sodja A (2012) Simulation modelling practice and theory. Elsevier p. article in press (2012)

18. Fantoni I, Lozano R (2002) Non-linear control for underactuated mechanical systems. Springer Verlag, Berlin
19. Leontaritis IJ, Billings SA (1985) Int J Control 41(1):303
20. Billings SA, Chen S, Korenberg MJ (1988) Int J Control 49(6):2157
21. Billings SA, Korenberg MJ, Chen S (1989) Int J Syst Sci 19(8):1559
22. Luo W, Billings S (1995) Sig Procces 46(1):179

[illegible]

Multi-robot Exploration and Mapping Strategy in Underground Mines by Behavior Control

Antoni Mauricio, Ayrton Nieves, Yomar Castillo, Kenji Hilasaca, Christian Fonseca, Jhair Gallardo, Ricardo Rodríguez and Glen Rodríguez

Abstract Exploration in high-risk areas is a topic that has motivated the development of mobile robotics in recent years. Moreover, the incursion of multi-agent systems in this field has opened a lot of solutions and applications. In this paper we propose a strategy of exploration and mapping for multi-robot systems in underground mines where toxic gases concentration (ex. : CO_2, CO, Sb) is unknown. The principal algorithm is the behavior control which evaluates the status of each agent and makes decisions that maximize system performance and minimize the cost of them. We will use scanning algorithms based on dynamic graph to reduce bandwidth consumption and use of memory. The system has been tested by simulating several situations such as partial loss of communications or agents.

Keywords Multi-robot systems · Exploration in high-risk areas · Behavior control · Mapping · Localization · Path planning · Motion planning

1 Introduction

Nowadays the necessity to solve complex problems at low cost has encouraged the development of multi-agent systems. Cooperative robotics has shown strong growth since its inception in the 1980s as evidenced by [1, 2], while the work developed in [3, 4] shows us the high performance of this kind of systems. There are many applications of multi-robot systems, such as space exploration, object transportation or risk areas exploration.

Mining is one of the sectors that has driven the development of the robotics, such as robotic arms for industrial processes or robots for specific tasks, for instance the extraction of minerals. Thrun et al. [5] show that it is possible to include autonomous

A. Mauricio · A. Nieves · Y. Castillo · K. Hilasaca · C. Fonseca · J. Gallardo · R. Rodríguez · G. Rodríguez (✉)
Universidad Nacional de Ingeniería, Lima, Peru
e-mail: grodriguez@uni.edu.pe

M. Ceccarelli and E.E. Hernández Martinez (eds.), *Multibody Mechatronic Systems*, Mechanisms and Machine Science 25, DOI 10.1007/978-3-319-09858-6_10

robots in high noise environments such as underground mines. However, the complex workspace of the mine remains the main problem for designing a robot. It has been pointed in [6] that a solution is feasible by applying a multi-robot cooperative system.

One of the jobs that best explains the integration of the principal algorithm is [7], from which we conclude that to complete the objective of this work we will have to use the theory of probabilistic robotic and graph theory for exploration process, besides the behavior control, which will deal with inspection and search.

The required location algorithm will need a "initial" map, which is feed into the robots from the start, and the initial distribution of particles in a circular area (radius equal to the width of the mine tunnel), close to the belief of the initial position of the robot, because if some problems occur with the system, for instance a non-planned reset, then the particles have a new initial distribution close to the last position of the robot.

2 Algorithms Design

In this section we present the algorithms that make up the main algorithm, these are: SLAM [Simultaneous Localization and Mapping], Behavior control, Path and Motion planning algorithms. The Fig. 1 shows the way those algorithms are connected in the context of a block diagram.

The system inputs are the initial position (X_o), positions of landmarks ($L_{i:n}$) and data from laser sensors (Z). These measures are inputs of the SLAM algorithm, the outputs are the local map and the next position for each robot. Slam outputs feed the "map rebuilder" and together with the information coming from other robots,

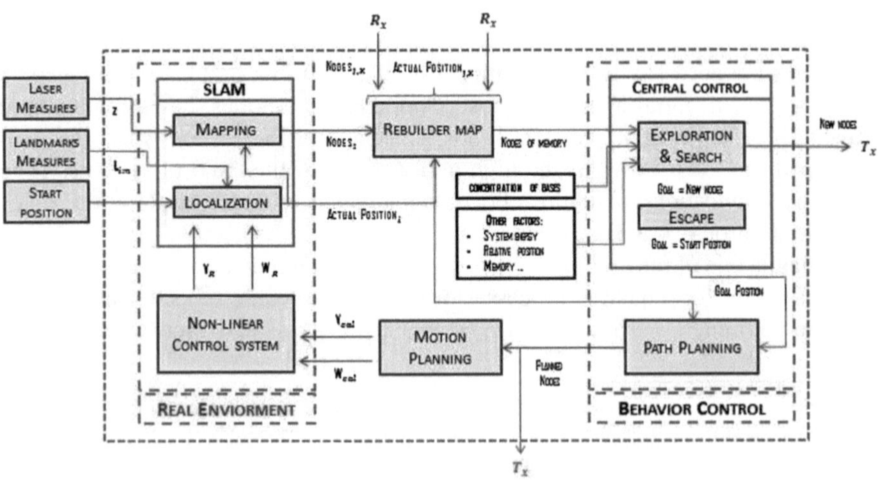

Fig. 1 Block Diagram

location is more precisely estimated and stored in memory. Each robot executes its behavior control based on gas sensors, energy level, memory and position of peers. The motion planner decides the linear and angular speed for the non-linear controller.

2.1 Behavior Control

Behavior control is based on decomposing the problem of autonomous control by task rather than by function. Traditional functional decomposition attempts to construct general purpose functional modules such as world modelers and planners, and advocates connecting them in a serial fashion [8]. Behavior control advocates the construction of special purpose task achieving modules that are connected directly to sensors and actuators and operate in parallel. In order to distinguish between the technical meaning of the word referring to a task achieving module in Behaviors control architecture, and the more generic meaning referring to the physical actions produced by such a module on a real robot, we shall capitalize the former. Thus, a set of Behaviors produces an emergent behavior when running on a real robot. One set of Behaviors might run on two different robots, or on one robot in two different circumstances, and produce two different behaviors [9].

There are two significant advantages to behavior control over traditional functional decomposition. First, because Behaviors operate in parallel rather than serially, fast Behaviors need not be delayed by slower ones, allowing the robot to respond to contingencies in real time. Second, because Behaviors are task-specific rather than general-purpose, Behavior designers can take advantage of the structure of the task in order to simplify the Behavior. This is the underlying reason why Behaviors tend to be simple [10].

Each robot has his own local behavior control algorithm (Fig. 2), with various behaviors for different states and situations [10], therefore, the decisions are taken based on their weights, with weight related to importance or urgency. Later, the weights are evaluated by a fuzzy function or a set of rules to choose the best decision.

Global behavior control is based in cooperative control theory, which relies in the agents' ability to coordinate their actions (Fig. 3). Advantages are reliability, robustness and better distribution of sensing and actions [11]. In this research project, underground exploration has 3 phases: entering the tunnel, exploration and gas search, and exit or escape (due to some emergency or low battery).

2.2 Robot Model

Figure 4 presents a diagram of the kind of robot we consider in this work. The robot has four wheels, driven by four independent electric motors. Two wheels are placed at each side of the robot. The robot configuration is represented by the position of

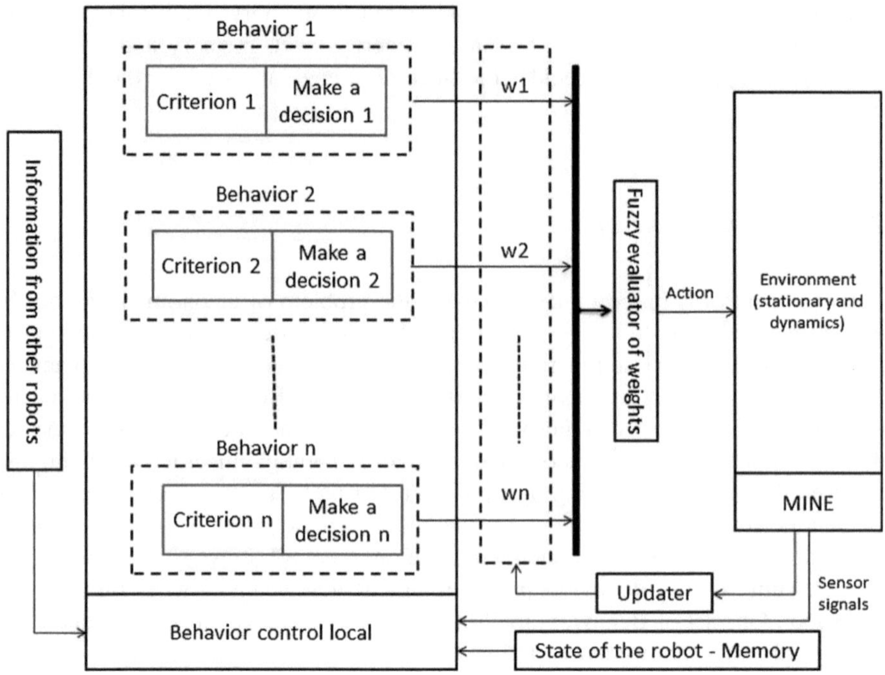

Fig. 2 Local behavior control

the center of the rectangular; dimensions are L (length) and H (width), in the Cartesian space (X and Y) and by its orientation α (angle between the vector of the robot orientation and the x-axis). Linear velocity is V, angular velocity is W, wheels radius is r and wheels speeds are $\dot{\theta}1, \dot{\theta}2, \dot{\theta}3$ y $\dot{\theta}4$.

2.3 Simultaneous Localization and Mapping [SLAM]

SLAM is considered a complex problem, because for obtain a good map is necessary a correct location, and vice versa. To solve the problem of SLAM exist various methods as explained by Thrun [12], in our case we use the Fast-SLAM method for positioning and occupancy grid mapping. The SLAM problem is based on calculating the next probability distribution:

$$p(\Theta, s^t | z^t, u^t, n^t) \tag{1}$$

Where: $z^t = z_1 \ldots z_t$ is a sequence of sensor measurements, Θ is the set of landmarks, u^t is the action taken, $s^t = s_1 \ldots s_t$ is the estimate of the current state and $n^t = n_1 \ldots n_t$ are variables associated to landmarks.

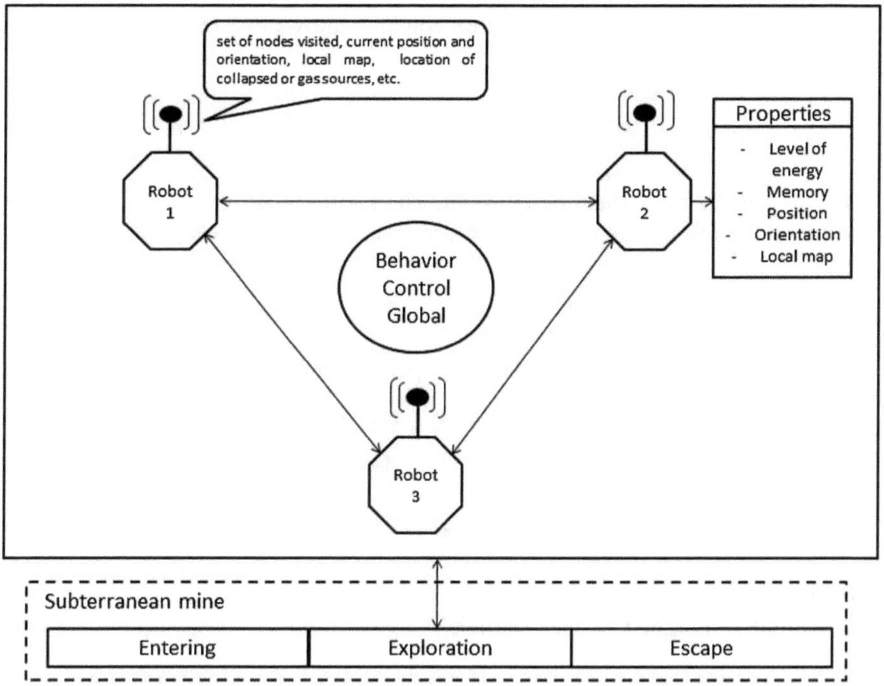

Fig. 3 Global behavior control

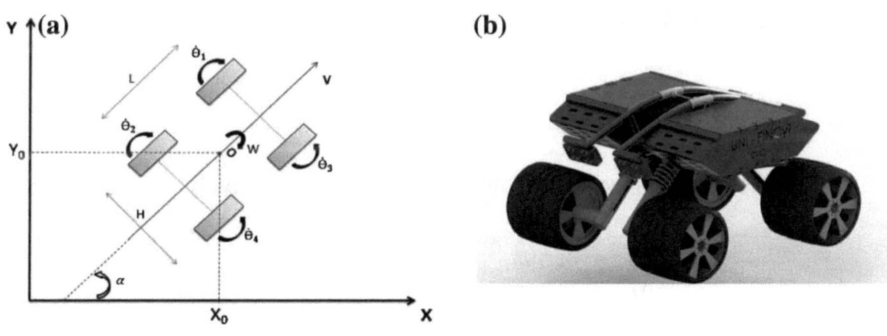

Fig. 4 **a** Mobile robot, **b** mechanical design of the robot

In the case of Fast-SLAM, there is known data association [14]; it is assumed that the landmarks are independent. It follows that:

$$p(\Theta, s^t|z^t, u^t, n^t) = p(s^t|z^t, u^t, n^t) \prod_n p(\theta_n|s^t, z^t, n^t) \quad (2)$$

The following probability distribution equation allows to estimate the location of the landmarks:

$$p(\theta_n|s^t, z^t, n^t) = \eta p\left(z_t|\theta_n, s_t^{[m]}, n_t\right) p(\theta_n|s^{t-1,[m]}, z^{t-1}, n^{t-1}) \quad (3)$$

Where η is the normalization constant. The particle filter is used to estimate the position of the robot. The probabilities are evaluated as follows [12]:

$$w_t^{[m]} = \eta \int p\left(z_t|\theta_{n_t}, s_t^{[m]}, n_t\right) p\left(\theta_{n_t}|s^{t-1,[m]}, z^{t-1}, n^{t-1}\right) d\theta_{n_t} \quad (4)$$

The Fast-SLAM algorithm updates the particles in the following way:

(1) Sampling: is randomly generated on a set of particles $\{x_t^{(i)}\}$ and is obtained from the previous generation $\{x_{t-1}^{(i)}\}$ Following the proposal distribution $\pi(.)$

$$x_{1:t}^{(i)} \sim \pi(.|z_{1:t}, u_{1:t-1}) \quad (5)$$

(2) Importance weight:
Weight is assigned to each particle

$$w_t^{(i)} = \frac{p(x_{1:t}^{(i)}|z_{1:t}, u_{1:t-1})}{\pi(x_{1:t}^{(i)}|z_{1:t}, u_{1:t-1})} \quad (6)$$

In many slam algorithms the proposal distribution π (.) is recursive

$$\pi(x_{1:t}|z_{1:t}, u_{1:t-1}) = \pi(x_t|x_{1:t-1}, z_{1:t}, u_{1:t-1}).\pi(x_{1:t}|z_{1:t-1}, u_{1:t-2}) \quad (7)$$

Therefore, importance weight can be calculated recursively:

$$w_t^{(i)} \propto \frac{p(z_t|m_{t-1}^{(i)}, x_t^{(i)})p\left(x_t^{(i)}|x_{t-1}^{(i)}, u_{t-1}\right)}{\pi\left(x_t|x_{1:t}^{(i)}, z_{1:t}, u_{1:t-1}\right)}.w_{t-1}^{(i)} \quad (8)$$

(3) Resampling: the particles are resampled considering the weights, after resampling all particles will have a weight proportional to $p(x_{1:t}^{(i)}|z_{1:t}, u_{1:t-1})$.
(4) Map estimation: for each particle the estimated map is calculated as follows:

$$p(m^{(i)}|x_{1:t}^{(i)}, z_{1:t}) \quad (9)$$

After the prediction step, a new set of points is obtained who is reapplied the update stage.

2.4 Occupancy Grid Mapping Algorithm

The main objective of this algorithm is to calculate the map, considering some environmental data:

$$p(m|z_{1:t}, x_{1:t}) \quad (10)$$

Where: m is the map, $z_{1:t}$ is sensor measurements from time 1 up to time t, and $x_{1:t}$ is the trajectory of the robot defined by a sequence of positions.

Let be m_i a grid cell with index I; the map "m" is divided into those cells. Each cell m_i has a binary value, 1 if it is occupied, 0 if it is free [12].

$$m = \{m_i\} \quad (11)$$

The estimation of each cell is now a binary problem with estimation and steady states. At the algorithm start, each cell has a probability of 0.5, the probabilities will updated according to the measure of the sensor through the Bayes filter [14].

$$p(m|z_{1:t}, x_{1:t}) = \prod_i p(m_i|z_{1:t}, x_{1:t}) \quad (12)$$

Each cell holds a probability value which indicates if that cell is occupied or not. The Occupancy Grid Mapping Algorithm is shown below:

```
Algorithm Occupnacy_Grid_Mapping(x_t, m_{t-1} , z_t):
// inputs: states of the robot from particle
//filters, Occupancy map, sensor range
  For all cell m_i do

     If m_i in perceptual field of z_t then

         If z_t observe an obstacle
              CellOcc = p(obst|occup)*p(m_i)
              CellUnocc = p(obst|unoccup)*(1-p(m_i))
              p(m_i) = CellOcc/(CellOcc + CellUnocc)
         Else
              CellOcc = (1-p(obst|occup))*p(m_i)
              CellUnocc = (1-p(obst|unoccup))*(1-p(m_i))
              p(m_i) = CellOcc/(CellOcc + CellUnocc)
         Endif
      Endif
   Endfor
Return {m_t}
```

2.5 Path Planning and Motion Planning

Path planning and motion planning allow each robot decides where to go and how to get there. The motion planner is developed based on the kinodynamic constraints for each robot [15], however the path planner is designed according to the workspace conditions.

Dijkstra's algorithm is a path planner algorithm based on graph theory for determining the shortest path from a source vertex to all other in a graph with weights of each edge. It works by visiting locations in the map starting with the ones closest to the start position, and expands outwards from the start point until it reaches the goal. For the implementation of this algorithm is necessary to have knowledge of the workspace, which is possible by running the algorithm with the reconstructed global map stored in memory; its main advantage is the low computational cost [16].

3 Simulations and Results

See Figs. 5 and 6.

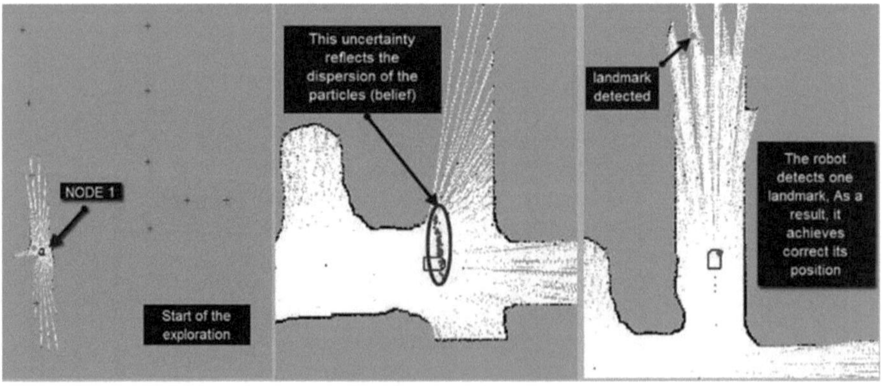

Fig. 5 Partial results of joining the path and motion planning with the simultaneous localization and mapping (SLAM)

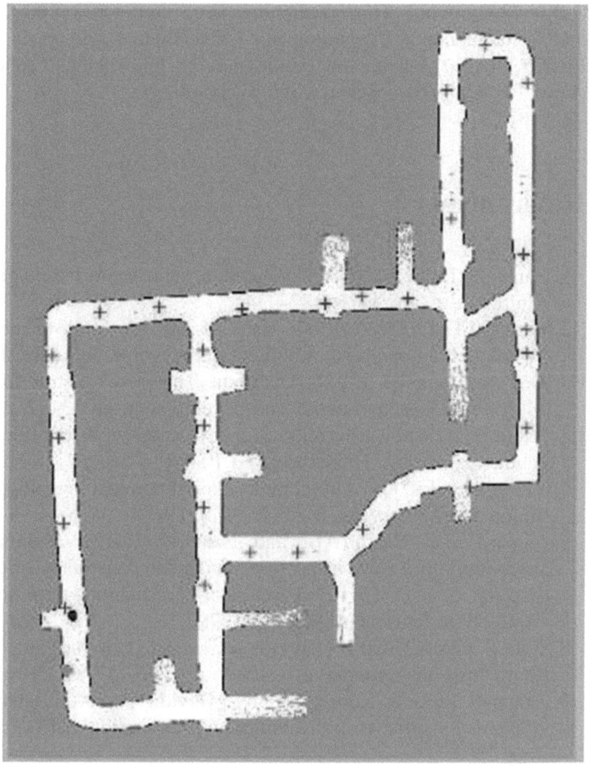

Fig. 6 Final result of the exploration

4 Conclusions

The local behavior control and global behavior control allow to increase the reach in the area to be mapped, and as a result, to increase the ability to find more toxic gases' sources. The probabilistic Occupancy Grid Mapping Algorithm has the advantage of storing the global map and the robot location; also, it takes into account the environmental noise of the tunnel, caused by leaking water from the roof and high concentration of dust and toxic gases.

The integrated algorithms solve the main problem when the any robot fails and has to reset as long as it has stored its last known position. The algorithms will be able to resample several particles in the area close to the last saved position, correct it estimated position when it detects a landmark and recalculate its motion planning to the next node. In the simulations, the integration of algorithms shows it is computationally expensive to analyse each cell map and calculating the probability of cell occupancy. Therefore, it could be necessary to use two microprocessors for a future implementation of the algorithms presented.

Acknowledgments This research has been supported by the CTIC-UNI (Center for Information and Communication Technologies—Universidad Nacional de Ingeniería) and by the grant 136-FINCYT-IA-2013 from the Fondo para la Innovación, Ciencia y Tecnología—FINCyT Peru (Fund for Innovation, Science and Technology).

References

1. Panait L, Luke S (2005) Cooperative multi-agent learning: the state of the art. Auton Agent Multi-Agent Syst 11(3):387–434
2. Parker LE (2000) Current state of the art in distributed autonomous mobile robotics. Distributed autonomous robotic systems, vol 4. Springer, Japan, pp 3–12
3. Das A, Frank LL (2011) Distributed adaptive control for networked multi-robot systems. Multi-robot systems, trends and development, pp 33–52
4. Şahin E (2005) Swarm robotics: from sources of inspiration to domains of application. Swarm robotics. Springer, Heidelberg, pp 10–20
5. Thrun S et al (2004) Autonomous exploration and mapping of abandoned mines. Robotics and Automation Magazine, IEEE 11.4, pp 79–91
6. Yinka-Banjo, Chika AB, and Isaac O (2013) Osunmakinde. Autonomous multi-robot behaviours for safety inspections under the constraints of underground mine terrains
7. Cepeda JS et al. (2012) A behavior-based strategy for single and multi-robot autonomous exploration. Sensors12.9:12772–12797
8. Reid S, Enk K, John B (1991) A six-legged rover for planetary exploration. In: Proceedings of Computrny in Aerocpuc e 8, 4IAA. pp 739–747, Oct 1991
9. Erann G, Rajiv D, Robert I, John L, David PM (2002) Behavior control for robotic exploration of planetary surfaces. In: Robotics and automation. IEEE Transactions, vol 10, #4, pp 490–503 Aug 2002
10. Zhiqi L, Naoyuki K (2007) Hybrid learning approach based on multi-objective behavior coordination for multiple robots. In: Proceedings of the 2007 IEEE International Conference on Mechatronics and Automation
11. Umit YO, Oguz HD, Umit O (2004) Cooperative control of mobile robots for target search. Proceedings of the IEEE International Conference on Mechatronics
12. Sebastian T, Wolfram B (2005) Probabilistic robotics. The MIT Press
13. Montemerlo, Michael, "FastSLAM (2003) A factored solution to the simultaneous localization and mapping problem with unknown data association. Doctoral dissertation, tech. report CMU-RI-TR-03-28, Robotics Institute, Carnegie Mellon University
14. Mejdi Ben Ardhaoui (2011) Implementation of autonomous navigation and mapping using a laser line scanner on a tactical unmanned aerial vehicle. Naval Postgraduate School, California
15. López Garcia Diego A (2011) New contributions in scheduling algorithms for executing maneuvers in autonomous robots non-holonomic. Electronic Engineering Department, Huelva University, July 2011
16. Guivant JE, Brett S, Mark AW (2012) 2D path planning based on Dijkstra's algorithm and pseudo priority queues, SOCS

Modeling Online via Clustering and Fuzzy SVM

J.C. Tovar, C.R. Mariaca and I. Álvarez Villalobos

Abstract In this article online clustering and modification of kernel Support Vector Machines (SVM's) is presented by a fuzzy modelling for nonlinear kernel plant. The structure identification was carried out by online clustering and fuzzy Support Vector Machines (FSVM's). When the process is slow, the fuzzy rules can be obtained automatically. Identification parameters of the fuzzy neural network are used. The range of variation in learning time ensuring the stability of the modelling error.

Keywords Clustering online SVM's · Kernel · Membership function · Fuzzy rules

1 Introduction

Both neural networks and fuzzy logic are universal estimators. The process of fuzzy rule extraction for nonlinear systems modeling is called structure identification. A common method is to partition the input and the output data, it is also called fuzzy grid [1]. Most of structure identification approaches are based on off-line data clustering, such as fuzzy C-means clustering [2], mountain clustering [2], and subtractive clustering [3]. These approaches require that the data is ready before the modeling. There are a few of on-line clustering methods in the literature. A combination of on-line clustering and genetic algorithm for fuzzy systems is proposed in [4]. In [5] the input space was automatically partitioned into fuzzy subsets by adaptive resonance theory. On-line clustering with a recursively calculated spatial

J.C. Tovar (✉) · C.R. Mariaca · I. Álvarez Villalobos
ESIME Zacatenco IPN Av. IPN s/n Lindavista, Gustavo A. Madero, 07738 Méxio, México
e-mail: jctovar77@hotmail.com

M. Ceccarelli and E.E. Hernández Martinez (eds.), *Multibody Mechatronic Systems*, Mechanisms and Machine Science 25, DOI 10.1007/978-3-319-09858-6_11

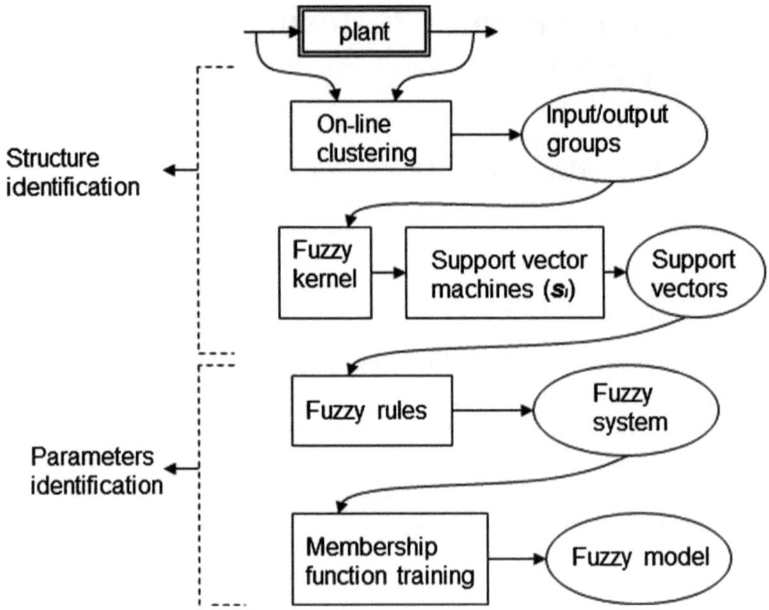

Fig. 1 The scheme of the modeling via on-line clustering and fuzzy support vector machines

proximity measure was given in [6]. There is one weakness for the above on-line clustering methods: the partitioning of the input (precondition) and the output (consequent) do not take into account time mark. They use all data to train each rule. In this paper, a novel on-line clustering approach is proposed. The time relationship in the input and the output spaces is considered.

The purpose of FSVMs (fuzzy support vector machines Fig. 1) is trained to treat the data with different importance in the training process. The lack of the term of the cost function is minimized by reformulating the construction of the optimization problem, and the construction of the Lagrangian so that solutions to the optimal hyper plane in the first form can be found in the dual form.

In this work use SVM for calculating estimation function. To find the support vectors in the group j, the input/output [y(k), x(k)] are used to approximate a nonlinear function. Consider the regression of a set of nonlinear functions

$$f(x) = w^T \varphi(x) + b \tag{1}$$

where $K(x, x_k) = \varphi(x)^T \varphi(x_k)$.

2 Clustering Online

The following state-space discrete-time smooth nonlinear system with fuzzy rules is described as

$$x(k+1) = f[x(k), u(k)],\ y(k) = h[x(k)] \tag{2}$$

where $u(k) \in \Re^m$ is the input vector, $x(k) \in \Re^n$ is the state vector, $y(k) \in \Re^q$ is the output vector, f and h are smooth nonlinear functions. (1) Can be written as

$$\begin{aligned} y(k) &= h[x(k)] = F_1[x(k)], \\ y(k+1) &= h[f[x(k), u(k)]] = F_2[x(k), u(k)] \\ y(k+n-1) &= F_n[x(k), u(k), \ldots u(k+n-2)] \end{aligned} \tag{3}$$

Denoting $y(k) = [y(k)\ and\ (k + 1), \ldots, y(k + n{-}1)]^T$, $U(k) = [u(k), u(k + 1), \ldots, (k + n{-}2)]^T$, and $(k) = F\ [x(k), U(k)]$, $F = [F_1, \ldots, F_N]^T$. From (2) is a smooth non-linear system, (3) it can be expressed as $x(k + 1) = g[y(k + 1), u(k + 1)]$. This leads to a multivariable model NARMA

$$y(k) = h[x(k)] = \Psi[X(k)] \tag{4}$$

where

$$X(k) = [y(k-1), y(k-2), \ldots u(k-d), u(k-d-1), \ldots]^T \tag{5}$$

$\psi(\cdot)$ is a string representing the dynamics of the plant, $u(k)$ and $y(k)$ is unknown nonlinear function measurable scalar inputs and outputs, d is a time delay.

The purpose of the identification of structures is to partition the input and output data $[y(k), x(k)]$ of the nonlinear system and fuzzy rules extracted. The following example to explain the importance of clustering online at the same time is used indexed. Nonlinear function is deemed to:

$$y(k) = f[x(k)] \tag{6}$$

For the norm of line grouping method proposed in [1] the inlet and outlet can be partitioned into four groups, as can be seen in Fig. 2. These groups can be formed within four rules like "*IF x(k) is Aj THEN y(k) is Bj*", $j = 1 \ldots 4$. Obviously, for the 3rd rule: "*IF x(k) is A3 THEN y(k) is B3*", this does not satisfy the relationship (6), because the precondition $x(k)$ and the result $y(k)$ does not occur in simultaneously.

The basic idea of clustering online is that the input and output spaces are partitioned hauled out at the same time indexed. If the distance from the point to the center is less than the required length, the point is within that group. When new data arrives, the center and the group should be changed according to new data. The following algorithm is given.

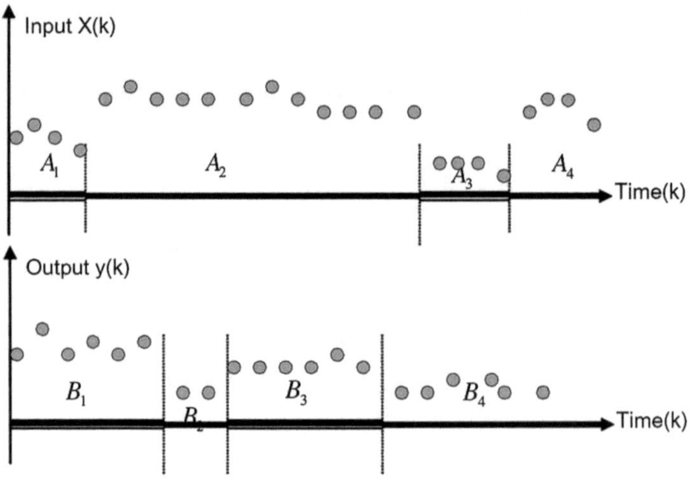

Fig. 2 Partitioning space input/out

The Euclidean distance at time k is defined as

$$d_{k,x} = \left(\sum_{i=1}^{n} \left[\frac{x_i(k) - \bar{x}_i^j}{x_{i,max} - x_{i,min}} \right]^2 \right)^{1/2}$$
$$d_{k,x} = \left| \frac{y(k) - \bar{y}^j}{y_{max} - y_{min}} \right| \tag{7}$$
$$d_k = \alpha d_{k,x} + \beta d_{k,y}$$

where $x_{i\_max} =$ max$k$ $\{x_i\}(k)\}$, $x_{i\_min} =$ min $\{x_i(k)\}$, $y_{max} =$ max $\{y(k)\}$, $y_{min} =$ min $\{y(k)\}$, $\bar{x}_i(k)$ and $\bar{y}(k)$ are the centers $\bar{x}_i(k)$ and $\bar{y}(k)$ while k, α and β are positive factors, typically $\alpha = \beta = (1/2)$ can choose. For group j, the centers are updated as follows

$$\bar{x}_i^j = \frac{1}{l_2^j - l_1^j + 1} \sum_{l=l_1^j}^{l_2^j} x_i(l)$$
$$\bar{y}^j = \frac{1}{l_2^j - l_1^j + 1} \sum_{l=l_1^j}^{l_2^j} y_i(l) \tag{8}$$

where l_1^j is the first issue of Group j, l_2^j is the last issue of Group j. The length of group j is $m^j = l_2^j - l_1^j + 1$. The time interval Group j is $[l_1^j, l_2^j]$. The process identification structure can be formed with the following steps

1. For the first data G_1, $k = 1$. $x_i(1)$, $y(1)$ are the centers of the first group, $\bar{x}_i^1 = x(1)$, $\bar{y}^j = y(1)$, $l_1^j = l_2^j = 1$.
2. If a new data $[y(k), xi(k)]$ arrives, $l_2^j = l_2^j + 1$, is used (7) and (8) for calculating d_k. If no new data arrives, go to 5.
3. If $d_k \le L$ then $[y(k), x_i(k)]$ is still in the group G_j, go to 2
4. If $d_k > L$ then $[y(k), x_i(k)]$ is a new set $j = j + 1$, the center of G_j there is $\bar{x}_1^j = x(k)$, $\bar{y}^j = y(k)$, $l_1^j = l_2^j = k$, go to 2
5. Check the distances between all centers $\bar{x}^j$, $\bar{y}^j$, if $\sum_{i=1}^{n} [\bar{x}_i^p - \bar{x}_i^q]^2 + |\bar{y}_i^p - \bar{y}_i^q \le L|$ the two groups G_p and G_q are combined into one group.

3 Fuzzy Kernel

There are many ways to choose the kernel $K(x, x_k)$ only requires that $K(x, x_k)$ satisfies the Mercer condition [2]. For example, the linear kernel $K(x, x_k) = x_k{}^T x$, the MPL kernel $K(x, x_k) = \tanh(k_1 x_k^T x + k_2)$, RBF kernel $K(x, x_k) = exp(-||x-x_k|^2/\sigma^2$ In this paper, the fuzzy kernel $K(\hat{x}, \hat{z})$ is used which is defined as

$$K(x, x_k) = \begin{cases} \prod_{i=1}^{M} u_i(x_k) \cdot u_i(x_i)\, x_k & \text{and } x \text{ both are in } jth \text{ group} \\ 0 & \text{otherwise} \end{cases}$$

where $\hat{x} = [x_1, x_2, \ldots, x_M] \in \mathbb{R}^M$ and $\hat{z} = [z_1, z_2, \ldots, z_M] \in \mathbb{R}^M$ are trained any two samples. $u_i(x_k)$ is the membership function of the jth cluster.

Let the group trained be $S = \{(x_1, y_1), (x_2, y_2), \ldots, (x_v, y_v)\}$ with interpretative variables x_i and tagged classes corresponding y_i, for all $i = 1, 2, \ldots, v$ where v is the total number of trained samples. It is assumed that the trained samples are partitioned into the clusters. You can run the following permutation of trained samples:

$$\begin{aligned} \text{cluster } 1 &= \{(x_{k_1}^1, y_{k_1}^1), \ldots, (x_{k_1}^1, y_{k_1}^1)\} \\ \text{cluster } 2 &= \{(x_1^2, y_1^2), \ldots, (x_{k_2}^2, y_{k2}^2)\} \\ &\vdots \\ \text{cluster } l &= \{(x_1^l, y_1^l), \ldots, (x_{k_l}^l, y_{k_l}^l)\} \end{aligned} \tag{9}$$

where k_g, $g = 1, 2, \ldots, l$ is the number of quality points for clustering ght, so he has $\sum_{g=1}^{l} k_g = v$. Then the kernel can be calculated using fuzzy set trained on (9), and obtaining kernel matrix K can be rewritten as follows:

$$K = \begin{bmatrix} K_1 & 0 & \cdots & 0 \\ 0 & K_2 & \cdots & 0 \\ \vdots & \ddots & \ddots & \vdots \\ 0 & \cdots & 0 & K_l \end{bmatrix} \in R^{v \times v} \tag{10}$$

How to choose the membership function $u_j(x_i)$ is another problem. The Gaussian function and the delta function are the most popular function of membership functions of fuzzy systems. When $u_j(x_i)$ is a Gaussian function, the kernel function is

$$K(x_k, x_j) = \varphi(x_k) \cdot \varphi(x_j) = \exp\left(\frac{\|x_k - x_j\|}{2\sigma^2}\right) \tag{11}$$

It is assumed that the trained fuzzy set is already obtained; the next step is to formulate the FSVM. Begins with the construction of a cost function, the FSVM also want to maximize the margin of separation and minimize the classification error such that good generalization ability can be performed.

Construction of FSVM optimization problem is formulated as

$$\begin{aligned} &\min \Phi(w, \xi, \mu) = \frac{1}{2} w^T w + C \sum_{i=1}^{n} \mu_i^m \xi_i \\ &subject\, y_i(w^T x_i + b) \geq 1 - \xi_i,\ i = 1, 2, \ldots, n \\ &\xi_i \geq 0,\ i = 1, 2, \ldots, n \end{aligned} \tag{12}$$

where m influences in the cost function. Now the Lagrangian is

$$\begin{aligned} Q(w, b, \xi, \alpha, \beta, \mu) = &\frac{1}{2} w^T w + C \sum_{i=1}^{n} \mu_i^m \xi_i \\ &- \sum_{i=1}^{n} \alpha_i \left[y_i(w^T x_i + b) - 1 + \xi_i \right] - \sum_{i=1}^{n} \beta \xi_i \end{aligned} \tag{13}$$

where α_i and β_i are non-negative Lagrange multipliers. Differentiating Q with respect to w, β and ξ_i, and equating the results to zero the three following optimality conditions are obtained

$$\begin{aligned} \frac{\partial Q(w, b, \xi, \alpha, \beta, \mu)}{\partial w} &= w - \sum_{i=1}^{n} \alpha_i y_i x_i = 0 \\ \frac{\partial Q(w, b, \xi, \alpha, \beta, \mu)}{\partial b} &= \sum_{i=1}^{n} \alpha_i y_i = 0 \\ \frac{\partial Q(w, b, \xi, \alpha, \beta, \mu)}{\partial \xi} &= C\mu_i^m - \alpha_i - \beta_i = 0 \end{aligned} \tag{14}$$

Each group has a fuzzy model. Using the idea Takagi-Sugeno model [9], we can combine the local models in each group within a global model. The fuzzy rules have the form

$$\begin{aligned} &R^j : IF\, x_{l_1^j} \leq x_1 \leq x_{l_2^j}\ and\ \ldots x_{l_1^j} \leq x_n \leq x_{l_2^j} \\ &THEN\, y(k) = f_i[X(k)] \end{aligned}$$

where $j = 1 \ldots p$, p is the number of the cluster group online. Membership functions for xi. The final fuzzy model is

$$y = \frac{\sum_{i=1}^{p} f_i[X(k)] \left[\prod_{j=1}^{n} \left(\mu_{A_i^j} \right) \right]}{\sum_{i=1}^{p} \left[\prod_{j=1}^{n} \left(\mu_{A_i^j} \right) \right]} \tag{15}$$

The following theorem provides a steady downward gradient algorithm for neuro-fuzzy modeling.

Theorem 1 If the fuzzy system to identify the nonlinear plant (1) in group j, the following gradient descent algorithm with a range of time-variant learning can identify the error $e(k)$ is used bounded [7]

$$W(k+1) = W(k) - \eta_k e(k) \Phi^T[X(k)] \tag{16}$$

where $\eta_k = \eta \big/ 1 + \|\Phi[X(k)]\|^2$, $0 < \eta \leq 1$. The standard error identification

$$e_N(k) = \frac{e(k)}{1 + \max_k \left(\|\Phi[X(k)]\|^2 \right)} \tag{17}$$

satisfies the following embodiment average

$$\lim_{T \to \infty} \sup \frac{1}{T} \sum_{k=1}^{T} \|e_N(k)\|^2 \leq \overline{\mu} \tag{18}$$

where $\overline{\mu} = max_k\}[\|\mu\|^2]$.

Remark 1 In general, a fuzzy system cannot match exactly any nonlinear system. The parameters of fuzzy system will not converge to the optimal values. The idea is to force online identification of the fuzzy system output to follow the output of the plant. Although the parameters may not converge to the optimal values (18) shows that the normalization of identification converge to the ball of radius $\overline{\mu}$. If the fuzzy system can match the nonlinear plant (4) exactly $(\mu(k) = 0)$, i.e. can find the best role membership $\mu_{A_i^j}$ and $W*$ such that the nonlinear system can be written as $y(k) = W^*\Phi\left[\mu_{A_{ji}}\right]$. Since $\| e(k) \| > 0$, the same law of learning (16) will asymptotically stable misidentification $\| e(k) \| > 0$.

Remark 2 Normalizing learning range (16) is time-varying in order to ensure the stability of the identification error. These learning ranges are easier to decide that in [8] (for example by selecting $\eta = 1$), without any a priori information. The ranges of variations in learning time can be found in the same standard adaptive scheme [3]. But these changes need to ensure stability robust identification. The algorithm is derived from the analysis of the stability (or ISS—Lyapunov function) algorithm [5] was obtained by minimizing the cost function. The approach to the identification error bound [5] focused on the analysis of convergence. Interestingly, two different methods can achieve similar results.

Now if trained membership functions of consequence and premise. The initial conditions are $c_{ji}(1) = \overset{*}{x_i}$, $w_j(1) = \overset{*}{y_i}$,$\sigma_{ji}(1)$, is random in (0, 1). Since the membership functions are Gaussian functions, the output of fuzzy system can be expressed as

$$y = \frac{\sum_{i=1}^{sv_j} w_i \left[\prod_{j=1}^{n} \exp\left(-\frac{\left(x_j - c_{ji}\right)^2}{\sigma_{ji}^2}\right)\right]}{\sum_{i=1}^{sv_j} \left[\prod_{j=1}^{n} \exp\left(-\frac{\left(x_j - c_{ji}\right)^2}{\sigma_{ji}^2}\right)\right]} \tag{19}$$

Theorem 2 For the neuro-diffuse TSK model, A_{ji} is selected as Gaussian functions [7, 8]. The output of the fuzzy logic system can be expressed as

$$y = \frac{\sum_{i=1}^{sv_j} \left(\sum_{k=0}^{n} p_k^i x_k\right) \left[\prod_{j=1}^{n} \exp\left(-\frac{\left(x_j - c_{ji}^*\right)^2}{\sigma_{ji}^{*2}}\right)\right]}{\sum_{i=1}^{sv_j} \left[\prod_{j=1}^{n} \exp\left(-\frac{\left(x_j - c_{ji}^*\right)^2}{\sigma_{ji}^{*2}}\right)\right]} \tag{20}$$

where c_{ji}^* and σ_{ji}^{*2} are unknown parameters which minimize the unmolded dynamics μ, y $x_0 = 1$.

4 Simulations

The nonlinear function which has the form [7, 8] is considered

$$f(x_1, x_2) = 0.52 + 0.1x_1 + 0.28x_2 - 0.6x_1x_2.$$

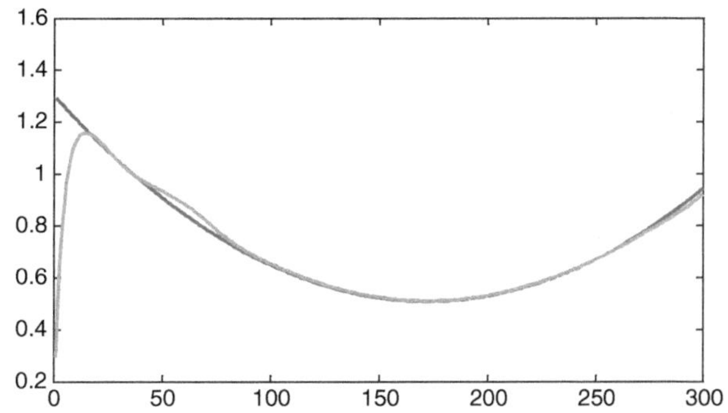

Fig. 3 Tested signal (*blue*) versus tested result (*green*)

The training data are selected as $x_1(k) = -1 + ((2k)/T)$, $x_2(k) = 1 - ((2k)/T)$, $k = 1, 2, \ldots, T$. Is chosen $\alpha = 0.4$ and $\beta = 0.6$. It is observed that the maximum changes in the input and output are about 1 and 2 so $\alpha\|x_{max} - x_{min}\| + \|y_{max} - y_{min}\|\beta = 16$, L could be chosen as $L < 1.2$, in this application is selected $L = 0.5$. The center as *"o"* for each group is represented, and *"+"* is the border between the groups. Can see that there are 7 groups. For example, the time intervals the 4th group are $l_1^2 = 94$, $l_1^2 = 238$.

Membership functions trained for each group are seven functions of membership for x_1. Then set the parameters of the membership functions other 300 data for testing the model used. The tested data $x_1(k) = 1 - ((2k)/T)$, $x_2(k) = -1 + ((2k)/T)$, $k = 1, 2, \ldots, T$. The end result of the modeled after local models is shown in Fig. (3).

5 Conclusions

This paper presents an efficient approach for modeling of nonlinear systems using FSVM with extraction rules is proposed. Some techniques are used to this new approach. First a clustering online which divides the data space Enter /output groups in the same time interval is proposed. After FSVM approach is used to obtain the support vectors in each group. With these support vectors are constructed fuzzy rules to obtain the corresponding fuzzy system. Subsequently identification structure, variation in time interval is applied to obtain the identification parameters. The contributions of this work are: (1) the online clustering method and the FSVM approach are used to extract rules, (2) the upper bound of modeling error and stability are tested for fuzzy modeling.

Such work can be used in robotic or mechatronic systems, to model the behavior or dynamic trajectories of nonlinear systems, when the uncertainties are unknown.

Acknowledgments The support given by COFAA, EDI ESIME Zacatenco National Polytechnic Institute and CONACyT.

References

1. Angelov P (2004) An approach for fuzzy rule-base adaptation using online clustering. Int J Approximate Reasoning 35(3):275–289
2. Cristianini N, Shawe-Taylor J (2000) An introduction to support vector machines. Cambridge University Press, Cambridge
3. Ioannou PA, Sun J (1996) Robust adaptive control. Pretince-Hall, Upper Saddle River
4. Juang CF (2005) Combination of on-line clustering and Q-value based GA for reinforcement fuzzy system design. IEEE Trans Fuzzy Syst 13(3):289–302
5. Mandic DP, Hanna AI, Razaz M (2001) A normalized gradient adaptive step size. IEEE Signal Process Lett 8:295–297
6. Takagi T, Sugeno M (1985) Fuzzy identification of systems and its applications to medeling and control. IEEE Trans Syst Man and Cybern 1:116–132
7. Tovar RJC (2008) Fuzzy SVM modeling with automated structure selection, thesis. CINVESTAV IPN, México
8. Tovar JC, Yu W, Ortiz F, Mariaca CR, José de Jesús R (2011) Modeling via on-line clustering and fuzzy support vector machines for nonlinear system. In: 50th IEEE Conference on Decision and Control and European Control Conference, 12–15 Dec 2011, Orlando, FL
9. Wang LX (1994) Adaptive fuzzy systems and control. Prentice-Hall, Englewood Cliffs

Cylindrical Contact Force Models for the Dynamics of Roller Chain Drives

J. Ambrosio, C. Malça and A. Ramalho

Abstract A multibody dynamics methodology for the study of roller chain drives based on the contact between its mechanical elements is proposed here. In particular, the chain drive mechanisms are described as planar multibody systems in which the connection between each pair of links is modelled as a multiple revolute clearance joint, instead of the traditional kinematic constraint, and the meshing between the roller and sprocket teeth is also treated as a generalized revolute clearance joint. The contact force model includes the geometric contact detection and the contact force evaluation. Recognizing the numerical and physical limitations of the existing cylindrical contact models, a new enhanced model is applied here and demonstrated with the study of the dynamics of a roller chain drive with initial pre-tension. In the process, the energy dissipative effect of friction and coefficient of restitution on the chain dynamics is also studied.

Keywords Multibody dynamics · Contact mechanics · Elastic line contact · Clearance joints · Impact

1 Introduction

Real machine characteristics such as link flexibility, backlash or clearances lead to impacts that often degrade the machine dynamic performance, increase vibration and noise, shorten life and decrease precision [1—4]. Dynamic models to represent

J. Ambrosio (✉)
LAETA, IDMEC, Instituto Superior Tecnico, Universidade de Lisboa, Lisboa, Portugal
e-mail: jorge.ambrosio@tecnico.ulisboa.pt

C. Malça
Polytechnic Institute of Coimbra, Coimbra, Portugal
e-mail: candida@isec.pt

A. Ramalho
University of Coimbra, Coimbra, Portugal
e-mail: amilcar.ramalho@dem.uc.pt

M. Ceccarelli and E.E. Hernández Martinez (eds.), *Multibody Mechatronic Systems*,
Mechanisms and Machine Science 25, DOI 10.1007/978-3-319-09858-6_12

such non-ideal characteristics constitute challenges to find new and better machine designs. This work proposes an approach to the modelling of roller chain drives based on the use of a generalized clearance revolute joint formulation. Although focusing on roller chain drives, the range of application is wider.

Among the flexible element drives, chain drives are one of the most precise and effective forms of power transmission [5, 6]. They are characterized by the discrete nature of their chain links and sprocket teeth, which lead to one of the most significant sources of noise and vibration in roller chain drives, especially at high operating speeds. These undesirable phenomena motivated several researchers to investigate their influence on the dynamic behaviour and vibration of chain drive systems to obtain higher operating speeds, lighter weight and better quality [5, 7—10].

Integrated models to describe the dynamics of the roller chain drive, including the detailed description of the nontrivial geometry of the sprocket surface and of the dynamics associated with the polygonal effect, the compliance of the chain, i.e., the flexibility of links, the transversal and longitudinal vibrations, the friction effects, and the contact/impact forces between the chain and sprockets, have been proposed by Kim and Johnson [11], Pedersen et al. [12] or Veikos and Freudenstein [6]. However, an integrated model to accurately describe the full complexity of the chain drive dynamics has not yet been developed. In all these models the impact between the pin/bushing and bushing/roller pairs was not considered. In addition, the force model used to describe the contact between the roller/sprocket pair, generally used for the chain meshing dynamics represents the internal contact between cylindrical geometries that does not relate to the geometry of the sprocket teeth or roller.

Clearances lead to the deterioration of the dynamic performance of mechanisms insofar as internal impact forces and impulsive forces are generated. These internal impact forces increase the level of vibration, noise, energy dissipation, which intrinsically result in the amplification of internal forces and stresses resulting in wear and tear, and consequently, to the degradation of the system performance [1, 3, 4, 13]. This work attempts to improve the models recently proposed for the chain drives and for mechanical systems [11, 12, 14], in particular the description of the interaction between the different elements in the system, through the use of revolute clearance joint formulation in the multibody model development.

2 Multibody Dynamics Chain Drive Modelling

Chain drives are assemblages of large number of bodies interconnected by cylindrical revolute joints [15]. A chain drive, represented in Fig. 1a, is composed of sprockets, chain and eventually a tensioner and delimiters of the chain vibrations. The chain itself is composed of inner and outer links, articulated through a pin/bushing hinge, and eventually rollers hinged onto the inner links bushings, if the chain is a roller chain instead of a bushing chain, as seen in Fig. 1b.

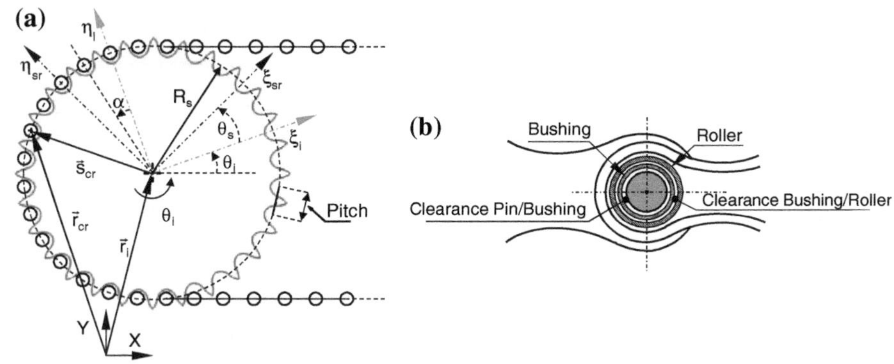

Fig. 1 Multibody representation of a chain drive. **a** individual mechanical components, **b** pin, bushing and roller assembly

Let the position and orientation of a bodies i of a multibody system be represented by $\mathbf{q}_i = [\mathbf{r}^T\theta]_i^T$ and the position and orientation of all bodies by $\mathbf{q} = [\mathbf{q}_1^T, \mathbf{q}_2^T \ldots \mathbf{q}_{nb}^T]^T$. The equations of motion of a multibody system are [16]

$$\begin{bmatrix} \mathbf{M} & \mathbf{\Phi}_\mathbf{q}^T \\ \mathbf{\Phi}_\mathbf{q} & \mathbf{0} \end{bmatrix} \begin{Bmatrix} \ddot{\mathbf{q}} \\ \boldsymbol{\lambda} \end{Bmatrix} = \begin{Bmatrix} \mathbf{g} \\ \boldsymbol{\gamma} \end{Bmatrix} \tag{1}$$

where $\mathbf{M}$ is the system mass matrix, $\mathbf{\Phi}_\mathbf{q}$ the Jacobian matrix associated to the kinematic constraints, $\ddot{\mathbf{q}}$ the system accelerations, $\boldsymbol{\lambda}$ a vector of Lagrange multipliers associated with the kinematic constraints, $\boldsymbol{\gamma}$ the vector with the velocity dependent on the terms of the kinematic acceleration constraints and $\mathbf{g}$ the vector with the forces applied to the rigid body of the system.

The contact force between two bodies i and j of the multibody system is the result of a normal force f_n, perpendicular to the contacting surfaces, and of a tangential force f_t, associated to friction phenomena, i.e.,

$$\mathbf{f}_i = -f_n\,\mathbf{n} + f_t\,\mathbf{t} \text{ and } \mathbf{f}_j = -\mathbf{f}_i \tag{2}$$

in which $\mathbf{n} = [n_x\ n_y]^T$ is the normal vector to the contact surfaces and $\mathbf{t} = [n_y\ -n_x]^T$ is the tangential vector. In the body forces description by Eq. (2), the contact mechanics problem is defined as the evaluation of the normal and tangential contact forces and the identification of their point of application in each body of the contact pair.

Based on multibody dynamics, two different formulations for modelling chain drive mechanisms can be used: (i) the revolute joints are considered as ideal or perfect joint constraints; (ii) a revolute clearance joint formulation is used considering the contact between elements [14, 15]. Figure 2 shows the connection between an inner link and an outer link where revolute joints are modelled as kinematic constraints and in which the kinematic constraints are removed and

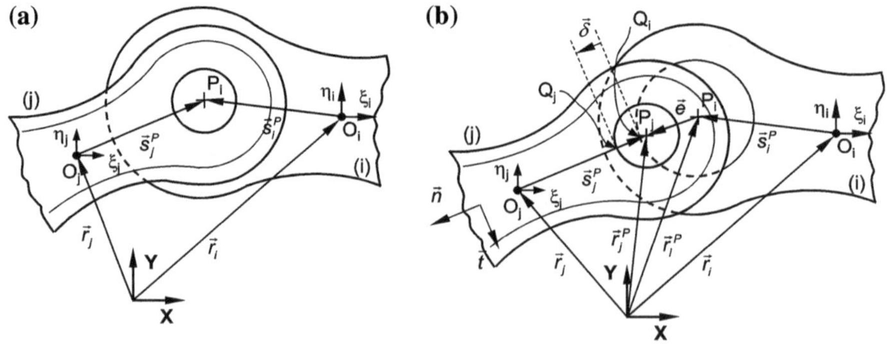

Fig. 2 Cylindrical revolute joint connecting pin link and bushing link. **a** perfect or ideal joint, **b** clearance joint

replaced by a pair of forces representing the contact between the connected bodies. While a perfect or ideal joint imposes two kinematic constraints, limiting the motion between bodies, the presence of clearance in a revolute joint implies replacing those kinematic constraints by a pair of forces representing the interaction between the connected bodies [1, 13].

2.1 Application of the Clearance Revolute Joint

Take the cylindrical revolute clearance joint shown in Fig. 2b representing the pin/bushing hinge connection. Let body i refer to the outer link that includes the bushing and body j be the inner link that includes the pin. When the centers of the bushing and pin, given by points P_i and P_j respectively, separate, there is the possibility for contact to take place if their distance exceeds the existing clearance of the joint denoted by c. With reference to Fig. 2b, let the eccentricity vector be defined as $\mathbf{e} = \mathbf{r}_j^P - \mathbf{r}_j^P$. A penetration between bodies i and j exist [13]:

$$\delta = \sqrt{\mathbf{e}^T\mathbf{e}} - c > 0 \tag{3}$$

When Eq. (3) is fulfilled, a normal contact force between the rigid bodies included in the contact pair must be evaluated, using a suitable cylindrical, and applied and added to the bodies force vector denoted by $\mathbf{g}$ in Eq. (1).

The methodology described for the contact between the pin links/bushing links is generalized for the bushing links/rollers and the rollers/sprocket teeth surface contact pairs. Regardless of whether the clearance revolute joint connects the elements roller/tooth or bushing/tooth, which depends on the chain type, the contact conditions are defined as a function of the tooth shape. A procedure for the accurate geometric description of the real tooth profile has been developed [15]. The strategy proposed starts by performing a discrete analysis of the real tooth profile geometry,

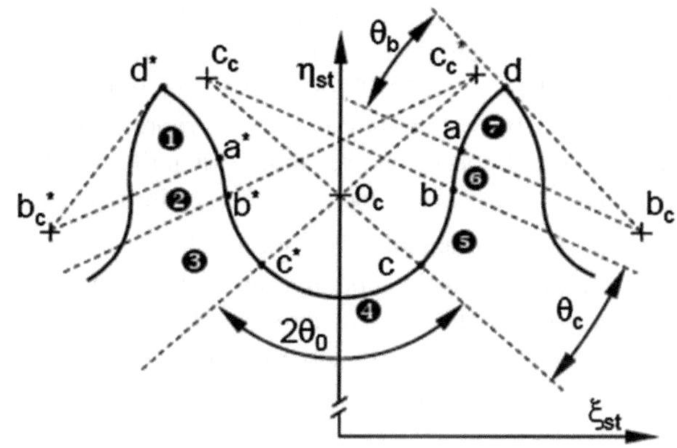

Fig. 3 Real tooth profile divided into seven contact regions

i.e. by subdividing the geometry into seven distinct contact areas, as shown by Fig. 3 where the roller/sprocket contact can occur and where internal and external contacts geometries can happen.

For each possible contact area, the geometric conditions that define whether or not the different elements that compose the revolute clearance joints are in contact have been established. By defining all the contact conditions in the tooth coordinate system, the number of transformations of coordinates is minimized and the detailed tooth geometry can be used with computational efficiency [15]. In this formulation, dry contact is assumed, i.e. contact without any interposition of a fluid layer between the contacting bodies. Furthermore, the joint components are considered as rigid bodies, in which relative penetration exists, even though without deformation.

2.2 Normal Contact Force for Cylindrical Geometries

Most of the cylindrical contact force models available are derived based on the Hertz pressure distribution exhibiting the same restrictions of the Hertz elastic contact theory, preventing them from being used with conformal contact often observed for low clearances. The existing cylindrical contact models represent the contact force as an implicit function of the penetration with logarithmic expressions, which pose some limitations in their use. The ability and accuracy of current cylindrical contact force models to describe the contact between bodies with cylindrical geometries are discussed by Pereira et al. [17]. In addition, it is demonstrated that purely elastic contact models, which do not account for energy dissipation, lead to unrealistic results. An enhanced cylindrical contact force model, including energy dissipation that characterizes the contact occurrence, has been proposed [18]

$$f_n = \frac{(a\ \Delta R + b)LE^*}{\Delta R}\delta^n\left[1 + \frac{3(1-c_e^2)}{4}\frac{\dot{\delta}}{\dot{\delta}^{(-)}}\right] \tag{4}$$

for internal cylindrical contact $a = 0.965$, $b = 0.0965$, $n = Y\Delta R^{-0.005}$ and $\Delta R = R_i - R_j$, being $Y = 1.51[\ln(1000\ \Delta R)]^{-0.151}$ if $\Delta R = [0.005, 0.34954[$ mm or $Y = 0.0151\Delta R + 1.151$ if $\Delta R = [0.34954,10.0[$ mm. L is the axial contact length dimension. For internal contact ΔR represents the clearance existing between impacting cylinders. The remaining quantities in Eq. (4), similar for internal and external cylindrical contact, are the restitution coefficient c_e, the relative impact velocity $\dot{\delta}^{(-)}$, the actual penetration velocity $\dot{\delta}$ and $E^* = E/2(1 - \nu^2)$ is the composite reduced elastic modulus, assuming materials with similar elastic modulus and Poisson coefficients denoted by E and υ, respectively. The direction of the application of the contact force f_n, given by Eq. (4), is the normal to the contact surfaces in the contact point given as $\mathbf{n} = \mathbf{e}/\sqrt{\mathbf{e}^T\mathbf{e}}$.

2.3 Tangential Contact Force

The normal force f_n is obtained using the enhanced cylindrical contact force model described by Eq. (5). The friction force is described by the modified Amontons-Coulomb friction force model [13]

$$f_t = \begin{cases} -c_f c_d f_n \frac{\mathbf{v}_t}{\mathrm{v}_t} & if \quad v_t \neq 0 \\ 0 & if \quad v_t = 0 \end{cases} \tag{5}$$

in which c_f is the friction coefficient, $\mathbf{v}_t$ the relative tangential velocity between the contact surfaces and v_t its magnitude. The dynamic correction coefficient c_d is

$$c_d = \begin{cases} 0 & if \quad \mathrm{v}_t \leq \mathrm{v}_0 \\ \frac{\mathrm{v}_t - \mathrm{v}_0}{\mathrm{v}_1 - \mathrm{v}_0} & if \quad \mathrm{v}_0 \leq \mathrm{v}_t \leq \mathrm{v}_1 \\ 1 & if \quad \mathrm{v}_t \geq \mathrm{v}_1 \end{cases} \tag{6}$$

in which v_0 and v_1 are given preset tolerances for the tangential velocity.

3 Demonstrative Examples

An ASA bushing chain n.º40 wrapping around two sprockets with 17 teeth each is used to demonstrate the methodologies presented here. Figure 4 shows the initial positions of chain drive multibody model developed, in which the initial distance between the sprockets is such that 38 links are required for the closure of the chain. The initial velocity of the chain links and sprockets is defined by the initialization procedure in order to ensure that the driving sprocket rotates at a constant angular velocity of 120 rpm. For the driving sprocket, a very large inertia moment is considered to ensure its rotation at a constant angular velocity. The bushing chain

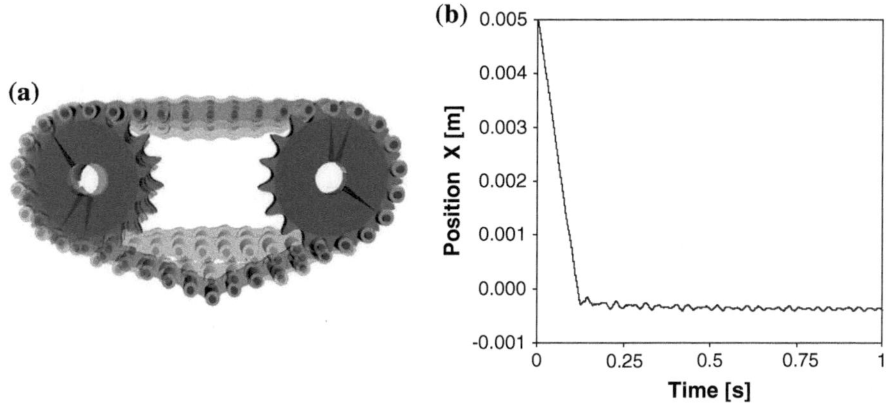

Fig. 4 Tensioner in a roller chain drive. **a** transient motion from initialization to final position, **b** displacement of one sprocket with respect to the other

considered in this application has a clearance of 10.0 μm on the pin/bushing clearance revolute joint. The chain drive is subjected to gravitational forces acting downwards. The dynamic analysis is carried out using the integrator DE/STEP [19] for which the numerical tolerances are set to 10^{-5} and the maximum time step allowed is 10^{-4}.

The chain transversal vibration is described for a case without pre-tension and another case with a pretension of 25 N. The trajectory of a typical bushing in steady state conditions is depicted in Fig. 5 for both cases.

The contact forces that develop between a typical tooth of a sprocket and a bushing are illustrated in Fig. 6. It is observed that as the chain strands change length, when a bushing starts or leaves the contact with sprockets, and the polygonal effect excites the transversal vibration of the chain, a particular bushing may hit a sprocket with a high normal impact velocity, thus leading to higher peaks for

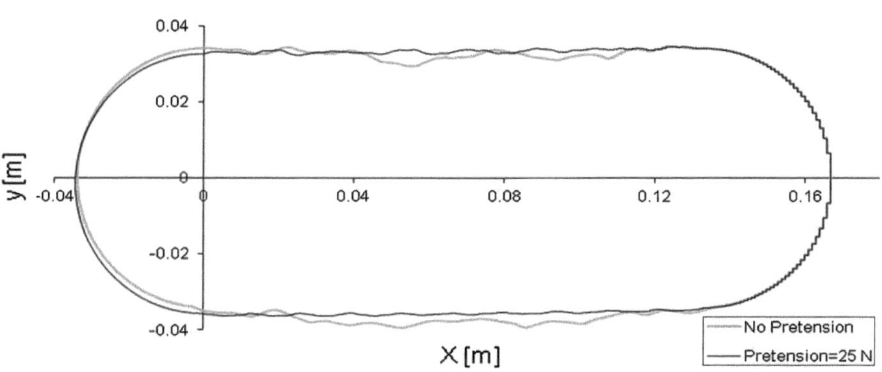

Fig. 5 Trajectory of a typical bushing for the two cases: without and with a pretension of 25 N

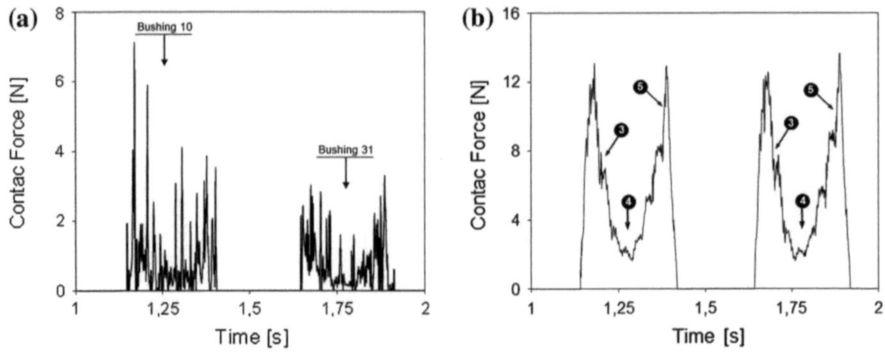

Fig. 6 Contact forces on a bushing/sprocket tooth for a $c_e = 0.7$. **a** $c_f = 0.0$, no pretension, **b** $c_f = 0.1$, 25 N pretension

initial and end contact. It is shown that a null pretension leads to multiple impacts instead of a continuous contact force, as shown in Fig. 6a. Even during the continuous contact of the bushing with the sprocket, for a pretension of 25 N, several force peaks are observed on the contact force. These peaks are due to the transition of the bushing contact between the different regions of the sprocket tooth, as labeled in Fig. 6b.

Figure 7 depicts the influence of pretension in the dynamics of cylindrical revolute joints composed of pin/bushing contact pairs. Figure 8 shows the effect of friction in the orbit of the pin with respect to the bushing, actually demonstrating the importance of the energy dissipation in the smoothing of the relative pin/bushing kinematics. Another consequence is that the numerical integrator time steps become larger, adding to the numerical efficiency.

When multiple simultaneous contacts occur, as in chain drive systems, the existence of friction forces and other energy dissipative sources, e.g. through the

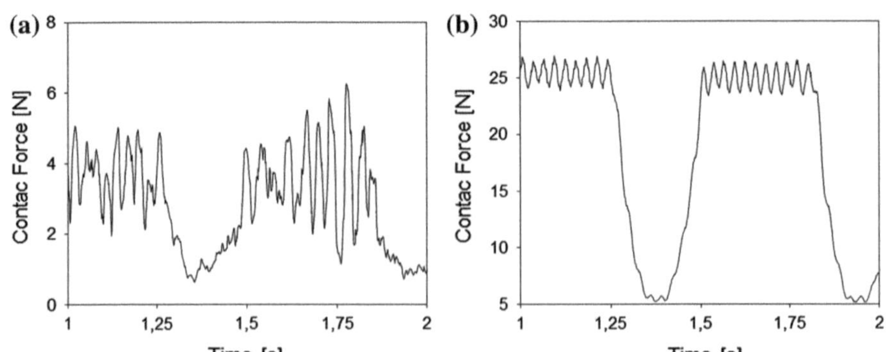

Fig. 7 Contact forces on a pin/bushing joint for a $c_e = 0.7$, a $c_f = 0.1$ and for: **a** no pretension; and **b** 25 N pretension

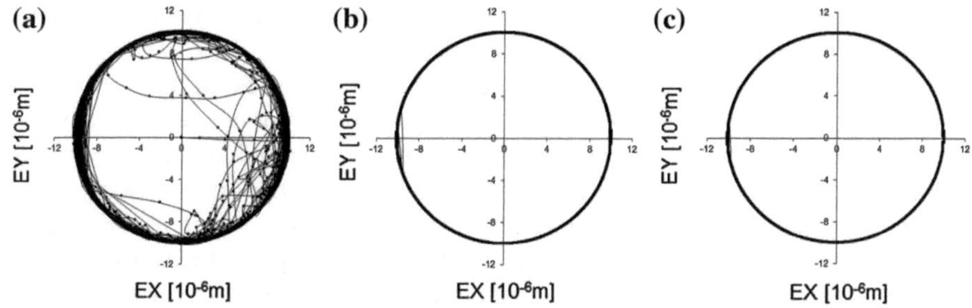

Fig. 8 Trajectory of the pin inside the bushing for c_e = 0.7 and: **a** c_f = 0.00, no pretension; **b** c_f = 0.00, 25 N pretension; **c** c_f = 0.10, 25 N pretension

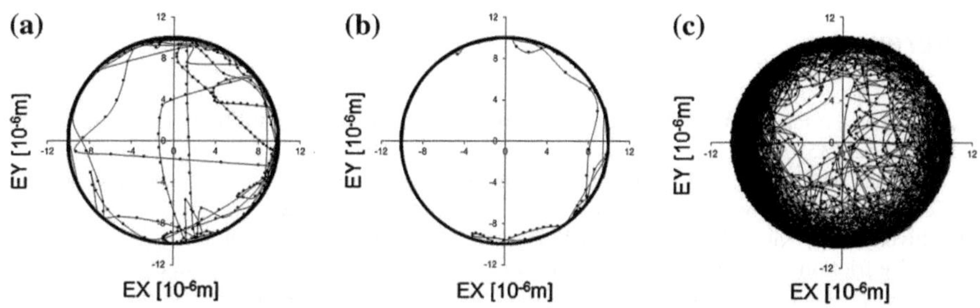

Fig. 9 Trajectory of the pin inside the bushing with no pretension, for c_f = 0.1 and: **a** c_e = 0.1; **b** c_e = 0.5; **c** c_e = 1.0

restitution coefficient, c_e, which reflects the type of collision, fully elastic if $c_e = 0.0$ or plastic contact if $c_e = 1.0$, and plays an important role on the control of the energy dissipation—are relevant for the dynamics of the system. The effect of restitution coefficient value on the dynamics of chain drive mechanism with a no pretension chain is depicted for the c_e of 0.1, 0.5 and 1.0, in Fig. 9.

The trajectories of the pins inside the bushings show that when the coefficient of restitution increases from 0.5 to 1.0 the contact becomes increasingly discontinuous, indicating small energy dissipation for the higher restitution coefficient. When the restitution coefficient increases from 0.1 to 0.5, the opposite trend is observed, the contact force peak values are reduced, indicating an increase of energy dissipation in the contact pairs. As a result, the pin is seated in the bushing during long periods leading to a smoother dynamic response of the system.

4 Conclusions

This work demonstrates that the dynamics of the clearance revolute joint is sensitive not only to the values assumed by the friction and the restitution coefficients, but also mainly to the pretension level. It is observed that if no pretension is applied, the contact is mostly discontinuous, and characterized by a large number of impacts and rebounds with a short duration. Regardless of contact pair it is demonstrated that the use of some pretension changes the characteristics of the contact by enforcing the continuous contact with only a small number of contact losses, which helps to achieve the required smoothness of the chain drive dynamic response. In addition, the stabilization of the dynamic response is reached more quickly.

The wide range application of roller chain drives means that the simulation program developed can be used as a general purpose program for analyzing very different chain drive systems. The methodologies adopted to develop the roller chain drive model can, with small changes, be used more generally for chain drives in industrial machines, marine engines, car engines, motorbikes and so forth.

References

1. Flores P, Ambrósio J (2004) Revolute joints with clearance in multibody systems. Comput Struct 82:1359–1369
2. Gummer A, Sauer B (2013) Modeling planar slider-crank mechanisms with clearance joints in RecurDyn. Multibody Sys Dyn. doi:10.1007/s11044-012-9339-2
3. Muvengei O, Kihiu J, Ikua B (2012) Dynamic analysis of planar multi-body systems with LuGre friction at differently located revolute clearance joints. Multibody Sys Dyn 28:369–393
4. Muvengei O, Kihiu J, Ikua B (2012) Numerical study of parametric effects on the dynamic response of planar multi-body systems with differently located frictionless revolute clearance joints. Mech Mach Theory 53:30–49
5. Chen CK, Freudenstein F (1988) Towards a more exact kinematics of roller chain drives. J Mech Transmissions Autom Des 110(3):123–130
6. Veikos NM, Freudenstein F (1992) On the dynamic analysis of roller chain drivers: Part I—Theory. Mech Des Synth 46:431–439
7. Ariaratnam ST, Asokanthan SF (1987) Dynamic stability of chain drives. J Mech Transmission Autom Des 109(3):412–418
8. Conwell JC, Johnson GE (1996) Design, construction and instrumentation of a machine to measure tension and impact forces in a roller chain drives. Mech Mach Theory 31(4):525–531
9. Ryabov GK (1968) Inertia effects of impact loading in chain drives. Russ Eng J 48(8):17–19
10. Turnbull SR, Nichol SW, Fawcett JN (1977) An experimental investigation of the dynamic behavior of a roller chain drive. In: Proceedings of the ASME Design Engineering Technical Conference, paper no. 77-Det-168
11. Kim MS, Johnson GE (1993) A general multibody dynamic model to predict the behavior of roller chain drives at moderate and high speeds. Adv Des Autom 1:257–268
12. Pedersen SL, Hansen JM, Ambrósio J (2004) A roller chain drive model including contact with guide-bars. Multibody Sys Dyn 12(3):285–301
13. Flores P, Ambrósio J, Claro J, Lankarani H (2008) Kinematics and dynamics of multibody systems with imperfect joints: models and case studies. Springer, Dordrecht
14. Pereira CM, Ambrósio J, Ramalho AL (2010) A methodology for the generation of planar models for multibody chain drives. Multibody Sys Dyn 24(3):303–324
15. Pereira C, Ambrósio J, Ramalho A (2011) Contact mechanics in a roller chain drive using a multibody approach. In: Proceedings of the 13th World Congress in Mechanism and Machine Science, paper no. IMD 123
16. Nikravesh P (1988) Computer-Aided analysis of mechanical systems. Prentice-Hall, Englewood Cliffs, New Jersey

17. Pereira CM, Ramalho AL, Ambrósio J (2011) A critical overview of internal and external cylinder contact force models. Nonlinear Dyn 63(4):681–697
18. Malça C, Ambrósio J, Ramalho A (2012) An enhanced cylindrical contact force model for multibody dynamics applications. In: Proceedings of the EUROMECH Colloquium 524—Multibody system modelling, control and simulation for engineering design. University of Twente, Enschede, Netherlands, pp 30–31
19. Shampine L, Gordon M (1975) Computer solution of ordinary differential equations: the initial value problem. Freeman, California

A Three-Dimensional Multibody Model of a Full Suspension Mountain Bike

B. Corves, J. Breuer, F. Schoeler and P. Ingenlath

Abstract To analyze the operating loads and also the handling qualities of full supscnsion mountain bikes in off-road use, a simulation model representing all important effects is required. Most bicycle models discussed in literature are quite elementary. These models are adequate for analysing the general dynamic behaviour of bicycles but they are not capable of showing all effects appearing in off-road operation of mountain bikes. In this paper a detailed three-dimensional multibody model of a system consisting of a full suspension mountain bike, a passive mass-spring-damper system for the rider, stochastic track excitation and a control system for the velocity and lean angle stabilisation is presented. Additionally first results for straight driving of geometrically different mountain bikes are presented and discussed. Although a experimental validation of the model has not been performed yet the first results indicate that the development of the presented model is a first step for simulation based evaluation of handling qualities and operating loads for mountain bikes.

Keywords Multibody dynamics · Bicycle dynamics · Matlab/adams cosimulation

---

B. Corves (✉) · J. Breuer · F. Schoeler · P. Ingenlath
Department of Mechanism Theory and Dynamics of Machines, RWTH Aachen University, Kackertstr. 16/18, 52072 Aachen, Germany
e-mail: corves@igm.rwth-aachen.de

J. Breuer
e-mail: jens.breuer@alumni.fh-aachen.de

F. Schoeler
e-mail: schoeler@igm.rwth-aachen.de

P. Ingenlath
e-mail: ingenlath@igm.rwth-aachen.de

M. Ceccarelli and E.E. Hernández Martinez (eds.), *Multibody Mechatronic Systems*, Mechanisms and Machine Science 25, DOI 10.1007/978-3-319-09858-6_13

1 Introduction

Bicycles are worldwide popular and technical advancements in cycling sports are made every year. Still it seems that simulation techniques and methods that are applied to analyze the dynamics of bicycles are less developed.

In scientific publications mostly elementary models like the Carvallo Whipple model, which consists of a frame with a rigidly attached rider, a front frame consisting of a fork and a handlebar and two knife edge wheels (some models have toroidal wheels instead), are used [1, 2]. Usually nonlinear or linearized equations of motion are derived and analyzed. With these simplified models the most significant characteristics of the steering behavior of bicycles can be investigated. Yet Sharp notes that due to possible variations in the design of a bicycle and its effect on the steering behavior no generalizations can be made [3]. Detailed reviews of the general bicycle dynamics and theoretical research can be found in [4] or [1] for instance.

However the intention of this paper is not to derive the equations for the bicycle rider system and to analyze it analytically. Furthermore a comprehensive model of a bicycle-rider-track system created in the commercial software MSC Adams is presented.

One important characteristic of the bicycle-rider-system is that the rider represents usually around 80 % of the total system mass. Therefore the rider has a big influence on the dynamics of the bicycle-rider-system [5]. To be able to analyze such systems within simulations Hull, Wilczynski and Wang developed a multibody rider model for the seated and for the standing position [6, 7]. Waechter, Riess and Zacharias showed that it is possible to predict vibrational stress acting on riders on full suspension bicycles [8]. As they used a two dimensional model cornering or handling qualities could not be investigated. Finally they investigated two different kinds of city bicycles which have certain differences from mountain bikes.

In mountain biking there are additional effects and operating conditions that distinguish mountain biking from other cycling sports. For example steep descents in combination with high speeds, vibrational excitation due to bad tracks or special conditions like jumps [5]. Because of these operating conditions it is difficult to derive loading assumptions theoretically and experimental testing is costly [9]. Hence having a multibody model that can predict the loads appearing in normal use and even in misuse cases is worthwhile.

Subsequently there are several nonlinear effects that have an influence on bicycle dynamics. For actual mountain bikes the wheels can travel up to 200 mm and more relative to the frame. Furthermore while cornering big roll angles are possible. From such large deflections result geometric nonlinearities [10].

An additional source of nonlinearities in mountain biking is the changing position of the contact point of the tire-ground-contact due to track obstacles (can be seen in Fig. 1). This changing position of the contact point is also significant to the handling of a bicycle since it causes a change of the trail N. Although Kooijman et al. showed in [11] that the trail is not necessarily needed for the self stability of a bicycle it

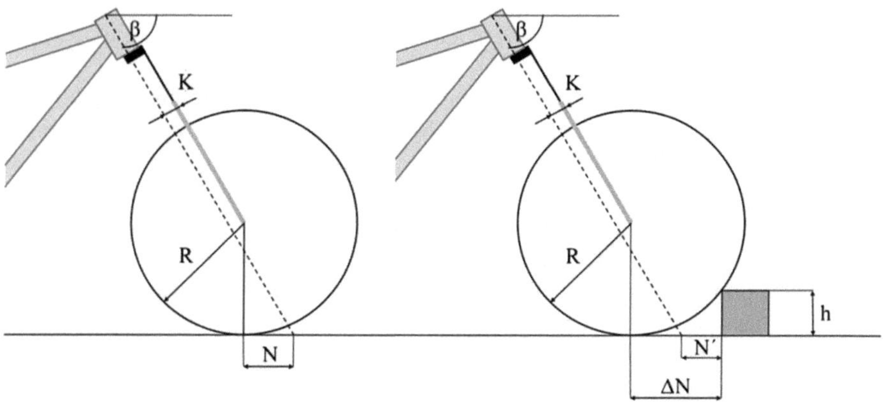

Fig. 1 Definition of trail and change of trail due to track obstacles following [5]

certainly has an influence on the handling qualities and steering characteristics since it directly affects the lever arm of the tire forces with respect to the steer axis.

The trail N at the front wheel is defined as the horizontal distance between the contact point of the wheel and the point where the steer axis passes through the ground [12].

$$N = \frac{R - \frac{K}{\cos(\beta)}}{\tan(\beta)} \tag{1}$$

in which R is the rolling radius of the tire, K is the fork offset and β is the steering angle. The change of trail ΔN caused by an obstacle with the height h can be expressed as

$$\Delta N = \sqrt{R^2 - (R - h)^2} \tag{2}$$

The change of trail can moreover lead to a negative trail N' like it is shown in Fig. 1. All these effects mentioned above are particular for mountain biking.

To predict the handling qualities, riding comfort and structural loads for mountain bikes a complex nonlinear model is needed. Basic steps for the creation of such a model are presented in this paper.

2 Materials and Methods

The model consists of four major components, the full suspension bicycle, the rider, the ground excitation and a control system for straight driving, cornering and lane change maneuver. These components are described in detail below.

2.1 Bicycle Model

Basis of the multibody model of the bicycle is the Nicolai Helius AC shown in Fig. 2 (also see www.nicolai.net). The Model is completely parameterized and implemented in MSC Adams. Different versions and sizes of this bicycle are available e.g. frame sizes from small to extra-large and three different wheel sizes 26, 27, 5 and 29 inch. In the following data for a medium size frame with 26 inch wheels is presented, data for other frame or wheel sizes are slightly different.

The bicycle model consists of nine rigid bodies for the frame and two more rigid bodies for the two wheels. These bodies are connected to each other through joints. At the rear damper and at the suspension fork linear spring damper elements are used. For the tire-ground-contact a penalty based contact definition is used. Thereby contact nonlinearities can be investigated within the model.

Mass and inertia properties of the frame are calculated within Adams assuming the constant crosssections for the tubes of the frame parts. The properties of the wheels are calculated through detailed 3D-CAD models. Other attached parts like brakes or derailleur mechanisms are considered as lumped masses. The mass properties of the model match with the manufacturer information. For the stiffness and damping constants needed for the tire-contact definition and for spring-damper elements in the fork and rear damper following assumptions are made: The spring constants of the rear damper and the suspension fork are adjusted in such a way that the sag is about 25 % of the possible spring deflection. That is in the range that is recommended by manufacturers of bicycles. The damping coefficients of the rear damper and the suspension fork are adjusted by driving tests within the simulation model. The stiffness coefficients for the tires are adjusted in such a way that a static tire deflection between 10 and 20 mm appears, which is common for mountain bike tires. The resulting values are in the scale of the values given in [5]. The tire damping coefficients are adjusted to gain a damping factor of 0.055 for the tires (value taken from [5]).

In Table 1 a summary of some of the physical properties of the bicycle is provided (the masses of the attached parts are included in the given data).

Fig. 2 A picture of the bicycle and the multibody model of the bicycle (*left* is courtesy of Meissner Raeder)

Table 1 Properties of the bicycle (frame size medium, wheel size 26 inch)

Component	Mass [kg]	Stiffness coefficient [$\frac{N}{m}$]	Damping coefficient [$\frac{Ns}{m}$]
Rear frame	4.45	–	–
Front tire	1.84	2.0×10^6	1.0×10^3
Rear tire	1.88	2.5×10^6	1.3×10^3
Rear damper	1.15	1.1×10^5	3.0×10^3
Front frame	5.55	9.0×10^3	3.7×10^2

2.2 Rider Model

The rider is modeled as a passive mass-spring-damper system in a standing position and consists of ten rigid bodies. The head and upper body (as one rigid body), the hip, two upper arms, two forearms, two lower and two upper legs (shown in Fig. 3).

Between shoulder and wrist as well as between hip and ankle two linear spring and two linear damper elements act on each side. Between upper body and hip one additional rotary spring-damper element acts. These Parameters shown in Table 2 are taken from [7]. The inertia properties of the rider are shown in Table 3.

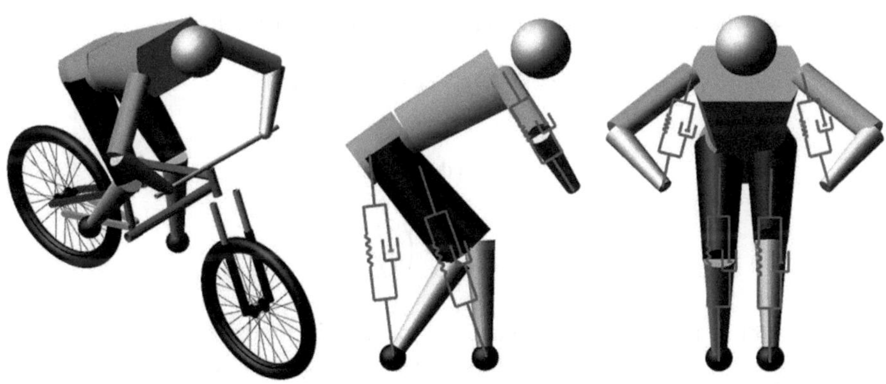

Fig. 3 The rider model with exemplarily shown spring-damper elements

Table 2 Stiffness and damping properties of the rider model [7]

Component	Stiffness coefficient	Damping coefficient
Arms	6356 N/m	392 Ns/m
Legs	15300 N/m	318 N/m
Hip	300 Nm/rad	100 Nm/rad

Table 3 Inertia properties of the rider model

Component	Mass [kg]	I_{xx} [kgm^2]	I_{yy} [kgm^2]	I_{zz} [kgm^2]
Body and head	33.01	0.943	0.886	1.611
Hip	12.3	0.112	0.103	0.130
Fore arm	1.45	$9.56e^{-3}$	$9.39e^{-3}$	$1.42e^{-3}$
Upper arm	2.5	$1.417e^{-2}$	$1.512e^{-2}$	$2.92e^{-3}$
Lower leg	4.04	$3.928e^{-2}$	$3.913e^{-2}$	$2.91e^{-3}$
Upper leg	10.34	0.178	0.169	$4.64e^{-2}$

2.3 Track Excitation

The track geometry is the most important excitation for bicycle dynamics. It is composed of a long-wave trace and short-wave irregularities. Especially the irregularities of mountain bike tracks can be rough [5]. It is common to model track irregularities as a stationary, ergodic and Gaussian stochastic process with zero mean [13, 14]. In this case power spectral density (PSD) may be used to describe measured tracks and to synthesize irregularities for simulations.

In this paper only vertical irregularities are considered. The vertical track $h_z(x)$ can be obtained using a harmonic series (3). The amplitudes $b_{z,i}$ are calculated from the PSD (4) whereas the phases ψ_i have to be chosen at random.

$$h_z(x) = \sum_i b_{z,i} \cdot \cos(\Omega_i \cdot x + \psi_i) \tag{3}$$

$$b_{z,i} = (\Phi_i \cdot \Delta\Omega)^{\frac{1}{2}} \tag{4}$$

$$\Phi(\Omega) = \Phi_0 \cdot \left(\frac{\Omega}{\Omega_0}\right)^{-w} \tag{5}$$

PSD for rails, roads as well as unpaved tracks (dirt roads) are presented in [15] (Fig. 4). For the investigations in this paper three spectra (good, bad and worst unpaved tracks) were chosen and approximated using Brauns formular (5). The coefficients are listed in Table 4. For good and bad tracks wavelengths between $L_{\min} = 0.2\,\text{m}$ and $L_{\max} = 20\,\text{m}$ were included, whereas only wavelengths smaller than $L_{\max} = 5\,\text{m}$ for the worst track. One track obtained is shown in Fig. 4.

2.4 Control Scheme

The controller for the model is made up in Matlab/Simulink. The linkage of this controller with the multibody model is created in a Co-simulation of Matlab/Simulink and Adams. For the model three different control schemes for straight driving,

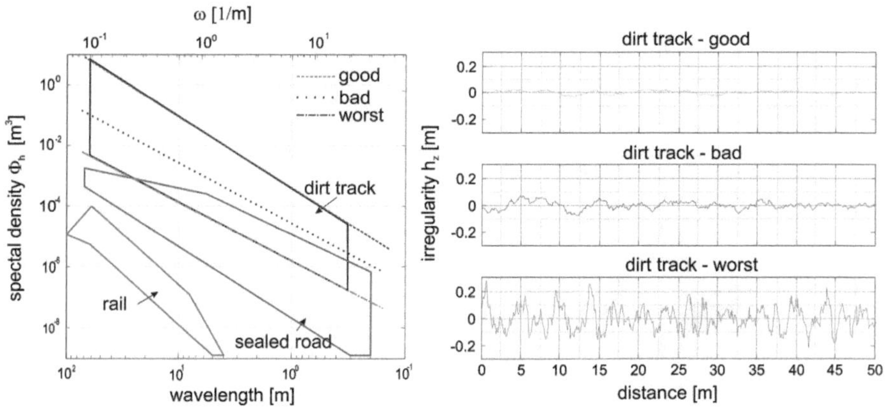

Fig. 4 PSD (following [15]) and irregularities for different track types

Table 4 Parameters of the different track types

Track type	Unevenness Φ_0 [m^3]	Waviness w [−]	Wavelengths L [m]
Good	0.0004	2	0.2—20
Bad	0.03	2	0.2—20
Worst	0.02	2	0.2—5

constant cornering and lane change were designed. In the following the control scheme for straight driving is explained. The scheme of the feedback control consists of two separated closed loops. In Fig. 5 these loops are shown.

The components of each loop are the controller, the controlled section and the feedback. A PID-module is used as controller. The controlled section is the multi-body model represented by the Bicycle-Rider-Model, which contains the Adams-model with all its components.

The upper feedback loop controls the velocity. As controlled variable the current speed of the rear wheels center of mass is used. The desired speed is the command variable. The deviation of the target velocity and the actual velocity is gained in the PID-module and is used to compute an actuating torsional moment acting on the rear wheel. Proportional, integral and derivative ratios are determined for the specific case of use. With the feedback of the controlled variable the loop is closed.

The lower feedback loop controls the rolling motion of the rider-bicycle-system. Controlling the rolling motion means to set the roll angle to a desired angle by applying a torsional moment between the handlebar and the main frame. As controlled variable the current roll angle is used whereas the desired roll angle represents the command variable.

For the stabilization of the rolling motion it is necessary to determine the exact current roll angle. The angle is measured referring to an initial coordinate system. In the inertial coordinate system it is straightforward to measure the degree of freedom

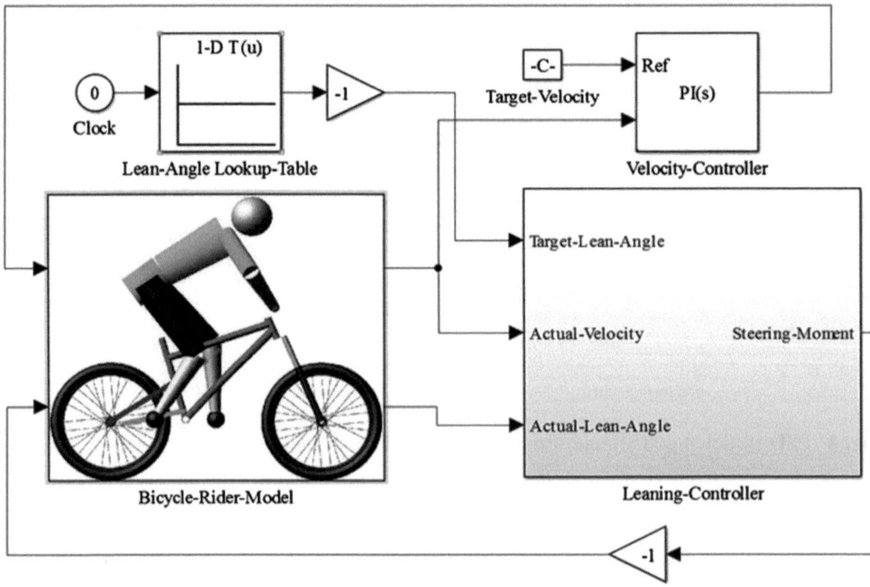

Fig. 5 Layout of the control sheme

(DOF) for rotation around the vertical axis of the bicycle. However due to the changing position and orientation of the bicycle the roll angle can not be measured in the inertial coordinate system. Thus a moving coordinate system is defined. The direction of the roll axis of that coordinate system is always equal to the vehicles roll axis. In this manner an exact roll angle measurement during cornering movements is secured. By adapting the "Lean-Angle Lookup-Table" in the Simulink model the different maneuvers like straight driving, steady state cornering and lane change are achieved.

3 Results

In this section results for two different bicycle-rider-systems are presented. The first system consists of a 26 inch bicycle and the other one of a 29 inch bicycle. All geometric data and mass properties are taken from the manufacturers data or calculated within Adams like described in 2.1. The rider is modelled like it is described in 2.2. The position of the rider is adjusted to the actual bicycle geometry. The track excitation that was used for the shown simulation results is the so called "bad" excitation shown in Fig. 4. In Fig. 6 the roll angle of the bicycle, that occurs while the controller tries to achieve straight driving, is shown.

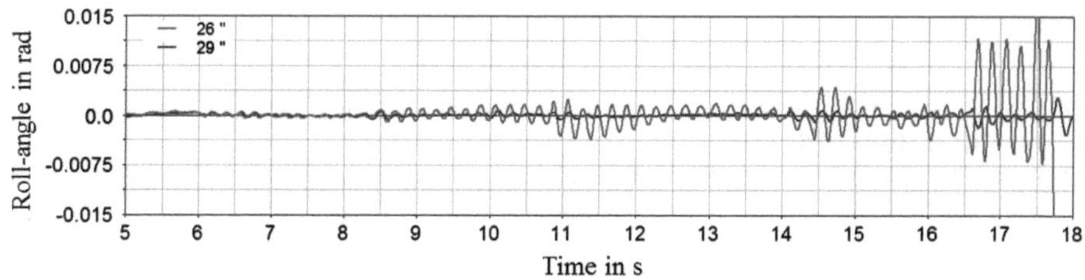

Fig. 6 Roll angle of the bicycle while straight driving at a speed of 30 km/h

4 Discussion

Although the control parameters have an influence on the dynamic behavior of the system, we assume that, since only geometric parameters have been changed in the model we can predict the influences of changes in geometry to the handling qualities of a bicycle.

The amplitude of the roll angle of the bicycle with 29 inch wheels are much smaller than the roll angle amplitudes of the bicycle with 26 inch wheels. So it seems that the bicycle with the 29 inch wheels is more stable.

The difference between the two investigated bicycles has also been studied by three test riders of a mountain bike magazine. The test riders all testified that the bicycle with 29 inch wheels has the highest running smoothness [16]. So the simulation result matches witch the subjective opinion of the test riders.

5 Conclusion

This paper provides a multibody model that consists of a full suspension bicycle, a rider, track excitation and a control scheme. Although the model is already quite sophisticated it can be extended with the consideration of elastic parts (e.g. the frame and the wheels). Like Waechter, Riess and Zacharias showed in [8] it is possible to predict riding comfort due to multibody models, Gross showed that it is also possible to calculate operation loads on bicycles [5]. Our results show that it is generally possible to evaluate handling qualities. We expect that in future it will be possible to use a model like this to aid the design process of mountain bikes what might lead to better bicycles in means of fatigue safety, handling qualities and riding safety.

References

1. Schwab AL, Meijaard JP (2013) A review on bicycle dynamics and rider control. Veh Syst Dyn 51(7):1059–1090
2. Frosali G (2012) Francesco Ricci. Kinematics of a bicycle with toroidal wheels. Commun Appl Ind Math 3(1)
3. Sharp RS (1976) The dynamics of single track vehicles. Veh Syst Dyn 5(1–2):67–77
4. Kooijman JDG, Schwab AL (2013) A review on bicycle and motorcycle rider control with a perspective on handling qualities. Veh Syst Dyn 51(11):1722–1764
5. Gross E (1997) Betriebslastenermittlung, Dimensionierung, strukturmechanische und fahrwerkstechnische Untersuchungen von Mountainbikes, volume 308 of Fortschritt Berichte VDI, Reihe 12. VDI-Verlag, Düsseldorf
6. Wilczynski H, Hull ML (1994) A dynamic system model for estimating surface-induced frame loads during off-road cycling. J Mech Des 116(3):816–822
7. Wang EL, Hull ML (1997) A dynamic system model of an off-road cyclist. J Biomech Eng 119(3):248–253
8. Waechter Matthias, Riess Falk, Zacharias Norbert (2002) A multibody model for the simulation of bicycle suspension systems. Veh Syst Dyn 37(1):3–28
9. Hölzel Christin, Hoechtl Franz, Senner Veit (2011) Operational loads on sport bicycles for possible misuse. Procedia Eng 13:75–80
10. Piersol AG, Paez TL, Harris CM (2010) Harris' shock and vibration handbook. McGraw-Hill handbooks, 6th edn. McGraw-Hill, New York
11. Kooijman JDG, Meijaard JP, Papadopoulos JM, Ruina A, Schwab AL (2011) A bicycle can be self-stable without gyroscopic or caster effects. Science 332(6027):339–342
12. ISO 8855:2011. Road vehicles—vehicle dynamics and road-holding ability—vocabulary
13. Knothe K, Stichel S (2003) Schienenfahrzeugdynamik, 3rd edn. Springer, Berlin
14. Dodds CJ, Robson JD (1973) The description of road surface roughness. J Sound Vib 31 (2):175–183
15. Braun H, Hellenbroich T (1991) Messergebnisse von straßenunebenheiten. In Unebenheiten von Schienen und Strassen als Schwingungsursache, pp 47–80. VDI-Verlag, Düsseldorf
16. Schmidt A (2012) Die 15 wichtigsten fragen zum thema 650b. Mountain BIKE, (8)

Gear Shifting Strategies Co-simulations to Optimize Vehicle Performance and Fuel Consumption

Jony J. Eckert, Fernanda C. Corrêa, Fabio M. Santiciolli, Eduardo S. Costa, Heron J. Dionísio and Franco G. Dedini

Abstract The vehicle longitudinal dynamics is responsible for calculating the vehicle power consumption to attend a specific route, estimating, by the equations, the forces acting on the system such as aerodynamic drag, rolling resistance, climbing resistance and the driver behavior. The gear shifting strategies influence significantly in the vehicle acceleration performance and fuel consumption because it changes the powertrain inertia and the engine speed. The literature presents gear shifting strategies based on the engine power and torque. A fuel economy strategy is more difficult to determine, because it depends on a large number of factors like the engine efficiency, vehicle speed, transition ratio and required acceleration. This paper presents a study based on the US06 standard velocity profile, in which the high speeds and acceleration stretches create a situation where the vehicle performance is limited by the engine available power and by the tire-ground traction limit. Because of the many factors involved in the vehicle behavior, it was developed an algorithm to optimize the gear shifting process to choose the more adequate strategy to each stretch. The analysis were performed by co-simulation between the multibody dynamics software Adams™ and Matlab/Simulink™, where is defined the vehicle power demand.

Keywords Gearshift strategies · Vehicle performance · Fuel consumption · Co-simulation · Vehicle longitudinal dynamics

J.J. Eckert (✉) · F.M. Santiciolli · E.S. Costa · H.J. Dionísio · F.G. Dedini
State University of Campinas - UNICAMP, 200 Mendeleyev Street, Campinas, SP 13083970, Brazil
e-mail: javorski@fem.unicamp.br

F.C. Corrêa
Federal Technological University of Paraná, Campus Ponta Grossa - UTFPR Monteiro Lobato Avenue - Jardim Carvalho Ponta Grossa, Ponta Grossa, PR 84016210, Brazil
e-mail: fernandacorrea@utfpr.edu.br

M. Ceccarelli and E.E. Hernández Martinez (eds.), *Multibody Mechatronic Systems*, Mechanisms and Machine Science 25, DOI 10.1007/978-3-319-09858-6_14

1 Introduction

The vehicle longitudinal dynamics evaluates the engine power required to remain at the desired speed, evaluating the forces acting on the system. The power demand is a function of the aerodynamic drag, tire-ground interaction, climbing resistance, powertrain inertia and driving behavior.

The gear shifting changes the powertrain inertia and the engine speed, influencing significantly the vehicle acceleration performance and fuel consumption. There are some gear shifting strategies or tactics based on the engine torque and power. Strategies based on fuel economy are more difficult to determine, because it depends on vehicle parameters like vehicle speed, acceleration, transition ratio and efficiency.

The shift control governs the transmission behavior, during a shift event, in function of the resistance forces and vehicle speed. The top gear is selected to get a maximum speed and it is limited by the engine power, speed and fuel economy [1]. The longitudinal vehicle dynamics models proposed by the literature do not define a standard gear shifting due this factor is represented through the driver behavior, depending on road and traffic conditions. According to [2] efficiency and shift quality are determined by the shift strategy, defining at what time a shift event is executed, and presents significant effects over the vehicular performance and fuel consumption [3].

As described by [4] the use of gear shifting indicators, composed of light and/or sound alarms connected to management software installed in the Electronic Control Unit (ECU), results in 3.6 % of fuel economy in a vehicle submitted to the New European Driving Cycle. Furthermore, [5] optimized the gearshift time in heavy vehicles as a function of the engine torque and clutch performance. A driving assistance system was developed by [6] to help drivers giving fuel efficiency guidelines that are suited for the current vehicle situation. With the same purpose of decreasing fuel consumption [7] evaluated experimentally the variability of driver behaviors, in relation with the engine speeds amplitude according to the driving style for 21 drivers. After that, the tests were reproduced using ecodriving techniques to reduce the fuel consumption.

This paper aims to optimize the vehicle acceleration performance and fuel consumption by co-simulations between Adams^{TM} where the multibody model based on a 1.0 L Brazilian vehicle operates, and Matlab/Simulink^{TM} where the vehicle dynamics equations and the gear shifting control are applied.

2 Longitudinal Vehicle Dynamics

In this paper it will be used the longitudinal vehicle dynamics methodology proposed by Thomas [8] where the model is based on the movement resistance forces.

The aerodynamic drag (D_A) is the air resistance imposed by the vehicle passage. Due to the complexity of the airflow outside the vehicle, this load is based on empirical constant known as drag coefficient (C_D), vehicle frontal area (A) [m^2], air density (ρ) [kg/m^3] and the vehicle speed (V) [m/s] Eq. (1).

$$D_A = \frac{1}{2}\rho V^2 C_D A \tag{1}$$

Rolling resistance is the energy loss in the tire, in function of the contact area and rubber damping properties. These lead to the transformation of mechanical energy into thermal [9]. At low speeds on hard pavement, rolling resistance is the primary resistance force. The rolling resistance R_x is calculated by the Eq. (2), where W is the vehicle weight [N].

$$R_x = 0.01\left(1 + \frac{0.62\,V}{100}\right)W \tag{2}$$

The climbing resistance is the vehicle weight force component resulting from the road grade. The effects of grade and longitudinal acceleration can be combined in finding the changes in front and rear loads [10].

The vehicle acceleration and the powertrain rotational inertia also generates resistance forces. The available traction force (F_x) in function of the engine torque (T_e) and the transmission ratio (N_{tf}) is given by Eq. (3).

$$F_x = \frac{T_e N_{tf}}{r} - ((I_e + I_t)N_{tf}^2 + I_d N_f^2 + I_w)\frac{a_x}{r^2} \tag{3}$$

a_x Vehicle acceleration [m/s^2]
r Tire external radius [m]
I_t Gearbox inertia [kgm^2]
I_e Engine inertia [kgm^2]
η_{tf} Transmission overall efficiency
I_w Wheels and tires inertia [kgm^2]
N_f Gearbox transmission ratio
I_d Differential inertia [kgm^2]

The vehicle acceleration performance is given by Eq. (4), where M is the vehicle mass [kg]. The climbing resistance is neglected because road grade is considered null in US06 cycle.

$$Ma_x = F_x - R_x - D_A \tag{4}$$

3 Reference Driving Cycle and Dynamic Co-simulation

To establish a benchmark, standard cycles were utilized to determine the reference vehicle speed behavior. A driving cycle represents the vehicle driver behavior during a trip, it is defined as a sequence of vehicle speed and therefore acceleration.

These driving cycles are designed to be representative of urban and extra-urban driving conditions and they reproduce measures of vehicle speed in real roads. In the simulations it was adopted the US06 cycle [11] that complements the U.S. test cycle FTP-75 because it represents high speeds and accelerations. Therefore, this cycle allows a better analysis of the vehicle performance in acceleration condition, to evaluate a critical operation in the proposed model.

The co-simulation technique is used in a development stage where the physical or mathematical mechatronic and control system are designed [12].

The simulations implemented in this paper were made via the multibody dynamic analysis program Adams™ (Automatic Dynamic Analysis of Mechanical Systems), where is implemented the vehicular model analysed. The control of variables related to longitudinal dynamics is done through the interface between Adams™ and Matlab/Simulink™.

The simulated vehicle was based on a compact hatchback equipped with 1.0 L engine (Table 1). The effects of vehicle suspension system were neglected to simplify the model according to the current literature.

The CAD model is based on a chassis dynamometer containing two rotating masses that simulate the vehicle displacement inertia. The model was exported to Adams™ (Fig. 1) where revolution joints were created to allow the movements. On the wheels were applied torques related to the powertrain and brake system. In the rotating masses were applied a resistance torque. The vehicle chassis was connected to the base to aligned wheels with the rollers.

To facilitate the implementation of the vehicle dynamics equations, it was used a Simulink™/Adams™ interface, generating a block of data from the dynamic model. The Simulink™ programmed algorithm works together with the Adams™ solver. The Simulink™ provides for Adams™ torque values from the engine model. To do this it adopts the engine torque curves values multiplied by the transmission ratio. The engine model also calculates the fuel consumption based on the fuel consumption map.

Table 1 Simulated vehicle parameters [17]

Components	Units	Speed				
		1st	2nd	3rd	4th	5th
Engine inertia	kgm^2	0.1367				
Transmission inertia	kgm^2	0.0017	0.0022	0.0029	0.0039	0.0054
Transmission ratio	–	4.27	2.35	1.48	1.05	0.8
Differential inertia	kgm^2	9.22E-04				
Differential ratio	–	4.87				
Wheels + tires inertia	kgm^2	2				
Vehicle mass	kg	980				
Tires	–	175/70 R13				

Fig. 1 Adams™ model

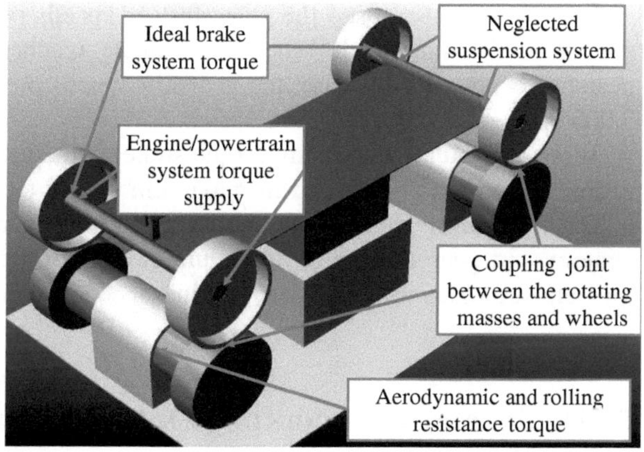

When the vehicle is braking, an ideal brake model applies the required torque at the Adams™ model wheels. After that, the wheels angular velocity is calculated by Adams™ in function of the applied torques. This angular velocity supplies the Simulink™ algorithm to it recalculates the required torque.

4 Engine Parameters and Gear Shifting Strategies

The simulated model compares the engine available torque curves with the required torque calculated by the longitudinal dynamics equations in function of the acceleration percentage and the engine speed. If the required torque exceeds the maximum torque available, there will be loss of performance. The fuel consumption is given by the specific consumption map as a function of torque and engine speed.

To represent the vehicle condition is important to define some specific parameters to avoid gear shifting instability. In the simulations performed were used a gear shifting time of 1 s as proposed by [13]. The time between two subsequent gear shifts is important to the stabilization, it avoids chattering and also it satisfies comfort conditions [14]. According to [15] the downshift occurs at 5 km/h below the upshift speed to prevent gearshift instability.

The maximum power and torque strategies consist in gearshift when the vehicle reaches the maximum engine power and torque point respectively. This tactics optimize the vehicle acceleration performance, usually acting at engine higher speeds, thereby generating a fuel consumption increase.

The fuel economy strategy is difficult to define because it depends on factors like the engine behavior, available gear ratios, required acceleration and fuel type. Taking into account these parameters, the tactics consist of keeping the engine running in higher efficiency regions.

In this paper they used the gear shifting speeds proposed by [16] for a vehicle similar to the simulated. When the vehicle reaches these standard speeds, the gearshift occurs. In this case, the gearshift speeds proposed are close to the engine 3,000 rpm.

When is used the maximum torque strategy, the gear shifting velocities are close to the engine 5,300 rpm and in maximum power tactics the upshift occurs in the 6,400 engine rpm. Due to high engine speed range available between fuel economy strategy and gear shifting at maximum engine torque, they were added two intermediate strategies to make gear shifting at 3,500 and 4,500 engine rpm.

5 Results and Gear Shifting Optimization Algorithm

Primarily, simulations for the gear shifting tactics mentioned above were performed and the results were compared with the US06 standard velocity profile, where it was observed that in high acceleration stretches, the vehicle did not have enough power to keep the required speed, staying at velocities lower than the cycle until the vehicle reaches lower acceleration stretches or braking sections.

To evaluate the difference between the simulation results, we used the linear correlation between the simulated results and the standard cycle. The term r^2 is the determination coefficient that measures the variability proportion between the two curves. It is a direct correlation function between the variables, showing the variance percentage. The r^2 consists of the sum of squares of prediction errors obtained as shown in Eq. (5), where x and y represent the curves values.

$$r^2 = \frac{\left(\sum(x_i - \bar{x})(y_i - \bar{y})\right)^2}{\sum(x_i - \bar{x})^2 \sum(y_i - \bar{y})^2} \tag{5}$$

The first simulations were performed considering the same gear shifting tactics throughout all cycle stretches. The results were compared with the US06 standard (Fig. 2) where it was observed that in high acceleration stretches, the vehicle did not have enough power to keep the required speed.

It was found a different performance between the simulated tactics, mainly because of tire-ground traction limit, where high acceleration strategies have a speed decrease due to weight transfer to the rear wheels, reducing the gravitational force applied to the vehicle front wheels responsible for traction torque.

With the speed increase, the maximum power and torque tactics perform better results. However, these strategies operate at lower efficiency engine regions, increasing fuel consumption. The results are shown in Table 2.

The regression r^2 is always close to 1 and it indicates a strong relationship between the results and the required velocity profile.

To maximize the simulated vehicle performance, and optimize the fuel consumption, it was developed an algorithm based on the results obtained by the

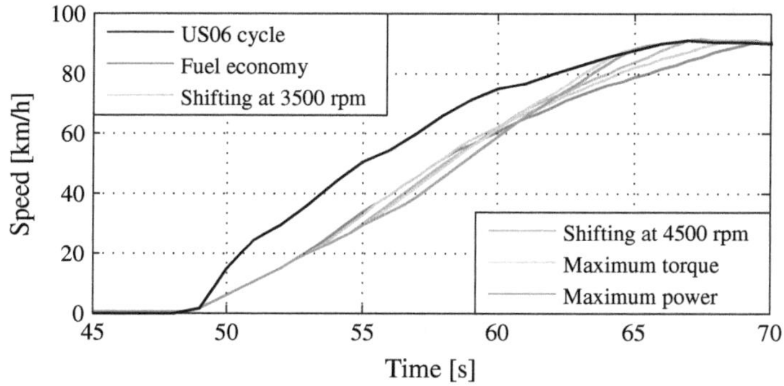

Fig. 2 Comparison between simulated tactics and the US06 standard

Table 2 Correlations and fuel consumption according to the gear shifting strategy

Gear shifting strategy	Linear correlation	Fuel consumption (mL)	Consumption average (km/L)
Fuel economy	0.9928	813.3	15.49
Gear shifting at 3,500 rpm	0.9929	833.8	15.12
Gear shifting at 4,500 rpm	0.9921	884.5	14.24
Maximum torque	0.9916	956.5	13.18
Maximum power	0.9907	1087.9	11.57

previously simulated strategies that select the most appropriate strategy for each cycle stretched. The algorithm selects the best performance tactics on a particular simulated stretch and, if the strategies show a similar performance, the algorithm select the lowest fuel consumption tactics. The algorithm creates a vector that controls the simulation gear shifting strategy. To avoid the gear shift instability, the strategies that are used for a short time are replaced to the previous used tactics in the control vector and thus a new simulation is performed. The results found are stored in a database used in the next optimization step. The database is limited to 20 simulation and when this limit is reached, the worst performance data is replaced by the last simulation result.

After 137 interactions in approximately 2 h of simulations applying the optimization algorithm, it was found a 0.9,937 correlation coefficient that is higher than the results obtained using the standard strategies previously simulated as shown in Fig. 3. The fuel consumption was 866.2 mL which represents a 14.87 km/L average fuel consumption. This average result between the fuel economy and 3,500 rpm gear shifting tactics.

The optimized strategy uses the most adequate tactics for each US06 cycle stretch, where the performance is limited by tire-ground traction limit and it enables

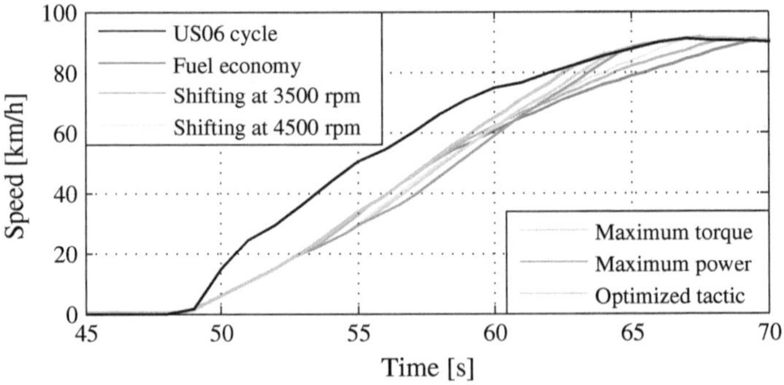

Fig. 3 Optimized tactics compared as the standard strategies

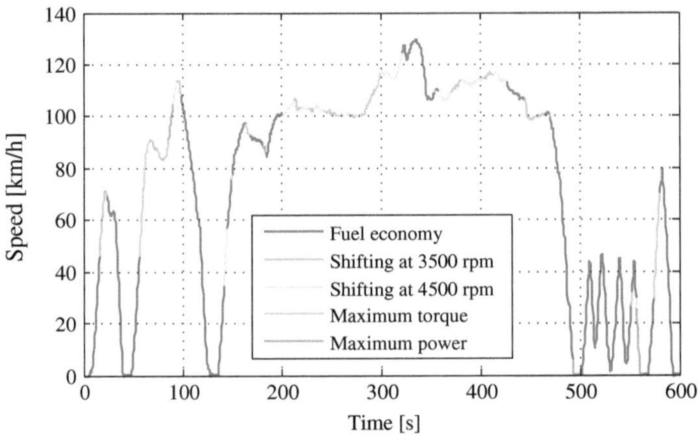

Fig. 4 Optimized gear shifting tactics utilized

the downshift process even in acceleration stretches to increase the vehicle wheels torque, improving the vehicle performance and the fuel consumption. The Fig. 4 shows the used tactics in the simulated cycle.

As can be seen, the optimized tactics utilized the fuel economy strategy, at the beginning of the acceleration stretches, to prevent a large load transfer to the vehicle rear wheels caused by the longitudinal acceleration that reduce the weight force applied at the front wheels, that propel the vehicle decreasing the transmitted torque in the tire-ground contact. With the speed increase, it was used another tactics that aims the vehicle acceleration performance based on the traction limit like the maximum engine torque as can be seen at the second acceleration stretch of the US06 cycle (Fig. 4).

In braking conditions are used the fuel economy strategy to decrease the fuel consumption. The maximum power strategy is used only at short high speed stretches, because this tactics provides the maximum vehicle acceleration, and when

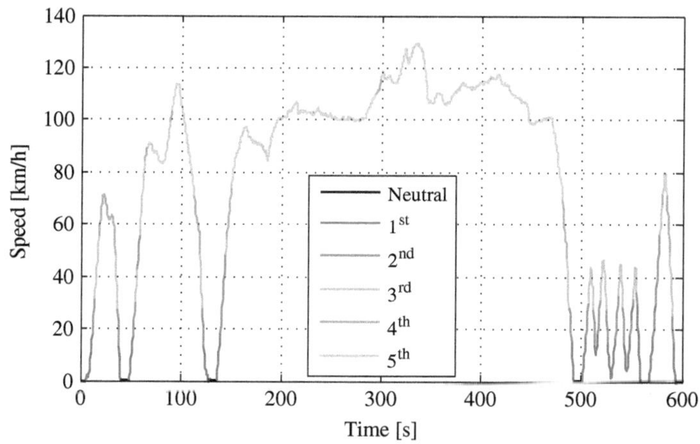

Fig. 5 Optimized strategy gear profiles

the vehicle is at lower speeds it causes a large load transference to rear wheels, therefore this tactics performance is the worst compared to the standard simulated strategies.

The gear shifting profile found in the optimized strategies is different when compared with the standard tactics, in which the gear shifting speeds are constant. The optimized tactics allows a downshift, even in acceleration process, to increase the available torque at the drive wheels when the vehicle performance is lower than objective velocity. When the vehicle reaches the required velocity it occurs, quickly, upshiftings to keep the engine running in the better efficiency regions. The optimized gearshift profile is shown in Fig. 5.

6 Conclusion

This paper has studied the gear shifting strategy influence in the vehicle longitudinal dynamics. The results were obtained by co-simulations performed in the multibody dynamics analysis Adams^{TMM} with Simulink/Matlab^{TM}.

The first simulations considered a standard gear shifting tactics and it was observed that the vehicle performance was lower than the expected in some accelerations stretches where the engine maximum power is required, but the traction is limited by tire-ground contact.

To optimize the simulated vehicle performance and fuel consumption it was developed an algorithm that merges the gear shifting tactics. It compares the previous simulated results and finally it chooses the best tactics for each cycle stretch.

The optimized gear shifting tactics profile results in a better correlation compared with the standard strategies. In this case, the fuel consumption is similar to the results when using the 4,500 rpm strategy. The optimized strategy makes the best representation of the driver behavior according to the required performance, which is inferred from the US06 standard velocity profile.

Finally the application of the gear shifting strategies is crucial in the vehicle dynamics, because they influence the vehicle performance and the fuel consumption.

Acknowledgments The authors thank CNPq and CAPES for financial support.

References

1. Jaideep S, Jagmindar S (2012) Selection of gear ratio for smooth gear shifting, SAE Technical Paper, No. 2012-01-2005
2. Kahlbau Sebastian, Bestle Dieter (2013) Optimal shift control for automatic transmission. Mech Based Des Struct Mach 41(3):259–273
3. Fabio MS, Eduardo dos SC, Jony JE, Heron JD, Franco GD (2013) Gear shifting optimization strategy for brazilian vehicles and traffic. In: Proceedings of the 22th international congress of mechanical engineering
4. Christopher V, Christian JB, Roshan W, Sam A, Lloyd A (2012) Development of a new method to assess fuel saving using gear shift indicators. In: Proceedings of the institution of mechanical engineers, journal of automobile engineering, pp 1630–1639
5. Bingzhao G, Yulong L, Anlin G, Hong C, Kazushi S (2011) Observer-based clutch disengagement control during gear shift process of automated manual transmission. Veh Syst Dyn 49(5):685–701
6. Tianyi G, Christian W (2012) Frey, fuel efficiency driver assistance system for manufacturer independent solutions, Intelligent Transportation Systems (ITSC). In: 15th International IEEE conference
7. Orfila Olivier, Guillaume SP, Cindie A (2012) Gear shifting behavior model for ecodriving simulations based on experimental data. Procedia-Soc Behav Sci 54:341–348
8. Thomas DG (1992) Fundamentals of vehicle dynamics. Society of Automotive Engineers, Warrendale, pp 11–28
9. Jörnsen R, Helmut S (1996) The automotive chassis: engineering principles. Society of Automotive Engineers, Warrendal USA, pp 121–128
10. William FM, Douglas LM (1995) Race car vehicle dynamics. Society of Automotive Engineers, Warrendale, pp 690–694
11. Barlow TJ, Latham S, McCrae IS, Boulter PG (2009) A reference book of driving cycles for use in the measurement of road vehicle emissions, 33TRL Limited
12. Tomas B, Zdenek H, Jan V (2011) Using of co–simulation ADAMS–SIMULINK for development of mechatronic systems. In: 14th international symposium IEEE
13. Xiaofeng Y, Dianlun X, Yun C (2007) Application of time-optimal strategy and fuzzy logic to the engine speed control during the gear-shifting process of AMT, In: IEEE Fourth international conference on fuzzy systems and knowledge discovery
14. Alessandro C, Giovanni P, Giuseppe R (2010) Efficient gear shifting strategies for green driving policies. In: IEEE American Control Conference (ACC)
15. Lu X, Xu X, Liu Y (2009) Simulation of gear-shift algorithm for automatic transmission based on MATLAB, software engineering. In: IEEE WRI World Congress on, vol 2
16. General Motors (2013) Owner Manual Chevrolet Celta 2013. General Motors Brazil Ltda
17. Jony JE, Eduardo dos SC, Fabio MS, Fernanda CC, Franco GD (2014) Influence of gearshift strategies in vehicle performance and fuel consumption. In: 14th Pan-American congress of applied mechanics

Vibration Analysis of an Offshore Platform-Like Structure Excited by Earthquake

J. Enríquez-Zárate and G. Silva-Navarro

Abstract This paper considers the vibration analysis of an offshore platform perturbed by seismic forces. The stability of these structures is related with their frequency response during seismic events. The mathematical model of the structure considers the excitation force provided by a real earthquake occurred and registered in Mexico City in the year 1985. The mechanical structure is modeled using the Euler-Lagrange methodology and its frequency response is validated by using modal analysis techniques. A Finite Element Model (FEM) analysis is performed to show the planar and torsional behavior of the structure into a bandwidth. The offshore platform-like structure is excited via an electromagnetic shaker coupled in the base of the structure. Some experimental and numerical results are presented to demonstrate the performance of the overall system using a passive vibration control scheme.

Keywords Natural frequency · Offshore platform · Earthquake · Modal analysis

1 Introduction

Offshore platforms are mechanical structures that are constantly exposed to environmental loads such as wind, waves, flows and earthquakes that may cause continuous vibrations [1, 2]. Acquiring information with respect to the natural frequencies of the overall system allows the analysis of problems like fatigue damage and risk of collapse of the structure. There are some techniques to mitigate or minimize the effects of vibrations in offshore platforms. These methods are divided in: passive control, semi-active control, active control and hybrid control of vibrations in mechanical structures.

J. Enríquez-Zárate (✉) · G. Silva-Navarro
Departamento de Ingeniería Eléctrica—Sección de Mecatrónica, Centro de Investigación de Estudios Avanzados del I.P.N, A.P. 14-740, C.P. 07360 Mexico, D.F., Mexico
e-mail: enriquezz@cinvestav.mx

M. Ceccarelli and E.E. Hernández Martinez (eds.), *Multibody Mechatronic Systems*, Mechanisms and Machine Science 25, DOI 10.1007/978-3-319-09858-6_15

Offshore platforms are divided into two main categories: Fixed platforms and Compliant platforms. In shallow waters the most common type of offshore platform is the fixed piled structure known as jacket. The fixed jacket structure consists of interconnected tubular members and this is suitable for construction in water depths from a few meters to more than 100 m. These structures usually have four to eight battered legs to achieve stability against toppling in waves [3]. The central structural components of the offshore platform are the jacket, the piles and the deck.

Currently the offshore industry needs to produce oil in geographical areas with higher seismic activity (ground motion), bigger waves and natural phenomena like hurricanes and winds. During the design and application of vibration control schemes to offshore platforms necessarily considers the solution of the eigenproblem and determination of the modal parameters of the overall system [4].

This paper describes the dynamic modeling, analysis and experimental modal analysis on a simple offshore platform-like structure. The results are considered to be used as parameters to design active or semi-active vibration isolation control schemes as Positive Position Feedback (PPF), by considering the frequency response and interaction with base motion due to seismic activity or waves [5, 6].

2 Scheme of the Offshore Platform-Like Structure

A representative model of an offshore platform-like mechanical structure is shown in Fig. 1a and the experimental platform designed is illustrated in Fig. 1b.

The offshore platform-like structure consists of one equivalent mass (m_1) interconnected by four cylindrical columns-like springs denoted by the equivalent stiffness (k_1); the mass is also coupled to (linear) viscous damping (c_1) as shown in Fig. 1a. The base of the structure is directly affected by a ground (force) motion $\ddot{z}(t)$ generated by an electromechanical shaker coupled to the structure. The acceleration measurements are obtained on the mass (m_1) using an accelerometer connected to a data acquisition system from *National Instruments*®. The displacements and velocities are numerically approximated in *Matlab/Simulink*®.

In this work we are interested in analyzing the frequency response of the structure considering the undesired vibrations that cause the displacements in the structure and which are generated by the action of a real seismic force $\ddot{z}(t)$ in terms of acceleration. In this case, we consider an earthquake occurred and registered in Mexico City on September 19, 1985 with a magnitude of 8.1 Mw [7, 17]. It was originated between the Michoacán and Guerrero coasts with the exact epicenter in the mouth of the Balsas River. Figure 2a, b depict seismic force and a component of the acceleration of such earthquake.

The main idea to consider is that a real seism excites the main natural frequencies of the overall structure and, hence, the frequency behavior can be obtained to be used as a preliminary result to synthesize proper passive and active vibration isolation systems.

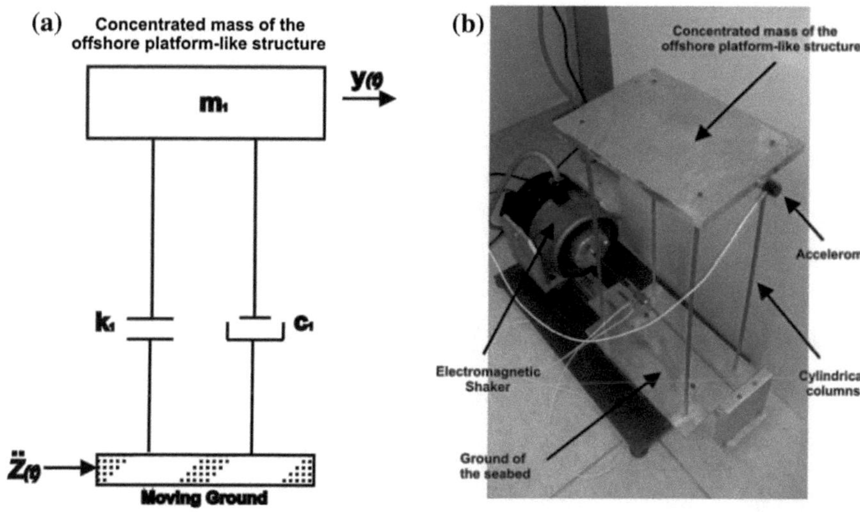

Fig. 1 **a** Schematic model of the offshore platform-like structure, **b** Experimental offshore platform-like structure

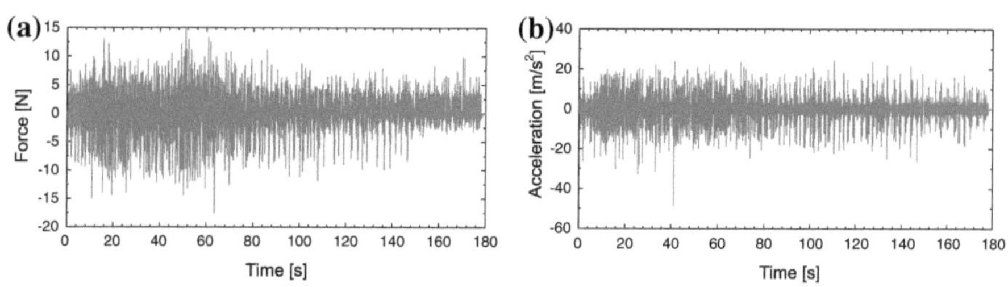

Fig. 2 **a** Seismic force occurred in 1985, **b** Seismic acceleration occurred in 1985

A seismic registration is constituted by three defining components, two in the horizontal plane, known as longitudinal and transverse and one in the vertical plane [8—10]. Figure 3 shows a frequency analysis of the three components of this earthquake in terms of acceleration. The frequency responses correspond to low frequency signals, which have a strong impact over tall structures like buildings, bridges or mechanical multi-story structures of high-rise, high-mass or weight, because they have long natural periods of vibrations [11].

In order to analyze the frequency response of experimental setup considering its mechanical design we decided to apply a frequency sweep in the range of 0—120 Hz with amplitude of 3.05 N, using an electromagnetic shaker to excite the overall mechanical structure (see Fig. 4a, b).

An experimental low rise and low mass structure was used, therefore, vibration modes were hardly excited by the real seismic record. It is then desirable to scale

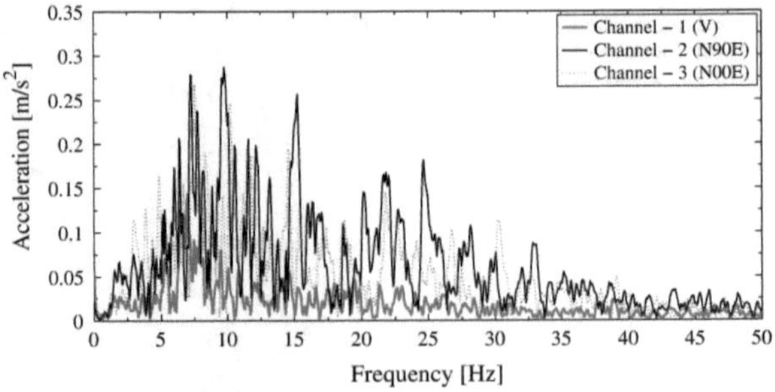

Fig. 3 Real components of the earthquake in terms of the frequency

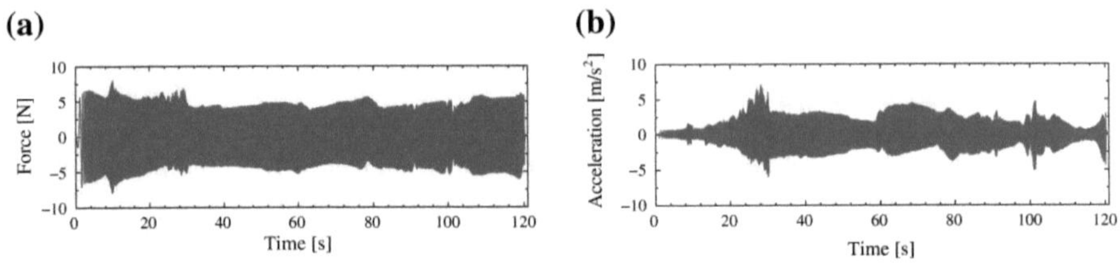

Fig. 4 **a** Ground force, **b** Ground acceleration

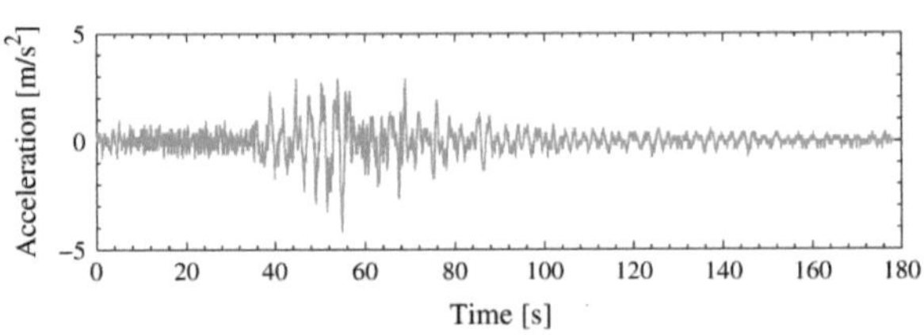

Fig. 5 Scaled vertical component of the earthquake

the vertical component of the seismic record, but only in amplitude to a value of 20 times its value, with the objective of exciting the first mode of vibration in the structure within the frequency range of the real seismic record (see Fig. 5).

The *Labworks Inc*© ET-139 electromagnetic shaker was used to excite the overall system with harmonic components containing low frequencies obtained directly from the seismic record. The mechanical design in the shaker has a limit career of ± 12.7 mm. Therefore, the vertical component of the seismic recording was considered because its acceleration range is lower relating to the horizontal components (see Fig. 3). However, the structural analysis is not restricted to the vertical component, that is, it is possible to perform an excitation of the offshore platform-like structure with the horizontal components of the seismic recording.

3 Model of an Offshore Platform-Like Structure

A simplified mechanical model of an offshore platform, subject to an external disturbance (ground motion) $\ddot{z}(t)$, is presented. The structural response is dominated mainly by the first mode of the system. The structure is modeled as a linear single degree-of-freedom (SDOF) structure extracting its first vibration mode. The offshore platform motion without any control scheme can be described as follows

$$M\ddot{y}(t) + C\dot{y}(t) + K(t) = -Me\,\ddot{z}(t),\ y \in R^1,\ z \in R \tag{1}$$

where $y = [y_1]^T$ is the generalized vector (relative displacement with respect to the main frame), $\ddot{z}(t)$ is the ground acceleration on the base and M, C, K are 1×1 mass, damping and stiffness matrices, respectively. The vector $e = [1]^T \in R^1$ is the influence vector representing the displacement of the mass by the effect of the ground motion. The corresponding mass, damping and stiffness matrices for the offshore platform are easily described by $M = [m_1]$, $C = [c_1]$ and $K = [k_1]$, respectively. For modal analysis purposes, the damping matrix C is assumed to be proportional, that is, $C_1 = a_0 M + b_0 K$ where $a_0 = \xi_i \times \frac{2\omega_i \times \omega_j}{\omega_i + \omega_j}$ and $b_0 = \xi_j \times \frac{2}{\omega_i + \omega_j}$, with ω_i and ω_j the structural modal frequencies and ξ_i and ξ_j the structural damping ratios for modes i and j, respectively [12].

It is important to note that more vibration modes can be modeled, including planar lateral modes or critical torsional modes on the structure. In such cases the model (1) can consider an arbitrarily large number of degrees-of-freedom into the platform, columns and other components, and as a consequence results more appropriate the application of finite element methods and experimental modal analysis techniques.

4 Validation Using Experimental Modal Analysis

The offshore platform-like structure was excited with a real seismic registration and a frequency sweep $\ddot{z}(t)$ in the range of 0—120 Hz using an electromechanical shaker and the resulting experimental Frequency Response Function (FRF) was analyzed using modal analysis techniques.

The corresponding experimental FRF is described in Fig. 6a, b, respectively. Note that, the first FRF (seismic excitation) exhibits only the first dominant mode at 9.189 Hz. The second FRF (frequency sweep), however, shows up to 6 modes, including more bending and torsional modes of the overall structure. The physical system parameters to be used in the simplified Euler-Lagrange model (1), without any TMD or active controller, are summarized in Table 1.

The modal parameters (resonant frequencies ω_i and modal dampings ξ_i) associated to the offshore platform-like structure were indirectly obtained, from the

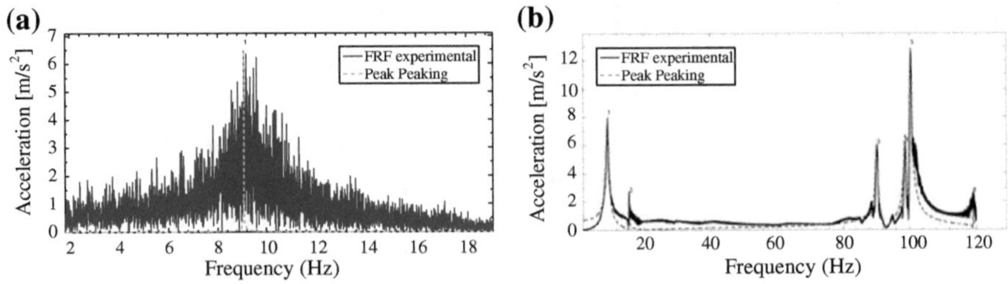

Fig. 6 **a** Experimental FRF of the offshore platform under realistic seismic motion, **b** Experimental FRF of the offshore platform submitted to frequency sweep

Table 1 Physical system parameters of the offshore platform-like structure

$m_1 = 1.3626\,\text{kg}$	$k_1 = 4834.1\,\text{N/m}$	$c_1 = 5.1921\,\text{N/(m/s)}$

Table 2 Modal parameters of the offshore platform-like structure using a frequency sweep

Mode	Resonant frequency ω_i (Hz)	Modal damping ξ_i
i	Experimental	Experimental
1	9.1887	0.0330
2	15.6450	0.0023
3	90.2415	0.0034
4	98.7721	0.0026
5	100.5031	0.0027
6	119.0807	0.0001362

experimental FRF in Fig. 6b, by using well-known modal analysis techniques like *Peak-Peaking* and *Curve Fitting* methods [13, 14]. These modal parameters, for the first 6 modes, are summarized in Table 2, where one can find the same dominant (bending) mode of the structure at 9.1887 Hz.

In order to be more descriptive, some modes shapes of the structure were also computed using finite element methods via the well-known *ANSYS*® software. The first 5 modes are illustrated in Fig. 7, where one can observe a similar dominant bending mode as well as mode shapes at higher frequencies. In particular, it results interesting that torsional mode at 23.735 Hz, because this is close enough to the higher excitation frequencies into the seismic record shown in Fig. 3.

In Table 3 the first vibration mode is evaluated by means of the experimental FRF of the structure obtained with seismic base excitation in Fig. 6a and compared

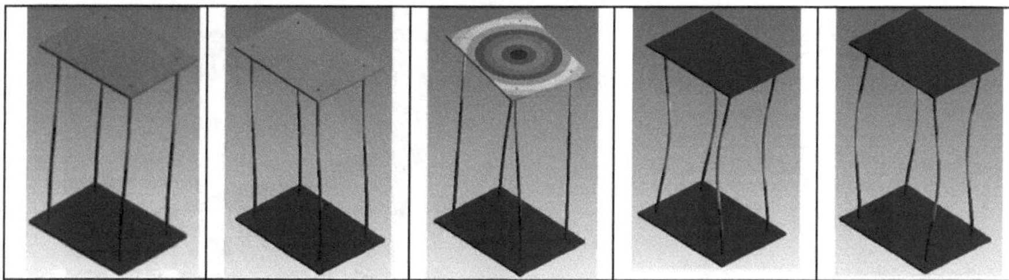

1st frequency [9.9128 Hz]	2nd frequency [10.142 Hz]	3rd frequency [23.735 Hz]	4th frequency [149.86 Hz]	5th frequency [152.1 Hz]

Fig. 7 Mode shapes of the offshore platform-like structure using ANSYS®

Table 3 Comparison of the first mode in the offshore platform-like structure

Mode	Resonant frequency ω_i (Hz)			Modal damping ξ_i experimental
i	Experimental	Numerical	FEM	
1	9.0695	9.4795	9.9128	0.2829

with previous numerical and FEM results. It is possible to check a reasonable matching to be used for structural analysis and vibration control purposes.

5 Vibration Isolation of an Offshore Platform-Like Structure with a TMD

Ground and base excitation on an offshore platform structure are common vibration problems to be solved by means of passive and/or active vibration control schemes. First of all, passive isolation systems (passive suspension) can be employed into the column-like (legs) of the structure in order to isolate the vibrations on the main platform. These can be implemented via shock absorbers or spring-damper suspensions for some specific design frequency band. However, this solution is not robust enough against high variations on the excitation frequency and, more importantly, when the excitation occurs in the main platform, which is the case on such mechanical structures. One solution consists in add Tuned Mass Dampers (TMD) or Active Tuned Mass Dampers (ATMD), which can dissipate high energy rates and reduce the dynamic amplification factor on the main platform [15, 16].

In our case study, a TMD is mounted and connected directly to the concentrated mass of the offshore platform-like structure, resulting in a two degrees-of-freedom mechanical system considering the design of the TMD to compensate the first vibration mode (see Fig. 8a, b).

The equations of motion of the offshore platform-like structure with TMD and ground (base) acceleration $\ddot{z}(t)$ are obtained as follows

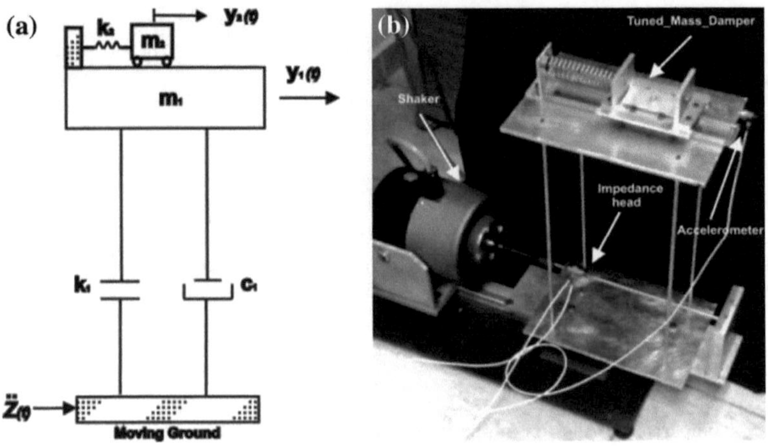

Fig. 8 **a** Schematic diagram of the offshore platform-like structure with a TMD, **b** Experimental offshore platform-like structure with a TMD

$$M_2\ddot{y}(t) + C_2\dot{y}(t) + K_2(t) = -M_2 e_2 \ddot{z}(t),\ y \in R^2,\ z \in R \qquad (2)$$

where $y = [y_1, y_2]^T$ is the generalized vector (relative displacement with respect to the main frame), $\ddot{z}(t)$ is the ground acceleration on the base and M_2, C_2, K_2 are 2×2 mass, damping and stiffness matrices, respectively. The vector $e = [11]^T \in R^2$ is the influence vector representing the displacement of the mass by effect of the ground motion. The corresponding mass, damping and stiffness matrices for the offshore platform are described by

$$M_2 = \begin{bmatrix} m_1 & 0 \\ 0 & m_2 \end{bmatrix},\quad C_2 = \begin{bmatrix} c_1 + c_2 & -c_2 \\ -c_2 & c_2 \end{bmatrix},\quad K_2 = \begin{bmatrix} k_1 + k_2 & -k_2 \\ -k_2 & k_2 \end{bmatrix}$$

The TMD is designed to passively minimize the j-th mode at $\omega_j \cong \sqrt{k_2/m_2}$, where ω_j is the j-th resonant frequency of the structure to be minimized and m_2 and k_2 are the mass and the stiffness of the TMD. In this case, the TMD is designed to minimize the first mode of the structure, that is, with $\omega_1 \approx 10.4173$ Hz. Note how this resonant frequency slightly changed because of the addition of the TMD on the main platform.

The approximated physical system parameters of the offshore platform-like structure with TMD are given in Table 4. These parameters were directly measured and/or computed by using experimental modal analysis techniques.

The overall passive offshore platform-like structure was excited with the earthquake ground motion described in Fig. 2. The experimental FRF is depicted in Fig. 9 and the modal parameters of the first two modes are summarized in Table 5.

Table 4 System parameters of the offshore platform-structure with a TMD

Primary structure	TMD
$m_1 = 1.3626\,\text{kg}$	$m_2 = 0.420\,\text{kg}$
$k_1 = 4834.1\,\text{N/m}$	$k_2 = 767\,\text{N/m}$
$c_1 = 8.3294\,\text{N/(m/s)}$	$c_2 = 17.1620\,\text{N/(m/s)}$

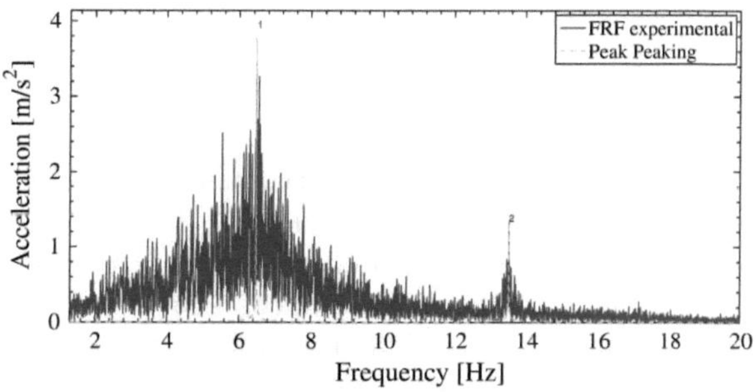

Fig. 9 Experimental FRF of the offshore platform-like structure with TMD

Table 5 Modal parameters of the offshore platform-like structure with TMD

Mode	Resonant frequency ω_i (Hz)		Modal damping ξ_i experimental
i	Experimental	Numerical	
1	6.4755	6.0418	0.1139
2	13.5231	10.6712	0.1425

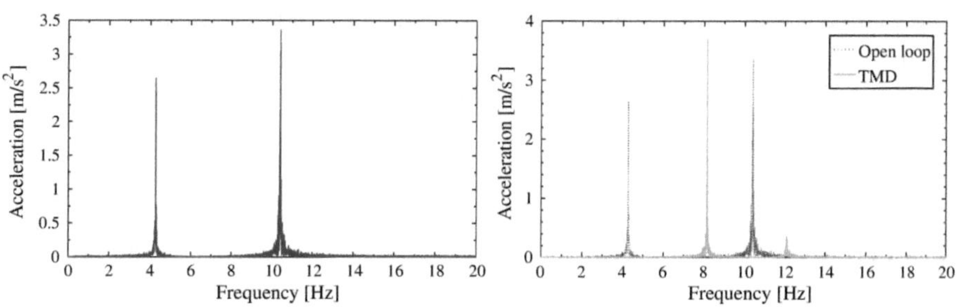

Fig. 10 **a** Numerical FRF of the 2DOF system and FRF response of TMD with $\omega_{TMD} \approx 6.8\,\text{Hz}$

Note that the numerical results are quite consistent with the modal parameters. The first mode shape of the structure was modified with the coupling of the TMD, thus shifting its resonant frequency to 6.4735 Hz.

For the system parameters given in Table 4 the designed tuned frequency of the TMD is computed as $\omega_{TMD} = \sqrt{k_2/m_2} \approx 6.8\,\text{Hz}$, therefore, the first mode can be passively attenuated. The numerical FRF responses of the offshore platform-like structure without/with TMD are shown in Fig. 10a, b.

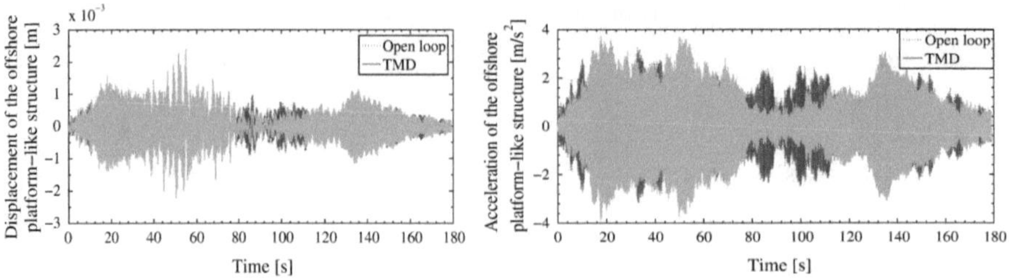

Fig. 11 Dynamic response of the offshore platform-like structure without/with TMD

The dynamic response (displacement and acceleration) of the offshore platform-like structure with TMD submitted to the earthquake ground motion in Fig. 2 is shown in Fig. 11a, b. It is important to observe how the first mode is attenuated up to 89 %, although the TMD resulted in one more resonance. This inconvenience can be overcome with the addition of a force actuator on the mass of the TMD and the synthesis of proper robust control laws in the active TMD.

6 Conclusions

In this work the modal and structural analysis of an offshore platform-like structure submitted to ground motion is presented. The experimental platform is tested when the ground motion corresponds to a realistic earthquake record. In order to attenuate the overall system response a TMD was designed and coupled to the main platform, thus reducing and shifting the first mode shape, but adding another resonance. Further work is being completed to get isolation and vibration control schemes with the application of an active TMD and a modal controller based on Positive Position/Acceleration Feedback.

References

1. Shu-Qing W, Li N (2013) Semi-active vibration control for offshore platforms based on LQG methode. J Mar Sci Technol 21(5):562−568
2. Zahariev EV (2013) Earthquake dynamic response of large flexible multibody systems. Mech Sci 4:131−137
3. Taghikhany T, Ariana Sh, Mohammadzadeh R, Babaei S (2013) The effect of semi-active controller in Sirri jacket seismic vibration control under Kobe earthquake. Int J Mar Sci Eng 3 (2):77−84 (Springer)
4. Mohammad Reza T, Younes K, Ali Akbar G (2012) Assessment and rehabilitation of jacket platforms. In: Earthquake resistant structures—design, assessment and rehabilitation, Chap. 15. Intech, Rijeka, pp 382−568

5. Ríos-Gutierrez M, Silva-Navarro G (2013) Active vibration control in building-like structures using piezoelectric actuator and positive acceleration feedback, DYNA, Year 80, 179th edn. Medellin, Colombia, pp 116–125 (In Spanish)
6. Gawronski WK (2002) Advanced structural dynamics and active control of structures. Springer-Verlag, New York
7. Centro de Instrumentación y Registro Sísmico, A.C. Data of the seismic registration of the 1985 earthquake in Mexico City. www.cires.org.mx
8. Schmidt Díaz V, Quirós Serrano C (2007) Caracterización de los registros acelerográficos obtenidos en el laboratorio de ingeniería sísmica de la Universidad de Costa Rica. San José, Costa Rica, Ingeniería, 17(1) 27–41. (In Spanish)
9. Ceccarelli M, Ottaviano E, Castelli G (2005) An application of a 3-DOF parallel manipulator for earthquake simulations. In: 22nd international symposium on automation and robotics in construction ISARC 2005. Ferrara, Italy
10. Selvi Ö, Ceccarelli M (2012) Interpretation of earthquake effects on mechanism operation: an experimental approach. J Naval Sci Eng 8(2):31–45
11. Meli R, Bazán E (2013) Diseño Sísmico de Edificios. Editorial LIMUSA, México (In Spanish)
12. Enríquez-Zárate J, Silva-Navarro G (2013) Vibration control in a building like structure using positive position feedback and sliding mode control. In: 20th international congress on sound and vibration. Bangkok, Thailand
13. Ewins DJ (1984) Modal testing: theory and practice. Research Studies Press Ltd., Wiley, New York
14. Hu J, Fu Z-F (2001) Modal analysis. Butterworth-Heinemann, Great Britain
15. Dongmei C, Dong Z, Zhaofu Q (2008) Vibration control simulation of offshore platform based on Matlab and ANSYS program. In: Global design to gain a competitive edge, Chap. 4. Springer, New York, pp 546–556
16. Patil KC, Jangid RS (2005) Passive control of offshore jacket platforms. Ocean Eng 32: 1933–1949
17. Auvinet G, Mendoza MJ (1986) Comportamiento de diversos tipos de cimentación en la zone lacustre de la Ciudad de México durante el sismo del 19 de septiembre de 1985, Proc. Symposium: Los Sismos de 1985; Casos de Mecánica de Suelos, México (In Spanish)

Dynamic Parameter Identification in the Front Suspension of a Vehicle: On the Influence of Different Base Parameter Sets

L.A. Mejía, V. Mata, F. Valero, J. Ros and X. Iriarte

Abstract This paper addresses the identification of the inertial parameters of a dynamic front suspension. The concept of inertia and mass transfer is used to obtain three different sets of base parameters, reducing them to the relevant parameters through the analysis of dynamic contribution. These models are compared with two models obtained by numerical methods, which have no physical meaning. The evaluation is performed by crossing data with different trajectories to those used in the identification process. The results show that the use of inertia and mass transfer in the selection of potential base parameters generates well-conditioned models that are very close to the original model behavior.

Keywords Inertia and mass transfer · Parameter identification · Front suspension · Vehicle dynamics

L.A. Mejía (✉)
Universidad Tecnológica de Pereira, Pereira, Colombia
e-mail: adriamec@utp.edu.co

V. Mata · F. Valero
Centro de Investigación de Tecnología de Vehículos, Universitat Politècnica de València, Valencia, Spain
e-mail: vmata@mcm.upv.es

F. Valero
e-mail: fvalero@mcm.upv.es

J. Ros · X. Iriarte
Universitat Pública de Navarra, Madrid, Spain
e-mail: jros@unavarra.es

X. Iriarte
e-mail: xabier.iriarte@unavarra.es

M. Ceccarelli and E.E. Hernández Martinez (eds.), *Multibody Mechatronic Systems*,
Mechanisms and Machine Science 25, DOI 10.1007/978-3-319-09858-6_16

1 Introduction

Unlike serial and parallel robots, the identification of dynamic parameters in vehicles has been little reported in the literature, despite the advantages of having an accurate dynamic model of the vehicle based on dynamic data obtained experimentally, which can be modified over the lifetime or service conditions of the vehicle. Among the published works are Chen and Beale [1], where a MacPherson type suspension is identified, Venture et al. [7] where procedures used in robotics were applied, and Valero et al. [4, 6] which is considered a double-wishbone suspension. Specific problems that arise when identifying vehicles are caused as described in [6], due to the difficulty of obtaining trajectories that are capable of exciting the various components of the equation of motion of the vehicle system in a suitable manner, leading to numerically ill-conditioned problems. One way to overcome this difficulty is to combine symbolic and numeric methodologies to obtain reduced models for describing the dynamic behavior of the vehicle accurately. Note that one of the advantages of symbolic methods is that they allow us to know the explicit relationships between the minimum set of parameters depending on the geometry of the mechanical system, which is a valuable tool in the optimization process of system performance. These symbolic procedures were applied to serial robots by Khalil and Dombre [3], where the parameters which must be grouped (base parameters) are specified based on the linear relationship between the energy functions of inertial parameters and the type of kinematic joint. For vehicle systems, Beale and Chen [1] identified a set of basic parameters for a MacPherson suspension mechanism and demonstrated the relationship between inertia—mass transfer and the identification process by comparison with the numerical model obtained. Ros et al. [5] established a fully symbolic methodology to obtain a set of base parameters using the concept of transfer of the inertial properties between adjacent bars, considering the type of kinematic pair.

In this work the concept of inertia and mass transfer is performed to obtain just a few base parameter sets with physical significance and its application is evaluated in vehicle suspension. From this point, we will compare the performance of the different sets of base parameters obtained when representing the dynamic behavior of the front suspension of a vehicle. The comparison process will also include different base parameter sets obtained by traditional numerical methods. The work is organized as follows. The characteristics of the model for identification are presented in Sect. 2. The method of inertia and mass transfer and its application in the reduction of base parameters for the front suspension of a vehicle is presented in Sect. 3, for which several models are obtained associated with different sets of base parameters. Then the models obtained are validated after comparison with a virtual model created using a commercial program ADAMS and the results of the identification process are presented.

2 Identification Model

The first step in the process of parameter identification is to obtain the dynamic model. If the reference systems used in the formulation are not centroidal systems, i.e. its location is such that it does not match the centers of mass of the elements, these equations of motion exhibit linearity with respect to the dynamic parameters [6]. The inclusion of equations for different positions of the mechanical system allows us to obtain a numerically overdetermined linear system. W is the coefficient matrix named observation matrix, $\vec{\tau}$ is the vector of generalized forces, and the dynamic model expressed in linear form with respect to the parameters can be expressed as follows

$$W_{k\times m} \cdot \vec{\Theta}_{m\times 1} = \vec{\tau}_{k\times 1} \tag{1}$$

where $\vec{\Theta}$ is the vector of dynamic parameters for n bodies

$$\vec{\Theta} = [m_1, mx_1, my_1, mz_1, Ixx_1, Ixy_1, Ixz_1, Iyy_1, Iyz_1, Izz_1, \ldots, Izz_n]^T \tag{2}$$

As mentioned above, this dynamic model includes all dynamic parameters, but there may be parameters with minimal or no contributions to system dynamics or parameters that due to the physical nature of the kinematic constraints act only as linear combinations. Therefore, the matrix W is usually rank deficient. However, the solution of the system expressed in (1) requires a full-rank observation matrix, which involves reducing the system and obtaining a new set of parameters called base parameters. Consequently, the matrix W is broken down into singular values as

$$W_{k\times m} = U_{k\times k} \times S_{k\times m} \times V^T_{m\times m} \tag{3}$$

where U and V are orthogonal matrices and S is a diagonal matrix with singular values $s_1, s_2, \ldots, s_r$ of W, r being the rank of the matrix. The V matrix could be expressed separately as

$$V = \begin{bmatrix} V1_{m\times r} & V2_{m\times (m-r)} \end{bmatrix} \tag{4}$$

An independent column in W corresponds to the null column of $V2$. Rearranging $V2$ so that full range submatrix obtained a permutation matrix P such that it is determined as

$$P^T \cdot V2 = \begin{bmatrix} V21_{r\times (m-b)} \\ V22_{(m-r)\times (m-r)} \end{bmatrix} \tag{5}$$

The determination of $V22$ is not unique, and therefore neither is the permutation matrix. While for serial and parallel robots it is usually obtained from the first or last row of the matrix $V2$, adding rows provided these increase the rank of the matrix, for vehicle systems this group has no physical significance nor is it an obvious set of parameters. In order not to leave the determination of the matrix $V22$ to chance, in this work the inertia and mass transfer process is used, referenced in the work of Valero et al. [6], to determine which parameters can be combined. The solution for base parameters is obtained as

$$\vec{\Theta}_{base} = \begin{bmatrix} I_{r\times r} & \beta_{r\times(m-r)} \end{bmatrix} P^T \vec{\Theta} \tag{6}$$

where $\beta = -V21 \cdot V22^{-1}$. Finally, the dynamic model can be represented by the expression

$$W1 \cdot \vec{\Theta}_{base} = \vec{\tau} \tag{7}$$

In parallel robots, Diaz et al. [2] use the relative standard deviation of the identified parameters to reduce the model by removing those base parameters with a high standard deviation to achieve an optimal set of base parameters. However, we must remember that in vehicle systems with ill-conditioning of the matrix W the quality of identification of the parameters evaluated by relative standard deviation does not necessarily allow us to obtain the best set of parameters. Another possibility for obtaining the set of base parameters is based on the dynamic contribution of each parameter to the total dynamic. The percentage contribution of each parameter to the system dynamics can be established by

$$\varsigma_i = \frac{\sqrt{\left(W_{(:i)} \cdot \Theta_i\right)^T \cdot \left(W_{(:i)} \cdot \Theta_i\right)}}{\sqrt{\vec{\tau}^T \cdot \vec{\tau}}} \cdot 100 \tag{8}$$

Having established the base parameters with low dynamic contribution, they are removed from the vector of parameters to give a new set called *relevant parameters*. The identification process is repeated for the new set of relevant parameters. To determine the contribution limit in order to eliminate parameters we must consider the condition number of the reduced observation matrix associated with new relevant parameters and the relative error achieved.

3 Inertia Transfer Models

The use of numerical methods such as SVD to reduce the model provides a set of parameters that may represent the dynamic behavior of the system analyzed, but are not necessarily a natural combination of parameters nor do they allow us to analyze the symbolic expressions that can be obtained from the method. Moreover, as the permutation matrix is not unique -in fact, it is an NP-hard problem-, it can give a high number of different linear combinations and therefore many sets of parameters representing the dynamic database system. Deciding a priori to represent the combination of the system can actually be a solution to the multiplicity of possible sets of base parameters. The concept of inertia transfer is based on the virtual redistribution of inertial properties of the elements of the mechanical system as long as the Lagrangian of the system is not changed. This redistribution results in equivalent inertial systems that depend on fewer parameters, but this reduction has no effect on the dynamics of the system and therefore on how the parameters are grouped to obtain a set of base parameters. The inertia and mass transfer methodology is presented in [5].

In this section the concept of inertia transfer is used to obtain several sets of basic parameters which can represent the dynamics of front suspension in order to compare whether the identified model is better than that obtained with the application of a numerical method such as SVD. The front suspension analyzed consists of a right and left side, each with upper and lower control arms, upright and wheel, spring-damper and track rod. Both sides are linked through the steering rack and the latter is attached to the chassis by a translational joint. Suspension topology is presented in Fig. 1.

Applying the concept of mass transfer requires prior identification of the dynamic parameters of the model that cannot be identified, as they do not show any dynamic influence. There are considerations for the type of kinematic joint that determine whether or not the associated parameters influence the system model [5]. Similarly, the observation matrix W allows us to know these parameters. If there are null columns in W, it means that the associated parameters are not included in the dynamic model and therefore cannot be selected in mass transfer. In the case of the suspension analyzed, these parameters are my_1, Ixx_1, Ixy_1, Ixz_1, Iyz_1, Izz_1,my_2, Ixx_2, Ixy_2, Ixz_2, Iyz_2, Izz_2, m_6, mx_8, my_8, mz_8, Ixx_8, Ixy_8, Ixz_8, Iyy_8, Iyz_8, Izz_8, my_9, Ixx_9, Ixy_9, Izz_9, Ixz_9, Iyz_9, my_{10}, Ixx_{10}, Ixy_{10}, Ixz_{10}, Izz_{10}, Iyz_{10}, m_{14}.

The next step in the process of obtaining the base parameters is to build a structure or branched open kinematic chain, to which cuts are made in the original topology of the system. Thus, in each branch a solid is chosen as convergent. First, inertial properties corresponding to the broken links will be transferred, trying not to delete parameters which can be transferred later in another kinematic chain.

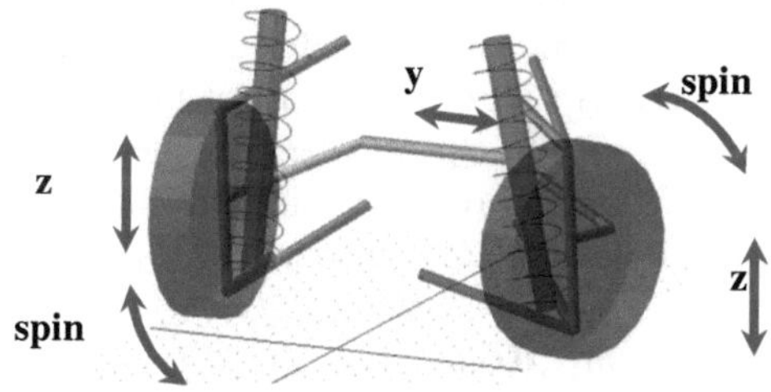

Fig. 1 Topology and degrees of freedom of the suspension analyzed

Then, the inertial properties are transferred from the farthest elements to the convergence element. Three mass transfer sequences are considered. In the first transfer, the convergence solids will be the wheels 5 and 13, but the steering rack will not be transferred; the second case will be the lower control arms 1 and 9 with the same consideration previous to the steering rack, and in the third the steering rack will be transferred.

T1 Model. Convergence solids are right and left wheels. Invariant steering rack. To clarify the procedure, the complete transfer of moment of inertia on the rotational axis of the revolute joint, where the body 2 (upper control arm) meets the chassis, is presented. One mass m_{0A} is added from the chassis to the body 2 at the point of action of revolute joint A, which gives $Iyy_2 = Iyy_2^{'} + m_{0A}\left(x_{AE}^2 + z_{AE}^2\right) = 0$. Where x_{AE}, y_{AE} and z_{AE} are the distances from the local reference system of the body 2 to the point of insertion of the mass in the local reference system. If after transfer $Iyy_2^{'} = 0$, it follows that the value of the transferred mass is $m_{0A} = \frac{Iyy_2}{x_{AE}^2 + z_{AE}^2}$. For the suspension, the distance z_{AE} is zero. Thus, the inertial parameters of link 2 after transfer are

$$m_2^{'} = m_2 + m_{0A} = m_2 + \frac{Iyy_2}{x_{AE}^2}, mG_2^{'} = mG_2 - m_{0A}\begin{bmatrix} x_{AE} \\ y_{AE} \\ z_{AE} \end{bmatrix} = \begin{bmatrix} mx_2 - \frac{Iyy_2}{x_{AE}} \\ my_2 - \frac{Iyy_2 \cdot y_{AE}}{x_{AE}^2} \\ mz_2 \end{bmatrix}$$

$$IG_2^{'} = IG_2 - m_{0A}\begin{bmatrix} y_{AE}^2 + z_{AE}^2 & x_{AE} \cdot y_{AE} & x_{AE} \cdot z_{AE} \\ x_{AE} \cdot y_{AE} & x_{AE}^2 + z_{AE}^2 & y_{AE} \cdot z_{AE} \\ x_{AE} \cdot z_{AE} & y_{AE} \cdot z_{AE} & x_{AE}^2 + y_{AE}^2 \end{bmatrix}$$

$$= \begin{bmatrix} Ixx_2 - \frac{Iyy_2}{x_{AE}^2} y_{AE}^2 & Ixy_2 - \frac{Iyy_2}{x_{AE}} \cdot y_{AE} & Ixz_2 \\ Ixy_2 - \frac{Iyy_2}{x_{AE}^2} \cdot y_{AE} & 0 & Iyz_2 \\ Ixz_2 & Iyz_2 & Izz_2 - \frac{Iyy_2\left(x_{AE}^2 + y_{AE}^2\right)}{x_{AE}^2} \end{bmatrix}$$

It is observed that the parameter Iyy_2 appears as a combination in the new set of parameters. Taking the new parameters for the body 2, their total mass can now be transferred to body 3 in the spherical joint. Since the reference system of the body 2 coincides with the spherical joint, mass transfer does not affect any other inertial parameters. Thus, the mass m_2 will become part of the body 3 and appears to be dependent with the mass m_3 as a base parameter. The inertia and mass transfer takes place throughout the entire branch of the chains until it reaches the convergence solids. After completing all the transfers, the dependent parameters are m_1, Iyy_1, m_2, Iyy_2, m_4, Iyy_4, m_5, mx_5, Izz_5, m_7, Ixy_7, Ixx_7, Ixz_7, Iyy_7, Iyz_7, Izz_7, m_9, Iyy_9, m_{10}, Iyy_{10}, m_{12}, Iyy_{12}, m_{13}, mx_{13}, Izz_{13}, m_{15}, Ixy_{15}, Ixx_{15}, Ixz_{15}, Iyy_{15}, Iyz_{15}, Izz_{15}.

The total number of parameters to be transferred in this model is 32. Eliminating the parameters without any dynamic influence, and after the combination of dependent parameters, the matrix $W1$ presents a rank of 83.

T2 Model. Invariant steering rack. In the second model convergence solids will lower control arms 1 and 9. The parameters to be transferred are Iyy_1, m_2, mx_2, m_3, m_4, mx_4, m_5, mx_5, Iyy_5, m_7, Ixy_7, Ixx_7, Ixz_7, Iyy_7, Iyz_7, Izz_7, Iyy_9, m_{10}, mx_{10}, m_{11}, m_{12}, mx_{12}, m_{13}, mx_{13}, Iyy_{13}, m_{15}, Ixy_{15}, Ixx_{15}, Ixz_{15}, Iyy_{15}, Iyz_{15}, Izz_{15}.

T3 Model. Convergence solids are right and left lower control arms. Transfer steering rack. In this model the steering rack is transferred. In this case, the mass m_8 is transferred to the right track rod. Possible transfers were then performed according to the kinematic joints to reach convergence solids 1 and 9. This generates the dependent parameters m_1, m_2, mx_2, m_3, m_4, m_5, mx_5, Iyy_5, m_7, Ixy_7, Ixx_7, Ixz_7, Iyy_7, Iyz_7, Izz_7, m_8, m_9, m_{10}, mx_{10}, m_{11}, m_{12}, mx_{12}, m_{13}, mx_{13}, Iyy_{13}, m_{15}, Ixy_{15}, Ixx_{15}, Ixz_{15}, Iyy_{15}, Iyz_{15}, Izz_{15}.

R4 Model. By random combination. In this model, the parameters are determined randomly, considering only the combination used to generate a full-rank matrix *V*22. Dependent parameters for this model are m_1, mx_1, m_2, mx_2, m_3, m_4, mx_5, Izz_5, Ixz_6, Izz_6, m_7, Ixx_7, Ixy_7, Iyy_7, Iyz_7, Izz_7, m_8, m_9, mx_9, Iyy_9, m_{10}, Iyy_{10}, m_{11}, m_{12}, mx_{12}, Ixz_{14}, Izz_{15}, Ixx_{15}, Ixy_{15}, Ixz_{15}, Iyy_{15}, Iyz_{15}, Izz_{15}.

R5 Model. By random combination. Dependent parameters for this model are Iyy_1, Iyy_2, m_3, mz_3, m_4, Iyy_4, m_5, mx_5, Izz_5, m_7, Ixx_7, Ixy_7, Ixz_7, Iyy_7, Iyz_7, Izz_7, Iyy_9, Iyy_{10}, m_{11}, mz_{11}, m_{12}, Iyy_{12}, m_{13}, mx_{13}, Izz_{13}, m_{15}, Ixx_{15}, Ixy_{15}, Ixz_{15}, Iyy_{15}, Iyz_{15}, Izz_{15}.

4 Reduced Models for Identification of Parameters in the Suspension Analyzed

The model used for identification depends on the permutation matrix used in Eqs. (5)–(6). Because the combinations of parameters are multiple, and for the case of the suspension analyzed this number is 2.81×10^{28} possibilities, we selected three models whose combinations are obtained from inertia and mass transfer (T Models) and two additional models obtained by random combinations (R Models). For the latter two models, their determination has no physical meaning. In order to perform the identification procedure, optimized trajectories are used to obtain a lower condition number of matrix *W*1. These have been parameterized as finite Fourier series

$$f(t) = A_0 + \sum_{j=1}^{NH} \left[\frac{a_j}{2 \cdot \pi \cdot f \cdot j} \cdot sen(2 \cdot \pi \cdot f \cdot j \cdot t) - \frac{b_j}{2 \cdot \pi \cdot f \cdot j} \cdot \cos(2 \cdot \pi \cdot f \cdot j \cdot t) \right] \quad (9)$$

where t is time, NH is the number of harmonics and f is frequency. The coefficients for the Fourier series for each variable are presented in Table 1.

The movement of the steering is given by the sine function $Y_8 = -0.1302611 - 0.01197389 \cdot \sin(16 \cdot \pi \cdot f \cdot t)$. For the spin of both wheels

Table 1 Coefficients of the Fourier series used in the trajectories of identification

Variable	A0	a1	a2	a3	b1	b2	b3
Z_7	1.828×10^{-1}	1.667×10^{-3}	-6.253×10^{-3}	6.734×10^{-3}	-1.193×10^{-3}	-2.681×10^{-3}	-6.583×10^{-1}
Z_{15}	1.799×10^{-1}	8.321×10^{-5}	-5.154×10^{-6}	-7.10×10^{-6}	-3.573×10^{-5}	-1.911×10^{-5}	-6.774×10^{-1}

$\phi_5 = 2\pi \sin(t)$ and $\phi_{13} = 2\pi \sin(t)$. Applying Eqs. (3)–(7) for different combinations, the corresponding reduced models are obtained for base parameters. The problem now is the high condition number of the observation matrix generated for each set of base parameters. This high condition number makes it impossible to think of an identification process in the presence of errors in the experimental data when the system arises in a real environment, hence this value should be momentous when determining which of the models obtained is best. A final reduction is carried out by means of dynamic analysis based on the contribution of each parameter to give rise to a so-called set of relevant parameters.

5 Results

Crossing data with different trajectories to that used in the identification process is performed for each model. This procedure is intended to evaluate the behavior of each model. A summary of the characteristics obtained for each model before reduction by dynamic contribution is presented in Table 2.

It can be observed that the numerically better conditioned models coincide with models that have a low condition number for the matrix $V22$. In the case of the three models obtained by mass transfer, they all exhibit well conditioning to $V22$. Nevertheless, the condition number of $W1$ remains high, so base parameters are eliminated by the analysis of dynamic contribution to obtain a set of relevant parameters that generate a well-conditioned model. Figure 2 presents the evolution of the condition number of W_{red} and the relative absolute error when base parameters are eliminated for the T2 model.

The figure shows that when base parameters are eliminated, the condition number of the reduced model decreases but the error obtained increases due to the small percentage of dynamic contribution for base parameters with respect to the original model. For different models, a considerable increase in the condition number above 10 relevant parameters is observed; however, for this number the fluctuation in error is not substantial. Secondly, over 90 % of the contribution of the suspension dynamics is achieved from these number of relevant parameters, thus reducing models to 10 relevant parameters in all cases.

Table 2 Characteristics obtained for each model before reduction by dynamic contribution

Model	Rank	Condition number $V22$	Condition number $W1$
T1	83	22.98	5.75×10^{10}
T2	83	4.80	5.75×10^{10}
T3	82	3.21	4.56×10^{15}
R4	83	7.43×10^{7}	1.00×10^{16}
R5	83	16.47	5.75×10^{10}

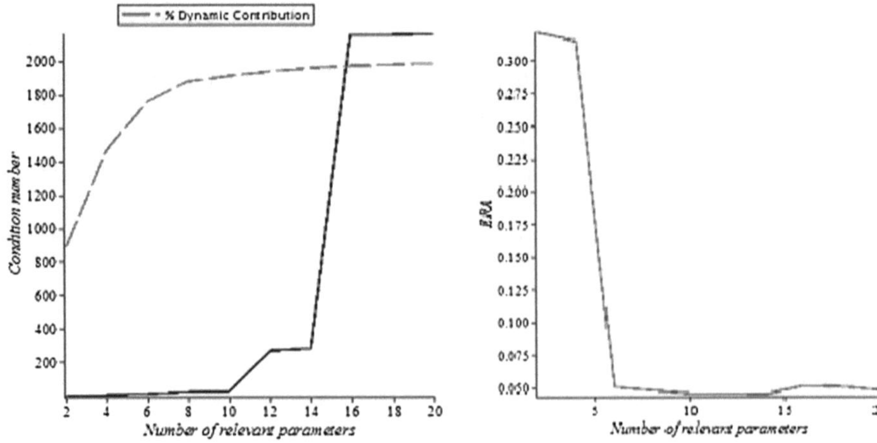

Fig. 2 Evolution of the condition number and ε_{RA} of relevant parameters for the T2 model

For crossing data a sine function with a variation in frequency from 1 to 8 Hz is used, expressed as $\phi_5 = 2\pi \sin(t)$, $\phi_{13} = 2\pi \sin(t)$, $Z_7 = 0.15 + 0.1 \cdot \sin(2 \cdot \pi \cdot f \cdot t) - 0.1 \cdot \cos(\pi \cdot f \cdot t)$, $Y_8 = 0.2 + 0.05 \cdot \sin(\pi \cdot f \cdot t)$, $Z_{15} = 0.2 + 0.1 \cdot \sin(2 \cdot \pi \cdot f \cdot t + \frac{\pi}{4})$. To quantify the accuracy of the identified model, the relative absolute error in each path is obtained by the expression

$$\varepsilon_{RA} = \frac{\sum_i \tau_{red_i} - \tau_{orig_i}}{\sum_i \tau_{orig_i} - \bar{\tau}} \tag{10}$$

Errors obtained for crossing data of different models and their condition numbers are presented in Table 3. All models have an acceptable behavior according to the error generated for the different frequencies analyzed. However, the differences are with respect to the condition number of the reduced models, which shows that all the models obtained by mass transfer are numerically well conditioned. Although a random pattern was obtained, the R5 model also has good conditioning, suggesting the importance of obtaining a combination that generates a good conditioning of the array *V*22.

Table 3 Error generated for different frequencies analyzed

Model	Condition number W_{red}	ε_{RA}				
		1 Hz	2 Hz	4 Hz	6 Hz	8 Hz
T1	157	0.0470804	0.0518912	0.0522014	0.0524212	0.0525480
T2	21.03	0.0458449	0.0554292	0.0569194	0.0570210	0.0570710
T3	22.69	0.0483099	0.0625443	0.0648960	0.0650802	0.0651611
R4	1.24×10^{15}	0.0468503	0.0851324	0.0935832	0.0936888	0.0938476
R5	38.58	0.0448046	0.0702265	0.0758308	0.0760716	0.0761545

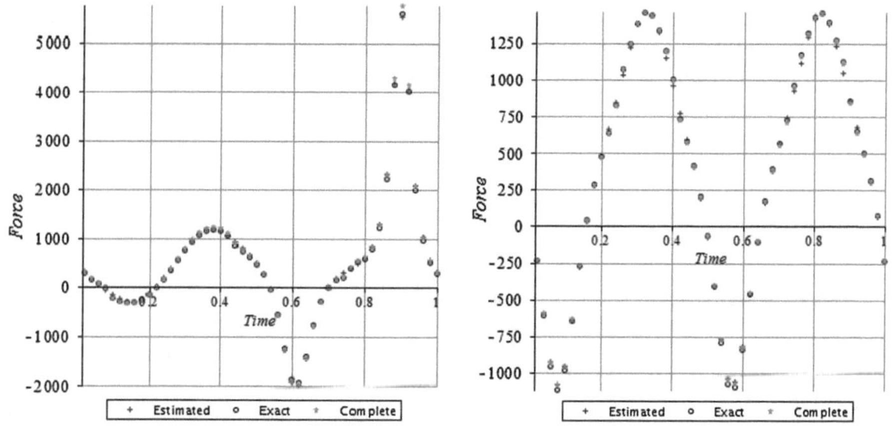

Fig. 3 Comparison between the estimated model, the exact model and the full original model for generalized forces Z_7, Z_{15} of the T2 Model for trajectories at a frequency of 2 Hz

The generalized forces that T2 obtained for a frequency of 2 Hz are presented in Fig. 3. These forces reproduce the response generated by the original model. Additionally, the forces for this model are very close to those obtained by the relevant parameters given the values obtained from Adams.

6 Conclusions

The application of traditional methods of identification of dynamic parameters produces reduced models that approximate the dynamic behavior of the original systems accurately. As some parameters have no influence on the dynamic behavior of the system and others influence it as a linear combination with other parameters, finally a set of parameters called base parameters is obtained. This set of parameters is not unique and depends on how the initial parameters are combined. The larger the system analyzed, as is the case with vehicle suspension, the greater the number of possible combinations that can occur. So far in the literature known to the authors no method has been reported to assess what the best set of parameters is based on all possible combinations. In this paper the concept of inertia and mass transfer was applied to obtain a reduced number of combinations, which, among other advantages, has physical significance because it is obtained from purely symbolic procedures.

The models obtained were observed to have better conditioning than are those in which the condition number of the associated matrix *V*22 is low.

The three models obtained by inertia and mass transfer for the front suspension analyzed complies with that feature. So it shows how the concept of mass transfer and inertia can be applied to determine a priori limited combinations that achieve better levels of numerical conditioning. A further reduction applied by analysis of the contribution of dynamic parameters yields even better conditional models with better dynamic response to the inclusion of errors in the measurements of forces and kinematic variables of the model.

References

1. Chen K, Beale D (2003) Base dynamic parameter estimation of a MacPherson suspension mechanism. Int J Veh Syst Dyn 39(3):227–244
2. Díaz M, Mata V, Valera A, Page A (2010) A methodology for dynamic parameters identification of 3-DOF parallel robots in terms of relevant parameters. Mech Mach Theory 45:1337–1356
3. Khalil W, Dombre E (2004) Modeling, identification and control of robots. Kigan Page Science, London
4. Mejía L, Valero F, Mata V (2013) Development of analytical models for the identification of dynamic parameters in a double wishbone front suspension. SAE Int J Passeng Cars—Mech Syst 6(1):231–240
5. Ros J, Iriarte X, Mata V (2012) 3D inertia transfer concept and symbolic determination of the base inertial parameters. Mech Mach Theory 49:284–297
6. Valero F, Iriarte X, Mata V, Ros J (2013) Identification of dynamic parameters in low-mobility mechanical systems: application to short long arm vehicle suspension, vehicle system dynamics. Int J Veh Mech Mobility
7. Venture G, Ripert P, Khalil W, Gautier M, Bodson P (2006) Modeling and Identification of Passenger Car Dynamics Using Robotics Formalism. IEE Trans Intell Transp Syst 7 (3):349–359

An Alternative Method for the Optimum Dynamic Balancing of the Four-Bar Mechanism

Mario Acevedo, Eduardo Haro and Félix Martínez

Abstract This article presents the optimum dynamic balancing of the four-bar mechanism, in particular the crank-rocker, by the addition of counterweights. This is done by imposing as little restrictive as possible constraints on the counterweights parameters. First the general analytical equations of motion of the crank-rocker four-bar mechanism are obtained, using natural coordinates. This model allows expressing the dynamic equations of the mechanism just in terms of the mass, as opposed to the need of using also the moment of inertia, and the coordinates of the center of gravity of the counterweights, that are used as optimization variables. This implies that no particular counterweight shape is assumed in advance. The only constraints imposed on these optimization variables are that masses must be non-negative. As a novelty, the most influencing variables in the optimization are identified using a global sensitivity analysis based on polynomial chaos. This allows to impose different constraints an also to reduce the total number of optimization variables without affecting the global results. The results obtained are validated by simulations, and compared to those expressed in representative papers obtained by other authors.

Keywords Dynamic balancing · Optimization · Four-bar · Natural coordinates

1 Introduction

Dynamic balancing of mechanisms consists on reducing (or in the best case eliminating) the dynamic reactions resulting from its motion. The reactions generally considered are the shaking force and the shaking moment transmitted to the supporting frame, as well as the input or driving torque required to drive the

M. Acevedo (✉) · E. Haro · F. Martínez
Faculty of Engineering, Universidad Panamericana, Augusto Rodin 498, 03920 Mexico, Mexico
e-mail: macevedo@up.edu.mx

M. Ceccarelli and E.E. Hernández Martinez (eds.), *Multibody Mechatronic Systems*, Mechanisms and Machine Science 25, DOI 10.1007/978-3-319-09858-6_17

mechanism with constant drive speed. The general motivations for dynamic balancing are to reduce noise, wear and fatigue caused by vibrations on the supporting frame, and to reduce flywheel or actuator torques requited to obtain the constant drive speed. Two general approaches are usually followed for dynamic balancing: springs addition, or counterweights and counter-rotary inertias addition. As an introduction to the topic the reader is referred to the representative papers like [4], where the so called linear independent vectors method is introduced, [6] where complete force and moment balancing of the in-line four-bar mechanism is presented, and [21] where the conditions for reducing the angular momentum fluctuations allowing for off-line mass distributions. For a complete reference of different methods the reader is referred to the excellent reviews done in [1–3, 17].

This work focuses on the counterweight addition for the optimum dynamic balancing of the planar four-bar mechanisms. This problem involves a trade-off between minimizing the mentioned dynamic reactions. A common way of handling these trade-offs is to incorporate them into an optimization problem, which can cope with more complicated mechanisms and allow a more flexible problem formulation. Different authors have solved this problem following different approaches. In [5] a least-square theory for the optimization of the shaking moment of fully force-balanced inline four-bar linkages is presented. Partial force balancing of four-bar linkages using the Lagrange Multiplier method is presented in [19]. In [12] improved methods for optimizing the root-mean-square shaking moment and driving torque of a fully forced-balanced four-bar mechanism is presented, while in [14] analytical mechanics and heuristic optimization techniques are used. Later this work is refined in [18] by a method by combining the use of Newton's equations and the principle of virtual work with the HOT code proposed in [15]. More recently in [8] the optimum balancing of four-bar linkages is solved based on the maximum recursiveness of the dynamic equations and in [7] the complete shaking force and shaking moment balancing of planar in-line four-bar linkages with constant input speed is achieved by defining the coupler link as a physical pendulum and by adding a class-two Assur group with prescribed geometrical and mass parameters. In [11] optimal balancing of a four-bar mechanism is formulated and a genetic algorithm is used to solve the optimization problem.

In this work the general analytical equations of motion of the crank-rocker four-bar mechanism are obtained, using natural coordinates, [9]. This model allows expressing the dynamic equations of the mechanism just in terms of the mass, as opposed to the need of using also the moment of inertia, and the coordinates of the center of gravity of the *ith* counterweight. All, the mass and the coordinates of the center of gravity of each counterweight, are used as optimization variables. This implies that no particular counterweight shape is assumed in advance. So the only constraints imposed on these optimization variables are that masses must be non-negative, without worrying about how these values correspond to a proper inertia moments.

The results obtained are validated by simulations and some are compared to those presented in the complete and detailed work in [10].

2 Modeling in Natural Coordinates, Inverse Dynamics

Figure 1a shows the kinematic scheme of a planar crank-rocker four-bar. All links are assumed to be rigid. Link 1, the crank *AB*, is assumed to rotate at constant speed ω rad/s, which determines the mechanism's period of motion $T = 2\pi\omega$ rad/s. Link 2, the coupler *BC*, connects the crank with link 3, the rocker *CD*. The latter performs an oscillating rocking motion. Link 4 is the fixed ground *AD*; ϕ_4 rad is a constant angle, which is of no influence if gravity is neglected. The classical assumption in the area of high-speed linkages can be made here. Finally the link lengths are denoted as $a_i\,[m]$, $i = \{1,2,3,4\}$.

Figure 1b shows the corresponding free-body diagram, based on which the dynamic reactions are defined. F^x_{ij} and F^y_{ij}, $[N], i = \{1,2,3,4\}, j = \{1,2,3,4\}$, denote the x and y components of the joint forces of the body i over the body j, while T^{drv}_1 N m is the driving torque applied to body 1, the crank. The x and y components of the shaking force acting on the supporting frame are equal, respectively to:

$$F^{\text{shak}}_x = \left(F^x_{14} + F^x_{34}\right);\; F^{\text{shak}}_y = \left(F^y_{14} + F^y_{34}\right) \tag{1}$$

The magnitude F_{shak} N of the shaking force hence is equal to:

$$F^{\text{shak}} = \sqrt{\left(F^x_{14} + F^x_{34}\right)^2 + \left(F^y_{14} + F^y_{34}\right)^2} \tag{2}$$

On the other hand the shaking moment can be calculated with respect to a point in the joint A as:

$$M_{\text{shak}} = -\left[T^{\text{drv}}_1 + a_4(\cos\phi_4)F^y_{34} - a_4(\sin\phi_4)F^x_{34}\right] \tag{3}$$

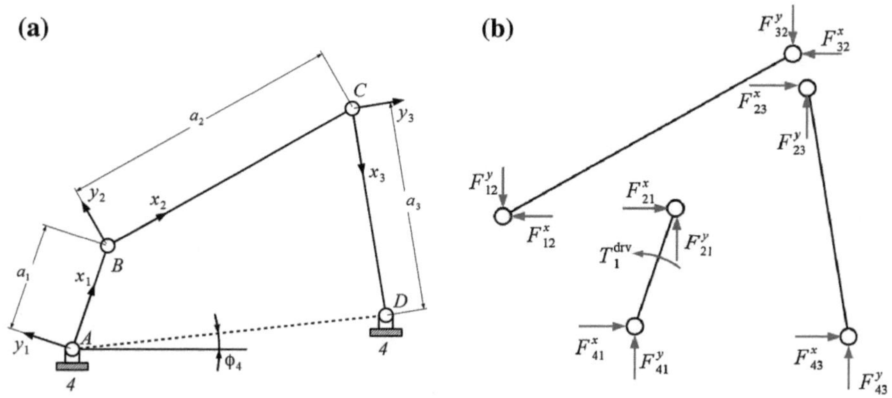

Fig. 1 Modeling of a generic crank-rocker four-bar mechanism. **a** Kinematic scheme of a planar crank-rocker four-bar mechanism: link 1 (crank) = *AB*; link 2 (coupler) = *BC*; link 3 (rocker) = *CD*. **b** Corresponding free-body diagram

All reactions can be calculated solving the inverse dynamics problem for the mechanism. In this case a global formulation based on natural coordinates is used. This has the advantage of using just the mass of each element in the equations of motion, inertia moments are taken into account in an indirect way. This implies that no particular counterweight shape is assumed in advance.

2.1 Kinematics

From the kinematics point of view the mechanisms has been defined with four points $P_j = (x_j, y_j)$, $j = \{A, B, C, D\}$, leading to a vector of eight natural coordinates:

$$\mathbf{q} = \begin{bmatrix} x_A & y_A & x_B & y_B & x_C & y_C & x_D & y_D \end{bmatrix}^{\mathrm{T}} \tag{4}$$

where $[\cdot]^{\mathrm{T}}$ denotes a matrix transpose. Vector $\mathbf{q}$ constitute a set of dependent coordinates related by a vector of nonlinear equations, $\boldsymbol{\Phi}(\mathbf{q}) = \mathbf{0}$, as:

$$\boldsymbol{\Phi}(\mathbf{q}) = \begin{bmatrix} x_A \\ y_A \\ x_D \\ y_D \\ (x_B - x_A)^2 + (y_B - y_A)^2 - a_1^2 \\ (x_C - x_B)^2 + (y_C - y_B)^2 - a_2^2 \\ (x_C - x_D)^2 + (y_C - y_D)^2 - a_3^2 \end{bmatrix} = [\mathbf{0}] \tag{5}$$

As defined, Eq. (5) depend only on the positions vector $\mathbf{q}$, but in general can also depend explicitly on time which for now is not the case. These equations are used to solve the positions problem (initial position and finite displacements) by the well known Newton-Raphson numerical method.

Equation (5) can be derived with respect to time to solve the velocities problem as:

$$\boldsymbol{\Phi}_{\mathbf{q}}\dot{\mathbf{q}} = \mathbf{0} \tag{6}$$

where $\boldsymbol{\Phi}_{\mathbf{q}} \equiv \partial\boldsymbol{\Phi}/\partial\mathbf{q}$ is the Jacobian matrix.

Equation (6) can in turn be derived with respect to time to solve the accelerations problem as:

$$\begin{aligned} \boldsymbol{\Phi}_{\mathbf{q}}\,\ddot{\mathbf{q}} + \dot{\boldsymbol{\Phi}}_{\mathbf{q}}\,\dot{\mathbf{q}} &= \mathbf{0} \\ \boldsymbol{\Phi}_{\mathbf{q}}\,\ddot{\mathbf{q}} &= -\,\dot{\boldsymbol{\Phi}}_{\mathbf{q}}\,\dot{\mathbf{q}} \end{aligned} \tag{7}$$

where $\dot{\mathbf{\Phi}}_{\mathbf{q}} \equiv d\mathbf{\Phi}_{\mathbf{q}}/dt$. Positions, velocities, and acceleration equations are solved for the mechanisms imposing a constant speed ω rad/s to the crank. These results from kinematics are used to solve the inverse dynamics problem.

3 Inverse Dynamics

To solve the inverse dynamics problem it is necessary to obtain the dynamic equations of motion of the system. In this case the equations of motion are obtained by the application of the virtual power principle, see [9], so the complete set of force equilibrium equations for the system can be written as:

$$\mathbf{M}\,\ddot{\mathbf{q}} + \mathbf{\Phi}_{\mathbf{q}}^{T}\boldsymbol{\lambda} = \mathbf{f} \tag{8}$$

where $\mathbf{M}$ is the system's mass matrix, $\mathbf{f}$ is the vector that includes the external forces and the velocity-dependent inertia forces (centrifugal and Coriolis), and $\boldsymbol{\lambda}$ is the corresponding vector of Lagrange multipliers as the system is expressed in a set of dependent coordinates. In Eq. (8) the first term, $\mathbf{M}\,\ddot{\mathbf{q}}$, represents the inertia forces, and $\mathbf{\Phi}_{\mathbf{q}}^{T}\,\boldsymbol{\lambda}$ is associated to the forces produced by the constraints. Then some terms of the vector of Lagrange multipliers are directly related to the forces at the kinematic joints, $\mathbf{f}_c$, so it can be said that $\boldsymbol{\lambda} \equiv \mathbf{f}_c$. In this way the equations can be organized as:

$$\mathbf{M}\,\ddot{\mathbf{q}} + \begin{bmatrix} \mathbf{\Phi}_{\mathbf{q}}^{T} & \mathbf{Q}_m \end{bmatrix} \begin{bmatrix} \mathbf{f}_c \\ \mathbf{f}_m \end{bmatrix} = \mathbf{f}_e \tag{9}$$

where $\mathbf{f}_m$ is the vector of driving forces (forces and/or torques), $\mathbf{f}_e$ is the vector of external forces (excluding driving forces) and velocity dependent inertia forces, and $\mathbf{Q}_m$ is a matrix that contains the terms associated (terms multiplying) to the driving forces. This matrix in general contains 1's and 0's when using Natural Coordinates.

Then considering that a constant angular velocity is imposed to the crank in the four-bar mechanism, the kinematics can be completely solved. So the Eq. (9) can finally be expressed in terms of the mass $m_{c,i}$ of the counterweights added to each i_{th}, $i = \{1, 2, 3\}$ link and its center of mass location $(x_{c,i}, y_{c,i})$. The mass, center of mass location, and inertia moments of all the original links in the mechanism are completely known.

4 Optimization Process

To reduce the dynamic reactions a counterweight is added to each moving link. The mass parameters of the counterweights constitute the optimization variables that can be grouped in a vector as:

$$\mathbf{p} = \begin{bmatrix} m_{c,1} & x_{c,1} & y_{c,1} & m_{c,2} & x_{c,2} & y_{c,2} & m_{c,3} & x_{c,3} & y_{c,3} \end{bmatrix}^{\mathrm{T}} \tag{10}$$

where $m_{c,i}$ kg is the mass of the i_{th}, $i = \{1,2,3\}$ counterweight. The center of mass coordinates $x_{c,i}$ m and $y_{c,i}$ m of the i_{th} counterweight are defined with respect to the corresponding link coordinate system (x_i, y_i), see Fig. 1a. Note that vector $\mathbf{p}$ has nine elements instead of twelve, the moments of inertia has been excluded by the use of natural coordinates. So in reference to the definition made in [10] for the ultimate limits for counterweight balancing, understood as that the constraints imposed on the counterweight parameters are as little restrictive as possible, the formulation proposed in this paper is even less restrictive than the one proposed in [10].

4.1 Objective Function

To define the objective function a dimensionless balancing effect index β_{shM} for the shaking moment is used. This kind of indexes have been used previously in [16, 19] and more recently in [10]. In this case this index is defined as the root mean square (rms, over one period T) of the optimized shaking moment with respect to the root mean square of the original shaking moment, indicated with a superscript $(\cdot)^{o}$:

$$\beta_{\mathrm{shM}} = \frac{\mathrm{rms}(M_{\mathrm{shak}})}{\mathrm{rms}\left(M^{o}_{\mathrm{shak}}\right)} \tag{11}$$

So the objective is to minimize β_{shM} subject to:

$$\beta_{\mathrm{shF}} \leq \beta^{\mathrm{max}}_{\mathrm{shF}}; \ \beta_{\mathrm{drvT}} \leq \beta^{\mathrm{max}}_{\mathrm{drvT}} \tag{12}$$

where β_{shF} and β_{drvT} are the corresponding dimensionless balancing effects for the shaking force and for the driving torque, while $\beta^{\mathrm{max}}_{\mathrm{shF}}$ and $\beta^{\mathrm{max}}_{\mathrm{drvT}}$ are the maximum allowed values. As pointed out in [10], the advantage of this approach is that the rms shaking moment is minimized while the designer directly controls the maximum allowed increase or the minimum wanted reduction of the rms shaking force and rms driving torque.

Apart from the constraints (12) additional constraints for the counterweights' mass parameters are required. Masses must be positive and bound constraints are necessary for the coordinates of the center of mass in order to get logical dimensions for the counterweights. The total mass added to the mechanisms by the counterweights is also constrained to a maximum $m^{\mathrm{max}}_{\mathrm{tot}}$. These constraints are expresses as ($i = \{1,2,3\}$):

$$m_{c,i} \geq 0 \tag{13}$$

$$x^{\mathrm{min}}_{c,i} \leq x_{c,i} \leq x^{\mathrm{max}}_{c,i} \tag{14}$$

$$y_{c,i}^{\min} \le y_{c,i} \le y_{c,i}^{\max} \quad (15)$$

$$m_c \le m_{\text{tot}}^{\max} \quad (16)$$

where $m_c = m_{c,1} + m_{c,2} + m_{c,3}$ is the sum of the masses of the counterweights. Then an additional useful index is defined as the total mass added by the counterweights in relation to the total original mass:

$$m_c^{\%} = \frac{m_c}{m_{\text{tot}}^o} \quad (17)$$

where m_{tot}^o is the total original mass of the linkage.

4.2 Sensitivity Analysis of the Optimization Parameters

In order to count on an additional source to understand the optimization process and lead the criteria for the optimization, a global sensitivity analysis based on polynomial chaos has been applied, see [13]. This is done, to the best of the authors' knowledge, by first time in this kind of systems.

The results from the application of this analysis are presented in Fig. 2. As can be seen, a counterweight added to the input link (the crank) has great influence on the shaking force while a counterweight at the coupler has almost null influence, besides a counterweight at the rocker has little influence. It is known by [4] that a general four-bar mechanism can be force balanced just by the addition of two counterweights. In [4] the counterweights are added to the crank and to the rocker. In spite of the results obtained by the sensitivity analysis done in this work, it is clear that this selection is the best. On the other hand the shaking moment and the

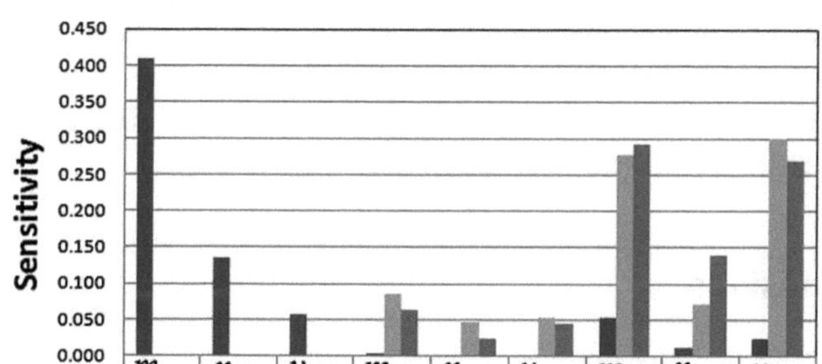

	$m_{c,1}$	$x_{c,1}$	$y_{c,1}$	$m_{c,2}$	$x_{c,2}$	$y_{c,2}$	$m_{c,3}$	$x_{c,3}$	$y_{c,3}$
Shaking Force	0.409	0.135	0.056	0.003	0.001	0.001	0.053	0.011	0.024
Shaking Moment	0.000	0.000	0.000	0.087	0.047	0.053	0.278	0.071	0.300
Driving Torque	0.000	0.000	0.000	0.063	0.024	0.046	0.291	0.139	0.270

Fig. 2 Results from the sensitivity analysis

Table 1 Crank-rocker four-bar mechanism used in the example

Parameters [units]	$i = 1$	$i = 2$	$i = 3$
a_i [m]	0.0508	0.1524	0.0762
m_i [kg]	0.0894	0.2394	0.1215
x_i [m]	0.0254	0.0762	0.0381
y_i [m]	0.0	0.0102	0.0
J_i [kg m^2]	0.0000198	0.0006792	0.0002198
a_4 [m]	0.1397	–	
ϕ_4 [°]	0.0		

driving torque can be greatly affected by a counterweight at the rocker, this information can be used to lead the optimization by controlling the increase or reduction of the driving torque.

5 Numerical Example

The method has been applied to a planar crank-rocker four-bar mechanism previously studied in [10, 14, 20], its parameters are defined in Table 1.

In the complete and detailed work presented in [10], the dynamic balancing of this mechanism is first approximated using punctual masses as counterweights. With the formulation used in this work it has been easy to replicate this approximation and obtain similar results which are presented in detail in Fig. 3.

In addition the dynamic balancing of the mechanisms has been defined as a design optimization problem where the dimensions (radius $R_{c,i}$, and thickness $t_{c,i}$, $i = \{1, 3\}$) of only two cylindrical counterweights, one attached to the crank and one attached to the rocker, are found. Just as considered for the calculation of the previous results, it was considered that $\beta_{\text{shF}} = 0.66$, $\beta_{\text{drvT}} = 1.2$, and $m_{c,1} + m_{c,3} = 0.8 \cdot m^o_{\text{tot}}$ (in this particular case $m^o_{\text{tot}} = 0.4503$ kg). The results are presented in Table 2 and compared to the those obtained in [10]. Simulation results validated with ADAMS are presented in Fig. 4.

6 Conclusions

An alternative method for the optimum dynamic balancing of the crank-rocker four-bar mechanism has been presented. From the ultimate limits for counterweight balancing, this method imposes the least restrictive parameters. Global or near global optimum has been obtained without reformulating the problem as a convex program. The application of this method to complex mechanism in plane remains open, as well as to mechanisms in space.

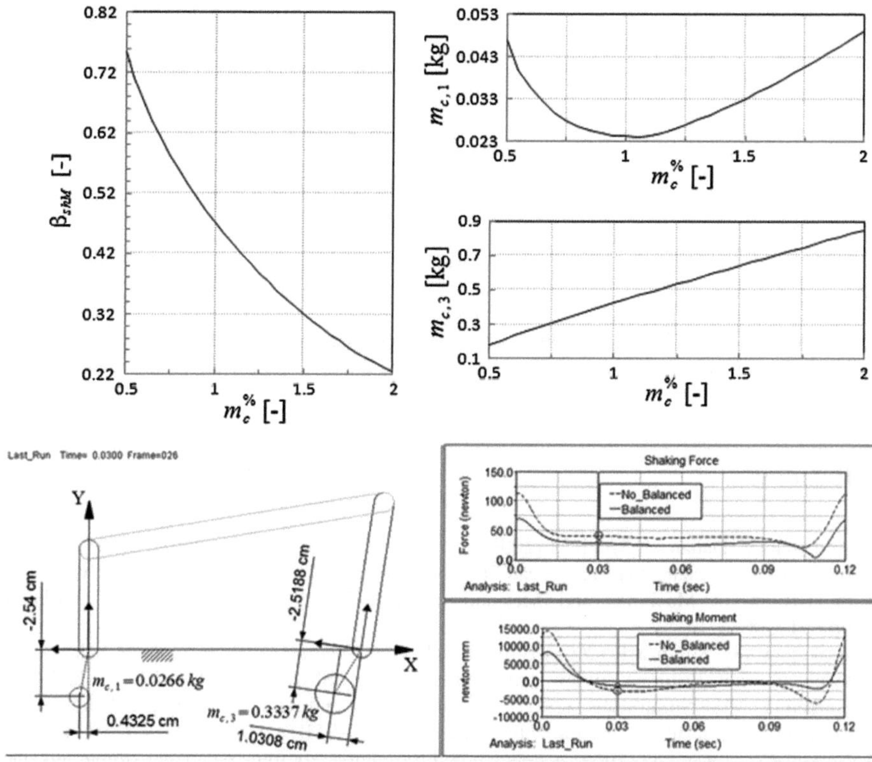

Fig. 3 Dynamic balancing of the four-bar mechanism of the example. At the *top*, three graphs present the variation of β_{shM}, $m_{c,1}$, and $m_{c,3}$, maintaining $\beta_{\text{shF}} = 0.66$, $\beta_{\text{drvT}} = 1.2$, while $m_c^{\%} = \{0.5\ldots2\}$. At the *bottom* simulation results validated with ADAMS are presented, masses location is indicated

Table 2 Results obtained in the design optimization problem and compared to those obtained in [10]

Results	$\frac{m_{c,1}+m_{c,3}}{m^o_{\text{tot}}}$	β_{shF}	β_{drvT}	β_{shM}	$R_{c,1}$ [mm]	$t_{c,1}$ [mm]	$R_{c,3}$ [mm]	$t_{c,3}$ [mm]
Results in [10]	0.8	0.66	1.2	0.650	26.45	2.83	26.37	18.20
This formulation	0.8	0.66	1.2	0.690	27.58	3.94	23.98	20.23

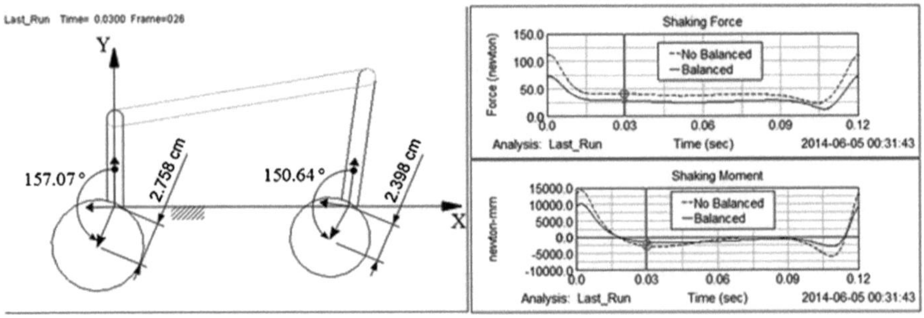

Fig. 4 Dynamic balancing of the four-bar mechanism of the example using two disks, one at the crank ($t_{c,1} = 3.94$ mm) and one at the rocker ($t_{c,3} = 20.23$ mm). Both disks made of steel ($\rho = 7833$ kg/m^3)

References

1. Arakelian V, Dahan M, Smith M (2000) A historical review of the evolution of the theory on balancing of mechanisms. In: Ceccarelli M (ed) Proceedings of the International Symposium on History of Machines and Mechanisms HMM2000. Kluwer Academic, Dordrecht, pp 291–300
2. Arakelian VH, Smith MR (2005) Erratum—shaking force and shaking moment balancing of mechanisms: A historical review with new examples. ASME J Mech Des 127(2):1034–1035
3. Arakelian VH, Smith MR (2005) Shaking force and shaking moment balancing of mechanisms: a historical review with new examples. ASME J Mech Des 127(2):334–339
4. Berkof RS, Lowen GG (1969) A new method for completely force balancing simple linkages. ASME J Manuf Sci Eng 91(1):21–26
5. Berkof RS, Lowen GG (1971) Theory of shaking moment optimization of force-balanced four-bar linkages. ASME J Manuf Sci Eng 93(1):53–60
6. Berkof RS (1973) Complete force and moment balancing inline four-bar linkages. Mech Mach Theory 8(3):397–410
7. Briot S, Arakelian V (2012) Complete shaking force and shaking moment balancing of in-line four-bar linkages by adding a class-two rrr or rrp assur group. Mech Mach Theory 57:13–26
8. Chaudhary H, Saha SK (2007) Balancing of four-bar linkages using maximum recursive dynamic algorithm. Mech Mach Theory 42(2):216–232
9. de Jalón JG, Bayo E (1994) Kinematic and Dynamic Simulation of Multibody Systems: the Real-Time Challenge. Springer, New York
10. Demeulenaere B, Aertbeliën E, Verschuure M, Swevers J, de Schutter J (2006) Ultimate limits for counterweight balancing of crank-rocker four-bar linkages. ASME J Mech Des 128 (6):1272–1284
11. Erkaya S (2013) Investigation of balancing problem for a planar mechanism using genetic algorithm. J Mech Sci Tech 27(7):2153–2160
12. Haines RS (1981) Minimum rms shaking moment or driving torque of a force-balanced 4-bar linkage using feasible counterweights. Mech Mach Theory 16(3):185–195
13. Haro Eduardo, Anstett-Collin Floriane, Basset Michel (2012) Sensitivity study of dynamic systems using polynomial chaos. Reliab Eng Syst Saf 104:15–26
14. Lee TW, Cheng C (1984) Optimum balancing of combined shaking force, shaking moment, and torque fluctuations in high-speed linkages. ASME J Mech Des 106(2):242–251
15. Lee TW, Freudenstein F (1976) Heuristic combinatorial optimization in the kinematic design of mechanisms, part 1: theory. ASME J Manuf Sci Eng 98(4):1277–1280

16. Lowen GG, Tepper FR, Berkof RS (1974) The quantitative influence of complete force balancing on the forces and moments of certain families of four-bar linkages. Mech Mach Theory 9:299–323
17. Lowen GG, Tepper FR, Berkof RS (1983) Balancing of linkages—an update. Mech Mach Theory 18:213–220
18. Qi NM, Pennestri E (1991) Optimum balancing of four-bar linkages. Mech Mach Theory 26 (3):337–348
19. Tepper FR, Lowen GG (1975) Shaking force optimization of four-bar linkage with adjustable constraints on ground bearing forces. ASME J Manuf Sci Eng 97(2):643–651
20. Tricamo SJ, Lowen GG (1983) A novel method for prescribing the maximum shaking force of a four-bar linkage with flexibility in counterweight design. ASME J Mech Des 105 (3):511–519
21. Wiederrich JL, Roth B (1976) Momentum balancing of four-bar linkages. ASME J Manuf Sci Eng 98(4):1289–1295

Behavior of Some Objects in Series with Dynamic Eliminators of Vibrations

Tadeusz Majewski

Abstract The paper presents dynamic eliminators of vibration which are located on some objects connected in series. It is shown that the vibrators are able to organize themselves in such a way that they may compensate the excitation. Depending on the magnitude of the excitation the vibration can be eliminated by the vibrators of the first object or the excitation enters deeper into the system and is compensated by the vibrators of the next objects. These phenomena can be observed for some parameters of the system.

Keywords Vibrations · Elimination of vibration · Self-organizing system

1 Introduction

In general, any excitation generates vibrations which are dangerous for mechanical systems and bothersome the personal who are close to the machine. Sometime it is not possible to eliminate excitation or decrease it. There exists a method of elimination or decreasing vibrations. The system can be modified in such a way to decrease the amplitude of vibrations. The suspension of the system may be made in such a way that at the point of work the force given by the suspension is almost a constant [1]. The dynamic absorber of vibration of Frahm may be used for some systems. A system is modified in such a way that an additional object is attached to the principle one whose vibration should be eliminated. The natural frequency of this subsystem should be equal to the frequency of excitation. The method is efficient for one frequency of excitation. To eliminate torsional vibration of shafts some pendulums are attached to the shaft. The centrifugal force acting on the pendulum results in its natural frequency increases with the spin velocity of the shaft. So the centrifugal pendulums may compensate for the torsional vibration of a

T. Majewski (✉)
Universidad de las Americas-Puebla, 72810 Puebla, Mexico
e-mail: tadeusz.majewski@udlap.mx

M. Ceccarelli and E.E. Hernández Martinez (eds.), *Multibody Mechatronic Systems*,
Mechanisms and Machine Science 25, DOI 10.1007/978-3-319-09858-6_18

shaft in greater range of spin velocity [2–4]. Vibratory equipment such as a pneumatic hammer should be equipped in a system for decreasing the transmission of vibration to an operator [5]. Introducing a damping may also help in decreasing of vibrations [6].

It can be observed in nature that some systems are able to self-organize in such a way as to adapt themselves to changing surroundings [7]. A man on a vibrating plane (factory, car, bus, train) changes the stiffness of his body or changes the position of his legs to decrease a transmission of harmful vibrations from the floor to his body. Some aquatic birds utilize the motion of the air caused by the oscillation of the ocean and they fly using less energy.

The author of this paper proposed a dynamic eliminator of vibration (DEV). The object in which vibration should be eliminated has a rotating drum with free elements inside it. If the vibrations of the object appear then the free elements in the drum reorganize themselves in such a way to generate a force opposite to the excitation of the object. The principle of the method was explained in paper [8]. The method was also applied to an object with more degrees of freedom [9, 10].

In this paper the author tries to apply the DEV to some objects which are arranged in series and they are connected by the viscous-elastic elements. The analysis should explain whether the method is effective or not and what parameters should have the DEV to eliminate vibrations.

2 Principle of the Method

Some objects are connected in a series by springs and dampers 3 as shown in Fig. 1. Each object has two pendulums rotating with the initial angular velocity ω. The pendulums can change their position in time depending on the inertial forces acting on them, i.e. they may change the position with respect to the reference frame C_ixy which rotates with a constant velocity ω. The position of the pendulums for the object *i*th are defined by the angles α_1^i, α_2^i—Fig. 2. It was assumed that all objects have the same mass *M*, all pendulums are the same with a mass *m* and at the distance *R* form the axis of rotation, the properties of all viscos-elastic elements are also the same; *k, c* are the coefficients of stiffness and damping.

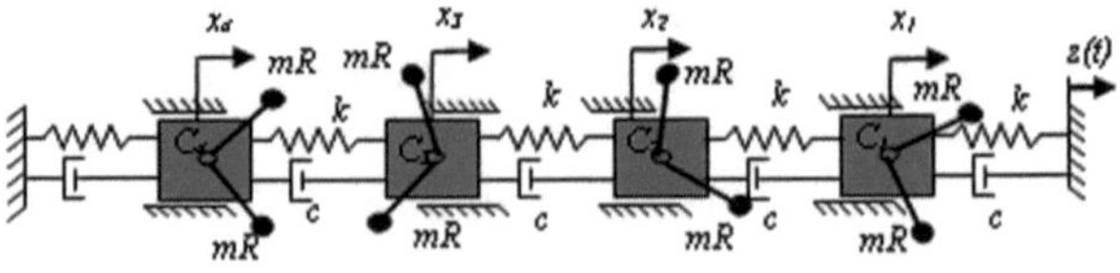

Fig. 1 Sistem with DEV

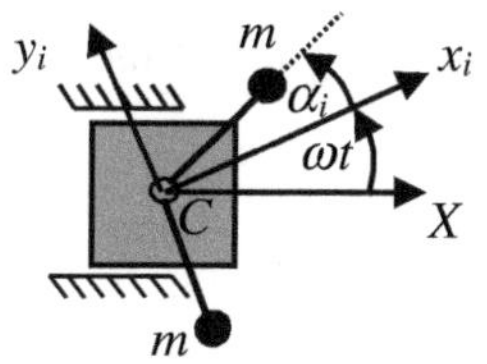

Fig. 2 Reference frame and position of the vibrator for the *i*th object

3 Mathematical Model

The differential equations defining the behavior of the system with kinematic excitation $z = z_o\cos(\Omega t)$ are obtained from the Lagrange equation. The motion of the *i*th object is governed by Eq. 1.

$$
\begin{aligned}
& M\underset{i}{\ddot{x}} + 2kx_i + 2c\dot{x}_i - kx_{i-1} - c\dot{x}_i - kx_{i+1} - c\dot{x}_{i+1} \\
& \quad = kz_i(t) + c_i\dot{z}_i(t) + m\sum_{j=1}^{2}\left[(\omega + \dot{\alpha}_j^i)^2\cos(\omega t + \alpha_j^i) + \ddot{\alpha}_j^i\sin(\omega t + \alpha_j^i)\right] \\
& i = 1, \ldots, N,
\end{aligned}
\tag{1}
$$

where N is the number of the objects, mR—is the static moment of the pendulum and $x_{N+1} = x_{-1} = 0$.

The motion of the first pendulum on the *i*th object is determine by

$$
mR\ddot{\alpha}_1^i = m\underset{i}{\ddot{x}}\sin(\omega t + \alpha_1^i) - c_\alpha mR\dot{\alpha}_1^i, \tag{2}
$$

where c_α is the coefficient of viscous damping for the pendulum. It was assumed that the viscous damping of the pendulum is proportional to its mass and relative velocity with respect to the reference frame *Cxy*. The differential equation for the second pendulum is similar to the first one with α_2^i.

It was taken for granted that kinematic excitation exists only for the first object. The right side of the Eq. (1) presents the excitation on *i*th object from the displacement of the wall and the inertial forces of two pendulums.

If the pendulums of the first object really are able to compensate the excitation then they should occupy a stable position α_{1f}, α_{2f} for which the resultant force

$$
\begin{aligned}
Q_1(t) &= z_o\sqrt{k^2 + (c\omega)^2}\cos(\omega t) - c\omega z_o\sin\omega t \\
&\quad + mR\omega^2\sum_{j=1}^{2}(\cos(\omega t + \alpha_{1f}) + \cos(\omega t + \alpha_{2f})) \equiv 0
\end{aligned}
\tag{3}
$$

It will be true if the angles α_{1f}, α_{2f} fulfill the following equations

$$z_o\sqrt{k^2+(c\omega)^2}+mR\omega^2\sum_{j=1}^{2}\cos\alpha_{jf}=0,\qquad \sum_{j=1}^{2}\sin\alpha_{jf}=0. \tag{4}$$

The first condition can be true only for $z_o\sqrt{k^2+(c\omega)^2}<2mR\omega^2$ and the second one shows that the positions of the pendulums are symmetrical with respect to the excitation given by the spring and damper.

4 Behavior of the System

The parameters of the system are: M = 2 kg; k = 2,500 N/m; c = 10 kg/s; m = 0.015 kg; R = 0.03 m. The diagrams of the objects vibration and the pendulums positions are given in a new time $\tau = \omega t$.

When the objects vibrate then there are inertial forces acting on the pendulums that force them to move to a new position. If the vibrations vanish vibratory forces also disappear. The behavior of the system for $kz_o<2mR\omega^2$ is presented in Fig. 3. The final positions of the pendulums for the first object are: 102°, −140° (the angle between the pendulums is ~120°) and the amplitude of this object is ~7 μm after the time 300/90 1/s = 3 s. As the excited vibration of the first object is eliminated so the pendulums of the second object are almost on one diameter with the positions 89.7°, −90.5°, there are very small free vibrations of the second, third and fourth objects and they vanish with time.

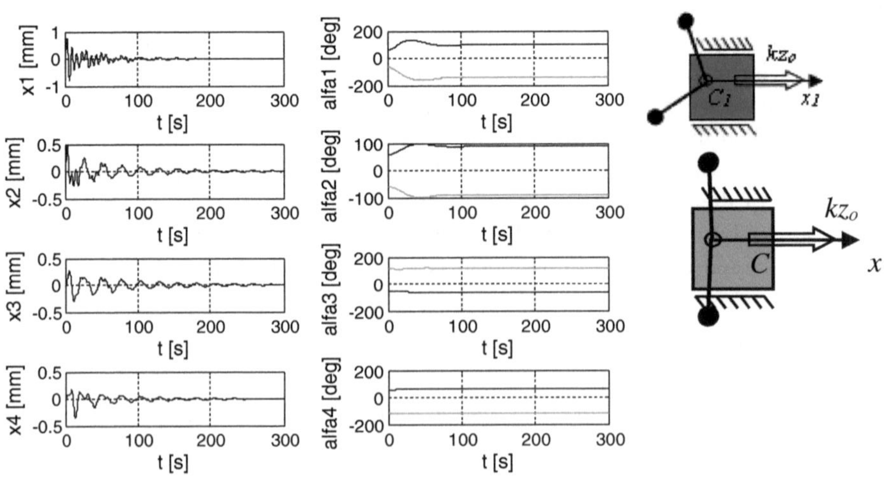

Fig. 3 Behavior of the system for the excitation z_o = 1.46 mm ($kz_o = mR\omega^2$) and ω = 90 rad/s

From this example one can come to the conclusion that the system is able to organize itself in such a way as to eliminate vibration. It should be explained that this is the reason for such behavior and when the pendulums compensate for the excitation.

The excitation for the *i*th object can be taken as

$$Q^i(t) \cong a_{i-1}\sqrt{k^2 + (c\omega)^2}\cos(\omega t) + \sum_{j=1}^{2}(\omega + \dot{\alpha}_j^i)\cos(\omega t + \alpha_j^i). \tag{5}$$

The pendulums change their position very slowly and the vibration of the *i*th object can be approximated as

$$x_i(t) \cong a_{io}\cos(\omega t - \phi) + \sum_{j=1}^{2} a_j^i\cos(\omega t + \alpha_j - \phi). \tag{6}$$

The principle force acting on the pendulum is shown in Fig. 4.

The inertia force $m\ddot{x}_i$ from the object's vibration gives a component F perpendicular to the pendulum and forces it to change position (Fig. 5).

$$\widehat{F}_j^i = m\ddot{x}_i \sin(\omega t + \alpha_j^i), \quad i = 1, \ldots N, j = 1, 2. \tag{7}$$

This force depends on the object's vibration and the position of the pendulum.

From the previous author's papers it is known that the behavior of the pendulum depends on the average of F. The vibratory force for the *jth* pendulum for the *i*th object is defined below as

$$F_j^i = -0.5m\omega^2\left[a_o^i\sin\left(\alpha_j^i + \phi\right) + \sum_{l=1}^{2} a_j \sin\left(\alpha_j^i - \alpha_l^i + \phi\right)\right] \tag{8}$$

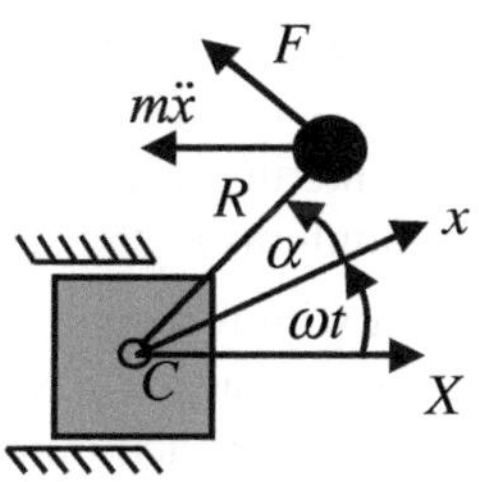

Fig. 4 Forces acting on the pendulum

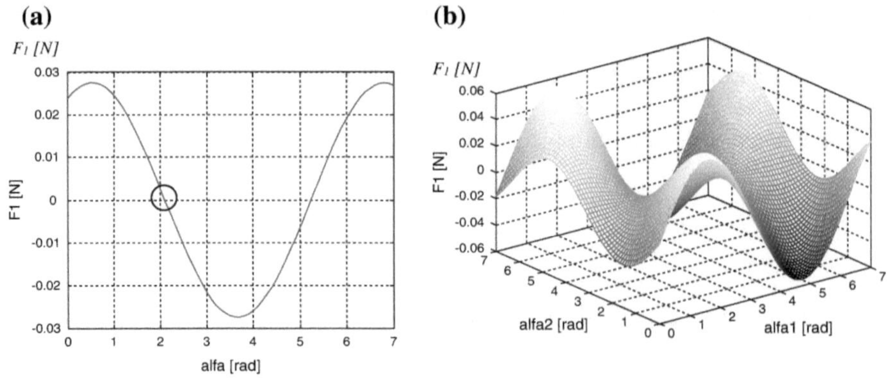

Fig. 5 Vibratory force **a** $F_1^1(\alpha_1, \alpha_{2f} = 240°)$, **b** $F_1^1(\alpha_1, \alpha_2)$

where a_o^i is the amplitude of vibration from given excitation ($z(t)$), a_1^i, a_2^i are the amplitudes of the object generated by the first and second pendulum. The amplitude of vibration of ith object depends also on the vibration of the adjacent objects.

If the excitation is small ($kz_o < 2mR\omega^2$) and the pendulums compensate the excitation then their final position are defined by

$$\cos_{1f} = \cos_{2f} = -z_o\sqrt{k^2 + (c\omega)^2}/2mR\omega^2 \tag{9}$$

Figure 5a presents the change of the force $F_1^1(\alpha_1)$ if $2mR\omega^2 = kz_o$ and the second pendulum is at its final position $\alpha_{2f} = 240^o$, Fig. 5b shows a change of the force $F_1^1(\alpha_1, \alpha_2)$ as a function of the position of two pendulums. It can be observed that the vibratory force of the first pendulum at its final position is $F_1(\alpha_{1f} = 120^o) = 0$—it is a position of equilibrium of this pendulum. When two pendulums simultaneously change their position then two vibratory forces F_1 and F_2 are zero only when the pendulums are at the positions $\alpha_{1f} = -\alpha_{1f} = 120^o$.

When the excitation is small, then the pendulums of the first object were able to compensate the excitation—Fig. 3. Excitation bigger than $2mR\omega^2$ cannot be compensated by the pendulums of the first object so it vibrates and transmits the vibration to the next object. The pendulums of the second object should compensate this excitation. For the excitation $kz_o = 3mR\omega^2$ the pendulums of the first object compensate for two thirds of the entering excitation and the rest of the excitation should be compensated by the pendulums of the second object. The final positions of the pendulums of the first object with respect to excitation $\overrightarrow{kz_o}$ are 164°, −196°, the pendulums of the second object are at the position 84°, −80°, and the pendulums of the other objects compensate for each other.

The amplitude of the first object a_1 depends on the resultant force of this object, the ratio of excited and natural frequencies of the system and the damping.

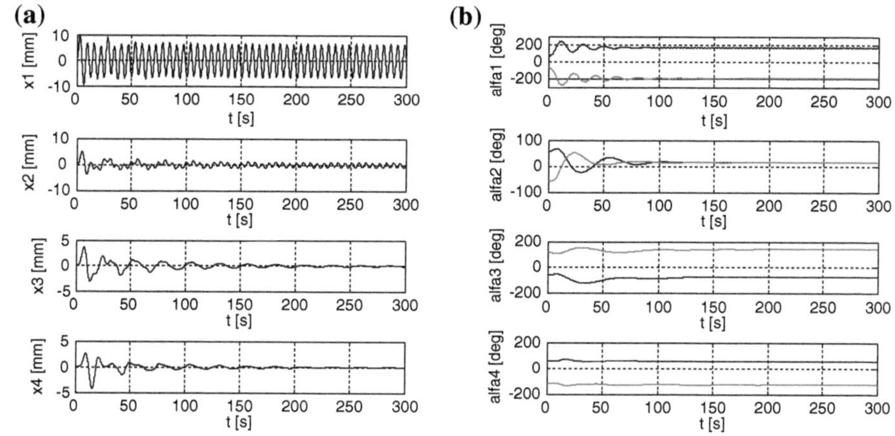

Fig. 6 Behavior of the system for the excitation $z_o = 30$ mm, $\omega = 90$ rad/s

$$a_1 = \frac{z_o\sqrt{k^2 + (c\omega)^2} - 2mR\omega^2}{\sqrt{(k^2 - M\omega^2)^2 + (0.5c\omega/M)^2}}. \tag{10}$$

The vibration of the first object is an excitation for the second one and the pendulums of the second object try to compensate its excitation.

The pendulums of the second object may compensate the excitation P_2 if

$$2mR\omega^2 > P_2 = a_1\sqrt{k^2 + (c\omega)^2}, \tag{11}$$

The equilibrium positions of the pendulums with respect to the excitation P_2 is defined as

$$\alpha_1^{(2)} = -\alpha_2^{(2)} = \arccos\left(-P_2/\left(2mR\omega^2\right)\right). \tag{12}$$

The vibratory forces acting on the pendulums of the second object are also determined in a similar way as (8) with excitation from the vibration of the first object.

$$F_1^{(2)} = -0.5m\omega^2\left[a_o^{(2)}\sin(\alpha_1^{(2)} + \phi) + \sum_{j=1}^{2} a_{1j}^{(2)}\sin(\alpha_1^{(2)} - \alpha_j^{(2)} + \phi)\right] \tag{13}$$

where $a_o^{(2)} = P_2\Big/\sqrt{(k - M\omega^2)^2 + (c\omega)^2}$,

$$a_1^{(2)} = a_2^{(2)} = mR\omega^2\Big/\sqrt{(k - M\omega^2)^2 + (c\omega)^2}$$

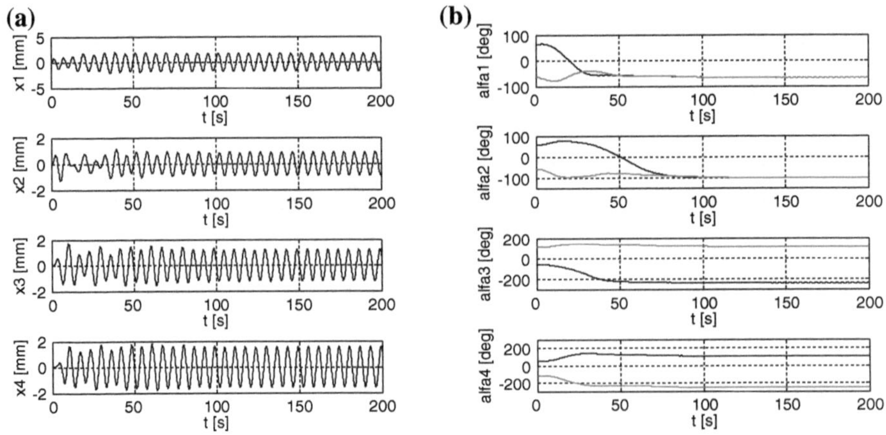

Fig. 7 Behavior of the system for the excitation z_o = 1 mm, ω = 35 rad/s

When the pendulums of second object are not able to eliminate its vibration then the second object is a source of vibration of the third object. Its pendulums try to compensate entering excitation—Fig. 6. The behavior of the third object is similar to the behavior of the first and second objects when the entering excitation kz_o is small.

This process repeats for the next objects of the system—the excitation can reach the third, fourth object and so on.

The natural frequencies of the system with four objects are $\omega_o = 21.8, 41.6, 57.2,$ and 76.3 rad/s. The behavior of the system presented in the above examples is for a spin velocity greater than the system's natural frequencies $(\omega > \omega_{omax})$ and for these velocities the pendulums do compensate the excitation. For lower spin velocities $(\omega < \omega_{omax})$ the pendulums do not compensate the excitation or even icrease the vibration of the system as it is demonstrated in Fig. 7.

5 Conclusions

It was shown that the pendulums of the objects which are connected in series are able to compensate the harmonic excitation which is applied at the borders of the system. The pendulums organize themselves in such a way as to obtain positions opposite to each object's excitation. Depending on the magnitude of the excitation it can be compensated at the first, second or the successive object. This property can be observed only for the frequencies of excitation greater than the largest natural frequency of the system. In this paper the frequency of the excitation is the same as the initial spin velocities of the pendulums. In the practice these frequencies can be different and also the excitation may be more complex than the harmonic one.

References

1. Cz. Cempel, Minimalization of Machine Vibration and Other Elements. Modern Problems of Dynamics. Ossolineum, Poland 1976
2. Curreri J, MacDuff J (1958) Vibration control. McGraw-Hill Book Company, New York, p 195
3. Korenev B, Reznikov LM (1993) Dynamic vibration absorbers. McGraw-Hill Book Company, New York
4. Den Hertog JP (2013) Mechanical vibrations. Courier Dover Publications, New York
5. Dobry MW (2001) Energy diagnostics and assessment of dynamics of mechanical and biomechatronics systems, machine dynamics problems, vol 25(3/4). Warsaw University of Technology, Warsaw, pp 35–54
6. Osinski Z, Bajkowski J (1998) Damping of vibrations. A. A. Balkema, Rotterdam, Brookfield
7. Blekhman II (2000) Vibrational mechanics. Nonlinear dynamic effects, general approach, applications. World Scientific, Singapore
8. Majewski T (1987) Synchronous vibration eliminator for an object having one degree of freedom. J Sound Vib 112(3):401–413
9. Majewski T (2000) Synchronous elimination of vibrations in the plane. Analysis of occurrence of synchronous movements. J Sound Vib 232(2):pp.553–570
10. Majewski T (2000) Synchronous elimination of vibrations in the plane. Method efficiency and its stability. J Sound Vib 232–2:571–584

Modeling, Analysis and Simulation of 3D Elastohydrodynamic Revolute Joints in Multibody Systems

P. Flores

Abstract In this work, a methodology for dynamic analysis of rigid-flexible multibody systems with elastohydrodynamic (EHD) lubricated joints is presented. The EHD lubricated cylindrical joint is formulated by the Natural Coordinate Formulation (NCF) and the twenty-node hexahedral element of Absolute Nodal Coordinate Formulation (ANCF), being the lubricant pressure determined through the resolution of the Reynolds' equation employing the finite difference method. The outcomes are validated with those obtained by using the commercial software ADINA. It is demonstrated that the bearing flexibility plays a significant role in the system responses, extends the lubricant distribution space and reduces the lubricant pressure.

Keywords Revolute joints · Elastohydrodynamic lubrication · Multibody dynamics

1 Introduction

A mechanical system usually consists of two major kinds of components, bodies and joints [6]. The bodies can be modeled as rigid or flexible elements, while the joints are represented by a set of kinematic constraints. The functionality of a mechanical joint relies upon the relative motion allowed between the connected components. This fact implies the existence of a clearance between the mating parts, and thus joint surfaces can contact each other or may be separated with a lubricant. It is of paramount importance to quantify the effects of both clearance joints and bodies flexibility on the global system response in order to define the minimum level of suitable tolerances that allow systems to achieve required performances.

P. Flores (✉)
University of Minho, Guimares, Portugal
e-mail: pflores@dem.uminho.pt

M. Ceccarelli and E.E. Hernández Martinez (eds.), *Multibody Mechatronic Systems*,
Mechanisms and Machine Science 25, DOI 10.1007/978-3-319-09858-6_19

In practice, lubricant is often utilized in mechanical joints to avoid the body-to-body (typically metal-to-metal) contact. This measure can reduce the level of impact and vibrations, and extends the joints lifetime. It is quite important to develop appropriate computational models that can account for the lubricant action in mechanical joints in the context of multibody system dynamics. For this purpose, two kinds of approaches can be found in the scientific domain of tribology, namely the hydrodynamic (HD) theory and the elastohydrodynamic (EHD) formulation. According to the HD theory, in the presence of dynamics of journal bearings, the hydrodynamic forces, which include both squeeze and wedge effects, generated by the lubricant fluid, oppose the journal motion. The hydrodynamic forces can be obtained by integrating the pressure distribution evaluated with the aid of Reynolds' equation established for the dynamic regime. Liu et al. [5] and Attia et al. [1] are among the very few authors who performed the EHD analysis for lubricated high-speed rotor-bearing systems by using the Fluid-Structure Interaction (FSI) analysis. These studies clearly demonstrated that the bearing deformations affect the pressure field in the clearance and increase the minimal film thickness. However, these works were performed only for isolated journal bearing systems.

2 Rigid-Flexible Multibody Formulation

In the present study, the flexible parts such as the flexible bearing and beams are modeled by using the finite elements of Absolute Nodal Coordinate Formulation (ANCF). In the ANCF, the location and deformation of a material point in a finite element are defined in a global coordinate system, such that no coordinate transformation is required and the mass matrix remains constant while the centrifugal and Coriolis forces in the finally derived dynamic equations vanish. The rigid bodies such as the journal in the cylindrical joint are described by NCF proposed by Jalón and Bayo [3]. It is known that NCF can also lead to a constant mass matrix for the rigid multibody system. The method that combines NCF describing the rigid bodies and ANCF describing the flexible bodies was named as the Absolute Coordinate Based (ACB) by Tian et al. [10], and has been widely adopted so that the mass matrix for the whole rigid-flexible system keeps constant and the system constraint conditions can be easily simplified. This approach is quite convenient from the computational point of view.

For a spatial rigid cylinder shown in Fig. 1, its motion can be defined, according to NCF, through two basic points and two unit vectors. Thus, the 12 global generalized coordinates of the rigid cylinder can be expressed as

$$\mathbf{q} = [\mathbf{r}_i^T, \mathbf{r}_j^T, \mathbf{u}^T, \mathbf{v}^T]^T \tag{1}$$

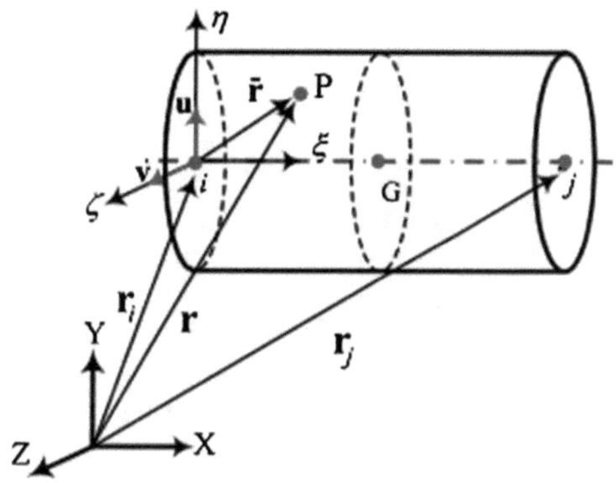

Fig. 1 Lubricant pressure distribution

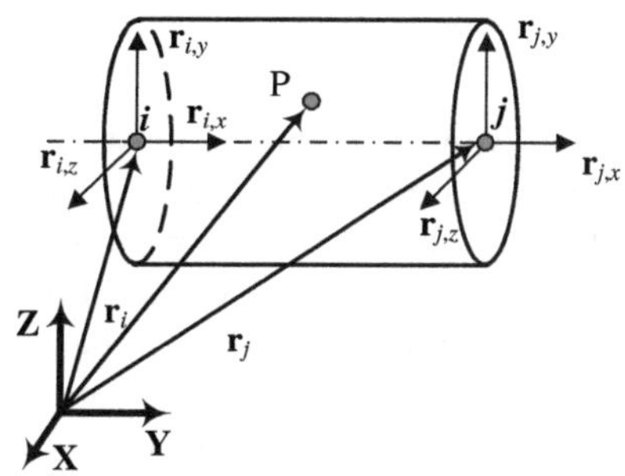

Fig. 2 Elastic deformation of bearing surface

where $\mathbf{r}_i$ and $\mathbf{r}_j$ are the position vectors of the basic points i and j, respectively. The vectors **u** and **v** are assumed to be unit and perpendicular vectors. The global position of an arbitrary point in the body can be written in the following form

$$\mathbf{r} = \mathbf{C}\mathbf{q} \tag{2}$$

where matrix **C** is determined by the local position ($\bar{\mathbf{r}}$) of point P defined in the body coordinate frame ξ-η-ζ, as Fig. 1 illustrates.

Different types of finite elements of ANCF have been proposed for modeling flexible parts undergoing both large overall motion and large deformation. In present study, the original two-node 3D beam element of ANCF developed by Shabana and Yakoub [8] is used to model flexible beams. There are a total of 24 nodal coordinates for each element, as depicted in Fig. 2.

3 EHD Model of Lubricated Cylindrical

In a broad sense, a suitable lubrication system can prevent body-to-body contact, reduce wear and, consequently, extend the service life of mechanical joints. With the intent to develop the EHD model of lubricated cylindrical joint with flexible bearing,

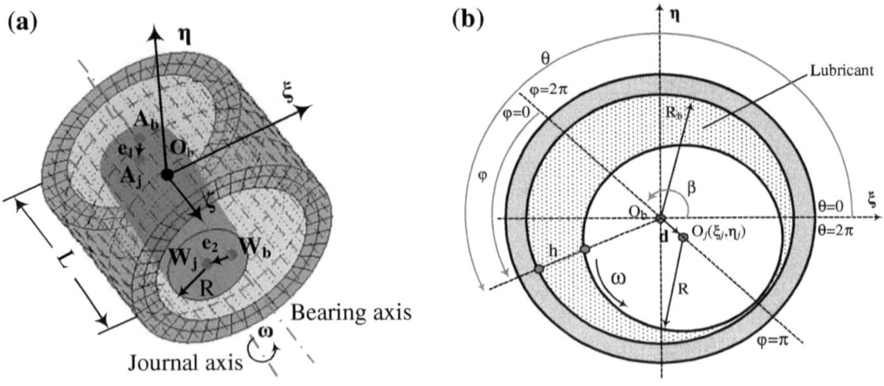

Fig. 3 **a** Generic configuration. **b** Arbitrary *cross section* along local axis ζ

the Reynolds' equations must be established. Figure 3a shows a generic configuration of a typical lubricated cylindrical joint, in which the journal misalignment is also represented. In the present study, the lubricated cylindrical joint is described by the ACB method. The center of mass of the bearing is denoted by point O_b and the journal bearing length by L. The local coordinate system is denoted by $\xi - \eta - \zeta$. In Fig. 3a, A_b and W_b are the bearing centers at end faces, while $A_j(\xi_1, \eta_1, -L/2)$ and $W_j(\xi_2, \eta_2, L/2)$ denote the journal centers at end faces. Figure 3b shows an arbitrary journal bearing cross section along the joint local axis ζ.

From Fig. 3a, when the journal misalignment is taken into account, the coordinates of an arbitrary journal cross section center O_j can be determined by the points $A_j(\xi_1, \eta_1, -L/2)$ and $W_j(\xi_2, \eta_2, L/2)$ from the following interpolation relations

$$\begin{aligned} \xi_j &= \xi_2 + \frac{(\xi_2 - \xi_1)(\zeta - L/2)}{L} \\ \eta_j &= \eta_2 + \frac{(\eta_2 - \eta_1)(\zeta - L/2)}{L} \end{aligned} \tag{3}$$

With regard to Fig. 3b, the general form of the isothermal Reynolds' equation can be expressed as [7]

$$\frac{\partial}{\partial \varphi}\left(h^3 \frac{\partial p}{\partial \varphi}\right) + R^2 \frac{\partial}{\partial \zeta}\left(h^3 \frac{\partial p}{\partial \zeta}\right) = 6\mu R^2 \omega \frac{\partial h}{\partial \varphi} + 12R^2 \mu \frac{\partial h}{\partial t} \tag{4}$$

where p denotes the lubricant pressure, μ is the dynamic lubricant viscosity and R is the journal radius. The lubricant film thickness can be calculated by

$$h = c + d\cos\varphi = c + d\cos(\theta - \beta) = c - \xi_j\cos\theta - \eta_j\sin\theta \tag{5}$$

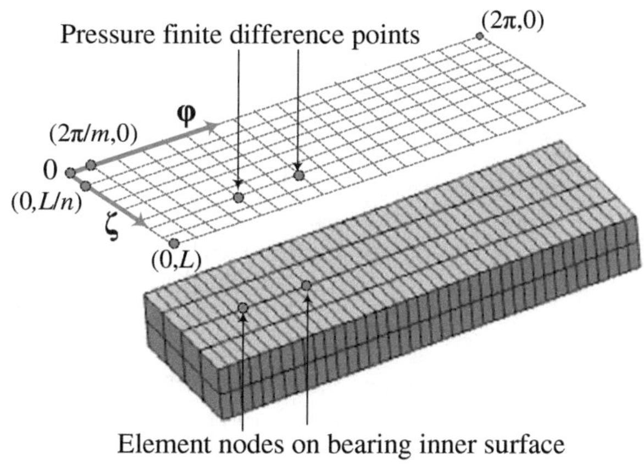

Fig. 4 EHD model of a lubricated cylindrical joint

in which c is the radial clearance, d represents the journal bearing eccentricity, θ is the angular coordinate and the variable β yields

$$\beta = \tan^{-1}\left(\frac{\eta_j}{\xi_j}\right) \tag{6}$$

When the elastic deformation of bearing is considered, the lubricant film thickness can be expressed in the following form

$$h = c - \xi_j \cos\theta - \eta_j \sin\theta + \delta \tag{7}$$

where δ denotes the elastic deformation of bearing.

As it is shown in Fig. 4, a lubricated cylindrical joint can be unfolded along with circumferential direction (ϕ). Then, the lubricant pressure field can be evaluated by imposing Eq. (8) to each calculation grid point of a finite-difference method [4]. The equal interval grid is adopted at both circumferential and axial directions. In Fig. 4, m and n represent the total number of finite difference points along circumferential direction (ϕ) and axial direction (ζ), respectively.

In order to evaluate the lubricant pressure, the Reynolds' equation (4) can be rewritten in the following form

$$h^3 \frac{\partial^2 p}{\partial \varphi^2} + 3h^2 \frac{\partial h}{\partial \varphi}\frac{\partial p}{\partial \varphi} + R^2 h^3 \frac{\partial^2 p}{\partial \zeta^2} + 3R^2 h^2 \frac{\partial h}{\partial \zeta}\frac{\partial p}{\partial \zeta} = 6\mu R^2 \omega \frac{\partial h}{\partial \varphi} + 12R^2 \mu \frac{\partial h}{\partial t} \tag{8}$$

According to the finite-difference method [7], the finite difference equation of the pressure can be expressed by

$$\left[\frac{-3h_{i,j}^2}{4(\Delta\varphi)^2}\left(h_{i+1,j}-h_{i-1,j}\right)+\frac{h_{i,j}^3}{(\Delta\varphi)^2}\right]p_{i-1,j}+\left[\frac{-3R^2h_{i,j}^2}{4(\Delta\zeta)^2}\left(h_{i,j+1}-h_{i,j-1}\right)+\frac{R^2h_{i,j}^3}{(\Delta\zeta)^2}\right]p_{i,j-1}$$
$$+\left[\frac{-2h_{i,j}^3}{(\Delta\varphi)^2}-\frac{2R^2h_{i,j}^3}{(\Delta\zeta)^2}\right]p_{i,j}+\left[\frac{3R^2h_{i,j}^2}{4(\Delta\zeta)^2}\left(h_{i,j+1}-h_{i,j-1}\right)+\frac{R^2h_{i,j}^3}{(\Delta\zeta)^2}\right]p_{i,j+1}$$
$$+\left[\frac{-3h_{i,j}^2}{4(\Delta\varphi)^2}\left(h_{i+1,j}-h_{i-1,j}\right)+\frac{h_{i,j}^3}{(\Delta\varphi)^2}\right]p_{i+1,j}=3\mu R^2\omega\frac{h_{i+1,j}-h_{i-1,j}}{\Delta\varphi}+12R^2\mu\left(\frac{\partial h_{i,j}}{\partial t}\right) \tag{9}$$

where $p_{i,j}$ is the pressure at the finite difference point (i, j), $i = 1, 2, ..., m$ and $j = 1, 2, ..., n$. Here, the pressure boundary conditions are $p(\varphi,0) = p(\varphi, L) = 0$, $p(\varphi_1, 0) = p(\varphi_2, L)$ and $p(\varphi_1, \zeta) = p(\varphi_2, \zeta)$, with φ_1 and φ_2 being the angles of the start and end point of a hydrodynamic film. In this work, the finite difference Eq. (9) is solved by using Successive Over Relaxation (SOR) method, that is

$$p_{i,j}=p_{i,j}+\lambda\left(a_0-a_1p_{i-1,j}-a_2p_{i,j-1}-a_3p_{i,j+1}-a_4p_{i+1,j}\right)/a_5 \tag{10}$$

where the six coefficients yield

$$\begin{aligned}
a_0 &= 3\eta R^2\omega\frac{h_{i+1,j}-h_{i-1,j}}{\Delta\varphi}+12R^2\mu\left(\frac{\partial h_{i,j}}{\partial t}\right)\\
a_1 &= \frac{-3h_{i,j}^2}{4(\Delta\varphi)^2}\left(h_{i+1,j}-h_{i-1,j}\right)+\frac{h_{i,j}^3}{(\Delta\varphi)^2}\\
a_2 &= \frac{-3R^2h_{i,j}^2}{4(\Delta\zeta)^2}\left(h_{i,j+1}-h_{i,j-1}\right)+\frac{R^2h_{i,j}^3}{(\Delta\zeta)^2}\\
a_3 &= \frac{3R^2h_{i,j}^2}{4(\Delta\zeta)^2}\left(h_{i,j+1}-h_{i,j-1}\right)+\frac{R^2h_{i,j}^3}{(\Delta\zeta)^2}\\
a_4 &= \frac{-3h_{i,j}^2}{4(\Delta\varphi)^2}\left(h_{i+1,j}-h_{i-1,j}\right)+\frac{h_{i,j}^3}{(\Delta\varphi)^2}\\
a_5 &= \frac{-2h_{i,j}^3}{(\Delta\varphi)^2}-\frac{2R^2h_{i,j}^3}{(\Delta\zeta)^2}
\end{aligned}$$

This procedure is performed until the pressure convergent criterion for the $(k + 1)$ step is reached, that is

$$\frac{\sum_{i=1}^m\sum_{j=1}^n\left|p_{i,j}^{(k+1)}-p_{i,j}^{(k)}\right|}{\sum_{i=1}^m\sum_{j=1}^n p_{i,j}^{(k+1)}}\le tol \tag{11}$$

where *tol* is a specified convergence tolerance. It must be stated that in the present work the tolerance considered was equal to 10^{-5}. The pressure field is calculated only over the positive part by setting the pressure in the remaining portion equal to zero. This boundary condition, associated with the pressure field, corresponds to Gümbel's boundary conditions or half Sommerfeld's conditions [7]. Once the lubricant pressures at the finite difference points are obtained, they are transformed into the corresponding generalized nodal forces for analyzing the elastic deformation of flexible bearing. Finally, the elastic deformation δ of flexible bearing described by Eq. (7) can be evaluated according to the classic finite element method.

4 Results and Discussion

The purpose of this section is to demonstrate that the twenty-node hexahedral element of ANCF can be applicable to a flexible hollow cylinder subjected to distributed forces as shown in Fig. 5. The distributed forces acting on the cylinder are assumed to be a linear function in time, defined as $F = 1000t$ N/m. The inner radius, length and thickness of the cylinder are equal to 0.225, 0.2, and 0.05 m, respectively. The Young's modulus of the bearing material is equal to 1×10^6 Pa.

Figure 6 shows the influence of mesh size on the displacement of point B in *Z*-direction and indicates that 4(axial direction) × 2(radial direction) × 40(circumferential direction) elements are enough to obtain the converged results.

In order to validate the obtained numerical results, the same model is also analyzed by using the commercial software ADINA [2]. The 3D 20-node brick solid element is used to mesh the model. From Fig. 7, the von Mises stresses at points A and B obtained by the numerical method are well agreement with those obtained by ADINA. Figure 8 shows the dynamic configurations and the von Mises stress contours of the hollow cylinder at different instants. Figure 9 shows the von Mises stress contour of the hollow cylinder obtained by using ADINA at the instant time $t = 1.00$ s. Also, after a careful analysis of the results represented in Figs. 8 and 9 leads to the assertion that the twenty-node hexahedral finite element of ANCF exhibits a very close response to the case of ADINA simulations. For this example

Fig. 5 A flexible *hollow cylinder* under uniform distribution forces

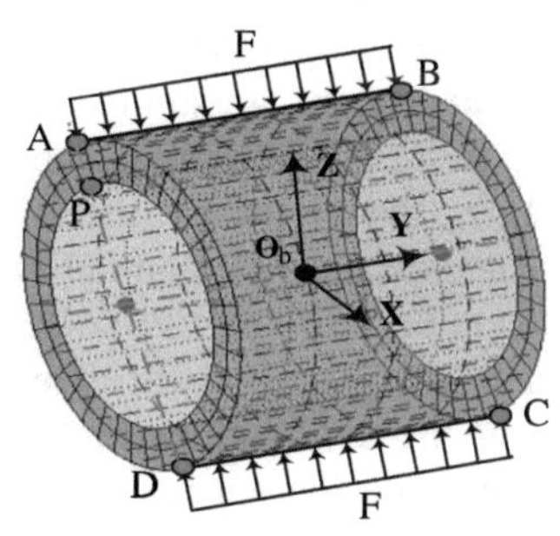

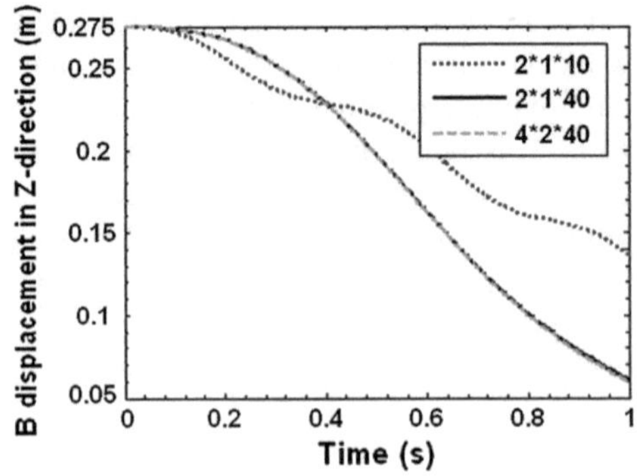

Fig. 6 Influence of the model mesh size on the displacement of point *B* in *Z-direction*

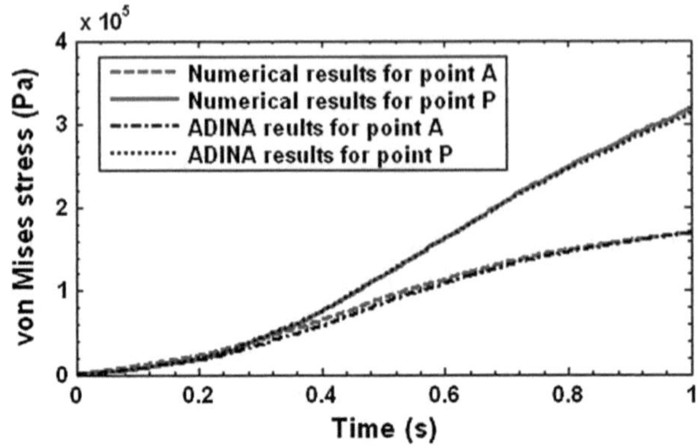

Fig. 7 Comparison of the von Mises stresses at *points A* and *P* numerical results versus ADINA results

the integration step is set to be 1e-3s, the cost computation time for the ADINA software and the proposed method are 616 and 878 s, respectively.

Finally, the EHD analysis of a lubricated cylindrical joint with rotating journal is presented. The length of the cylindrical joint is equal to 66 mm. The journal rotates around its axis A_jW_j at a constant angular speed of 3000 rpm. The dynamic lubricant viscosity is equal to 9 mPa s. The rigid journal is modeled by using the NCF, while the flexible bearing is modeled by the twenty-node hexahedral element of ANCF. The nodes on the outer bearing surface are assumed to be fixed in the space. The material density and the thickness of the bearing are set to be 7,800 kg/m^3 and 20 mm, respectively. As a comparison, the systems with and without journal misalignment are studied.

At first, the journal axis A_jW_j is assumed to be parallel with the bearing axis A_bW_b, that is, the rigid rotating journal is not misaligned being $e_1 = e_2 = 0.024$ mm. The clearance is equal to 0.03 mm. The Young's modulus of the bearing material is set to be 2.1×10^{12} Pa so that the results can be compared to those obtained by Sun and Gui [9], who considered a case of rigid bearing. Figure 10 shows the lubricant pressure distributions for this journal bearing system. As expected, Fig. 10b shows that if the journal is not misaligned, the lubricant pressure exhibits a symmetrical

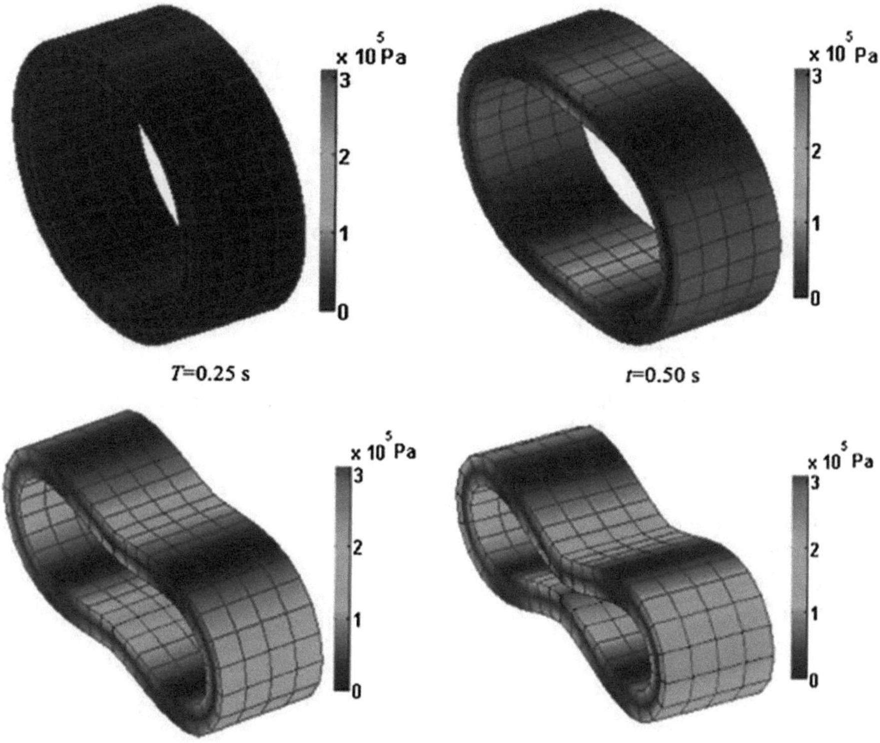

Fig. 8 von Mises stress contour of a *hollow cylinder* obtained using FEM of ANCF

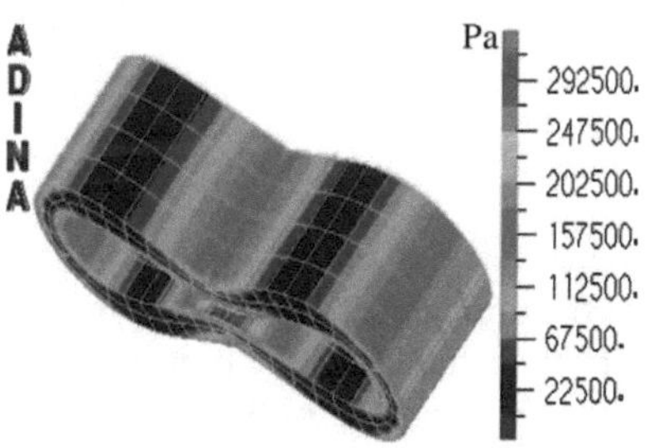

Fig. 9 The von Mises stress contour of a *hollow cylinder* obtained using ADINA (t = 1.00 s)

distribution about the plane ξ-C_j-η. Here, C_j is the mass center of the rigid journal. Figure 11 shows a scaled view of the deformation of the inner bearing surface. The figure indicates that for the system with a rotating journal without misalignment, the distribution of the deformation of the inner bearing surface also exhibits a symmetrical distribution about the plane ξ-C_j-η, which is consistent with the lubricant pressure distribution shown in Fig. 10b.

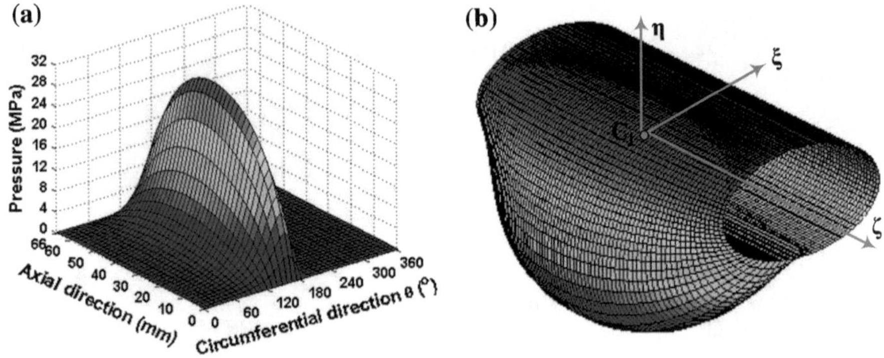

Fig. 10 Lubricant pressure distributions of a journal without misalignment ($E = 2.1 \times 10^{12}$ Pa). a pressure vs journal directions, b a 3D view

Fig. 11 Zoom view of the elastic deformation of inner bearing surface ($E = 2.1 \times 10^{12}$ Pa)

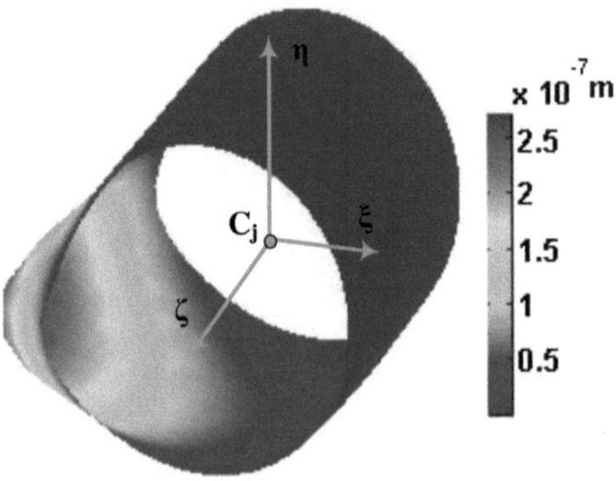

5 Conclusion

In the present study, a general and comprehensive methodology is proposed to integrate the EHD model of lubricated cylindrical joints into the flexible multibody system formulation. The EHD behavior of the lubricated cylindrical joints in flexible multibody system is studied by using the ACB method such that the flexible bearing is modeled via the twenty-node hexahedral elements of ANCF, while the rigid journal in the cylindrical joint is described via NCF. The lubricant pressure is evaluated by solving the Reynolds' equation via the finite difference method. The elastic forces and their Jacobian of ANCF finite elements are deducted through the definition of the Piola-Kirchhoff stress tensor of the first type in continuum mechanics. The numerical examples show that the bearing flexibility affects the system responses in a significant manner, since the flexible bearing will extend the lubricant distribution space, and then reduce the lubricant pressure. The methodology proposed in this study can be easily extended to studying the coupling dynamics of the lubricated rotor system with the bearing flexibility taken into account.

References

1. Attia HM, Bouziz S, Maatar M, Fakhfakh T, Haddar M (2010) Hydrodynamic and elastohydrodynamic studies of a cylindrical journal bearing. J Hydrodyn 22(2):155–163
2. Bathe KJ (2010) ADINA system. ADINA R&D Inc.
3. Jalón JG, Bayo E (1994) Kinematic and dynamic simulation of multibody systems: the real-time challenge. Springer, New York
4. Kumar MS, Thyla PR, Anbarasu E (2010) Numerical analysis of hydrodynamic journal bearing under transient dynamic conditions. Mechanika 2(82):37–42
5. Liu HP, Xu H, Ellision PJ, Jin ZM (2010) Application of computational fluid dynamics and fluid-structure interaction method to the lubrication study of a rotor bearing system. Tribol Lett 38(3):325–336
6. Nikravesh PE (1988) Computer aided analysis of mechanical systems. Prentice-Hall, New Jersey
7. Pinkus O, Sternlicht SA (1961) Theory of hydrodynamic lubrication. McGraw-Hill, New York
8. Shabana AA, Yakoub RY (2001) Three-dimensional absolute nodal coordinate formulation for beam elements: theory. J Mech Des 123:606–613
9. Sun J, Gui CL (2004) Hydrodynamic lubrication analysis of journal bearing considering misalignment caused by shaft deformation. Tribol Int 37:841–848
10. Tian Q, Liu C, Machado M, Flores P (2011) A new model for dry and lubricated cylindrical joints with clearance in spatial flexible multibody systems. Nonlinear Dyn 64:25–67

Analysis of Experimental Data from Complex Multibody System

J.C. Jáuregui-Correa, C.S. López Cajún and Mihir Sen

Abstract Complex mechanical systems are those that are made up of a large number of inter-connected individual components, such as a multi body system. Traditionally, interconnections among elements are assumed to be springs and dampers, in this case, weak connections are introduced to explain synchronization. The system itself is treated as a black box, but it is assumed that a number of sensors can be applied to it to obtain dynamical signals of quantities like acceleration, velocity or displacement. The methods with which these signals can be analyzed to display characteristics such as synchronization are described. Some of these methods have been applied to signals obtained in several different experiments with mechanical systems.

Keywords Complex system · Mechanical system · Kuramoto's order parameter

1 Introduction

The term *complexity* has been used in many different senses in engineering. A number, a measured signal, or even a problem may be *complex*. The theory of chaos has provided, for several decades, low-dimensional non-linear ordinary dimensional equations that have extremely complex solutions. These are all perfectly valid usages, as long as the term is precisely defined; sometimes it is, and sometimes it isn't. What we will consider here is something very specific, and is

J.C. Jáuregui-Correa (✉) · C.S. López Cajún
Universidad Autónoma de Querétaro, Santiago de Querétaro, Mexico
e-mail: jc.jauregui@uaq.mx

C.S. López Cajún
e-mail: cajun@uaq.mx

M. Sen
University of Notre Dame, Notre Dame, USA
e-mail: Mihir.Sen.1@nd.edu

M. Ceccarelli and E.E. Hernández Martinez (eds.), *Multibody Mechatronic Systems*,
Mechanisms and Machine Science 25, DOI 10.1007/978-3-319-09858-6_20

merely one of the ways in which the term is used. Our subject is the analysis of data obtained from *complex mechanical systems*. For this we will need to define what a complex system is, and then suggest methods of analysis of the experimental data obtained from them.

Of course, there are a large number of complex systems, including financial, social, political and biological. Here we will concentrate on mechanical multi-body systems. The reason is that, although many of the concepts that we discuss apply equally well to, say, social systems, we are first of all not qualified to carry the discussion to those areas. More importantly, we restrict ourselves to systems for which we can possibly write and solve equations. Our approach is thus conceptually equations-based, though the ideas themselves are indeed broader and the governing equations themselves will not be invoked.

On the other hand, synchronization is a phenomenon studied long time ago. However, this phenomenon as such, in mechanical engineering has been treated slightly and there are few works reported [2, 6, 7].

2 Complex Mechanical Systems

By *mechanical systems* we refer to machines and mechanisms with moving parts. We are interested in the *dynamics* of the system, i.e. its behavior with respect to time. Technical terms, as implied before, make sense only if they are defined more or less precisely. We will refer to an object as a *component* or *sub-system* if its dynamics can be represented by a small number of ordinary differential equations. A large number of components can be inter-connected in some way as a *complex system*. The inter-dependence is through *coupling* equations that connect the components. We have thus a large number of equations that describe the system, some of them describing the components in the system themselves, and the others the coupling between them.

A schematic of a complex system is shown in Fig. 1. A number of components (indicated by 1, 2, 3, ... are inter-connected. There is a single input, but many outputs. Typically, the input may be thought to be from some external stimulus, and the outputs are from sensors mounted at different locations. A linear mathematical model of component the motion u_i of component i may be of the form

$$L_i(u_i) + \sum_{j=1}^{N} k_{ij}\left(u_j - u_i\right) = f_i(t) \tag{1}$$

where L may be a differential operator, N is the number of components in the system, and k_{ij} are coupling coefficients, both, strong and weak, u_i is the dynamic response of each individual element, and $f_i(t)$ represents any excitation.

Usually in multi-body systems the strong coupling is the one considered in many models. The challenge is to determine, from the outputs alone, what the parameters of the above equation are.

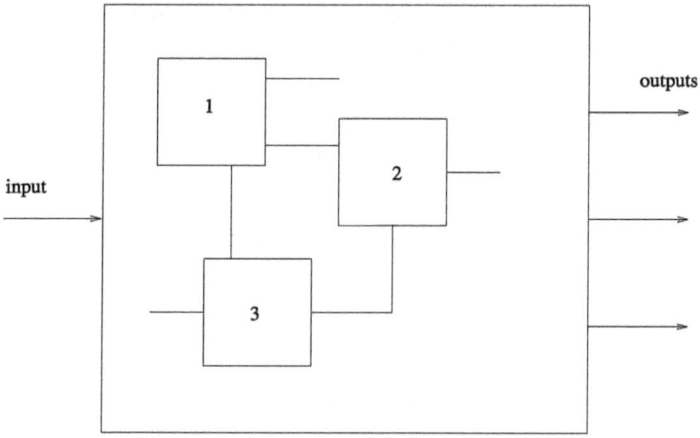

Fig. 1 A complex system

A forced system such as Eq. (1) will in the long run, as $t \to \infty$, reflect the characteristics of the external forcing $f(t)$, rather than those of the operator L_i. The complementary functions, on the other hand, will be due to the homogeneous operator. The way to perceive this is to carry out an impact test, the response to which will provide an indication of the solutions of the characteristics equation.

3 Methods of Data Analysis

3.1 Statistics

Statistics (regression, probability distribution, moments, auto-correlations, fractal dimensions) is a classical approach to the analysis of a single signal. Fourier and wavelet transforms also provide much information about the signals [7]. A small number of signals can be analyzed by cross-correlation. Measures have been proposed to quantify the unpredictability of a single time series, particularly for health monitoring. Among them are regularity [12] and various measures of entropy like measure-theoretic [4, 8], approximate [1] and multiscale [5]. In this work, we introduce the disorder parameter.

3.2 Disorder Parameter

This is of interest when a large number of components are in synchronization, or close to it. Synchronization is described in [10, 14]. Kuramoto's order parameter (KP) is a function that describes synchronization among several oscillators. The KP is in [9, 11, 13]. This parameter is based on the relative instantaneous phase that can be determined with the Hilbert's transform. This is given as

$$\widetilde{u}(t) = \frac{1}{\pi} \mathrm{PV} \int_{-\infty}^{\infty} \frac{u(s)}{t-s} ds \tag{2}$$

where PV is the principal value, and *u(s)* is a signal, so that the analytic signal is

$$\begin{aligned} U(t) &= u(t) + i\widetilde{u}(t) \\ &= A(t)e^{i\phi(t)} \\ \tan\phi &= \frac{\widetilde{u}(t)}{u(t)} \end{aligned}$$

Kuramoto's order parameter (KP), *r*, can be computed as

$$r = \frac{1}{n} \sum_{j=1}^{n} e^{i\phi_i(t)} \tag{3}$$

where $\phi_i(t)$, $i = 1, \ldots n$ represent the set of signals in synchronization (the instantaneous phase angle). Values of *r* close to 1, indicate full synchronization, whereas values close to zero indicate no synchronization at all.

In the discussion below, the proposed disorder parameter, *s*, is defined as

$$s \equiv 1 - r \tag{4}$$

4 Application to Simple Signals

The methodology presented in this paper is validated with numerical simulated signals, as well as experimental data. The simulated signals are of the form:

$$\begin{aligned} y_1 &= A\sin(\omega t) \\ y_2 &= A\sin(\omega t + \alpha_i) + g_i \end{aligned} \tag{5}$$

where $0 \le \alpha_i \le 1$, and g_i a random generated perturbation. The value of α_i was incremented from 0 to 1, the instantaneous KP was computed as well as the disorder parameter for comparing the effect of random variations, The results are shown in Fig. 2. The KP is affected by the amount of noise that the signal has. If the noise level is small, $\alpha_i \le 0.3$, synchronization is high ($r \ge 0.95$), whereas at higher values, KP falls to values less than 0.85, and remains around this value. The corresponding DP is shown in Fig. 2. This figure proves that two signals with a constant instant phase have a disorder parameter close to zero, when the signals cross each other, there is a change of sign that disturbs the solution.

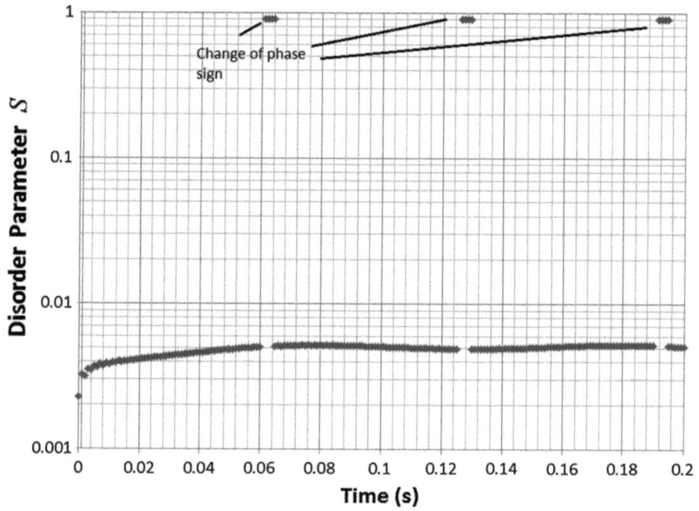

Fig. 2 Disorder parameter as a function of α_i

5 Experimental Applications

5.1 Electric Motor Shaft

Two accelerometers were placed at the end of an electric motor shaft. The accelerometers measured synchronous motion as a function of motor's speed. Measured signals are shown in Fig. 3. With this data, the DP varies as a function of time as shown in Fig. 4. The two accelerometers were placed at an arbitrary angle, in this way, the signals have a constant phase angle. It is clear that the DP value is always less than 0.011. Variations are due to the noise.

5.2 Blade Vibrations

An experimental analysis to show synchronization of blades (example of a multibody system with 22 individual elements), shown in Fig. 5 was carried out, and reported elsewhere [7]. Two types of tests were conducted, namely, impact, and an airflow. The airflow acts as a force excitation, and affects the response of each blade. Blades were fixed and the vibration of each blade was measured simultaneously with 22 MEMs accelerometers. With this example, it was possible to calculate the KP with 22 simultaneous signals and to compute the DP. Results are shown in Fig. 6. It is clear that the 22 blades vibrate synchronously since the DP value remained under 0.0025. Variations on the DP are due to signal noise and differences in blade dynamics.

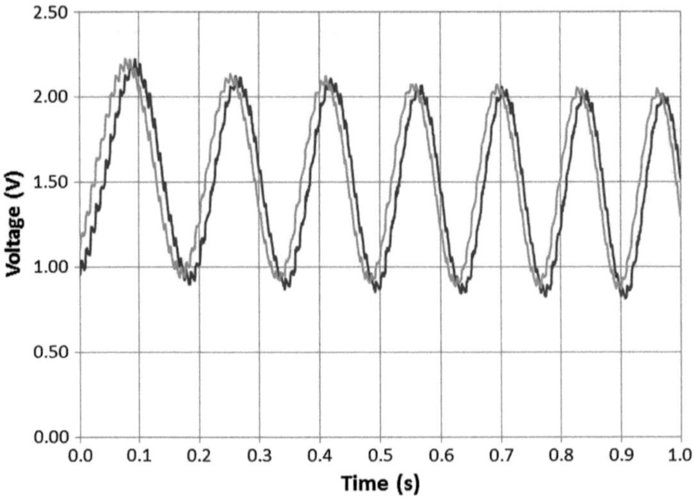

Fig. 3 Measured signals from shaft of electric motor

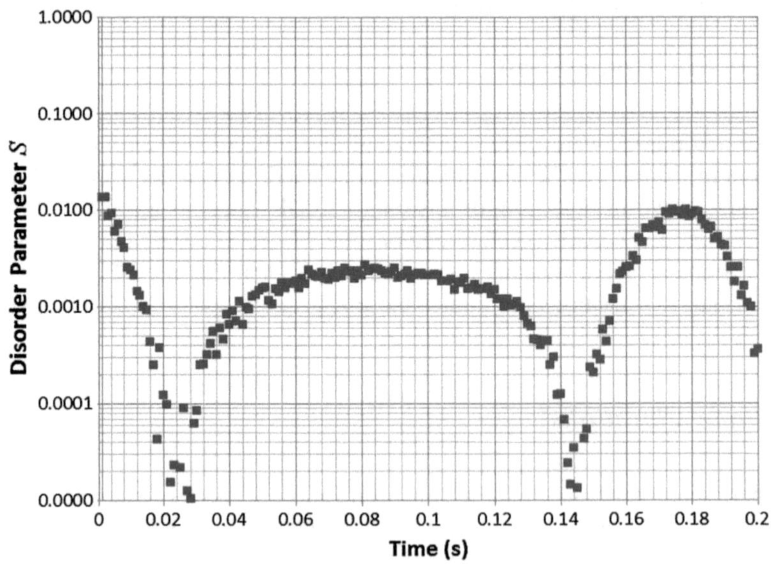

Fig. 4 DP variations for the electric motor

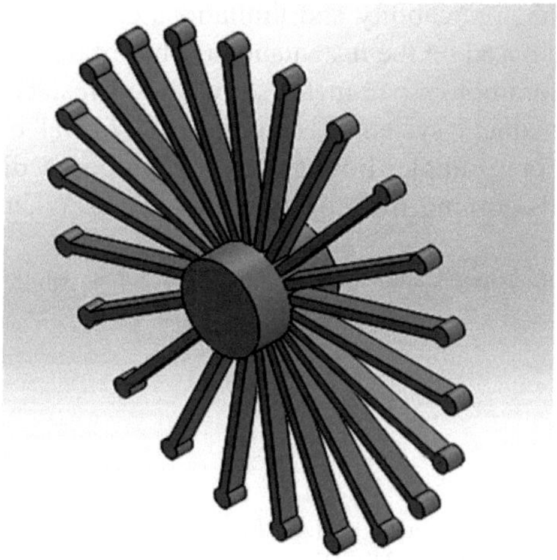

Fig. 5 Model of the rotor and blades

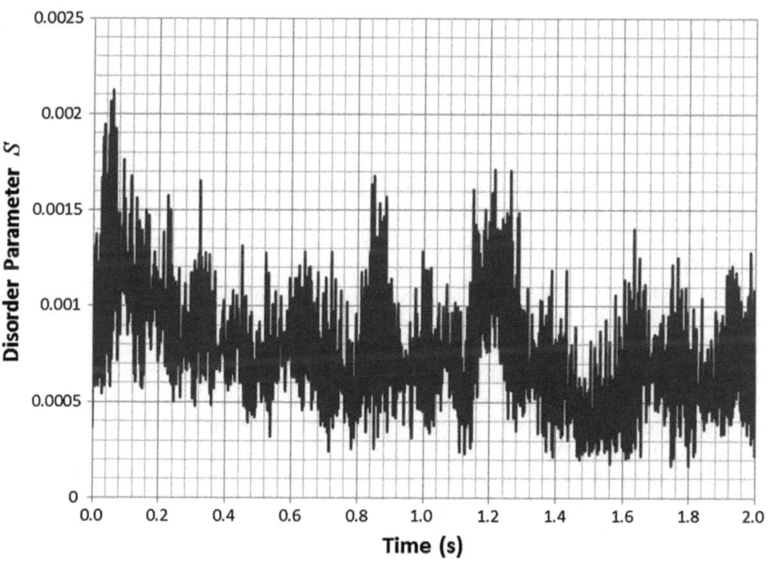

Fig. 6 Disorder parameter of 22 blades under a wind blow

6 Conclusions

Some methods of analysis of experimental data from complex mechanical systems have been presented. Among others, statistical methods and the proposed disorder parameter presented herein. Three cases related to synchronization were briefly discussed showing the applicability and limitations of each method. It was shown that the methodology based on the instantaneous phase calculated with the Hilbert's transform and the Kuramoto's parameter shows good results for identifying synchronization in mechanical systems. Nevertheless, several questions arose from these applications. For example, how to use the proposed disorder parameter in complex noisy signals coming from many components? What about randomness?

Acknowledgments Prof. López-Cajún thanks CONACyT for its support during his sabbatical leave.

References

1. Acharya U, Joseph K, Kannathal N, Lim C, Suri JS (2006) Heart rate variability: a review. Med Biol Eng Comput 44(12):1031–1051
2. Barrón MA, Sen M (2009) Synchronization of coupled self-excited elastic beams. J Sound Vib 324:209–220
3. Boetta G, Cencini M, Falcioni M, Vulpiani A (2002) Predictability: a way to characterize complexity. Phys Rep Rev Sect Phys Lett 356(6):367–474
4. Costa M, Goldberger A, Peng C (2002) Multiscale entropy analysis of complex physiologic time series. Phys Rev Lett 89:068102
5. González-Cruz CA, Jáuregui-Correa JC, López-Cajún CS, Sen M (2014) Dynamic behavior and synchronization of an automobile as a complex system. In: Proceedings of the ASME 2014 12th Biennial conference on engineering systems design and analysis, Paper no. ESDA2014-20037
6. Jáuregui-Correa JC, Sen M, López-Cajún C (2009) Experimental characterization of blade vibration synchronization. ASME Turbo Expo, Paper no. GT2011-46105
7. Katok A (2007) Fifty years of entropy in dynamics: 1958—2007—Preface. J Mod Dyn 1 (4):545–596
8. Kuramoto Y (1984) Chemical oscillations, waves, and turbulence. Springer, New York
9. Manrubia S, Mikhailov A, Zanette D (2004) Emergence of dynamical order: synchronization phenomena in complex systems. World Scientificc, New Jersey
10. Pikovsky A, Rosenblum M, Kurths J (2001) Synchronization: a universal concept in nonlinear sciences. Cambridge University Press, Cambridge
11. Pincus S, Goldberger A (1994) Physiological time-series analysis—what does regularity quantify. Am J Physiol 266(4, 2):1643–1656
12. Strogatz S (2000) From Kuramototo Crawford: exploring the onset of synchronization in populations of coupled oscillators. Physica D 143(1–4):1–20
13. Strogatz S (2003) Sync: the emerging science of spontaneous order. Theia, New York

Reduced Energy Consumption in Induction Motors with Possible Mechatronic Applications

G. Calzada-Lara and J. Álvarez-Gallegos

Abstract This document proposes an energetic output improvement of electric drives for induction motors. This modification pointing toward a better energetic output (improving the efficiency) according to the application, and can be made in the most common control algorithms as the Flux Oriented Control (FOC) and the Direct Torque Control (DTC). The experimental results show that the proposed controller ensures both, a good speed control and smooth torque response with current shapes with low THD, when these are compared with the conventional DTC scheme. Experimental results for a 1.1 kW induction motor are presented and analyzed using a dSpace system with DS1103 controller board based on the digital processor Texas Instruments TMS320F240. The obtained results showed that the proposed control scheme is able to obtain a high performance in mechatronic applications (robotics for example).

Keywords Direct torque control (DTC) · Efficiency improvement · Energetic output · Induction motor · Position control

1 Introduction

The annual incremental energy users while energy sources are more and more limited, requires research into methods that increase the systems efficiency. Lowering motor losses substantially will have a significant impact on the country's energy consumption [4]. More than 150 TWh of electricity consumed annually

G. Calzada-Lara (✉)
Universidad Autónoma de San Luis Potosí, Centro Histórico, México
e-mail: gabriel.calzada@uaslp.mx

J. Álvarez-Gallegos
Centro de Investigación y de Estudios Avanzados del Instituto Politécnico Nacional, Ciudad de México, México
e-mail: jalvarez@cinvestav.mx

M. Ceccarelli and E.E. Hernández Martinez (eds.), *Multibody Mechatronic Systems*, Mechanisms and Machine Science 25, DOI 10.1007/978-3-319-09858-6_21

could be saved in a cost-effective manner if drive systems were designed from an energy efficiency point of view [6]. Electric motors are one of the components involved that can contribute to energy savings. In the last decades this design problem has been tackled in different manners. Many changes in terms of materials and project criteria have been introduced in past years, with important results under the efficiency point of view [15]. Once more, this makes more evident the age-old problem of power losses in the electric motors. The problem was addressed in the 80s; good results came from the adoption of new magnetic materials on one side and of speed regulators (without dissipation) on the other [9]. Also, in recent decades have arisen several discussions concerning to the very low efficiency of the motors adopted for domestic appliances [3].

Recent studies are focused on small power rating machines; in fact, their distribution is very significant under the point of view of the total losses and of the costs of production with new technologies [2]. It is possible to estimate that today about 30 % of the electrical power is absorbed by machines with rated power lower than 5 kW. If we assume an average equivalent efficiency of 80 % (very optimistic), this means that 20 % is dissipated. As a consequence, more than 6 % of the total electric power produced is dissipated in small electric motors. It is also evident that the possibility to further reduce the internal losses of the motors should permit an increment of efficiency, even if it is limited to 1—2 %. In practical applications, many motors are working significantly below their design capacity, resulting in low efficiency levels. It has been shown that can save significant amounts of energy monitoring and reducing the electric energy consumed by induction motors in order to match their actual energy need under varying load conditions. The present work shows that it is possible to achieve the monitoring and reduction mentioned above, retaining the original control scheme.

2 Development

There are applications where induction motors are used and it is interesting and advantageous to optimize the performance in steady state: paper machines, fans, pumps, centrifuges, electric traction, etc. However, there are many other applications where the transient is of equal or greater duration than the permanent state, such as: rolling mills, elevators, cranes, electric vehicles, machine tools, robots, textiles, etc. In some applications the possibility to vary the speed becomes a necessity, whereas in others, there is a great potential for energy savings when using a variable speed control. This is the case of applications with torque-speed quadratic loads, for example, air conditioners, pumps, and air compressors, where they work at a constant speed by controlling the power output of the process by mechanical means [5]. In applications with variable speed, the motor operates for long periods of time in conditions below the rated power, for example, in electric vehicles, where nominal power is used only in startups and slope increases [12].

The dynamic high-performance control systems used in industrial applications like Field Oriented Control (FOC) or Direct Torque Control (DTC) operate with constant flux norms fixed at nominal rate (and recently in domestic applications too). Under this situation efficiency is maximal only when the system operates at its nominal torque rate. Away from this operating point, the machine dissipates a considerable part of the injected electrical power in terms of core losses, and it may inefficiently store too much energy in the coil inductances. In most of the applications, these machines do not operate at the nominal rate since the desired torque may change online or may be dependent on other system states such as position or speed (this is the case in electrical transportation systems) [1]. It is then technically and economically interesting to investigate other modes of flux operation seeking to optimize system performance. A traditional control with a possible modification that tries to solve this problem can be appreciated in Fig. 1. The gain in efficiency obtained because of the optimal flux application in comparison with the conventional flux, that is constant, is significant, especially at low load and also low speed.

There are some contributions about FOC modification taking into account machine performance optimization according to the connected load, however about modifications carried out to DTC there is very little reported [8]. This work shows that it is possible to make changes to the conventional DTC with the objective of improving its output from an energetic standpoint, principally based on the use of three previous ideas: the first one regarding to conventional DTC modifications made by [10] and [11], where an improvement in motor energy efficiency through

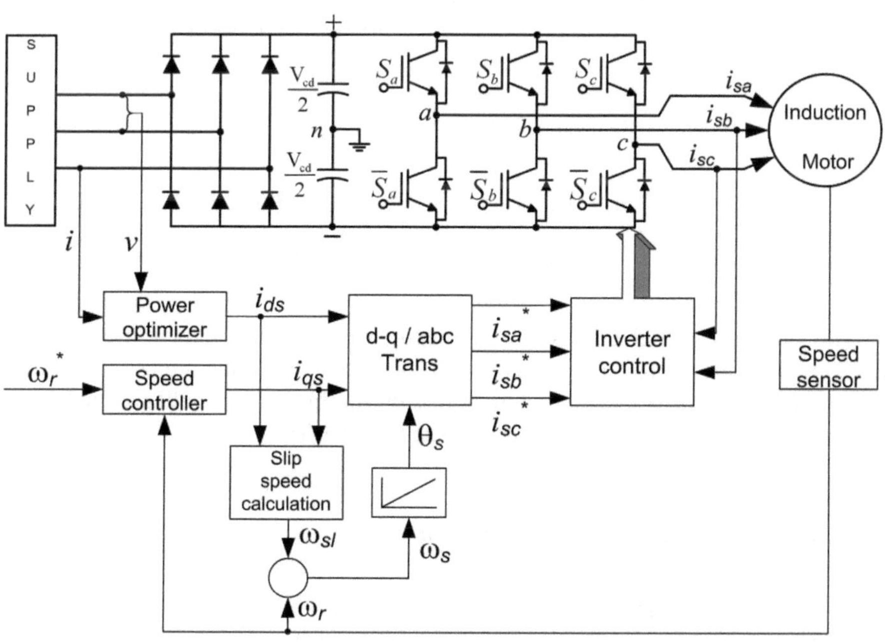

Fig. 1 FOC with output optimization based on the measure of consumed active power

searching an optimal flux is performed, the second one based on changing hysteresis controllers for flux and torque made by [17] and [19] to achieve a constant switching frequency; and the third one made in [7] regarding the introduction of new control elements that allow a better performance and greater robustness.

2.1 The Induction Motor Model

A natural approach to obtain the equations that govern the behavior of the three phase squirrel cage induction motor consists on considering such as an electric machine formed by coupled coils in relative movement, and then try to solve the state equations directly for each phase in the time domain. Although the approach is totally legitimate, an immediate difficulty would be founded; the states of the equations are coupled; besides the coefficients of these equations vary with time, which gives a group of nonlinear differential equations as a result with varying coefficients in the time [13, 14]. Finally, after simplifying to a great extent the model of squirrel cage induction motor, making a transformation from the three phase system to the coordinated system of axes α and β which stay fixed and are related with the stator, the following system of equations that is able to imitate the real behavior of the motor under normal conditions of operation is obtained:

$$\begin{bmatrix} \dot{i}_{S\alpha\beta} \\ \dot{\lambda}_{R\alpha\beta} \end{bmatrix} = \begin{bmatrix} -\frac{1}{\sigma\mathcal{L}_s}\left(R_s + R_R\frac{M^2}{\mathcal{L}_R^2}\right)\mathbf{I} & \frac{M}{\sigma\mathcal{L}_s\mathcal{L}_R}\left(\frac{R_R}{\mathcal{L}_R}\mathbf{I} - \mathbf{J}\omega\right) \\ \frac{MR_R}{\mathcal{L}_R}\mathbf{I} & -\frac{R_R}{\mathcal{L}_R}\mathbf{I} + \mathbf{J}\omega \end{bmatrix} \begin{bmatrix} i_{S\alpha\beta} \\ \lambda_{R\alpha\beta} \end{bmatrix} + \begin{bmatrix} \frac{1}{\sigma\mathcal{L}_s}\mathbf{I} \\ \mathbf{0} \end{bmatrix} \boldsymbol{v}_{S\alpha\beta} \tag{1}$$

where,

$$\mathcal{L}_s = L_s - M_s,\ \mathcal{L}_R = L_R - M_R,\ M = {}^3\!/_2 M_{SR},\ \sigma = 1 - \left(M^2/\mathcal{L}_s\mathcal{L}_R\right)$$

$$\boldsymbol{i}_{S\alpha\beta} = \begin{bmatrix} i_{s\alpha} \\ i_{s\beta} \end{bmatrix}, \boldsymbol{\lambda}_{R\alpha\beta} = \begin{bmatrix} \lambda_{R\alpha} \\ \lambda_{R\beta} \end{bmatrix}, \boldsymbol{v}_{\alpha\beta} = \begin{bmatrix} v_{s\alpha} \\ v_{s\beta} \end{bmatrix}, \mathbf{I} = \begin{bmatrix} 1 & 0 \\ 0 & 1 \end{bmatrix}, \mathbf{J} = \begin{bmatrix} 0 & -1 \\ 1 & 0 \end{bmatrix}, \mathbf{0} = \begin{bmatrix} 0 & 0 \\ 0 & 0 \end{bmatrix}$$

The previous model allows a relationship between the stator currents of the machine ($\boldsymbol{i}$) in the coordinates α and β with rotor fluxes ($\boldsymbol{\lambda}_{\mathbf{R}}$), having as control input the voltage applied to the stator winds ($\boldsymbol{v}$), that is in fact the voltage that will be controlled directly in the inverter by means of the DTC.

2.2 DTC Principles and Operation

The basic idea of the DTC is to calculate torque and flux instantaneous values from machine stator variables. Torque and flux are controlled direct and independently through optimum selection of inverter commutation states and limiting the errors of

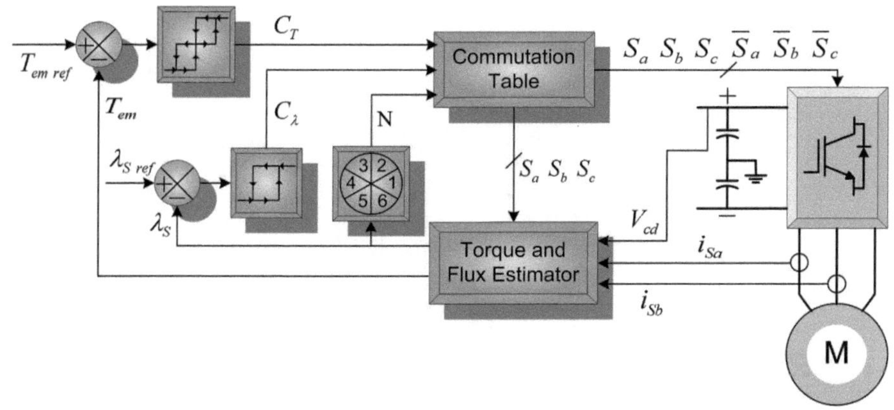

Fig. 2 DTC general structure

these through hysteresis controllers [18]. In the Fig. 2 a basic DTC blocks diagram is shown. By means of DTC the motor torque is effectively controlled with stator voltage vector rotation using appropriate commutation states. At the same time the magnitude of stator flux vector is controlled, that means, with use of inverter commutation states. This value can be changed according to requirements of flux and torque reference. The commutation table algorithm uses the discrete values encountered before in addition with the numerical value of one of the six sectors in which the flux vector is encountered. Thus, the algorithm chooses one of eight possible states to generate the spatial vector of inverter output voltage [16].

2.3 Experimental Results of High Performance DTC

Describing shortly the used equipment in the experimental tests, a three phase tension inverter was used to feed a squirrel cage induction motor controlled through the "*dSpace*" system. The power system consists of a three phase rectifier of bridge type at the input (built with diodes), a pair of capacitors which form the direct current bus, and finally the three phase tension inverter of half bridge type of six power switches, in this case IGBT's. The system is designed to provide a maximum of 30 A_{rms} per phase, with a three phase maximum feeding voltage of 480 V_{rms}, and a switches commutation frequency of 20 kHz with the capacity to give a three phase output power of 20 kVA. Some parameters of the used motor are: $R_S = 7.75\ \Omega$, $R_R = 2.95\ \Omega$, $L_{S,R} = 0.47$ H, $M = 0.44$ H, $\sigma = 0.123$, $J = 0.03$ kg-m^2, $p = 4$, nominal speed 1750 rpm, feeding voltage 220 V_{RMS} to 60 Hz of frequency, full load current of 4.2 A_{RMS} per phase, full load torque of 6.1 N-m, nominal power of 1.5 hp (Fig. 3).

Experimental results for speed and position tracking were obtained during the development of the investigation. The tracking of speed is considered very important because most of the commercial variable speed drives realize the tracking

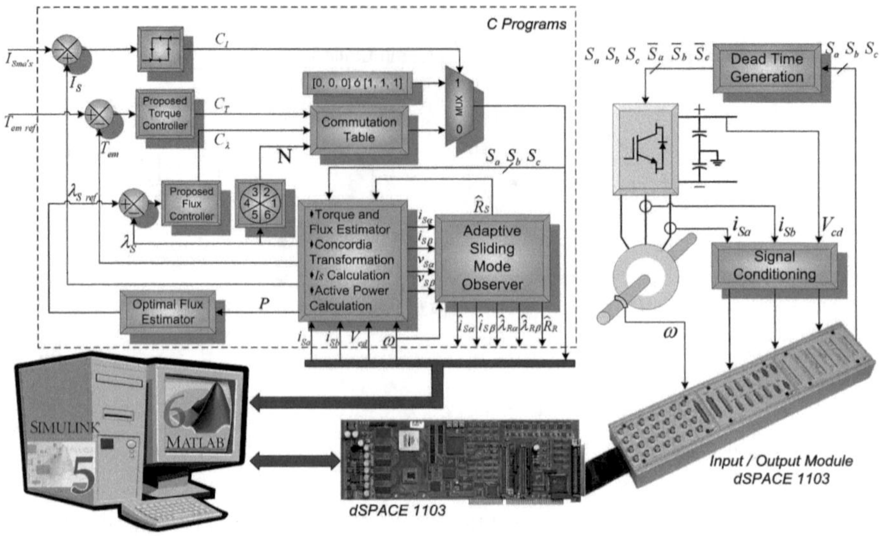

Fig. 3 General diagram of the high performance DTC

of a reference speed (function that can not be realised with the conventional DTC). For this purpose the speed measurement can be performed indirectly via an encoder (as in this case) and should be compared with a reference speed; finally, the error signal after the comparison is regulated through a proportional-integral type controller (PI), which sends its output to the torque reference in the DTC scheme. In the case of position tracking, a PID is needed to follow the reference position; this is because the original DTC only can follows a torque position because the result of the operations in the controller is a torque signal and the commutation states for the inverter IGBT's.

Figure 4 shows the behavior of the system for tracking position, which is important today for the development of many applications in mechatronics where only expensive DC servomotors can be used (specially in robotics). The experimental results shows that a perfect tracking position can be achieved by the controller with and without the measurement of the real position. This can be seen in the first graph of the Fig. 4, where a positional reference is imposed to the induction motor shaft, as can be seen, the exact target is achieved. Derived from the reference position, the reference speed is obtained and therefore can be compared with the measure of the real speed reached by the machine; a small delay in the measured signal is observed (second graph of the Fig. 4) due to a digital filter is placed after getting the measured position by the encoder.

In the third graph of the Fig. 4 can be seen the load on the motor shaft and the torque variations due to the imposed position setpoint. This result is particularly interesting, since induction motors have a higher load capacity than the DC servomotors and it is shown that even under load, can perform precise movements to change its position, tracking the reference position.

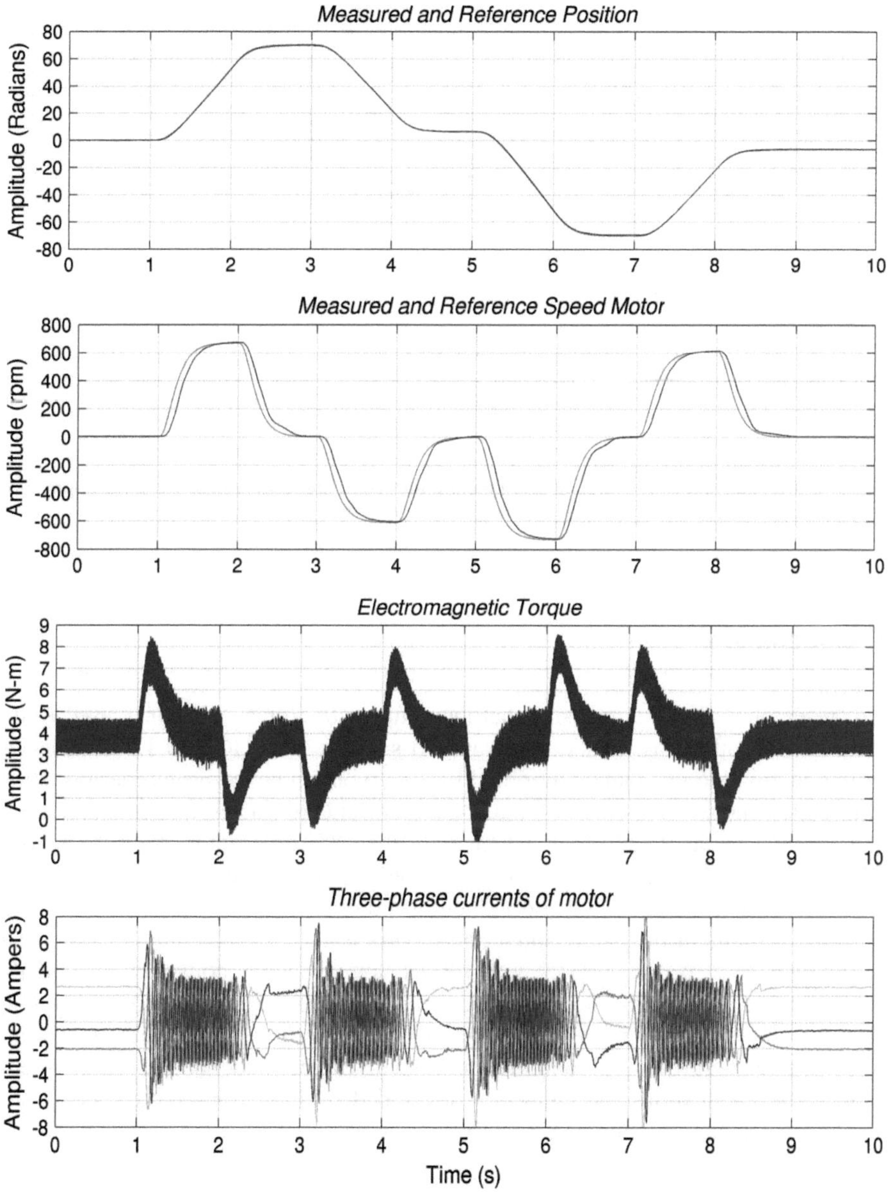

Fig. 4 Experimental results of the high performance DTC for position control

Finally in the fourth graph of the Fig. 4, the three currents consumed by the induction motor are shown. As can be seen, the currents remains constant under a constant load and they increase every time that the motor is forced to follow the reference position.

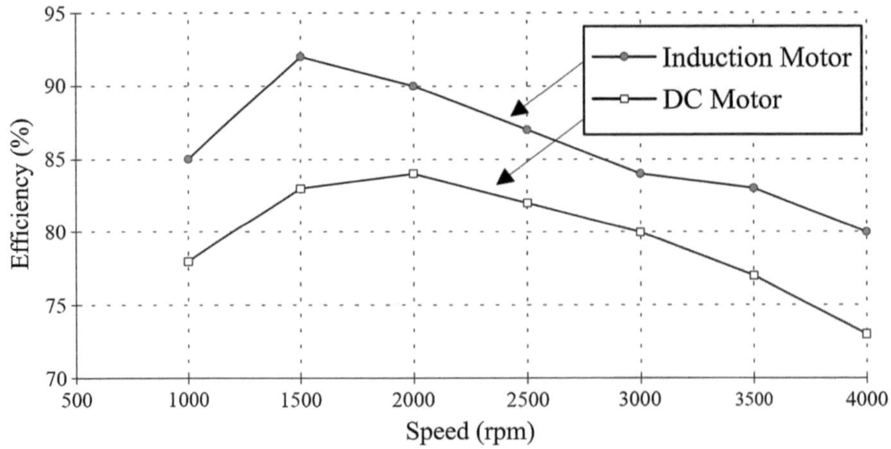

Fig. 5 Difference in the efficiency of an induction motor and a DC motor for the same application

3 Conclusion

This research seeks to emphasize that the control structures commonly used in the industry (recently in domestic applications) such as the FOC and DTC, working with induction motors, can be modified further to obtain better system performance. This paper shows that changes made in this context carries a significant good performance for applications in which only the DC servomotors have been used (like is showed in Fig. 5). With the proposed control strategy, new mechatronic applications as robotics could use induction motors.

Some benefits of the use of induction motors in the mentioned applications can be a lower cost in comparison with servos, lower energy consumption and the greater torque generated. Then it is a contribution that points to the commissioning of more efficient systems to better exploit the energy consumed, thus generating considerable energy savings, and therefore have a direct impact on the environment.

In some applications the efficiency developed by inductions motors is considerably greater than the efficiency of DC motors (Fig. 5).

Appendix

Table 1 Nomenclature used in the paper

Symbol	Description
I, J, 0	Identity, Antisymmetric and Null matrix
i_{Sa}, i_{Sb}, i_{Sc}	Stator currents in the abc reference frame
$i_{S\alpha}$, $i_{S\beta}$	Stator currents in the α-β reference frame
I_S, $I_{S\ max}$	Stator current vector magnitude and its maximum
$v_{S\alpha}$, $v_{S\beta}$	Stator voltages in the α-β reference frame
$\lambda_{R\alpha}$, $\lambda_{R\beta}$	Rotor fluxes in the α-β reference frame
λ_S, $\lambda_{S\ ref}$	Stator flux vector and its reference
T_{em}, $T_{em\ ref}$	Electromagnetic torque and its reference
C_T, C_λ, C_I	Torque, Flux and Current discrete logic signal
ω, ω_r	Electrical, Rotor mechanical angular speed
ω_S	Electrical angular speed error
R_S, R_R	Stator and rotor resistance (per phase)
L_S, L_R	Stator and rotor inductance (per phase)
M, V_{cd}	Mutual inductance, DC-bus voltage
M_S, M_R	Stator and rotor mutual inductance
σ, p	Total dispersion coefficient, Number of machine poles
J, N	Inertia moment, Sector number of flux vector localization
S_a, S_b, S_c	Commutation States
i_{ds}, i_{qs}	Direct and quadrature current comps

References

1. Arif A, Betka A, Guettaf A (2009) A minimization of torque ripple of sensorless DTC for controlled induction motor used in electric vehicles. Int Rev Electr Eng (IREE) Praise Worthy Prize 4(5):837—843
2. Boglietti A, Cavagnino A, Ferraris P, Lazzari M (2008) Energy-efficient motors. IEEE Ind Electron Mag 2(4):32—37
3. Boglietti A, Lazzari M, Profumo F (1986) About the selection of induction motors samples having representative character for energetic considerations. In: Proceedings of international conference on electrical machines (ICEM), Munchen, Germany, pp 223—226
4. Bonnett AH (1994) An update on ac induction motor efficiency. IEEE Trans Ind Appl 30 (5):1362—1372
5. Bonnett AH (1996) Regulatory impact on the application of AC induction motors. IEEE Ind Appl Mag 2(2):4—15
6. Bonnett AH, Yung C (2008) Increased efficiency versus increased reliability. IEEE Ind Appl Mag 14:29—36
7. Calzada-Lara G (2012) Estudio de técnicas de control relacionadas con el aprovechamiento de la energía en motores de inducción. Ph.D. thesis, CIEP-FI, Universidad Autónoma de San Luis Potosí, S.L.P., México, February 2012

8. Casadei D, Serra G, Tani A, Zarri L (2013) Direct torque control for induction machines: a technology status review. In: Proceedings of the IEEE workshop on electrical machines design control and diagnosis (WEMDCD), Paris, France, pp 117–129
9. Fuchsloch JF, Finley WR, Walter RW (2008) The next generation motor. IEEE Ind Appl Mag 14(1):37–43
10. Kaboli S, Zolghadri MR, Haghbin S, Emadi A (2003) Torque ripple minimization in DTC of induction motor based on optimized flux value determination. In: Proceedings of IEEE industrial electronics conference, pp 431–435
11. Kaboli S, Zolghadri MR, Roye D, Emadi A (2004) Online optimal flux controller for dtc based induction motor drives. In: Proceedings of the 30th annual conference of the IEEE industrial electronics society 2004
12. Khoucha F, Marouani K, Kheloui A, Benbouzid MH (2009) Electric vehicle induction motor DSVM-DTC with torque ripple minimization. Int Rev Electr Eng (IREE) Praise Worthy Prize 4(3):501–508
13. Leonhard W (2001) Control of electrical drives. Springer, Berlin
14. Lyshevski SE (2000) Electromechanical systems electric machines and applied mechatronics. CRC Press, Boca Raton
15. Malinowski J, McCormick J, Dunn K (2004) Advances in construction techniques of AC induction motors: Preparation for super-premium efficiency levels. IEEE Trans Ind Appl 40 (6):1665–1670
16. Moustafa-Azab MA (2002) Estudio y realización del control directo del par (DTC) para accionamientos de motores de inducción con inversores de diferentes topologías, Ph.D. thesis, Universidad Politécnica de Cataluña
17. Nik NR, Toh CL, Elbuluk ME (2006) A new torque and flux controller for direct torque control of induction machines. IEEE Trans Ind Appl 42(6):1358–1366
18. Noguchi T, Takahashi I (1985) A new quick response and high eficiency control strategy of an induction motor. In: Proceedings of the IEEE annual industry applications society meeting, pp 496–502
19. Toh CL, Idris NR, Yatim AHM (2005) Constant and high switching frequency torque controller for DTC Drives. IEEE Power Electron Lett 3(2):76–80

Modeling and Control of a Pendubot with Static Friction

Sergio Sánchez-Mazuca, Israel Soto and Ricardo Campa

Abstract The pendubot is an underactuated mechanical system with two degrees of freedom but only one control input. Several control techniques have been applied for swinging up and balancing the pendubot, but it is common to neglect the effects of friction in the joints of the system. In this paper we take a different approach by considering that the first joint of the pendubot has a significant amount of friction and it can not be neglected. We first review some friction models, including one that has been recently proposed by the authors. Secondly, we use some simple swing up and balancing controllers which compensate the friction effects. At the end, the results allow us to validate not only the viability of this approach, but also the proper identification of the friction model.

Keywords Pendubot · Static friction · Modeling · Control

1 Introduction

The pendubot is an inverted-pendulum-type system which consist of a planar robot with the first joint (shoulder) actuated an the second joint (elbow) unactuated [1]; i.e., it is a system with two degrees of freedom but with only one control input, for this reason, the pendubot, is considered as an underactuated system.

The pendubot has four equilibrium points, termed here as down–down, down–up, up–down and up–up, depending on the vertical position taken by each of the two links. Only the down-down equilibrium is stable, and to reach the top-top one is usually the control aim.

The pendubot is also a second-order nonholonomic system [2] and it is considered a nonminimum-phase system [3]. Because of these features the pendubot is

---

S. Sánchez-Mazuca · I. Soto · R. Campa (✉)
Instituto Tecnológico de la Laguna, Blvd. Revolución y Cuauhtémoc, Torreón, Coah, Mexico
e-mail: recampa@ieee.org

M. Ceccarelli and E.E. Hernández Martinez (eds.), *Multibody Mechatronic Systems*,
Mechanisms and Machine Science 25, DOI 10.1007/978-3-319-09858-6_22

considered an underactuated benchmark system used for education and research in nonlinear control theory.

The problem of controlling the pendubot can be divided in two control objectives: the *swing–up control* and the *balancing control*. The swing–up control consists in applying to the actuated joint the energy required to make the second link (a.k.a. the pendulum) swing up in order to reach the upright vertical position (up–up equilibrium); and the balancing control is required to stabilize the pendulum at such unstable equilibrium.

A great variety of control algorithms have been proposed to make the pendubot swing up, and they are based on different control techniques, such as partial linearization control [1], intelligent control [4], hybrid control [5, 2], energy-based control [6, 7], and nonlinear control [8]. For the balancing control most of the researchers use the method based on a linear quadratic regulator (LQR).

Nevertheless, there are many factors that affect the behavior of these controllers, being one of the most important the unmodeled dynamics of the system, such as the friction effects. Therefore, in order to make the pendubot have a better performance under control, it is desirable to have a friction model that best describes the actual behavior of this phenomenon.

The aim of this paper is twofold. First we recall some of the friction models found in the literature, including a modification to some of them that has recently been proposed by the authors in [9]. Secondly, we show how the compensation of friction can be implemented to improve the performance of a direct-drive pendubot with significant friction effects (see Fig. 1).

It is worth mentioning here that, compared with [9], the main contribution of this paper is the application of friction compensation to a complete underactuated system, such as the pendubot, with significant static friction. Practically all of the papers dealing with the control of underactuated systems assume that joint friction is negligible.

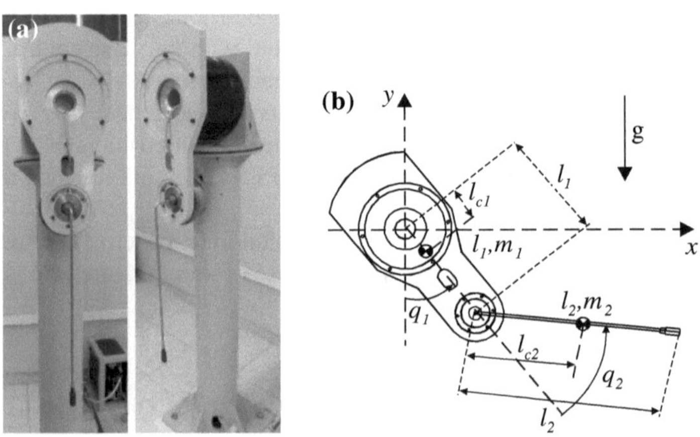

Fig. 1 Experimental direct-drive pendubot, **a** Actual prototype, **b** Definition of parameters

The remainder of this paper is organized as follow. The review of friction models is given in Sect. 2. Both the non-linear and linearized dynamic models of the pendubot are given in Sect. 3. The control techniques used to swing-up and balance the pendubot are mentioned in Sect. 4. The experimental results are presented in Sect. 4, while some concluding remarks are given in Sect. 5.

2 Friction Models

Friction is a force acting against the relative motion between two surfaces in contact. The phenomenon is present in all mechanical systems including rotating actuators. Throughout the time, several models have been proposed, each trying to capture the essence of the complicated phenomenon of friction; however, finding a model to describe all its effects is difficult.

Moreover, friction is generally divided into static friction and kinetic friction; the former corresponds to all the effects which are present without a visible motion, while the latter refers to everything what happens once that motion exists.

Consider a motor with friction torque f; let q, $\dot{q}$ and $\ddot{q}$ be, respectively, the angular displacement, velocity and acceleration of the motor shaft; also, let τ be the external torque applied to the motor to produce the motion. The relation among these quantities is given by the motor dynamics

$$J\ddot{q} + f = \tau \tag{1}$$

where J stands for the moment of inertia of the motor.

It is common to assume that the friction torque of a motor depends only on the velocity, i.e. $f(\dot{q})$. However, in 1833, Morin [10] established the first model for static friction, stating that such friction can not be described satisfactorily only as a function of velocity but also of the applied torque.

Morin's static friction model [10] is defined for $\dot{q} = 0$, and is given by

$$f(0,\tau) = \begin{cases} \tau, & \text{if} \quad |\tau| < f_s \\ f_s \operatorname{sgn}(\tau), & \text{if} \quad |\tau| \geq f_s \end{cases} \tag{2}$$

where the sign function is defined as

$$\operatorname{sgn}(x) = \begin{cases} 1 & \text{if} \quad x > 0 \\ 0 & \text{if} \quad x = 0 \\ -1 & \text{if} \quad x < 0 \end{cases}$$

and f_s is the maximum static friction coefficient, which is assumed to be constant.

With respect to kinetic friction, it is present when $\dot{q} \neq 0$ and does not depend on τ, so it is denoted as $f_k(\dot{q})$. One of the most common models for describing kinetic friction is the one combining the Coulomb friction and the viscous friction effects; this model can be written as:

$$f_k(\dot{q}) = f_c \operatorname{sgn}(\dot{q}) + f_v \dot{q} \tag{3}$$

where f_c and f_v are the Coulomb and viscous friction coefficients, respectively.

In 1902, Stribeck carried out some experiments at a constant velocity of low amplitude [11], showing the existence of the phenomenon now known as Stribeck effect, that describes the transition between the maximum static friction and the Coulomb friction. One of the expressions more used to model this phenomenon is the following exponential function known as the Armstrong model [12]:

$$f_k(\dot{q}) = \operatorname{sgn}(\dot{q})\left[f_c + (f_s - f_c)e^{-\left|\frac{\dot{q}}{v_s}\right|^{\delta_v}}\right] + f_v \dot{q} \tag{4}$$

which not only incorporates the static friction, Coulomb friction and viscous friction coefficients but also includes two more parameters: the Stribeck velocity coefficient v_s and a fitting parameter δ_v.

The static friction model (2) can be combined with kinetic friction in the following way,

$$f(\dot{q}, \tau) = \begin{cases} \tau, & \text{if} \quad \dot{q} = 0, |\tau| < f_s \\ f_s \operatorname{sgn}(\tau), & \text{if} \quad \dot{q} = 0, |\tau| \geq f_s \\ f_k(\dot{q}), & \text{if} \quad \dot{q} \neq 0 \end{cases} \tag{5}$$

where $f_k(\dot{q})$ can be (3, 4), or any other kinetic friction model (with $\dot{q} \neq 0$).

In 1968 Dahl proposed a dynamic friction model based on the behavior of the roughness between two surfaces in contact [13]. Dahl considers that this roughness can be modeled as the bristles of two brushes facing each other; thus, if z represents the average deflection of the bristles, then z times a stiffness coefficient σ_0 gives the static friction torque. If viscous friction effect is included then the Dahl model is given by,

$$f(\dot{q}, z) = \sigma_0 z + f_v \dot{q} \tag{6}$$

$$\dot{z} = -\frac{\sigma_0 |\dot{q}|}{f_c} z + \dot{q} \tag{7}$$

The so-called LuGre model [14] is a generalization of the Dahl model, and can also be expressed in terms of the velocity $\dot{q}$ and the average bristle deflection z, but also includes a term proportional to $\dot{z}$, that is

$$f(\dot{q}, z) = \sigma_0 z + \sigma_1 \dot{z} + f_v \dot{q}, \tag{8}$$

$$\dot{z} = -\frac{\sigma_0|\dot{q}|}{g(\dot{q})}z + \dot{q}, \tag{9}$$

$$g(\dot{q}) = f_c + (f_s - f_c)e^{-|\frac{\dot{q}}{v_s}|^{\delta_s}}; \tag{10}$$

where σ_1 is the damping coefficient of the bristles, and the function $g(\dot{q})$ allows to extend Dahl model to describe Stribeck effect, employing the same exponential function as in (4).

2.1 Improved Model for Static Fricción

Most of the friction models do not consider the variation of the break-away point; they only take into account the maximum static friction, given by f_s. But some well–documented observations such as those in [15] show that the break–away point is a function of the rate of change of the applied torque. Recently, in [9], an explicit expression for such a function (named $\phi_s(\dot{\tau})$) is given, and it is also proposed to replace f_s by $\phi_s(\dot{\tau})$ in those friction models including such parameter. As a result some new friction models are obtained, for which friction depends not only on $\dot{q}$ and τ, but also on $\dot{\tau}$.

Thus, the Armstrong model (4) becomes

$$f_k(\dot{q}, \dot{\tau}) = \text{sgn}(\dot{q})[f_c + (f_s - f_c)e^{-(|\frac{\dot{q}}{v_s}|^{\delta_v} + |\frac{\dot{\tau}}{t_s}|^{\delta_t})}] + f_v\dot{q}. \tag{11}$$

Moreover, the general static model (5) can be rewritten as

$$f(\dot{q}, \tau, \dot{\tau}) = \begin{cases} \tau, & \text{if} \quad \dot{q} = 0, |\tau| < \phi_s(\dot{\tau}) \\ \phi_s(\dot{\tau})\text{sgn}(\tau), & \text{if} \quad \dot{q} = 0, |\tau| \geq \phi_s(\dot{\tau}) \\ f_k(\dot{q}, \dot{\tau}), & \text{if} \quad \dot{q} \neq 0 \end{cases} \tag{12}$$

Now, the LuGre model (8–10) with $\phi_s(\dot{\tau})$ instead of f_s, results in

$$f(\dot{q}, z, \dot{\tau}) = \sigma_0 z + \sigma_1 \dot{z} + f_v\dot{q}, \tag{13}$$

$$\dot{z} = -\frac{\sigma_0|\dot{q}|}{g(\dot{q}, \dot{\tau})}z + \dot{q}, \tag{14}$$

$$g(\dot{q}, \dot{\tau}) = f_c + (f_s - f_c)e^{-(|\frac{\dot{\tau}}{t_s}|^{\delta_t} + |\frac{\dot{q}}{v_s}|^{\delta_v})} \tag{15}$$

Models (11, 13–15) are the improved versions of the Armstrong model (4) and LuGre model (8–10), respectively.

3 Pendubot System

In this section we describe the pendubot system that we have built at the Mechatronics and Control Laboratory of the Instituto Tecnológico de la Laguna. The prototype is shown in Fig. 1a. The first joint is a direct-drive brushless DC servoactuator, model DM1200A from Parker Compumotor, with a high–resolution encoder of 10,24,000 pulses per revolution, which is mounted on a pedestal; the first link is made out of aluminum while the pendulum is a steel rod; the second joint is an industrial encoder from Precizike Metrology, which was configured to have 14,40,00 pulses per revolution.

3.1 Dynamic Model

Considering the Lagrange's equations of motion the dynamic model of the pendubot is given by

$$M(\boldsymbol{q},\boldsymbol{\theta}_I)\,\ddot{\boldsymbol{q}} + C(\boldsymbol{q},\dot{\boldsymbol{q}},\boldsymbol{\theta}_I)\dot{\boldsymbol{q}} + \boldsymbol{g}(\boldsymbol{q},\boldsymbol{\theta}_I) + \boldsymbol{f}(\dot{\boldsymbol{q}},\boldsymbol{\tau},\dot{\boldsymbol{\tau}},\boldsymbol{\theta}_f) = \boldsymbol{\tau} \tag{16}$$

where $\boldsymbol{q}$, $\dot{\boldsymbol{q}}$, $\ddot{\boldsymbol{q}} \in \mathbb{R}^n$ are respectively the vectors of joint displacements, velocities and accelerations, $M(\boldsymbol{q},\boldsymbol{\theta}_I) \in \mathbb{R}^{n\times n}$ is the inertia matrix, $C(\boldsymbol{q},\dot{\boldsymbol{q}},\boldsymbol{\theta}_I) \in \mathbb{R}^{n\times n}$ is the matrix of Coriolis and centripetal torques, $\boldsymbol{g}(\boldsymbol{q},\boldsymbol{\theta}_I) \in \mathbb{R}^n$ is the vector of gravitational terms, $\boldsymbol{f}(\dot{\boldsymbol{q}},t,\dot{t},\boldsymbol{\theta}_f) \in \mathbb{R}^n$ is the vector of friction terms, and $\tau \in \mathbb{R}^n$ is the vector of external torques applied to the pendubot. $\boldsymbol{\theta}_I$ is the vector of inertial parameters, which is taken as

$$\boldsymbol{\theta}_I = \begin{bmatrix} \theta_1 \\ \theta_2 \\ \theta_3 \\ \theta_4 \\ \theta_5 \end{bmatrix} = \begin{bmatrix} m_1 l_{c1}^2 + m_2 l_1^2 + I_1 \\ m_2 l_{c2}^2 + I_2 \\ m_2 l_1 l_{c2} \\ m_1 l_{c1} + m_2 l_1 \\ m_2 l_{c2} \end{bmatrix}$$

where m_i, l_{ci}, l_i and I_i for $i = 1, 2$. Correspond to the mass, the distance to center of mass, the total length and the moment of inertia of the link i, respectively. $\boldsymbol{\theta}_f$ is the vector of friction parameters whose number depends on the chosen friction model. Then the elements of model (16) are given by

$$M(q,\theta_I) = \begin{bmatrix} \theta_1 + \theta_2 + 2\theta_3\cos(q_2) & \theta_2 + \theta_3\cos(q_2) \\ \theta_2 + \theta_3\cos(q_2) & \theta_2 \end{bmatrix}$$

$$C(q,\dot{q},\theta_I) = \theta_3\sin(q_2)\begin{bmatrix} -\dot{q}_2 & -(\dot{q}_1 + \dot{q}_2) \\ \dot{q}_1 & 0 \end{bmatrix}$$

$$g(q,\theta_I) = g\begin{bmatrix} \theta_4\sin(q_1) + \theta_5\sin(q_1+q_2) \\ \theta_5\sin(q_1+q_2) \end{bmatrix}$$

$$\tau = \begin{bmatrix} \tau_1 \\ 0 \end{bmatrix}, \qquad f(\dot{q},\tau,\theta_f) = \begin{bmatrix} f_1(\dot{q}_1,\tau_1,\theta_{f1}) \\ f_2(\dot{q}_2,\theta_{f2}) \end{bmatrix}$$

notice that τ_1 corresponds to the unique external torque applied to the actuated joint.

3.1.1 Linearized Model

In this section the pendubot's nonlinear model (16) is linearized around the up-up equilibrium position.

Considering that the states of the system are defined by $x = [q_1 \quad q_2 \quad \dot{q}_1 \quad \dot{q}_2]^T \in \mathbb{R}^4$ and neglecting the friction term of (16), it is possible to rewrite the dynamics in state space as the following nonlinear model

$$\dot{x} = \phi(x) + \gamma(x)\tau_1 \tag{17}$$

where

$$\phi(x) = \begin{bmatrix} x_3 \\ x_4 \\ M^{-1}(x)\left(-C(x)\begin{bmatrix} x_3 \\ x_4 \end{bmatrix} - g(x)\right) \end{bmatrix} \quad \text{and} \quad \gamma(x) = \begin{bmatrix} 0 \\ 0 \\ M^{-1}(x)\begin{bmatrix} 1 \\ 0 \end{bmatrix} \end{bmatrix}$$

Nonlinear system (17) can be linearized around any point $x = x_e$ to get

$$\dot{x} = Ax + b\tau_1 \tag{18}$$

where

$$A = \left.\frac{\partial\phi(x)}{\partial x}\right|_{x=x_e} \quad \text{and} \quad b = \gamma(x)|_{x=x_e}$$

by taking $x_e = [\pi \quad 0 \quad 0 \quad 0]^T$ we get

Table 1 Estimated friction parameters

Parameter	Value (+)	Value (−)	Unit	Parameter	Value (+)	Value (−)	Unit
f_c	8.955	10.01	Nm	δ_v	1.032	1.286	
f_v	5.234	5.335	Nms/rad	δ_t	0.172	0.287	
f_s	17.617	23.813	Nm	σ_0	17260		Nm/rad
v_s	0.145	0.078	rad/s	σ_1	66.708		Nm/rad
t_s	0.888	1.954	Nm/s				

$$A = \begin{bmatrix} 0 & 0 & 1 & 0 \\ 0 & 0 & 0 & 1 \\ \frac{(\theta_2\theta_4-\theta_3\theta_5)g}{\theta_1\theta_2-\theta_3^2} & -\frac{\theta_3\theta_5 g}{\theta_1\theta_2-\theta_3^2} & 0 & 0 \\ \frac{(\theta_1+\theta_3)\theta_5 g-(\theta_2+\theta_3)\theta_4 g}{\theta_1\theta_2-\theta_3^2} & \frac{(\theta_1+\theta_3)\theta_5 g}{\theta_1\theta_2-\theta_3^2} & 0 & 0 \end{bmatrix}, \text{ and } b = \begin{bmatrix} 0 \\ 0 \\ \frac{\theta_2}{\theta_1\theta_2-\theta_3^2} \\ -\frac{(\theta_2+\theta_3)}{\theta_1\theta_2-\theta_3^2} \end{bmatrix}$$

3.2 Parameter Identification

The inertial parameters were identified by using standard least-square procedures. After that, the estimation of the five inertial parameter resulted in $\theta_1 = 1.059$ [kg·m^2], $\theta_2 = 0.053$ [kg·m^2], $\theta_3 = 0.037$ [kg·m^2], $\theta_4 = 2.401$ [kg·m^2] and $\theta_5 =$ 0.117, [kg·m]. Regarding the friction term we employed the procedure proposed in [9] to obtain the nine parameters of the modified Lugre friction model (13–15) for the actuated joint of the pendubot. The estimated friction parameter are given in Table 1. The friction effect is neglected for the second link.

4 Experimental Evaluation

4.1 Swing–Up Controller

Due to significant friction effects in the first joint of our pendubot it was not possible to use known swing-up controllers such as those reported in [6, 1]. Thus, in order to make the pendubot swing up and reach the desired up-up equilibrium we tried a heuristic approach which employ a PD-like controller given by

$$\tau_1 = k_p \tanh(q_{d1} - q_1) - k_v\dot{q}_1 \tag{19}$$

to take the first link to its desired upright position (where $q_{d1} = \pi$). After that, in order to make the pendulum start an increasing oscillation, we switched the value of q_{d1} according to the following rule

$$q_{d1} = \begin{cases} \pi + \alpha, & \text{if} \quad q_2 \geq -\pi \\ \pi - \alpha, & \text{if} \quad q_2 < -\pi \end{cases}$$

where α initially starts with a value of $\frac{7\pi}{180}$, but once the condition $|q_2 - \pi| > \frac{\pi}{3}$ is reached by the first time, then $\alpha = \frac{2.3\pi}{180}$. For the experiments presented in this section we chose $k_p = 58$ [Nm], $k_v = 15$ [Nms].

4.2 Balancing Controller

Once the swing–up controller takes the pendulum near the up-up equilibrium (by entering a window given by $|q_1 - \pi| < \frac{\pi}{18}$ and $|q_2| < \frac{\pi}{15}$), the balancing controller starts its operation trying to establish the pendubot around the equilibrium.

Following the custom we decided to use an LQR controller which was designed employing the traditional procedure on the linearized dynamic model (18) and adding a term for friction compensation. The control law is given by

$$\tau_1 = -Kx + f_1(\dot{q}_1, \tau_1, \theta_{f1}) \tag{20}$$

After some tests we decided to use $K = \text{diag}\{-267.262, -267.282, -51.5, -35.0\}$.

4.3 Experimental Results

In order to verify the benefit of including friction models in the control law of the pendubot we performed the following experiments, intended to compare similar friction models:

1. The static + Armstrong friction model (4–5) versus the improved static + Armstrong friction model (11–12).
2. The LuGre friction model (8–10) versus the improved LuGre friction model (13–15).

Figure 2 shows the time evolution the joint displacements q_1, q_2, and the applied torque τ_1 for the experiments considering the static + Armstrong (S + A) friction model and the improved static + Armstrong (I(S + A)) friction model.

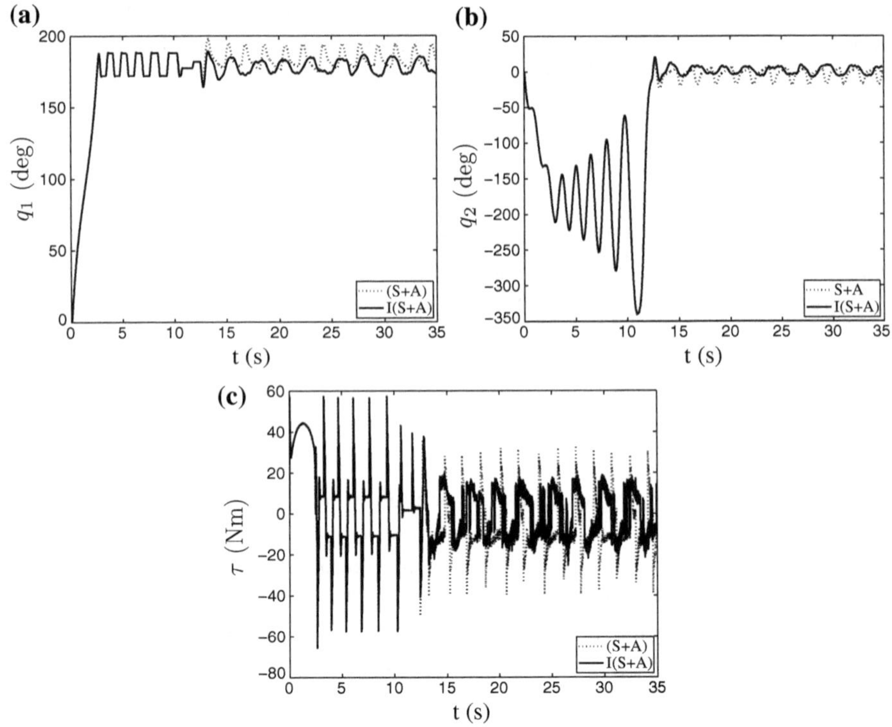

Fig. 2 Experiments using Armstrong model, **a** Position of the joint q_1, **b** Position of the joint q_2, **c** Torque of the joint q_1

Regarding the experiment that considers the the LuGre (LG) friction model and the improved LuGre (I(LG)) friction model, Fig. 3 shows the corresponding graphs.

It is noticed that for the improved friction models the position error is smaller and the applied torque has a lower amplitude than for those not having the improvement.

In order to quantify the position error and compare the different friction compensation methods, we employed the standard root mean square (RMS) value of the norm of the position error (defined as $e = \sqrt{(\pi - q_1)^2 + q_2^2}$) as a performance index. Table 2 show the values of $RMS(e)$ for the four controllers tested. We notice that the errors are smaller when we use the improved versions of the friction models.

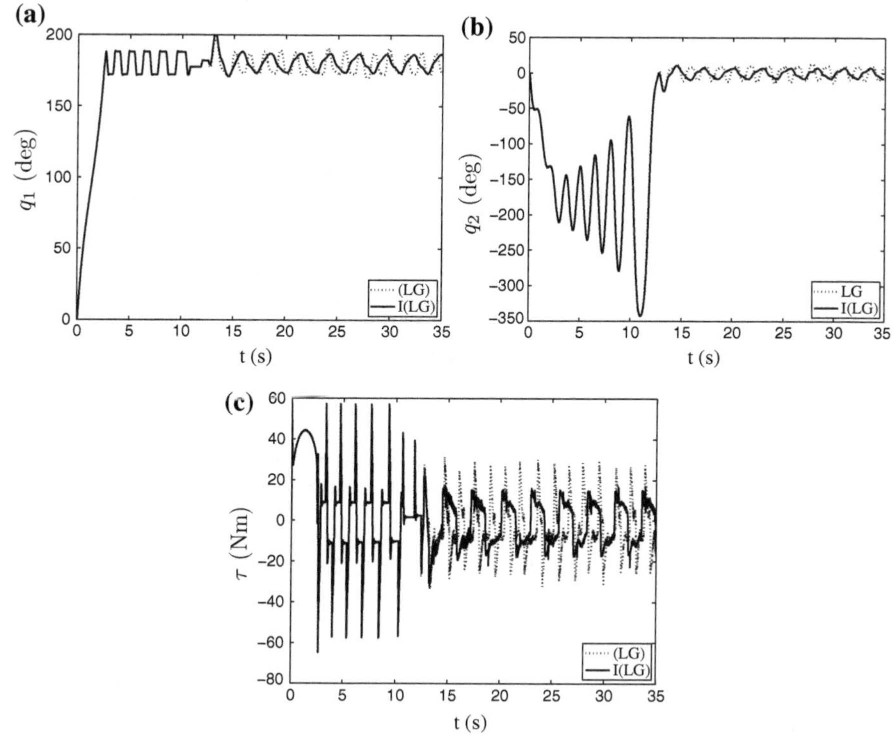

Fig. 3 Experiments using Lugre model, **a** Position of the joint q_1, **b** Position of the joint q_2, **c** Torque of the joint q_1

Table 2 Performance considering different friction models

Friction model	S + A	I(S + A)	LG	I(LG)
RMS(*e*)	7.7771	4.9842	7.3153	4.8543

5 Conclusions

This paper shows the performance of a direct–drive pendubot with significant friction effects in its first joint. After a swing–up phase, which employs a simple control law, the estabilization of the system around the up-up unstable equilibrium is achieved using an LQR controller with friction compensation. Experimental results show the viability of this approach, and also validate the improvement to the static friction model recently proposed in [9].

Acknowledgments This paper was partially supported by PROMEP, DGEST and CONACYT Mexico.

References

1. Spong MW, Block DJ (1995) The pendubot: a mechatronic system for control research and education. In: Proceedings of the conference on decision and control
2. Zhang MJ, Tarn TJ (2002) Hybrid control of the pendubot. IEEE Trans Mechatron 7:79–86
3. Guemghar K (2001) On the use of input-output feedback linearization techniques for the control of nonminimum-phase systems. Ph.D. thesis, École Polytechnique Fédérale de Lausanne, Lausanne, Switzerland
4. Ma XQ, Su CY (2002) A new fuzzy approach for swing up control of pendubot. In: Proceedings of the American control conference. Anchorage, AK
5. O'Flaherty RW, Sanfelice RG, Teel AR (2008) Hybrid control strategy for robust global swing-up of the pendubot. In: Proceedings of the American control conference
6. Fantoni I, Lozano R, Spong MW (2000) Energy based control of the pendubot. IEEE Trans Autom Control 45:79–86
7. Xin X, Kaneda M, Oki T (2002) The swing-up control for the pendubot based on energy control approach. In: Proceedings of the 15th IFAC world congress, pp 461–466
8. Wang W, Yi J, Zhao D, Liu X (2004) Adaptive sliding mode controller for an underactuated manipulator. In: Proceedings of the international conference on machine learning and cybernetics
9. Sanchez-Mazuca S, Campa R (2013) An improvement proposal to the static friction model. In: Proceedings of Mathematical problems in engineering 2013
10. Morin A (1833) New friction experiments carried out at metz in 1831–1833. Proc French R Acad Sci 4:1–128
11. Stribeck R (1902) The key qualities of sliding and rolling bearings. Z Ver Dtsch Ing 46:1342–1348
12. Armstrong-Helouvry B (1991) Control of machines with friction. Kluwer Academic Publishers, Boston
13. Dahl P (1968) A solid friction model. Technical report TOR-0158(3107-18)-1, The Aerospace Corporation
14. Canudas C, Olson H, Astrom K, Lischinsky P (1995) A new model for control of system with friction. IEEE Trans Autom Control 40:419–425
15. Johannes VI, Green MA, Brockley CA (1973) The role of the rate of application of the tangential force in determining the static friction coefficient. Wear 24:381–385

Design of Electronic Control Board to Obtain the Photovoltaic Module Power Voltage Curve as Temperature Function

J. Vega-Pérez, S. Vega-Pérez and L. Castañeda-Aviña

Abstract The test, construction and design of an electronic circuit in order to obtain the photovoltaic module power voltage curve as a function of temperature is reported. The electronic circuit permits to capture the current and voltage variables, processing the electrical signals and determines the power. After using a commercial plotter, the characteristic curve of photovoltaic module is drawn as a function of temperature level. This information is used for the engineers who design a photovoltaic system, to supply electric energy from solar energy, for activating any electrical, electronic or electro mechanical equipment. The electronic circuit was made using electronic and electrical components bought in Mexico. Experimental results indicate that the developed electronic circuit works fine and agrees with the technical design, and it can measure photovoltaic modules from 1 to 50 W from the short circuit condition to the open circuit condition at 23 V with the current level varying between 0—2 A, with a measurement error of about 1 %.

Keywords Photovoltaic module · PV module power voltage curve · PV module current voltage curve

1 Introduction

The photovoltaic (PV) module is an electronic device which transforms the visible light energy into direct current electrical energy [1]. The output power of the PV module depends on its area, incident light level and conversion efficiency basically,

J. Vega-Pérez (✉) · S. Vega-Pérez · L. Castañeda-Aviña
Escuela Superior de Ingeniería Mecánica y Electrica Ticoman—IPN, Mexico City, Mexico
e-mail: jvegap@ipn.mx

S. Vega-Pérez
e-mail: svega@ipn.mx

L. Castañeda-Aviña
e-mail: otilioh@yahoo.com

M. Ceccarelli and E.E. Hernández Martinez (eds.), *Multibody Mechatronic Systems*, Mechanisms and Machine Science 25, DOI 10.1007/978-3-319-09858-6_23

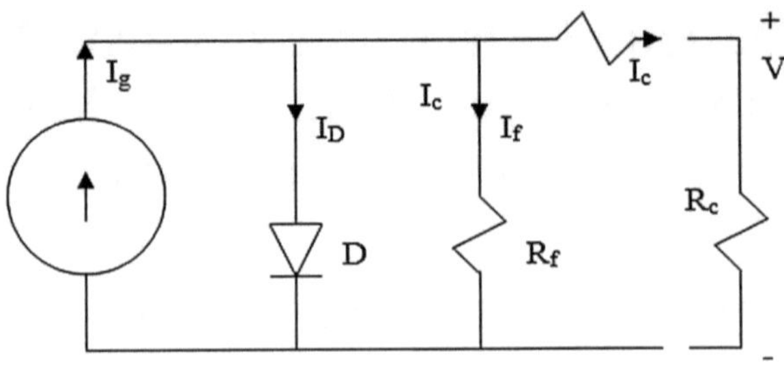

Fig. 1 Equivalent circuit of PV cell

but it is reduced by increase in temperature [2]. The PV module is represented in terms of an electrical circuit as shown in Fig. 1.

From analysis of Fig. 1 circuit, the output power of PV module is obtained using the generated current (Ig), output voltage (Vc) serial resistance (Rs), leakage resistance (R_f), output current (Ic), the leakage current (I_0), electron charge (q), Boltzmann constant (K), and output power (Pc), where the supply power is mathematically expressed by Eq. (1) [2].

$$P_c = V_c\left[I_g - I_0\left(e^{\frac{q(V_c+I_cR_s)}{KT}} - 1\right) - \frac{I_cR_s + V_c}{R_f}\right] \qquad (1)$$

The experimental work was done using the PV module which was composed by 36 PV cells of 3.5 diameter inches, connected in serial and made in I. P. N-Mexico. The current is the same as that of a PV cell with short circuit current 1.1 A but the module voltage increases which is given by 36(0.53) = 19 V for the open circuit [2]. It is important to indicate that it agrees with experimental work and the module temperature increase around 1.75 times the local environmental temperature [3]. So its Power-Voltage characteristic curve is necessary as it shows us the current voltage and power, which is useful for the design engineer [4].

When Eq. 1 is solved changing the voltage "Vc" from open circuit condition to short circuit condition, all the values of power and voltage are obtained, in order to get the power—voltage (P—V) curve of the PV module (Fig. 2).

From Fig. 2 it is observed that the PV module has only a maximum power point, and so in case of open circuit as well as short circuit the output power supplied by PV module is zero. It is necessary to detect and capture the maximum power point for taking advantage of the power of the PV module.

Some researchers have developed electronic systems in order to locate the power of PV modules. Henry Shu-Hung Chung et al. [1], detected a maximum power point for solar panels using a SEPIC or Cuk converter. Miguel Goncalves Wanzeller et al. [5] developed a Current control loop for the tracking of maximum power point supplied by photovoltaic array, Munji et al. [6] also worked on this topic with mathematical models to analyze electrical circuit of PV module and employed Newton's iterative method to get the I—V curve. Quaschning et al. [7] started from a numerical algorithm to get the current and voltage from PV cell. Analyzing theses methods it was found that Henry Sha-Hungchung et al., located the maximum power

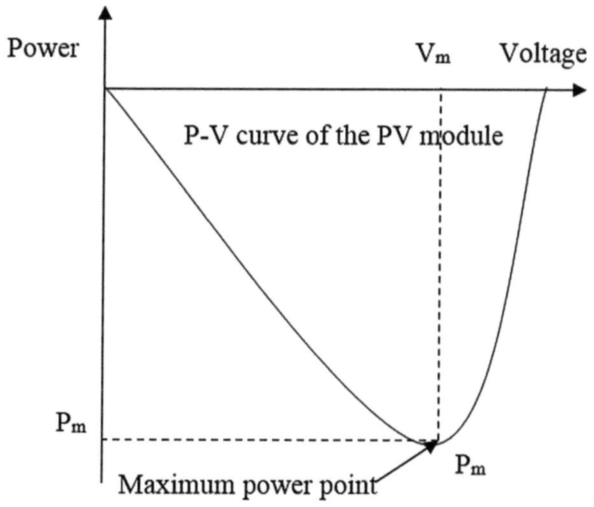

Fig. 2 Power-voltage theoretical curve of the PV module

point of the solar panel connecting a pulse width modulate (PWM) and a converter direct current to direct current (DC/DC) between solar panel and battery. Mungy et al. measured electrical parameter of the photovoltaic concentrator cell under illumination but he did not get the electronic circuit design. Miguel Goncalves Wanzeller et al. located the maximum power of the solar array, using a three phase static direct current converter to alternate current (DC/AC), but he did not consider the temperature. Quasching et al. proposed a mathematical model for description of a photovoltaic generator using a numerical algorithm, which was simulated on computer, but he did not design any electronic circuit.

In this work we developed and experimental method designing an analogical electronics circuit to get power—voltage and current—voltage (I—V) curves of the PV module.

2 Proposal Technique

Our proposal takes advantage of the property of electrical auto regulation of the PV module to get its power and voltage when the module is lighted. An electronic system of variable resistance is connected to PV module as it is shown in the Fig. 3. The electric signals of current and voltage are captured by varying the resistance values. After the current and voltage were processed to obtain the electrical power, the power and voltage signals are sent to plotter to get the characteristic curve of power versus voltage. The variable resistance is controlled by an electronic circuit, which was designed to work from the short circuit condition to the open circuit condition of the PV module. The electronic control circuit starts by fixing the resistance values to zero so that the PV module can be fixed in the short circuit

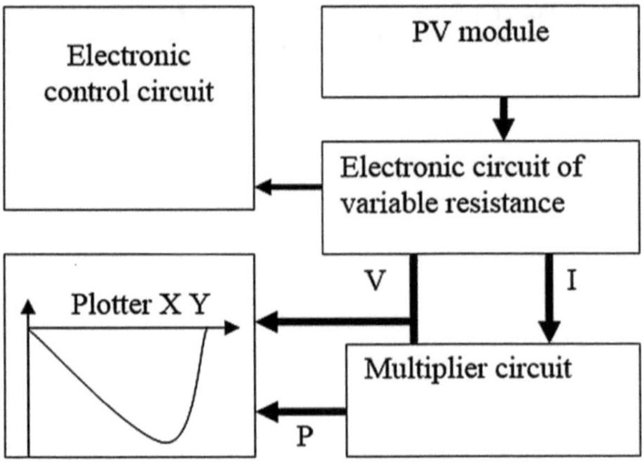

Fig. 3 Block diagram of proposed electronic system

point. Later the resistance is increased and at the same time the values of current and voltage are captured. This process is repeated until the resistance reaches high value (~10 KΩ). In this situation the solar module is fixed in the open circuit condition, and the I—V characteristics curve of the PV module is obtained.

2.1 Electronic Design

An electronic circuit which uses a MOSFET power transistor and FET integrated circuit (IC) was designed to reflect variable resistance (Fig. 4). The transistor drains the current of the PV module but it is driven from IC, then the power transistor works as an extension of integrated circuit itself. With this form, it is possible to sweep the current of the module from short to open circuit controlling the input voltage (Ve). From electronic circuit as shown in Fig. 4, the current from PV module can be determined, by using the equation given by Eq. 2

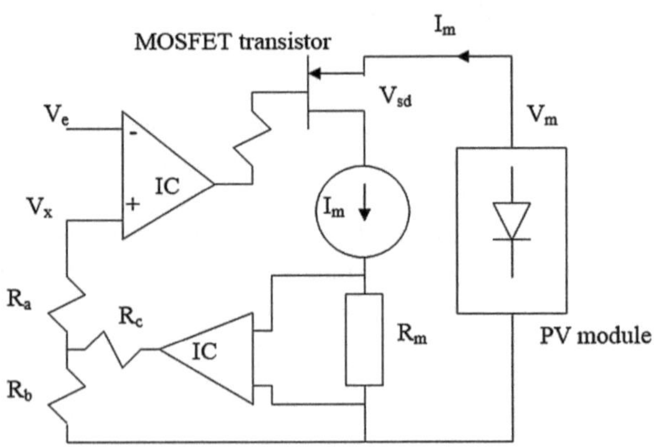

Fig. 4 Block diagram of current voltage electronic circuit

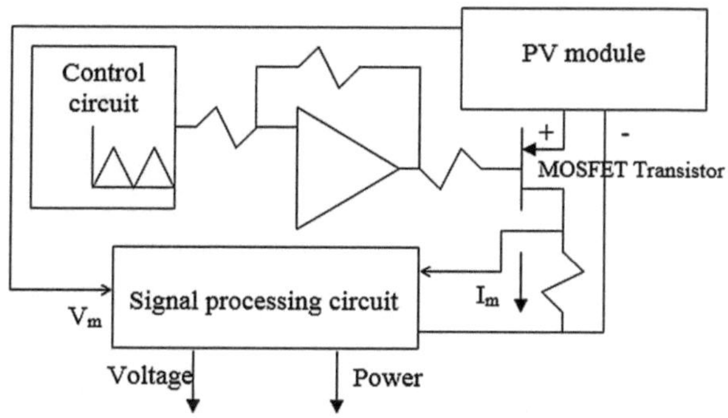

Fig. 5 Block diagram of proposed electronic circuit for obtain the power voltage curve of PV module

$$I = \frac{Vm - Vsd}{R_m} \quad (2)$$

Sawtooth oscillator circuit (Fig. 5) was designed with low frequency of 0.25 Hz using analogical and digital integrated circuits [8] to synchronize the I—V curve of the PV module.

3 Development

An integrated circuit for instrumentation was used to amplify the current and voltage. A logarithmic integrated circuit to processing the current and voltage for determining the power also was used. A MOSFET [8] power transistor polarized in active region reflects a variable electrical resistance, and so it was used to work as a amplifier. Comparator integrated circuit and transistors of low level signals were used to generate Sawtooth signal to change the resistance level of the MOSFET transistor. Printed circuit was designed using Protel software; all the electronics devises were purchased in Mexico. FET operational amplifiers were used to avoid information loss. The power transistor was installed on a heat sink to avoid the damage to the transistor as shown below in Fig. 6.

Fig. 6 Developed electronic circuit

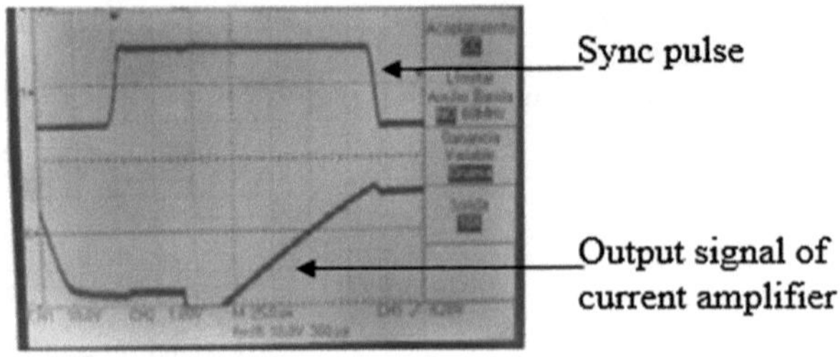

Fig. 7 Waveforms of developed electronics circuit

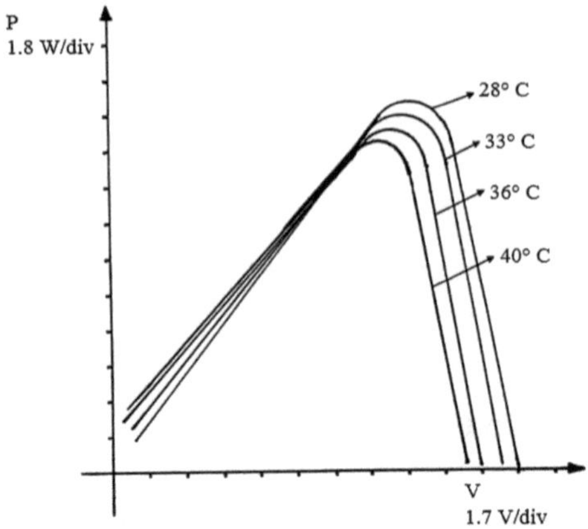

Fig. 8 P-V curves of the PV module measured from the developed electronic system

4 Measurements and Results

Electrical measurement were done on a PV module of 20 W having a PV cell of 3.5 inch diameter manufactured in CINVESTAV-IPN Mexico, which supplies 20 W, 1.1 A and 17 V at the maximum power point. It uses a developed electronic card, a plotter and a solar simulator calibrated to 960 W/m^2 approximately. The waveforms were obtained from developed electronic circuit (Fig. 7).

The power-voltage curves of PV module of 20 W were obtained from the developed electronic system for different temperature and the lighting level was calibrated on 960 W/m^2. The experimental results are shown in Fig. 8.

A solar simulator developed in laboratory was used to make the measurements. The light level can be calibrated by changing the supply voltage of the solar simulator. The module temperature increased only because of the effect of infrared light on the solar simulator. After the PV module was put into simulator it was necessary to wait 12 min approximately to increases the PV module temperature by 10 °C over the room temperature. This phenomenon permits us to get the measurements of I—V and P—V curves of the PV module.

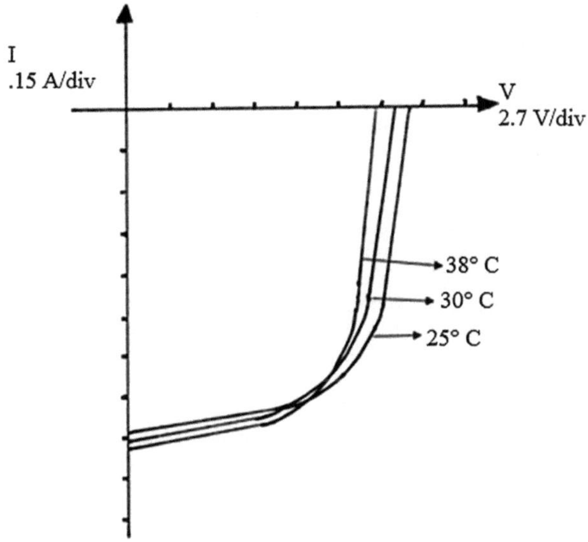

Fig. 9 PV module I—V curves to different temperature levels obtained with the designed electronic circuit

Also the current-voltage curve of the PV module for different temperatures levels were measured from the designed electronic system using the same conditions. The results are shown in Fig. 9.

5 Conclusions

From the experimental results, the proposed electronic circuit agrees with the technical design criteria. An electronic control board for measuring the characteristic curve of the PV module was designed to obtain the power voltage curves and the current voltage curves with different temperature levels, having a measurement error of ±1 % approximately. From the power voltage curves as shown in the Fig. 8, it is observed that the PV module power reduces with increase in temperature. As shown in Fig. 9, the current voltage curve of the PV module demonstrates a decrease in voltage with the temperature which is in contrary to the behavior of the current. As future research work, an electronic circuit can be designed to measure the power of the PV module.

Unlike the techniques reported by other researchers regarding photovoltaic cell and modules, here we propose a different technique for getting the power—voltage curve of the PV module, using an electronic system based on the use of MOSFET power transistor, connected to the photovoltaic module. It has an electronic control circuit for supplying the signal to control the transistor, so that it reflects a variable resistance at the PV module, and also an electronic circuit to process the electrical signal and supply the power signal to be plotted.

Acknowledgments The authors thank to National Polytechnic Institute from México, by the financial support to the project No. 20141121 SIP-IPN.

References

1. Shu-Hung Chung H, Tse KK, Ron Wui SY, Mok MC, Ho MT (2003) A novel maximum power point tracking technique for solar panels using a SEPIC or Cuk converter. IEEE Trans Power Electron 18(3):717–724
2. Vega-Pérez J, Ponomaryov V (2001) Maximum power detection electronic circuit for photovoltaic module. J Telecommun Radio Eng 56(1):126–134
3. Vega-Perez J, Volodymyr P, Nino-de-Rivera L (2003) Electronic system for power optimization of photovoltaic generator. Int J Electromagn Waves Electron Syst 8(7–8): 12–15. ISSN 1560-4128
4. Vega-Perez J, Volodymyr P, Nino-de-Rivera L (2002) Transfering electronic system of solar module power. Proceedings of the IEEE, Guanajuato Mexico Section, 2nd International Workshop on Random Fields Modeling and Processing in Inhomogeneous Media. Guanajuato, Mexico, 27–29 Nov 2002, pp 67–69
5. Wanzeller MG, Alves RNC, Joao Viana da FN, dos Santos Fonseca WA (2004) Current control loop for tracking of maximum power point supplied for photovoltaic array. IEEE Trans Instrum Meas 53(4):1304–1310
6. Munji MK, Okullo W, Van Dyk EE, Voster FJ (2010) Local devise parameter extraction of a concentrator photovoltaic cell under solar spot illumination. Sol Energy Mater Sol Cell 4: 2129–2136
7. Quaschning V, Hanitch R (1996) Numerical simulation of current voltage characteristics of photovoltaic system with shaded solar cell. Sol Energy 56(6):513
8. Rashid MH (1999) Electrónica de potencia. Prentice Hall, New Jersey

A Robust Control Scheme Against Some Parametric Uncertainties for the NXT Ballbot

R.A. García-García and M. Arias-Montiel

Abstract In this work a Linear Matrix Inequalities (LMIs) approach to provide robustness to an optimal control scheme for the NXT ballbot is proposed. The mathematical model of the system is obtained taking into account the actuators dynamics and in this overall model uncertain parameters appear. These uncertainties can affect (in a negative form) the control algorithm performance, so we define a polytopic model considering some parameters vary within some bounded sets. This polytopic model is used to synthesize a robust control scheme against parameters variation using LMIs. Numerical results show a significant improvement in the closed loop system in comparison with a classic Linear Quadratic Regulator (LQR) control. Finally, some experimental results are presented to show the viability of implementing the control strategy proposed.

Keywords NXT ballbot · Linear matrix inequalities (LMIs) · Uncertain parameters · Student paper

1 Introduction

A ballbot is a thin, agile and omnidirectional robot that balances on, and is propelled by, a single spherical wheel. The system relies on active balancing and thus is "dynamically stable" [1]. Dynamical characteristics and potential applications have made ballbot an interesting object of study for engineers and researchers in automatic control, mechatronics and robotics areas [2—7].

R.A. García-García (✉) · M. Arias-Montiel
Institute of Electronics and Mechatronics, Universidad Tecnológica de La Mixteca, Oaxaca, Mexico
e-mail: rafagarcia.0810@gmail.com

M. Arias-Montiel
e-mail: mam@mixteco.utm.mx

M. Ceccarelli and E.E. Hernández Martinez (eds.), *Multibody Mechatronic Systems*,
Mechanisms and Machine Science 25, DOI 10.1007/978-3-319-09858-6_24

An experimental platform used in order to study the dynamical behavior of the ballbot is the *LEGO Mindstorms NXT*® [8, 9]. *LEGO Mindstorms NXT*® is a programmable robotics kit with appropriate characteristics to be used in education and research activities [10—12].

Yamamoto [9] developed a version of a ballbot, called NXT ballbot, based on a *LEGO Mindstorms NXT*® kit. He presented a linearized model of the NXT ballbot and he proposed a Linear Quadratic Regulator (LQR) in order to achieve the stabilization of the system in a vertical position. Prieto et al. [8] built a ballbot using a *LEGO Mindstorms NXT*® kit with light structural modifications in relation to Yamamoto's. In addition, they proposed some procedures in order to obtain actuators parameters and to characterize gyro sensors used to measure angular rate. The control algorithm used to stabilize the ballbot was a LQR.

In order to control the NXT ballbot, actuators dynamics must be included in the model. Some authors have reported different values for physical parameters for NXT servomotors based on experimental tests [8, 9, 13, 14], so the parametric uncertainties are present and they can affect in a negative form the control algorithm performance. In this work an LMIs approach to provide robustness against parametric uncertainties to a LQR control scheme for the balancing problem in a NXT ballbot is proposed. A polytopic model considering variations within some bounded sets for inertial and friction parameters for system actuators is defined. This polytopic model is used to synthesize a robust control scheme against these parameters variations using LMIs. Numerical results are presented to show the significant improvement in the closed loop system using robust control in comparison with a classic LQR control. Some experiments were carried out and the obtained results demonstrate the viability of implementing the robust control strategy proposed.

2 System Description and Modeling

2.1 The NXT Ballbot

In Fig. 1a the mechanical structure for the NXT ballbot is shown. It consists of standard structural parts, a plastic spherical ball, two wheels and two servomotors, all of them included in the basic *LEGO Mindstorms NXT*® kit, and in addition, two *Hitechnic*® gyrosensors. The rubber wheels called driven wheels transmit the motion from the servomotors to the spherical ball in order to vary its rotation axis and thus, equilibrate the system. Gyrosensors measure pitch and yaw angles rate and this information is used by the controller to actuate the servomotors providing the control forces to the closed loop system.

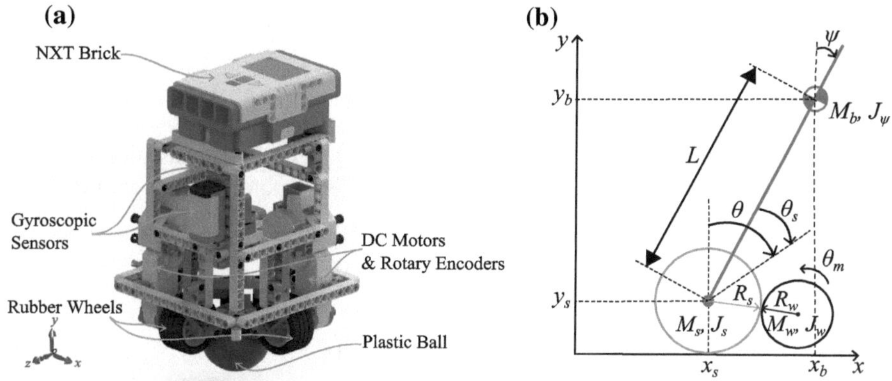

Fig. 1 **a** Render view of the NXT ballbot; **b** Schematic diagram for NXT ballbot

2.2 The NXT Ballbot Dynamic Model

Dynamic model for NXT ballbot is based on a spherical wheeled inverted pendulum considering that the motions in *xy* plane and *zy* plane are decoupled (see Fig. 1b). Under this assumption, two identical motion equations are obtained, so only one plane of motion is analyzed. In Fig. 1b NXT ballbot body is modeled as an inverted pendulum with mass center coordinates x_b, y_b, mass M_b and mass moment of inertia J_ψ. M_s, J_s and R_s are the mass, moment of inertia and radius of the spherical ball, and M_w, J_w and R_w are the mass, moment of inertia and radius of driven wheels, respectively.

The variables to describe the motion in *xy* plane are

- ψ: body inclination angle in relation to vertical axis.
- θ: spherical ball angle in relation to vertical axis.
- θ_s: spherical ball angle due to the contact with the driving wheel.
- θ_m: driving wheel angle imposed by the actuator.

By inspection of Fig. 1b, the next relationships can be obtained

$$\theta = \psi + \theta_s \tag{1}$$

$$R_\omega \theta_m = R_s \theta_s \tag{2}$$

Considering θ and ψ as the generalized coordinates and applying the Euler-Lagrange method, motion equations are given as

$$\left[(M_s + M_b)R_s^2 + \frac{R_s^2}{R_\omega^2}(J_m + J_\omega) + J_s\right]\ddot{\theta} + \left[M_b R_s L \cos\psi - \frac{R_s^2}{R_\omega^2}(J_m + J_\omega)\right]\ddot{\psi} \\ - M_b R_s L \dot{\psi}^2 \sin\psi = F_\theta \tag{3}$$

$$\left[M_bR_sL\cos\psi - \frac{R_s^2}{R_\omega^2}(J_m + J_\omega)\right]\ddot{\theta} + \left[M_bL^2 + J_\psi + \frac{R_s^2}{R_\omega^2}(J_m + J_\omega)\right]\ddot{\psi} - M_bgL\sin\psi = F_\psi \tag{4}$$

with J_m = moment of inertia of servomotor and g = gravity acceleration constant.

The generalized forces F_θ y F_ψ correspond to servomotor torques and they can be modeled as

$$F_\theta = \alpha e_a - \beta(\dot{\theta} - \dot{\psi}) \tag{5}$$

$$F_\psi = -\alpha e_a + \beta(\dot{\theta} - \dot{\psi}) \tag{6}$$

with $\alpha = \frac{k_b}{R_a}$; $\beta = k\left(\frac{k_pk_b}{R_a} + b_e\right)$, $k = \frac{R_s}{R_w}$, k_p, is the torque constant, b_e is the friction coefficient inside the motor, R_a is the armature resistance, k_b is the back electromotriz force constant and e_a is the input voltage.

As we can see, system model expressed by Eqs. (3) and (4) is non linear. In order to synthesize a linear control algorithm, this model is linearized about an equilibrium point [15], i.e. when the NXT ballbot is in vertical position ($\psi = 0$).

Linearized model is

$$\left[(M_s + M_b)R_s^2 + \frac{R_s^2}{R_\omega^2}(J_m + J_\omega) + J_s\right]\ddot{\theta} + \left[M_bR_sL - \frac{R_s^2}{R_\omega^2}(J_m + J_\omega)\right]\ddot{\psi} = F_\theta \tag{7}$$

$$\left[M_bR_sL - \frac{R_s^2}{R_\omega^2}(J_m + J_\omega)\right]\ddot{\theta} + \left[M_bL^2 + J_\psi + \frac{R_s^2}{R_\omega^2}(J_m + J_\omega)\right]\ddot{\psi} - M_bgL\psi = F_\psi \tag{8}$$

2.3 The NXT Ballbot Model in State Space

Equations (7) and (8) can be rewritten in matricial form as

$$\mathbf{E}\ddot{\phi} + \mathbf{F}\dot{\phi} + \mathbf{G}\phi = \mathbf{H}e_a \tag{9}$$

where

$$\phi = \begin{bmatrix}\theta \\ \psi\end{bmatrix}$$

$$\mathbf{E} = \begin{bmatrix}(M_s + M_b)R_s^2 + k^2(J_m + J_\omega) + J_s & M_bR_sL - k^2(J_m + J_\omega) \\ LM_bR_s - k^2(J_m + J_\omega) & M_bL^2 + J_\psi + k^2(J_m + J_\omega)\end{bmatrix}$$

$$\mathbf{F} = \begin{bmatrix} \beta & -\beta \\ -\beta & \beta \end{bmatrix}, \quad \mathbf{G} = \begin{bmatrix} 0 & 0 \\ 0 & -M_b g L \end{bmatrix}, \quad \mathbf{H} = \begin{bmatrix} \alpha \\ -\alpha \end{bmatrix}.$$

By defining the state vector $\mathbf{x} = [\psi \quad \theta \quad \dot{\theta} \quad \dot{\psi}]^T$ and the input $u = e_a$, the NXT ballbot dynamics (9) can be described in state space form as follows

$$\dot{\mathbf{x}} = \mathbf{A}\mathbf{x} + \mathbf{B}u \tag{10}$$

with

$$\mathbf{A} = \begin{bmatrix} 0 & 0 & 1 & 0 \\ 0 & 0 & 0 & 1 \\ 0 & \frac{E_{12}G_{22}}{E_{11}E_{22}-E_{12}^2} & \frac{E_{12}F_{21}-E_{22}F_{11}}{E_{11}E_{22}-E_{12}^2} & \frac{E_{12}F_{22}-E_{22}F12}{E_{11}E_{22}-E_{12}^2} \\ 0 & \frac{-E_{11}G_{22}}{E_{11}E_{22}-E_{12}^2} & \frac{E_{21}F_{11}-E_{11}F_{21}}{E_{11}E_{22}-E_{12}^2} & \frac{E_{21}F_{12}-E_{11}F_{22}}{E_{11}E_{22}-E_{12}^2} \end{bmatrix}, \quad \mathbf{B} = \begin{bmatrix} 0 \\ 0 \\ \frac{E_{22}H_{11}-E_{12}H_{21}}{E_{11}E_{22}-E_{12}^2} \\ \frac{E_{11}H_{21}-E_{21}H_{11}}{E_{11}E_{22}-E_{12}^2} \end{bmatrix}.$$

3 NXT Ballbot Polytopic Modeling and Control

The LMIs approach is used in system and control theory in order to solve a wide variety of problems (i.e. construction of quadratic Lyapunov functions for stability and performance analysis of linear differential inclusions, optimization over an affine family of transfer matrices, including synthesis of multipliers for analysis of linear systems with unknown parameters, inverse problem of optimal control, multicriterion LQG/LQR, etc.). The use of LMIs is based on polytopic models which can be defined if model matrices include terms with linear dependence of uncertain parameters and these parameters vary within some bounded sets [16, 17].

Different values for the NXT ballbot actuators (NXT servomotors) have been reported in literature; some of these values are presented in Table 1.

According to data presented in Table 1, moment of inertia J_m and friction coefficient b_e are the parameters with the widest rank of variation, so the polytopic

Table 1 Physical parameters for servomotors reported in literature

Parameter	Reference [8]	Reference [14]	Reference [13]	Reference [9]
$R_a[\Omega]$	4.6	6.85	5.0012	6.69
$L_a[\text{H}]$	3.7×10^{-3}	Neglected	1×10^{-3}	Neglected
$J_m[\text{kg} \cdot \text{m}^2]$	45×10^{-9}	– – – –	2.4589×10^{-6}	10×10^{-6}
$b_e[\text{N} \cdot \text{m} \cdot \text{s}]$	1.85×10^{-3}	1.1278×10^{-3}	38.745×10^{-6}	2.2×10^{-3}
$K_p[\frac{\text{N}\cdot\text{m}}{A}]$	0.48	0.3179	0.5246	0.317
$K_b[\frac{\text{V}\cdot\text{s}}{rad}]$	0.48	0.46389	0.5246	0.468

model for the NXT ballbot will be obtained in terms of these two parameter. The uncertain parameters vector is defined as

$$\rho = (b_e \quad J_m) \tag{11}$$

with the vertices number of resultant polytope $L = 4$, limited by the minimum and maximum values

$$b_e \in [b_{e\min} \quad b_{e\max}], \quad J_m \in [J_{m\min} \quad J_{m\max}] \tag{12}$$

These uncertain parameters are present in matrices **A** and **B** of the Eq. (10), and these matrices depend on the uncertain parameters in a linear form, so it is possible to define a polytope with 4 vertices which contains all the possible values for the uncertain parameters within their minimum and maximum values. The 4 vertices of polytopic model for matrices **A** and **B** can be defined as

$$\mathbf{A}_i = \begin{bmatrix} 0 & 0 & 1 & 0 \\ 0 & 0 & 0 & 1 \\ 0 & \frac{E_{Jm\,12}G_{22}}{E_{Jm\,11}E_{Jm\,22}-E_{Jm\,12}^2} & \frac{E_{Jm\,12}F_{be\,21}-E_{Jm\,22}F_{be\,11}}{E_{Jm\,11}E_{Jm\,22}-E_{Jm\,12}^2} & \frac{E_{Jm\,12}F_{be\,22}-E_{Jm\,22}F_{be\,12}}{E_{Jm\,11}E_{Jm\,22}-E_{Jm\,12}^2} \\ 0 & \frac{-E_{Jm\,11}G_{22}}{E_{Jm\,11}E_{Jm\,22}-E_{Jm\,12}^2} & \frac{E_{Jm\,21}F_{be\,11}-E_{Jm\,11}F_{be\,21}}{E_{Jm\,11}E_{Jm\,22}-E_{Jm\,12}^2} & \frac{E_{Jm\,21}F_{be\,12}-E_{Jm\,11}F_{be\,22}}{E_{Jm\,11}E_{Jm\,22}-E_{Jm\,12}^2} \end{bmatrix}$$

$$\mathbf{B}_i = \begin{bmatrix} 0 \\ 0 \\ \frac{E_{Jm\,22}H_{11}-E_{Jm\,12}H_{21}}{E_{Jm\,11}E_{Jm\,22}-E_{Jm\,12}^2} \\ \frac{E_{Jm\,11}H_{21}-E_{Jm\,21}H_{11}}{E_{Jm\,11}E_{Jm\,22}-E_{Jm\,12}^2} \end{bmatrix}, \quad i = 1, \ldots, 4 \tag{13}$$

where values of J_m and b_e depend on vertex of polytopic model

$$\begin{aligned} [J_{m\min} \quad b_{e\min}] &\in \mathbf{A}_1, \mathbf{B}_1, \\ [J_{m\min} \quad b_{e\max}] &\in \mathbf{A}_2, \mathbf{B}_2, \\ [J_{m\max} \quad b_{e\min}] &\in \mathbf{A}_3, \mathbf{B}_3, \\ [J_{m\max} \quad b_{e\max}] &\in \mathbf{A}_4, \mathbf{B}_4. \end{aligned}$$

In order to be able to solve the tracking problem and to assure the error in steady state converges to zero, another state variable ξ is added to the polytopic model which corresponds to integral of position error for variable θ. The extended polytopic model is expressed by

$$\dot{\mathbf{x}}_a = \mathbf{A}_a\mathbf{x}_a + \mathbf{B}_a u + \begin{bmatrix} 0 \\ 1 \end{bmatrix} \mathrm{r} \tag{14}$$

where r is the reference value to track and

$$\mathbf{x}_a = [\mathbf{x} \quad \xi]^T, \quad \mathbf{A}_a = \begin{bmatrix} \mathbf{A}_i & \mathbf{0} \\ -\mathbf{C} & 0 \end{bmatrix}, \quad \mathbf{B}_a = \begin{bmatrix} \mathbf{B}_i \\ 0 \end{bmatrix}, \quad \mathbf{C} = [1 \quad 0 \quad 0 \quad 0]$$
$$i = 1, \ldots, L.$$

The extended polytopic model (14) is used to synthesize a LQR control law to stabilize the NXT ballbot in spite of the uncertainties in inertia and friction parameters for the actuators by LMI formulations. The augmented gain vector obtained will be $\mathbf{K}_a = [\mathbf{K} \quad K_\xi]$ with $\mathbf{K} \in \mathbb{R}^{m \times n}$ and $K_\xi \in \mathbb{R}$.

In order to compare the performance of the LMI-LQR control with the classical LQR, values for **Q** and **R** given by Yamamoto [9] were considered. To obtain the classical LQR controller gains, *Matlab*® *lqr(A, B, Q, R)* function is used.

LQR formulation can be extended to polytopic systems by LMI constrains in order to provide robustness to the controller against uncertainties in parameters. For the system described by Eq. (14) the problem consists of finding a control law by state feedback $\mathbf{u} = -\mathbf{K}\mathbf{x}$, so that the next cost function in closed loop can be minimized

$$J = \int_0^\infty \left(\mathbf{x}^T \left(\mathbf{Q} + \mathbf{K}^T \mathbf{R} \mathbf{K}\right)\mathbf{x}\right) dt \tag{15}$$

Using the trace operator $\mathbf{Tr}(\cdot)$, which satisfies $\mathbf{Tr}(A^T BC) = \mathbf{Tr}(BCA^T)$, cost function (15) can be rewritten as

$$J = \int_0^\infty \mathbf{Tr}\left(\left(\mathbf{Q} + \mathbf{K}^T \mathbf{R} \mathbf{K}\right)\mathbf{x}\mathbf{x}^T\right) dt = \mathbf{Tr}\left(\left(\mathbf{Q} + \mathbf{K}^T \mathbf{R} \mathbf{K}\right)\mathbf{P}\right)$$

where $\mathbf{P} = \int_0^\infty (\mathbf{x}\mathbf{x}^T) dt$ is a symmetric positive definite matrix which satisfies the constrain

$$(\mathbf{A}_i - \mathbf{B}_i\mathbf{K})\mathbf{P} + \mathbf{P}(\mathbf{A}_i - \mathbf{B}_i\mathbf{K})^T + \mathbf{I} = 0, \quad i = 1, \ldots, L. \tag{16}$$

where subindex i represents each vertex of the polytpic model.

The optimal feedback gain can be obtained by

$$\min_{\mathbf{P},\mathbf{K},\mathbf{X}} \quad \mathbf{Tr}(\mathbf{QP}) + \mathbf{Tr}(\mathbf{X})$$

subject to

$$\begin{aligned} &\mathbf{P} > 0, \\ &\mathbf{A}_i\mathbf{P} + \mathbf{P}\mathbf{A}_i^T - \mathbf{B}_i\mathbf{Y} - \mathbf{Y}^T\mathbf{B}_i^T + \mathbf{I} < 0, \\ &\begin{bmatrix} \mathbf{X} & \mathbf{R}^{\frac{1}{2}}\mathbf{Y} \\ \mathbf{Y}^T\mathbf{R}^{\frac{1}{2}} & \mathbf{P} \end{bmatrix} > 0. \end{aligned} \tag{17}$$

When the constrains given by (17) are solved, the optimal feedback gain for the robust controller is obtained by $\mathbf{K} = \mathbf{Y}\mathbf{P}^{-1}$ [16—18].

The numerical problem described by constrains in Eq. (17) is solved by *Matlab*® LMI Toolbox [19]. Obtained gains controllers are

$$\mathbf{K}_{\mathrm{lqr}} = [-1.504 \times 10^{-2} - 1.569 - 2.700 \times 10^{-2} - 2.325 \times 10^{-1} - 7.125 \times 10^{-3}] \tag{18}$$

$$\begin{aligned} \mathbf{K}_{\mathrm{lmi-lqr}} = &[-7.241 \times 10^{-3} - 1.385 - 1.808 \times 10^{-2} - 2.298 \times 10^{-1} \\ &- 2.549 \times 10^{-3}] \end{aligned} \tag{19}$$

The control law is based on full state feedback and it is proposed as follows

$$u = -\mathbf{K}\mathbf{x_a} \tag{20}$$

where $\mathbf{K}$ is the gains vector given by Eq. (18) or (19) and $\mathbf{x_a}$ is the extended vector state.

4 Numerical and Experimental Results

In order to get some numerical results and to compare the performance of classical LQR with the LMI-LQR rosbust control scheme, nominal parameters values for the system given by Yamamoto [9] are considered. First, the extended polytopic model given by Eq. (14) with nominal parameters values and with the control input given by (20) is numerically solved. In Fig. 2a—b, a comparison of closed loop system response for classical non robust LQR and for the robust LMI-LQR control is presented. As we can observe, in this case there is not any significant difference in the response of the four state variables. Something similar can be noted in the control effort shown in Fig. 2c.

Now, we explore the variation in control performance when inertial and friction parameters of the actuators change using some vertices of the extended polytopic model which represent the extreme values (minimum or maximum) for the uncertain parameters, given in Table 1. In Fig. 3a—b closed loop system response for vertice 3 (i.e. matrices $\mathbf{A}_3$ and $\mathbf{B}_3$) of the extended polytopic model is depicted. The comparison between the non robust LQR and the robust LMI-LQR control

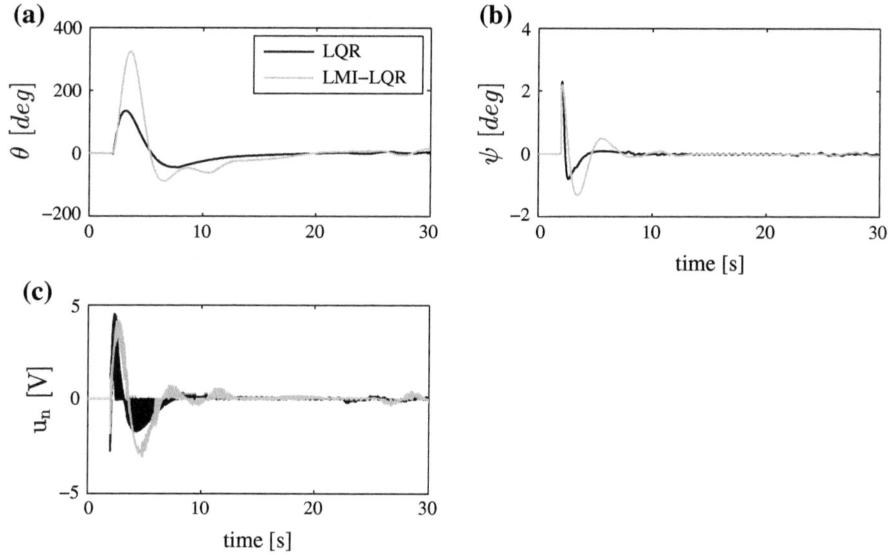

Fig. 2 State variables and control effort of the closed loop system with nominal parameters

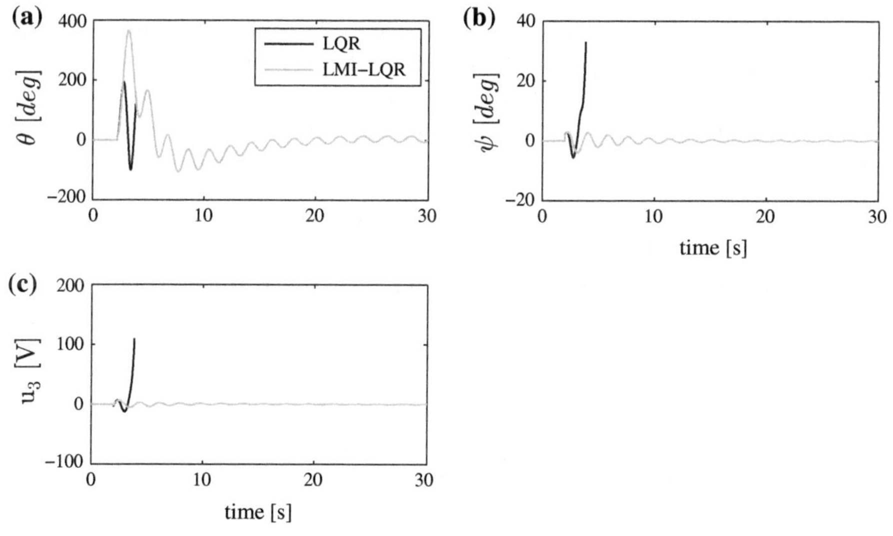

Fig. 3 State variables and control effort of the closed loop system for vertice 3 of the extended polytopic model

shows the first one is not able to stabilize the NXT ballbot, at around 4*s* the system falls down losing the equilibrium. In Fig. 3c the voltage input obtained from each controller is shown. A similar behavior is obtained at the other three vertices.

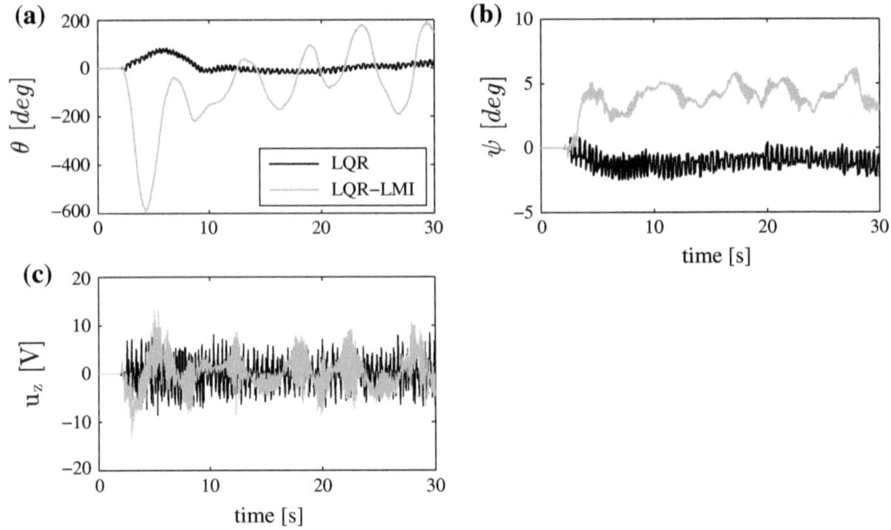

Fig. 4 Experimental response in the *xy* plane

In order to demonstrate the viability of practically applying the robust control scheme proposed, some experiments were carried out. Data were acquired by gyrosensors in both planes of motion. Data acquisition was carried out by transfer information in real time via bluetooth from NXT brick to a personal computer equiped with *Simulink® Support Package for LEGO MINDSTORMS NXT® hardware*.

Experimental results for the state variables and for the measured input voltage are presented in Fig. 4.

It is important to mention that implemented controllers were tuned with gains given by (18) and (19) and we were not able to physically vary the inertial and friction parameters for the actuators, so that the experimental results presented are equivalent to numerical results for the nominal model shown in Fig. 2. In Fig. 4a we can observe the response oscilates around $\theta = 0$. This means ballbot needs displacements in zx plane in order to get the vertical equilibrium. This effect can be reduced by increasing the available control effort limited by LMI constraints.

5 Conclusions

Because of its potential applications, ballbot system is an interesting object of study for engineers and researchers in automatic control, mechatronics and robotics areas. In this paper a robust control scheme against parametric uncertainty for a ballbot system was proposed. Robust controller is based on a optimal linear quadratic regulator (LQR) in combination with linear matrix inequalities (LMIs) theory which uses a polytopic model in order to consider the rank variation for some parameter values. Numerical results show an important improvement in the closed loop system behavior of the robust controller in comparison with the classical LQR reported in literature when the system parameters values change. In order to prove the dynamic behavior of the proposed control scheme, we use the experimental platform *LEGO Mindstorms NXT®* kit to built a ballbot system. Experimental results show the viability of practical implementation of the robust control based on LMIs approach. This control scheme can be extended in order to include uncertainties in other system parameters and even to take into account the disturbance rejection problem.

References

1. Hollis R (2006) Ballbots. Sci Am 295(4):72–77
2. Asgari P, Zarafshan P, Moosavian SAA (2013) Dynamics modelling and stable motion control of a ballbot equipped with a manipulator. In: Proceedings of the 2013 international conference on robotics and mechatronics, pp. 259–264
3. Fankhauser P, Gwerder C (2010) Modeling and control of a ballbot. Bachelor Thesis, Autonomous Systems Lab, Swiss Federal Institute of Technology Zurich
4. Fong J, Uppill S (2009) Design and build a ballbot, Bachelor Thesis, Faculty of Engineering, Computer and Mathematical Sciences, University of Adelaide
5. Mampetta A (2006) Automatic transition of ballbot from statically stable state to dynamically stable state. Master's Thesis, Robotics Institute, Carnegie Mellon University
6. Nagarajan U, Kantor G, Hollis R (2013) The ballbot: an omnidirectional balancing mobile robot. Int J Robot Res 1–14. Published online 13 November 2013
7. Skrabel C (2013) Mechanical design of a ballbot platform, Bachelor Thesis, Autonomous Systems Lab, Swiss Federal Institute of Technology Zurich)
8. Prieto S, Navarro T, Plaza M, Polo O (2012) A monoball robot based on lego mindstorm. IEEE Control Syst 32(2):71–83
9. Yamamoto Y (2009) NXT ballbot model-based design—control of self-balancing robot on a ball, built with Lego Mindstorm NXT, 1st edn. Cynernet Systems CO., LTD
10. Bradley P, De la Puente J, Zamorano J, Brosnan D (2012) A platform for real-time control education with lego mindstorms. In: Proceedings of the 9th IFAC symposium advances in control education, pp. 112–117
11. Kim Y (2011) Control systems lab using a lego mindstorms nxt motor system. IEEE Trans Educ 54(3):452–461
12. Pinto M, Moreira AP, Matos A (2012) Localization of mobile robots using an extended Kalman filter in a lego nxt. IEEE Trans Educ 55(1):135–144
13. Pinto-Salamanca ML, Bermúdez-Bohórquez GR (2007) Parameters determination of a mobil robot of lego minstorms. Ingeniería Investigación y Desarrollo 5(2):7–13
14. Hurbain P, Hurbain P, Gasperi M (2007) NXT motor. The International Fan-Created *Lego* Users Group Network, Extracted from World Wide Web: http://web.archive.org/web/20110328142611/http://web.mac.com/ryowatanabe/iWeb/RyosHoliday/NXTMotor.html
15. Vukic Z, Kuljaca L, Donlagic D, Tesnjak S (2003) Nonlinear control systems. Marcel Dekker Inc, Basel

16. Boyd S, El Ghaoui L, Feron E, Balakrishnan V (1994) Linear matrix inequalities in system anc control theory. Society for Industrial and Applied Mathematics SIAM, Philadelphia
17. Ostertag E (2011) Mono and multivariable control and estimation—linear, quadratic and LMI methods. Springer, Berlin
18. Olalla C, Leyva R, El-Aroudi A, y Queinnec I (2009) Robust LQR control for PWM converters: an LMI approach. IEEE Trans Indus Electron 56(7):2548–2558
19. Erkus B, Lee YJ (2004) Linear matrix inequalities and MATLAB LMI toolbox. University of Southern California Group Meeting Report, Los Angeles

Fuzzy Logic Control on FPGA Using LabVIEW

Juan Carlos García-Montalva, Jesús de la Cruz-Alejo and Jorge Díaz-Salgado

Abstract We implement a fuzzy logic control algorithm on a Nexys 3 FPGA board that uses coding look-up tables on VHDL which is generated by an application developed in LabVIEW. The algorithm used in this work is based on alpha-levels. This technique reduces execution time and simplifies VHDL programming. In addition, it is possible to use any shape of membership sets, although only triangular trapezoidal and gaussian are tested. Also, LabVIEW application helps in verifying each block of the system through boolean, numeric and graph indicators. Further, using look-up tables avoids the necessity of VHDL coded while loops. The system uses a max-min Mamdani inference mechanism. In Nexys 3 FPGA board COSAA defuzzification method [2] is implemented. Crisp output values are compared with those obtained with MATLAB Fuzzy Logic Tool Box. Results have very little variation.

Keywords VHDL · Fuzzy logic · Labview · FPGA · Look-up tables · Mamdani

1 Introduction

Currently, the use of fuzzy logic based controllers have been spreading significantly in mechatronic systems, especially in robotics, due to their effectiveness for modeling human expertise, using common sense rules [1, 2]. Fuzzy Logic Systems

J.C. García-Montalva (✉) · J. de la Cruz-Alejo · J. Díaz-Salgado
Tecnológico de Estudios Superiores de Ecatepec, Ecatepec de Morelos, Mexico
e-mail: jcparanoid79@hotmail.com

J. de la Cruz-Alejo
e-mail: jesus_ch517@hotmail.com

J. Díaz-Salgado
e-mail: xorge@hispavista.com

J.C. García-Montalva
Universidad Tecnologica de Tecamac, Tecamac, Mexico

M. Ceccarelli and E.E. Hernández Martinez (eds.), *Multibody Mechatronic Systems*,
Mechanisms and Machine Science 25, DOI 10.1007/978-3-319-09858-6_25

(FLS) are capable to provide robust controllers for mathematically ill-defined systems even if they are nonlinear, time varying and complex [3, 4]. Fuzzy logic systems provides advantages over traditional methods, in the sense of mathematical dispense systems analysis [5]. On the other hand, most of the architectures for the design of fuzzy systems are implemented in digital mode through dedicated fuzzy circuits or Application Specific Integrated Circuits, (ASICs). FPGAs are hardware devices used as user-programmable ASICs [6]. They are able to deal with complex engineering problems in real-time operation. However, because the systems necessary to implement fuzzy operations in programmable logic devices consume a lot of hardware (multiplication, division, powers), programming in hardware description language is a major challenge [1]. An alternative to the programming, is using an application developed in LabVIEW. The description can be step by step approach using programming structures and existing data types in LabVIEW. In this paper, an FPGA implementation of digital fuzzy logic system is reported with modular design concept based on LabVIEW [7]. Also, the use of lookup tables simplifies programming in addition to avoiding the use of while loops that do not adequately synthesized into FPGA devices. Another advantage is the reduction of the execution time by not having to calculate each value of membership each time an input value is obtained. The system is capable of 8 bits resolution with total 256 alpha levels. In Sect. 2, a fuzzy logic-based unified theory and LabVIEW is introduced for the system. In Sect. 3, experimental results are reported. Conclusions are described in Sect. 4.

2 Fuzzy Logic System

The block diagram of a Mamdani fuzzy system is illustrated in Fig. 1 [1]. Fuzzification stage is responsible for converting the crisp input data into a fuzzy data with a membership degree in one or more membership functions. Inference step is responsible for determining, using the rule base, the conclusion of the input rule, while defuzzification stage is responsible for converting the result to a continuous value.

Figure 2 shows the block diagram of the overall system implemented in the FPGA board. Since both inputs are DC voltage signals, it is necessary to pass them through an ADC which provides addresses for the look-up tables that contain the membership values of each fuzzy set. Values found are stored in weight vectors and subsequently implicated with the max-min mechanism. Later, alpha-levels are taken from another look-up table and aggregation is performed. Finally, the rigid output value is obtained by center of slice area average (COSAA) method and turned into DC voltage using an external DAC.

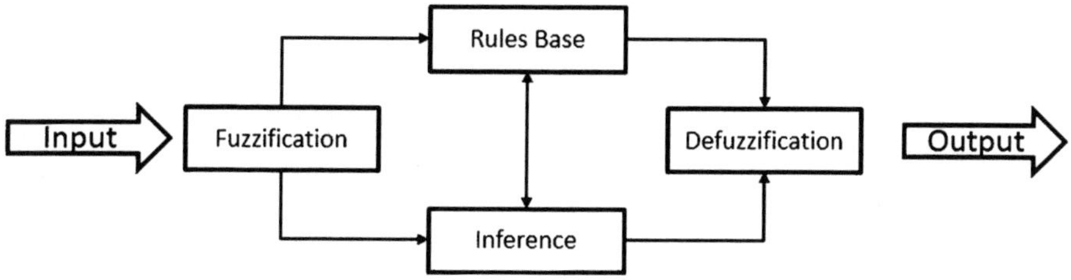

Fig. 1 Block diagram of Mamdani fuzzy system

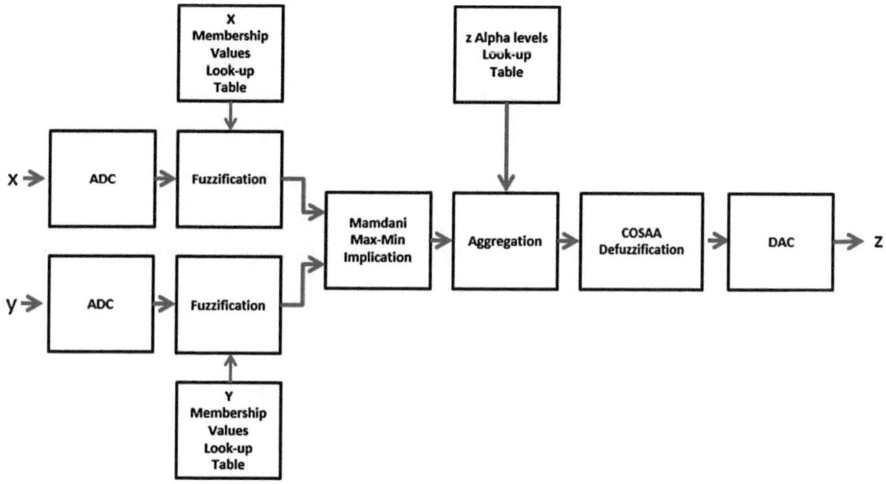

Fig. 2 Block diagram of the Mamdani FLS algorithm implemented on Nexys 3 FPGA board

2.1 Fuzzification

VHDL coded look-up tables of membership values for each set are generated by LabVIEW based application which is capable to obtain values for three types of sets: triangular, trapezoidal and gaussian (although any type function can be used).

An iterative structure increases a variable from zero to 2^(n − 1) (n is the number of bits used to represent the universe of discourse); which is used as an argument of the selected membership function. Figure 3a shows the flowchart of the process for a trapezoidal set and Fig. 3b shows this algorithm implemented on the LabVIEW based application.

In fuzzification step, α-levels represent the membership values of each function. Each α-level is defined by four parameters: M is the name of the membership function; i is the height, represents the membership value in M, whose maximum value is $I = 2^m - 1$; α_0^i and α_f^i are the initial and the final points over the universe of discourse for the membership value i. The α-levels for each continuous value input,

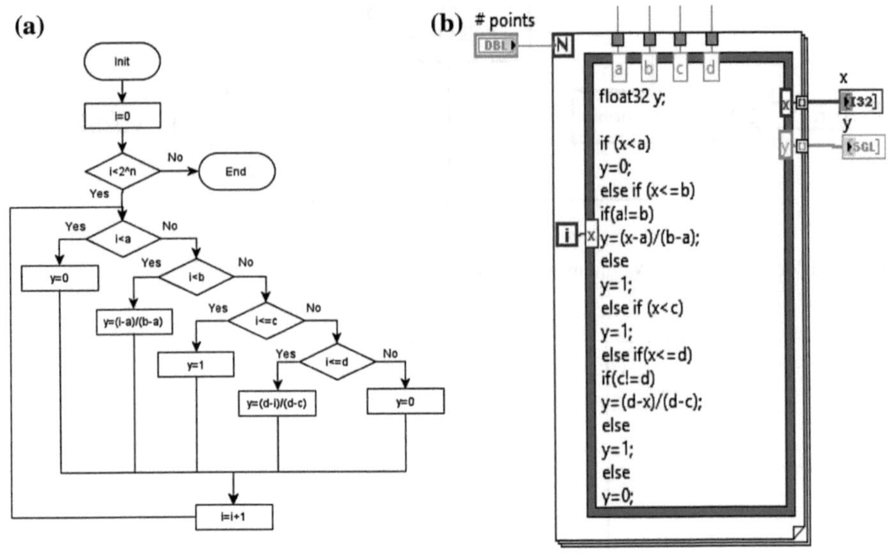

Fig. 3 **a** Flow chart for the generation of trapezoidal MF. **b** Loop in LabVIEW to generate membership functions

Fig. 4 Alpha-levels of a fuzzy set

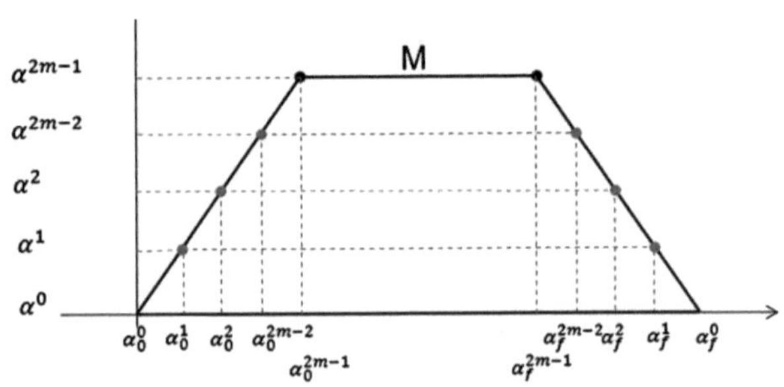

provide the points where the input value falls in one or more membership functions (Fig. 4). The base of each fuzzy set is found by α_0^i and α_0^f, whose membership value is zero. This values, are especially useful because they allow to know if the value of the input signal belongs to a given fuzzy set [1].

Weight vectors w_1 and w_2 which store the membership grades corresponding to every input fuzzy set and its length corresponds the number of sets in the universe of discourse of each input [1].

LabVIEW application converts membership values for each possible entry, in strings of binary numbers and exports data to a text file (Fig. 5a). Data in the text file contains binary words of 40 bits (8 bits for each set) as shown in Fig. 5b.

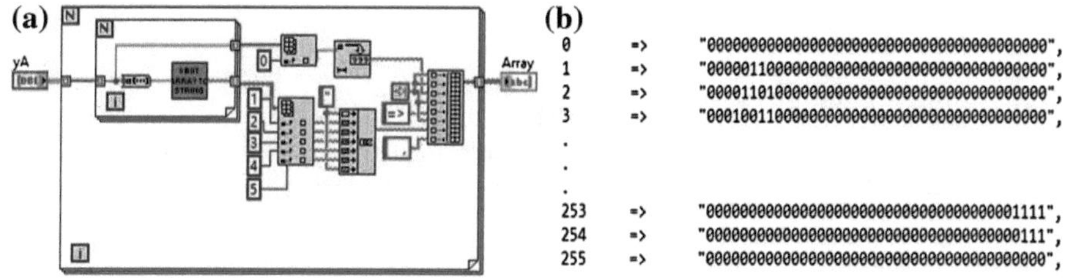

Fig. 5 VHDL coded look-up tables generated by LabVIEW application. **a** LabVIEW code. **b** Text file

2.2 *Inference*

In this fuzzy logic implementation a Mamdani max-min implication mechanism is used for two inputs (denoted by x and y) and one output (denoted by z) (Fig. 6a).

The inference matrix is built by interpolating weight vectors w_1 and w_2, with min operator as Eq. (1) shows.

$$\text{inference matrix}_{i,j} = T(w1, w2) \tag{1}$$

The vector below in Fig. 6b stores maximum values in each column. These values are necessary to build the output aggregated function.

To avoid while loops in VHDL programming, look-up table containing α-levels are used, thus only defuzzification and aggregation stages need to be calculated. This way all iterations with fixed number are performed, (for loops) [8].

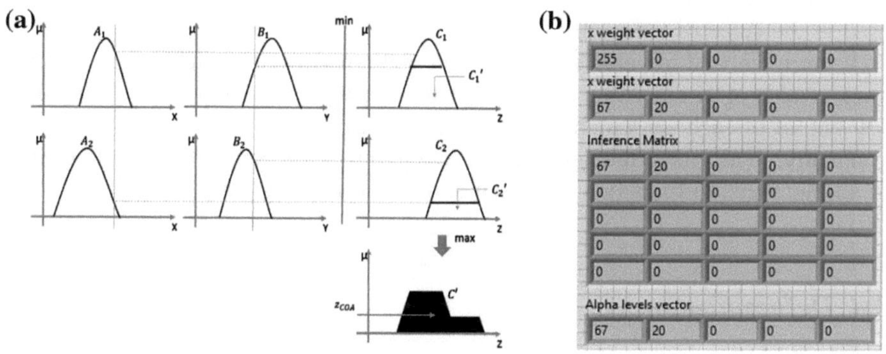

Fig. 6 **a** Mamdani inference system using the max-min implication. **b** Matrix built implicating w_1 and w_2 vectors with max-min operators

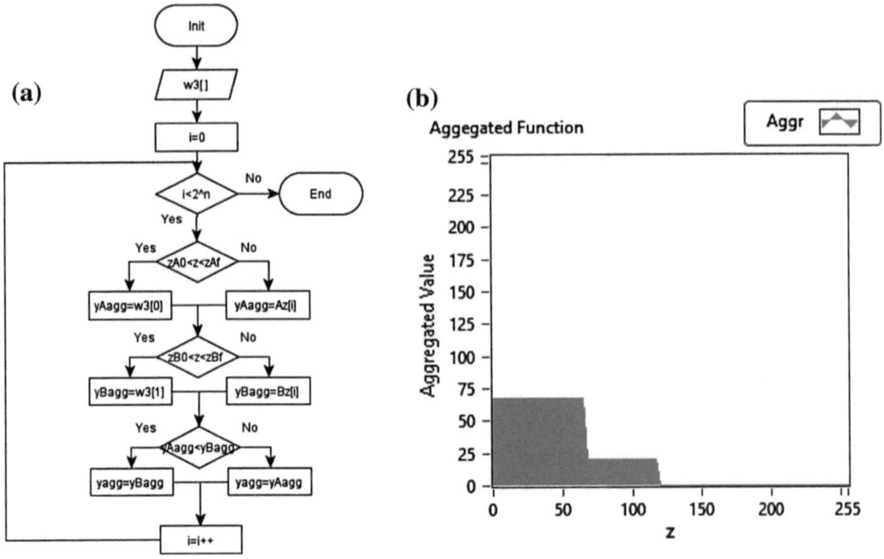

Fig. 7 **a** Algorithm used to build aggregated function. **b** Plot in LabVIEW

2.3 Aggregation

To make a comparison between aggregation points where the membership function is flat (α-levels), the output is obtained through a sweep from left to right in each set, as shown in the flowchart in Fig. 7a.

When it is greater or equal to α-level, the value is stored in zA0 for the α-level of the initial value and zAF for the α-level of final value (Fig. 7a). These values are used in the stage of defuzzification.

If the membership value is in the range delimited by the α-levels, the MF will have a value of 1, otherwise the value of the aggregate function is the value of the MF output of each set which is contained in *yAagg* array.

Finally, to assemble the aggregate function values are compared and the greatest in all membership functions is chosen and stored in *yagg* array (Fig. 7b).

2.4 Defuzzification

Rigid value on FPGA is obtained using COSAA (Center of Slices of Area Average) method which is faster than COG (Center of Gravity), but is less exact. It is described by Eq. (2).

Table 1 Parameters of the fuzzy sets used for experimental tests

Conj	Trapezoidal	Triangular	Triangular	Triangular	Trapezoidal
X	0,0,50,70	65,90,115	110,140,170	165,190,215	200,225,255,255
Y	0,0,45,68	60,85,120	115,150,180	170,200,220	210,235,255,255
Z	0,0,55,69	60,85,120	120,150,170	160,200,230	212,231,255,255

$$\mathrm{COSAA} = \frac{\sum_{i=0}^{\alpha_{max}} \left[\frac{x_f^{\alpha i} - x_0^{\alpha i}}{2} + x_0^{\alpha i}\right]}{\alpha_{max}} \quad (2)$$

3 Experimental Results

To test the performance of the fuzzy logic algorithm implemented on the FPGA board we use the application developed in LabVIEW as well as Matlab Fuzzy Toolbox as references. Table 1 values were used to set the parameters of the inputs and output fuzzy sets.

The response for each combination of inputs (x and y) is plotted in a tridimensional graph. Figure 8 shows the graph plotted with Matlab Fuzzy Toolbox while Fig. 9 shows the plot in LabVIEW application. It can be seen that both graphs have very similar shapes. LabVIEW application front panel (Fig. 10) simulate the x and y inputs. Each control ranges from 0 to 255, assuming 8 bits are used for each binary input word. The output is represented by a boolean array of 8 LEDs, Front panel shows the output crisp value obtained by two defuzzification methods: COG and COSAA. Table 2 shows a comparison between crisp values obtained in the FPGA board, LabVIEW application and Matlab Fuzzy Toolbox. To test the COSAA performance on FPGA, input x is fixed in 176 and y input is swept from 0 to 255. It can be seen that COSAA algorithm running on FPGA board provides crisp values very approximated to those ones obtained by the LabVIEW application and Matlab Fuzzy Logic Toolbox using COG defuzzification method.

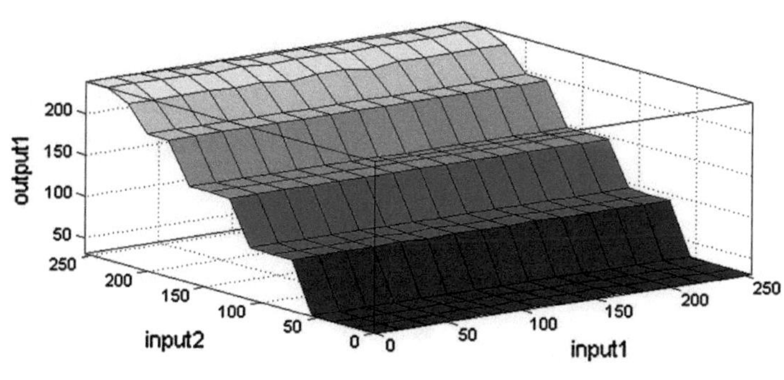

Fig. 8 3D graph plotted in Matlab

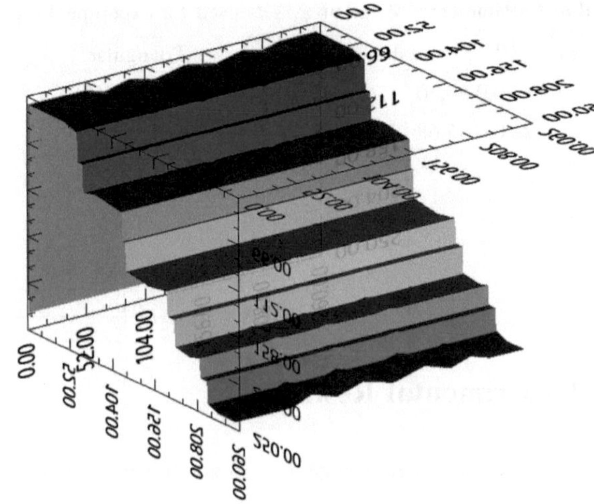

Fig. 9 3D graph plotted in LabVIEW

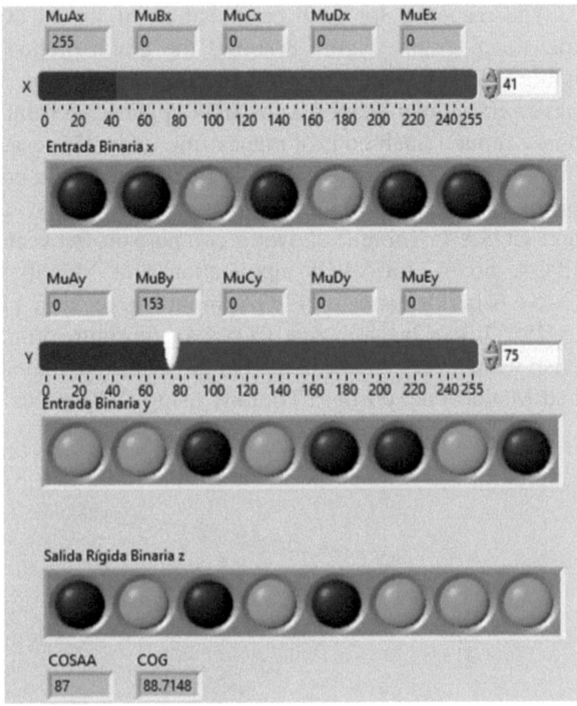

Fig. 10 Front panel in LabVIEW based application that helps in verifying the output crisp value for each inputs combination

Table 2 COSAA versus COG method comparison

Y	COSAA FPGA	COSAA LV	COG LV	COG MLTB
0	31	31	32.729	32.7
10	31	31	32.729	32.7
20	31	31	32.729	32.7
30	31	31	32.729	32.7
40	31	31	32.729	32.7
50	31	31	32.729	32.7
60	32	32	33.0369	33.1
70	88	88	89.0837	89.1
80	88	87.5	89.0023	89.0
90	88	87.5	89.0023	89.0
100	88	87.5	89.0023	89.0
110	88	89	89.3332	89.3
120	145	146	145.338	145.0
130	147	147	145.959	146.0
140	147	147.5	145.983	146.0
150	147	147.5	145.983	146.0
160	147	147.5	145.983	146.0
170	146	146.5	145.763	146.0
180	196	197	195.791	196.0
190	197	197	196.008	196.0
200	197	197	196.008	196.0
210	197	197	196.008	196.0
220	237	237.5	235.596	236.0
230	237	238	235.77	236.0
240	237	238	235.77	236.0
250	237	238	235.77	236.0
255	237	238	235.77	236.0

This FPGA implementation uses only 1 % of the available slices, 1 % of the available lookup tables (LUTs) and 3 % of input-output blocks (IOBs), see Table 3.

Also, mean square error was obtained applying (5) to data shown in Table 2.

$$MSE = \sqrt{\frac{\sum_{i=1}^{N} (COSAA_{FPGA} - COG_{MLFTB})^2}{N}} \tag{5}$$

Table 3 Utilization summary

Item	Available	Used	Percentage (%)
Slices	2278	34	1
Slice flip flops	90	42	46
Slice LUTs	9112	85	1
Bonded IOBs	232	9	3

Thus, MSE for the data in Table 2 is 1.66 which is very acceptable considering that COSAA method do not use iterations to obtain the crisp value. The average processing time is 52 us. This value was obtained using a triangular low frequency signal and taking the time it takes to generate an output FLC.

4 Conclusions

This work presents an approach for the implementation of a fuzzy logic controller on an FPGA board using VHDL coded look up tables which simplifies the VHDL programming.

The application developed in LabVIEW effectively generates look-up tables for Mamdani fuzzy model implementation in FPGAs. Moreover it was very helpful to verify each stage of the system.

Although COSAA defuzzification method (Center of Slice Area Average) provides a less-exact rigid value than COG (Centroid of Gravity), it requires fewer computations, and therefore improves the execution time, so it is a good choice when a hardware fuzzy system implementation is needed.

Crisp values obtained using COSAA method in FPGA are according to those ones obtained using COG in Matlab Fuzzy Toolbox.

Since this fuzzy logic system approach is thought as a general purpose controller, can have a lot of applications in many fields, especially in mechatronics due to their intrinsic real-time operation requirements.

References

1. Hernandez Zavala A (2009) High Performance architecture for fuzzy processors. Ph.D thesis, Instituto Politécnico Nacional. México, D. F., Jan 2009
2. Passino K, Yurkovich S (1998) Fuzzy control, 1st edn. Addison Wesley, California
3. Li Z, Gu J, Ming A, Xu C, Shimojo M (2006) Intelligent complaint force/motion control of non-holonomic mobile manipulator working on the non-rigid surface. Neural Comput Appl 15(3–4):204–216
4. Miyashitaa T, Ishiguroa H (2004) Human-like natural behavior generation based on involuntary motions for humanoid robots. Robot Auton Syst 48:203–212

5. Fung EHK, Wong YK, Zhang XZ, Cheng L, Yuen CWM, Wong WK (2011) Fuzzy logic control of a novel robotic hanger for garment inspection: Modeling, simulation and experimental implementation ser. Expert systems with applications, vol 38. Elsevier, Hong Kong, p 9929
6. Robin JP (2005) Design and Implementation of a Fuzzy Inference Engine on an FPGA. M.Sci thesis, Indian Institute of Technology, Delhi, India, Mar 2005
7. Ponce-Cruz P, Ramírez-Figueroa FD (2010) Intelligent control systems with LabVIEW, 1st edn. Springer, London
8. Yi J, Yubazaki N (2000) Stabilization fuzzy control of inverted pendulum systems. Artif Intell Eng 14:153–163 (Elsevier, Mar 2000)

High-Order Sliding Modes Based Linearization: An Application to Roll Autopilot

J. Davila, A. Monsivais and A. Mosqueda

Abstract The high-order sliding modes are applied for system linearization of a class of nonlinear systems. The nonlinearites of the system are treated as bounded external disturbances acting on a linear system. A standard controller is designed for the linearized system. The robust controller is composed by two terms: first, a linear controller, that provides desired performance is designed for the linear system in the absence of perturbations; second, a nonlinear compensation term, designed using the high-order sliding-modes techniques, is used to compensate the perturbations that affect the linear system. The workability of the proposed controller is illustrated with the design of a roll autopilot for a missile.

Keywords Roll control · High-order sliding-modes · Sub-optimal control

1 Introduction

The design of controllers for nonlinear systems is limited for the impossibility of integrate performance restrictions given the nonlinear nature of the system. This important restriction is very important for the aerospace applications. Sliding mode control is known as an effective technique to deal with perturbed or uncertain systems, the application of sliding mode control techniques is restricted by the appearance of the chattering effect.

The High-Order Sliding-Modes techniques mitigates the application problems related to standard sliding-modes by reducing the switching frequency needed to maintain the sliding motion. The High-order sliding-modes techniques have already been applied to flight control problems with satisfactory results. In [1] a sliding modes based control is designed using two control loops, the proposed control

J. Davila (✉) · A. Monsivais · A. Mosqueda
Section of Graduate Studies and Research, National Polytechnic Institute, ESIME-UPT, Av. Ticoman 600, Col. San Jose Ticoman, Gustavo A. Madero, Mexico
e-mail: jadavila@ipn.mx

M. Ceccarelli and E.E. Hernández Martinez (eds.), *Multibody Mechatronic Systems*, Mechanisms and Machine Science 25, DOI 10.1007/978-3-319-09858-6_26

ensures asymptotic tracking of the command deflections. The smooth second-order sliding-modes is applied for missile guidance in [2]. The dynamic sliding-manifolds technique is applied in combination with a transformation for stabilization of nonminimum phase aircrafts in the work by [3]. Recently [4] developed a second order sliding-modes based black-box control for signal tracking, the proposed controller is tested in simulations with a 6-DOF UAV model.

An alternative to obtain the characteristic robustness of the sliding-modes, without applying directly the discontinuous control signals on the system, is the use of these techniques for the design of estimation algorithms. The main advantages of the observers basing on high-order sliding-modes is their robustness against external perturbations [5, 6], and that they bring the possibility of exploit the equivalent output injection for the designing of the identification algorithms for the disturbances.

The combination of sliding-mode control techniques with conventional control has allowed the development of robust control algorithms that are capable to solve the stabilization problem under uncertain conditions. In [7] a Linear Quadratic Regulator is applied to stabilize a nonlinear affine system using a compensation term generated by the use of integral sliding modes. In [8] the high-order sliding mode based hierarchical observer is applied to identify disturbances acting on a perturbed system, the identified signal is used to add robustness to the smooth control signal generated by standard feedback control. A backstepping design that combines the high-order sliding modes differentiator and the feedback linearization is proposed in [9].

In [10] the general model of an autopilot for tactical missiles is proposed. In this article, the dynamic of the missile is spliced into two decoupled dynamics. In one side the first order rigid body effect is considered; while by the other side, the dynamic corresponding to the flexible body dynamics is considered. Using this model, in [11] a robust autopilot is proposed using a Linear Quadratic Regulator with a compensation term designed using the methodology proposed in [12].

In this paper, a robust controller is designed for a class of nonlinear systems. The proposed controller is composed by two terms: A linear controller capable to satisfy the desired performance for the a linearized version of system. A nonlinear compensation term, designed using the high-order sliding-mode observer, that is used to compensate the nonlinear part of the system seen as an external disturbance affecting the linear system.

The structure of the paper is the following: In Sect. 2, the class of systems under study is defined. A linear representation of the system is presented in Sect. 3. The proposed controller is presented in Sect. 4. A linear controller is proposed, in particular a Linear Quadratic Regulator is designed in Sect. 4.1. The high-order sliding-modes based compensation term is designed in Sect. 4.2. The proposed methodology is applied to the roll autopilot design in Sect. 5. Section 6 provides conclusions to this study.

2 Problem Statement

Let us consider the class of nonlinear systems given in the following form:

$$q^{(n+1)} = F(q^{(n)}, q^{(n-1)}, \ldots, q) + u \tag{1}$$

where $q \in \mathbb{R}$ is the generalized coordinate and $u \in \mathbb{R}$ is the control the nonlinear function $f : \mathbb{R}^n \to \mathbb{R}$ is a bounded and Lebesgue measurable function. It is important to remark that given the consideration on the function f, the considered class of nonlinear systems are Bounded Input—Bounded State systems.

Differential equations are understood in the Filippov sense (see [13]) in order to provide for the possibility to use discontinuous signals in controls. Filippov solutions coincide with the usual solutions, when the right-hand sides are Lipschitzian. It is assumed also that all considered inputs allow the existence of solutions and their extension to the whole semi-axis $t \geq 0$.

The aim of this article is to provide regulation for the variable q of the nonlinear system (1) by means of a linear controller ensuring satisfaction of linear performances after a finite-time transient.

3 Linear Representation

Let us consider the following state-space form

$$\dot{x} = Ax + Bu + Df \tag{2}$$

$$y = Cx \tag{3}$$

where $x \in \mathbb{R}^n$ is the state vector, $y \in \mathbb{R}$ is the system output and $u \in \mathbb{R}$ is the control signal. The nonlinear function $f \in \mathbb{R}$ is composed by the nonlinearities of the system and is defined as:

$$f = F(x_n, x_{n-1}, \ldots, x) - Ax \tag{4}$$

The matrices B, C, and D are defined as

$$C = \begin{bmatrix} 1 & 0_{1\times(n-1)} \end{bmatrix} \quad B = \begin{bmatrix} 0_{(n-1)} \\ 1 \end{bmatrix} \quad D = \begin{bmatrix} 0_{(n-1)} \\ 1 \end{bmatrix}$$

The matrix A is chosen arbitrarily such that the following assumption are satisfied.

The Rosenbrock matrix of the triplet $\{A, C, D\}$ is defined as:

$$R(s) = \begin{bmatrix} sI_n - A & -D \\ C & 0 \end{bmatrix} \quad (5)$$

The invariant zeros of the triplet $\{A, C, D\}$ are given by the points s_0 for which the Rosenbrock matrix $R(s_0)$ loses rank.

Assumption 1 The triplet {A, C, D} does not have invariant zeros.

Assumption 2 The perturbation signal f satisfies

$$||f||_\infty \leq f^+$$

for a known scalar $f^+ > 0$, here $|| \cdot ||_\infty$ denotes the infinite norm.

Notice that the last assumption is always satisfied given the considered class of nonlinear systems.

Assumption 3 The control distribution matrix B and the disturbance term Df satisfy:

$$Df \in span\ B$$

Differential equations are understood in the Filippov sense [13] in order to provide for the possibility to use discontinuous signals in controls. Filippov solutions coincide with the usual solutions, when the right-hand sides are Lipschitzian. It is assumed also that all considered inputs allow the existence of solutions and their extension to the whole semi-axis $t \geq 0$.

The aim of this paper is designing a Linear Quadratic Regulator algorithm that can stabilize the state of the system (2)—(3) even in the presence of the disturbances vector f.

4 Robust Linear Quadratic Regulator Design

The proposed controller is composed by two signals. The first one, is used to provide for optimal stabilization the nominal system, while the second one is designed to guarantee robustness against the perturbation f. The control takes the following form

$$u = u_1 + u_2 \quad (6)$$

where u_1 and u_2 will be designed below.

4.1 Linear Quadratic Regulator

The Linear Quadratic Regulator is designed to minimize a quadratic performance index of the form

$$J = \int_0^\infty \left(x^T Q x + u_1^T R u_1\right) dt$$

where $Q \geq 0$ and $R > 0$ are weights to be chosen. The resulting is an optimal control law given by:

$$u_1 = -Kx \tag{7}$$

where the gain K is computed as $K = R^{-1}B^T P$, where P is computed as the solution of the matrix algebraic Riccati equation

$$PA + A^T P - PBR^{-1}B^T P + Q = 0$$

By an appropriate selection of matrixes Q and R, one can obtain the desired performance. This optimal controller is usually applied to solve a wide variety of problems, as for example, this controller is an usual tool for the design of autopilots.

4.2 High Order Sliding Modes Based Compensator

Under Assumption 1 the system (2), (3) is strongly observable (to see a deeper study about strong detectability, the reader can refer, for example, to the tutorial book [14]). This assumption allows us to reconstruct exactly and in a finite-time the state, even in the presence of the disturbance f [15].

With this aim, an observer which is based on the high-order sliding modes is proposed as:

$$\begin{aligned} \dot{z} &= Az + Bu + L(y - y_z) \\ e_y &= y - y_z \\ \hat{x} &= z + U^{-1}\upsilon(e_y) \end{aligned} \tag{8}$$

where the matrix U takes the form

$$U = \begin{bmatrix} C \\ C(A-LC) \\ \vdots \\ C(A-LC)^{n-1} \end{bmatrix}$$

the compensation term $\upsilon(e_y)$ is composed by the variables

$$\upsilon(e_y) = \begin{bmatrix} \upsilon_1 & \upsilon_2 & \cdots & \upsilon_n \end{bmatrix}$$

where the components of the vector υ_i $i = 1, \ldots, n$, and the additional variable υ_{n+1} are taken from the high order sliding mode differentiator [16] given by:

$$\begin{aligned} \dot{\upsilon}_1 &= w_1 \\ w_1 &= -\alpha_{n+1} M^{1/(n+1)} |\upsilon_1 - e_y|^{n/(n+1)} sign(\upsilon_1 - e_y) + \upsilon_2 \\ \dot{\upsilon}_2 &= w_2 \\ w_2 &= -\alpha_n M^{1/n} |\upsilon_2 - w_1|^{(n-1)/n} sign(\upsilon_2 - w_1) + \upsilon_3 \\ &\vdots \\ \dot{\upsilon}_n &= w_n \\ w_n &= -\alpha_2 M^{1/2} |\upsilon_n - w_{n-1}|^{1/2} sign(\upsilon_n - w_{n-1}) + \upsilon_{n+1} \\ \dot{\upsilon}_{n+1} &= -\alpha_1 M \, sign(v_{n+1} - w_n) \end{aligned} \tag{9}$$

where the parameter M is chosen sufficiently large, in particular $M > |d|f^+$, where $d = C(A-LC)^{n-1}D$. The constants α_i are chosen recursively sufficiently large as in [16]. In particular, one of the possible choices is $\alpha_1 = 1.1$, $\alpha_2 = 1.5$, $\alpha_3 = 2$, $\alpha_4 = 3$, $\alpha_5 = 5$, $\alpha_6 = 8$, which is sufficient for $n \leq 6$. Note that (9) has a recursive form, useful for the parameter adjustment. In any computer realization one has to calculate the internal auxiliary variables υ_j and w_j, $j = 1, \ldots, n$, using only the simultaneously-sampled current values of e_y and υ_j.

The auxiliary output estimation error e_y and its first n derivatives take the following form

$$\begin{aligned} e_y &= y - Cz = C(x - z) \\ \dot{e}_y &= C(A - LC)(x - z) \\ &\vdots \\ e_y^{(n)} &= C(A-LC)^n (x - z) + C(A-LC)^{n-1} Df \end{aligned}$$

On the other hand, the high order sliding mode differentiator (9) brings an estimation of the derivatives up to order $n-1$. Hence, after the convergence of the differentiator, the derivative of order n satisfies:

$$- \alpha_2 M^{1/2} |\upsilon_n - w_{n-1}|^{1/2} sign(\upsilon_n - w_{n-1}) + \upsilon_{n+1} =$$
$$C(A - LC)^n (x - z) + C(A - LC)^{n-1} Df$$

Thus, the following equality holds after a finite time transient

$$\upsilon_{n+1} = C(A - LC)^n (x - z) + C(A - LC)^{n-1} Df. \tag{10}$$

The Eq. (10) is called the equivalent output injection. Given the properties of the differentiator (9), υ_{n+1} is a continuous term.

The perturbation f can be identified through the equivalent output injection as:

$$\hat{f} = \left(C(A - LC)^{n-1} D \right)^{1} \left(\upsilon_{n+1} - C(A - LC)^n U^{-1} \upsilon(e_y) \right)$$

Notice that $C(A - LC)^{n-1} D \neq 0$ otherwise no perturbations affects the system. The control term that provides robustness against the disturbance f is proposed as

$$u_2 = H\hat{f} \tag{11}$$

where the matrix H is computed as $H = B^+ D$, the matrix B^+ is the Moore-Penrose left pseudoinverse of B, i.e., $B^+ = (B^T B)^{-1} B^T$.

Theorem 1 *Be the system* (2)—(3). *Under Assumptions* 1—3 *the controller* (6) *provides suboptimal exact regulation under the presence of external perturbations.*

5 Application to the Robust Roll Autopilot Design

A wide variety of missiles possesses a cruciform configuration which brings to them a high accuracy and quick manoeuvering in any direction. However, the inherent instability of the roll yields in undesirable rolling motions that degrades the performance. To overcome this problem, the roll autopilots are proposed (see [10, 11]). The main objective of the above mentioned controllers is to maintain the attitude of the missile under system variations and external disturbances.

The block diagram of the missile roll dynamic is shown in Fig. 1. describe external perturbations.

The airframe flexibility is considered in the flexible body dynamics block. External disturbances d_{ext} are used to

The control is designed disregarding the flexible body effects and the external disturbances. In this sense, for analysis proposes, the flexible body dynamics, external perturbations and any other coupling effect derived from the pitch and yaw motions are concentrated in a single term, f. The state equations of the system can be written as:

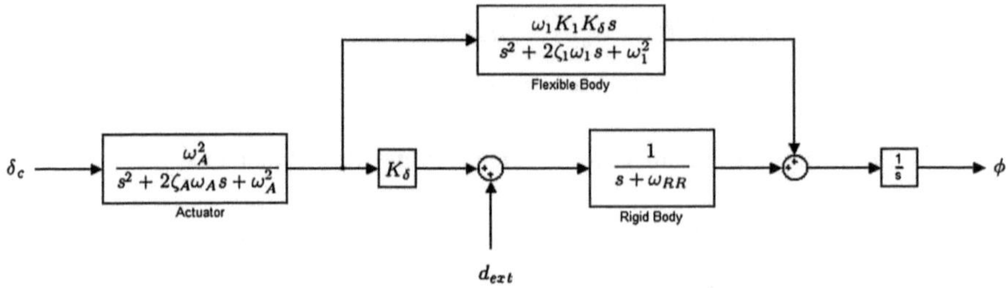

Fig. 1 System block diagram

$$\dot{x} = Ax + Bu + Df$$
$$y = Cx$$

where the matrices A, B and C take the following form

$$A = \begin{bmatrix} 0 & 1 & 0 & 0 \\ 0 & 0 & 1 & 0 \\ 0 & 0 & 0 & 1 \\ -a_3 & -a_2 & -a_1 & -a_0 \end{bmatrix}, \quad B = \begin{bmatrix} 0 \\ 0 \\ 0 \\ b \end{bmatrix}, \quad D = \begin{bmatrix} 0 \\ 0 \\ 0 \\ 1 \end{bmatrix}$$
$$C = \begin{bmatrix} 1 & 0 & 0 & 0 \end{bmatrix}$$

where $a_0 = 2\zeta_A\omega_A + \omega_{RR}$, $a_1 = 2\zeta_A\omega_A\omega_{RR} + \omega_A^2$ $a_2 = \omega_A^2\omega_{RR}$, $a_3 = 0$, $b = \omega_A^2 K_\delta$.

The parameters are given in the (Table 1):

For simulation purposes, the perturbation is given by:

$$d_{ext} = 54600 + 50000 \sin(\sin(t)\sin(0.1t) + 0.2)$$

The weighting matrices Q and R are the same as in [10]:

$$Q = \begin{bmatrix} \frac{1}{\phi_{max}^2} & 0 & 0 & 0 \\ 0 & \frac{1}{\dot{\phi}_{max}^2} & 0 & 0 \\ 0 & 0 & 0 & 0 \\ 0 & 0 & 0 & 0 \end{bmatrix} \quad R = \begin{bmatrix} \frac{1}{\delta_{c(max)}^2} \end{bmatrix}$$

where ϕ_{max}, $\dot{\phi}_{max}$, $\delta_{c(max)}$ are the maximal permissible values of the respective variables.

Using the solution of the algebraic Matrix Riccati equation, the gain K for the controller (7) is given by

$$K = \begin{bmatrix} 3 & 0.1286 & 0.001 & 0 \end{bmatrix}$$

The eigenvalues of matrix A are $0, -2, -65 + 75.9934i, -65 - 75.9934i$, notice that the system is marginally stable. The Luenberger gain of the observer is

Table 1 System parameters

Symbol	Variable	Value
ω_{RR}	Roll rate bandwidth	2 rad/s
K_δ	Fin effectiveness	9000 1/s^2
ω_A	Actuator bandwidth	100 rad/s
ζ_A	Actuator damping	0.65
ω_1	Torsional mode frequency	250 rad/s
ζ_1	Torsional mode damping	0.01
K_1	Torsional mode gain	−0.0000129
ϕ_{max}	Maximum desired roll angle	10
$\dot{\phi}_{max}$	Maximum desired roll rate	300 deg/s
$\delta_{c(max)}$	Maximum desired fin deflection	30°

designed to obtain a stable estimation error. The gain L is chosen to place the roots of the estimation error dynamics matrix $(A - LC)$ in −40, −41, −42, −43, as

$$L = \begin{bmatrix} 34 \\ -4417 \\ 499890 \\ -18385220 \end{bmatrix}$$

Using the compensated system matrix $(A - LC)$, the matrix U is given by:

$$U = \begin{bmatrix} 1 & 0 & 0 & 0 \\ -34 & 1 & 0 & 0 \\ 5573 & -34 & 1 & 0 \\ -839550 & 5573 & -34 & 1 \end{bmatrix}$$

The high-order sliding mode differentiator takes the form:

$$\begin{aligned}
\dot{\upsilon}_1 &= w_1 = -\alpha_5 M^{1/5}|v_1 - e_y|^{4/5}\text{sign}(\upsilon_1 - e_y) + \upsilon_2 \\
\dot{\upsilon}_2 &= w_2 = -\alpha_4 M^{1/4}|\upsilon_2 - w_1|^{3/4}\text{sign}(\upsilon_2 - w_1) + \upsilon_3 \\
\dot{\upsilon}_3 &= w_3 = -\alpha_3 M^{1/3}|\upsilon_3 - w_2|^{2/3}\text{sign}(\upsilon_3 - w_2) + \upsilon_4 \\
\dot{\upsilon}_4 &= w_4 = -\alpha_2 M^{1/2}|\upsilon_4 - w_3|^{1/2}\text{sign}(\upsilon_4 - w_3) + \upsilon_5 \\
\dot{\upsilon}_5 &= -\alpha_1 M\ \text{sign}(\upsilon_5 - w_4)
\end{aligned}$$

where the gains are chosen as $\alpha_1 = 1.1$, $\alpha_2 = 1.5$, $\alpha_3 = 2$, $\alpha_4 = 3$, $\alpha_5 = 5$ and $M = 2000$.

The control signal (6) is given by $u = -K\hat{x} + H\hat{f}$, where

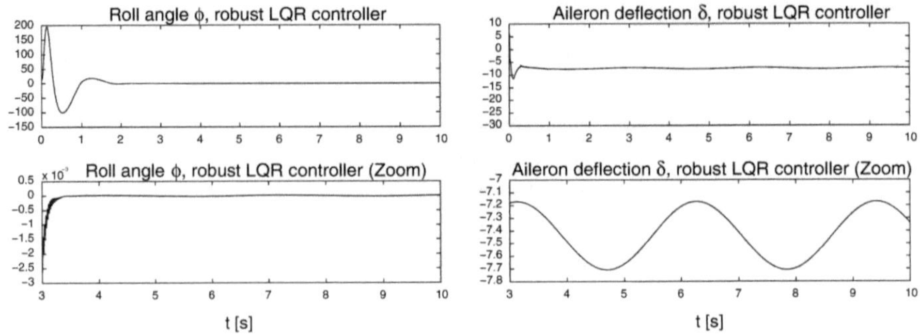

Fig. 2 *Left* Roll angle ϕ (*above*) and a zoom on the image (*below*) for the robust LQR controller. *Right* Aileron Deflections δ for the robust LQR controller

Fig. 3 Estimated disturbances

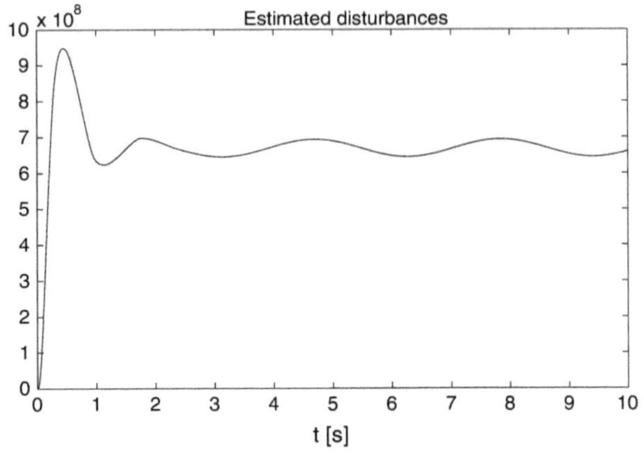

$$H = \begin{bmatrix} 0 & 0 & 0 & 0.1111 \times 10^{-7} \end{bmatrix}$$

The convergence of the roll angle ϕ to zero after a finite time transient is shown in the Fig. 2, left. Deflection of the ailerons δ and its ratio are presented in the Fig. 2, right. The perturbation identification $\hat{w}$ is shown in Fig. 3.

The results obtained with the proposed methodology are compared with a standard Linear Quadratic Regulator. With this aim, the control signal takes the form:

$$u = -Kx$$

The roll angle and a zoom on the graphic using the standard Linear Quadratic Regulator are shown in the Fig. 4, left. In comparison with the standard Linear Quadratic Regulator, the robust Linear Quadratic Regulator is exact with respect to the coordinate ϕ. The aileron deflection δ for the standard Linear Quadratic

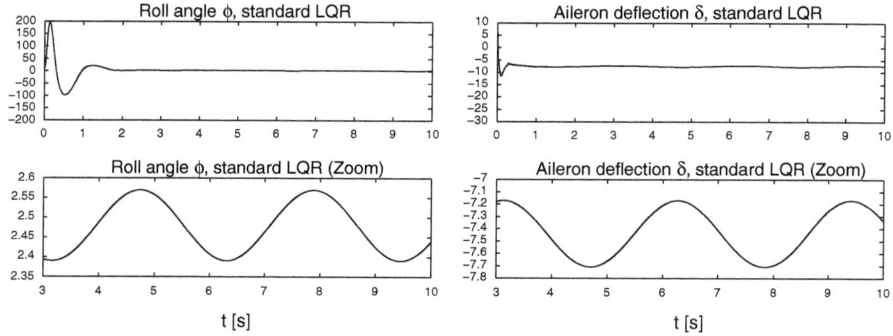

Fig. 4 *Left* Roll angle ϕ (*above*) and a zoom on the image (*below*) for the standard LQR controller. *Right* Aileron Deflections δ (*above*) and a zoom on the image (*below*) for the standard LQR controller

Regulator is shown in the Fig. 4, right. Notice that after 2.5 s, the deflections obtained for both controllers are very similar, then the most important contribution of the nonlinear compensation term takes place during the transient.

6 Conclusions

In this paper a controller for a class of nonlinear systems is proposed. The controller is composed of two terms. The first term is a linear regulator that allows the stabilization of the system satisfying linear performance criterions, while the second term is defined using high-order sliding-mode based compensation providing the cancelation of the nonlinear behavior of the system after a finite-time transient. The proposed method is illustrated by means of the design of a roll autopilot for a missile model. The robustness of the proposed controller is shown in simulations.

References

1. Shtessel Y, Buffington J, Banda S (1999) AIAA J Guid Control Dyn 22(6):873
2. Shtessel Y, Shkolnikov I, Levant A (2007) Automatica 43(8):1470
3. Shkolnikov I, Shtessel Y (2001) AIAA J Guid Control Dyn 24(3):566
4. Bartolini G, Pisano A (2010) Int J Robust Nonlinear Control 20(14):1594
5. Bejarano F, Fridman L, Poznyak A (2007) Int J Robust Nonlinear Control 17(18):1734
6. Davila J, Fridman L, Pisano A, Usai A (2009) Int J Control 82(8):1564
7. Pang H, Chen X (2009) J Syst Eng Electr 20(4):838
8. Ferreira A, Bejarano F, Fridman L (2011) IEEE Trans Contr Syst Technol 19(5):969
9. Davila J (2013) IEEE Trans Automat Contr 58(8):2077
10. Nesline F, Wells B, Zarchan P (1981) AIAA J Guid Control Dyn 4(3):316
11. Talole S, Godbole A, Kolhe J (2011) AIAA J Guid Control Dyn 34(1):107

12. Wang W, Gao Z (2003) In: Proceedings of the american control conference 2003, Cleveland State University, OH, USA, pp 4754–4759
13. Filippov A (1988) Differential equations with discontinuos right-hand sides. Kluwer Academic Publishers, Dordrecht
14. Trentelman HL, Stoorvogel AA, Hautus M (2001) Control theory for linear systems. Springer, London
15. Fridman L, Levant A, Davila J (2007) Int J Syst Sci 38(10):773
16. Levant A (2003) Int J Control 76(9–10):924

Proposal of Automated Inspection Using Camera in Process of VIN Validation

L.R.S. Souza, R.M.M. Oliveira and M.H. Stoppa

Abstract After recording the Vehicle Identification Number (VIN) on the chassis, a procedure of validation is indispensable, to ensure that the code be correctly recorded. The automotive sector utilizes this constantly, trying to eliminate mistakes, once that the VIN is utilized in world scale and these errors bring many troubles to the consumers, and consequently for the brand. In some cases, it is not available, the minimum requirements of reliability necessary for the inspection process. In cases where the inspection is performed manually, with an operator making a visual conferencing, there is risk of commercialize a vehicle with a different VIN of the one contained in the documents, owing a human mistake. This paper proposes the automation of the inspection process, using computational vision in data validation transcribed to the chassis, and low-cost components to read the VIN recorded in the chassis, comparing it to the previously authorized code, to increasing the quality control and avoiding future problems. We subject the system to experimental tests and find efficiencies in the inspection recording VIN. We increased reliability in the process, including a pre-validation of the code to be recorded, where the machine is not authorized to perform recording without requirements validated.

Keywords Computer vision · Image processing · Automated inspection · Pattern recognition · VIN recording

L.R.S. Souza (✉) · R.M.M. Oliveira · M.H. Stoppa
Universidade Federal de Goiás, Goiânia, Brazil
e-mail: leandrorodrigues.s@gmail.com

R.M.M. Oliveira
e-mail: romulomuriel@gmail.com

M.H. Stoppa
e-mail: mhstoppa@pq.cnpq.br

M. Ceccarelli and E.E. Hernández Martinez (eds.), *Multibody Mechatronic Systems*, Mechanisms and Machine Science 25, DOI 10.1007/978-3-319-09858-6_27

1 Introduction

Globally is used the standard VIN (Vehicle Identification Number) for vehicle identification [1]. The sequence of numbers is constituted by seventeen characters, which determines the manufacture country, year of production, model, and others vehicle informations [2].

Handled with great care by the manufacturers, the VIN registered on the chassis is one of the most important information of vehicle. Through it, government can emit all necessary documents for the vehicle carryover on the urban roads. Once the VIN numbers is compromised, the owner of a vehicle needs to re-write it, in an appropriated and authorized government agency. The procedure is bureaucratic and not simple to be performed, always demanding reports and authorizations. Adulterate the VIN is a criminal business, however there are effective methods to inhibit this crime, such as based on microdots, which uses nanotechnology to authenticate the code, avoiding falsifications [1].

Buying a vehicle with a compromised VIN is so bad as buying one with VIN recorded incorrectly. The paper here presents a proposal that avoids this last type of incident. Currently, many companies adopt conventional methods for inspect the recorded VIN, especially the visual conferencing by human. In the traditional process, after the VIN recording, a person responsible for the code conference, compares the barcode label and the code recorded on the chassis, copying the VIN using a graphite over a sheet of paper. After the first hours of inspection, the level of reliability decreases significantly, certainly due to fatigue, stress, physical, emotional problems, and the repeatability [3]. Thus, the conference proceeding becomes inefficient. There are approaches that consider only the VIN image analysis [4]. However, this paper proposes a system which besides to do this, also implements the system validation, integrating different software.

Using a conventional camera (webcam), of low cost, and including concepts of image processing and computer vision, the paper presents a method of automated visual inspection for validate the process of chassis recording. Besides ensuring reliability and efficiency, replacing the manual procedure, the proposal also includes a pre-conference of code to be written on the chassis, preventing erroneous records.

2 Method Description

Usually recording the VIN is made by a punching machine, that receives the data and performs the markings on the vehicle chassis. With a barcode's reader, the data are inputted on the machine. The reader streamlines to the recording process, however can infer information that are not expected, due to printing problems, standardization, dirt or wrinkled label. Any problem in the barcode label causes variation in the reading moment and so an incorrect recording. Generally the labels are printed in fragile materials, and not much safe [5].

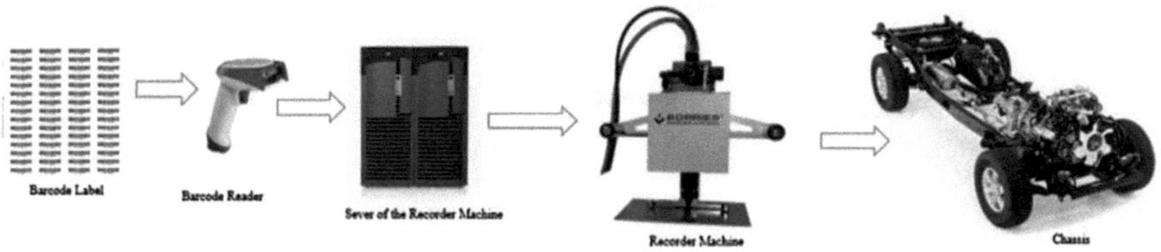

Fig. 1 Current process of VIN recording

Identify variations or errors in the process of VIN recording, prevents the use the chassis, which must be destroyed. There are no possibilities for markdowns or corrections. There are cases where the human limitations, prevents him or her to detected the error, causing major problems. Once identified the error in chassis, the consumer may go through embarrassing situations, for example, explain the legality of the vehicle. Any evidence of adulteration of the VIN is considered a crime and to prove to a court that the vehicle was acquired under these conditions, is rather difficult.

The companies did not disclose these types of recording errors and their consequences, yet when this becomes public, many internal and external disorders are generally accompanied by substantial financial compensation to the consumer.

2.1 Current Process

In the usual process (Fig. 1), as soon as the operator receive the list of VIN sequence to be recorded, he verifies codes which were been recorded. After read the barcode label and record the VIN on the chassis, the label is attached to the vehicle inspection formulary, accompanying it over the assembly line. When the label is read, the barcode reader sends the information directly to the punching machine, which performs recording when the operator authorizes it. Completed the recording, the operator verifies visually the sequence of seventeen digits, comparing the label and the VIN recorded. After each recording, the operator catch a VIN's image is recorded over the chassis, using a piece of paper and rubbing the graphite over the VIN. This is necessary for future conferences.

2.2 Proposal Process

The devices used in this experiment were a notebook with 4 GB of RAM, Intel core i3 M350 processor with frequency of 2.27 GHz and a hard drive of 640 GB. The operating system used was a Windows® 8 Pro with 64 bits. Also were used a Microsoft webcam, with resolution of 640 × 480/1280 × 720. For the development

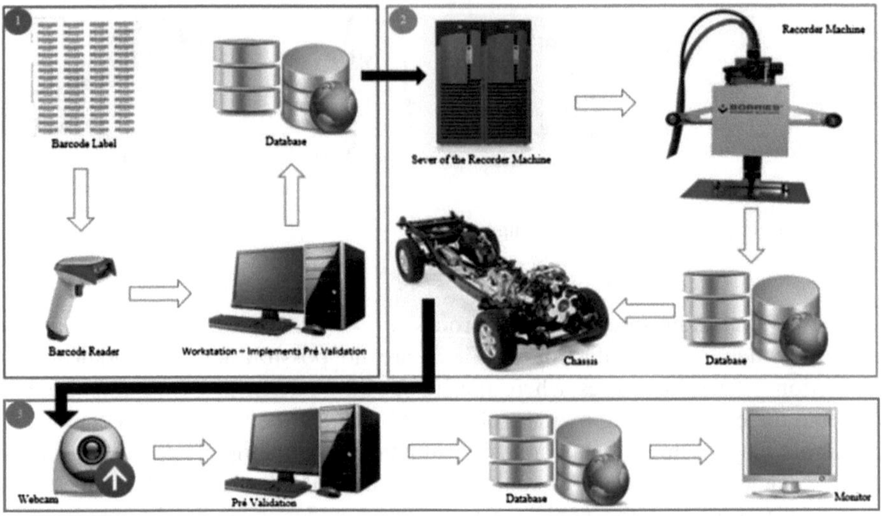

Fig. 2 Proposed diagram process

were used the following tools: Visual Studio 2012 Express Edition, for implementation of the source code, Emgu CV version 2.9.0, open source image processing and computer vision software. The Tesseract-OCR Engine, open source character recognition software. Finally, it is also used the MySQL 5.5.28 database, which allows traceability of the proposed process. All features operate in an integrated manner, providing support for the proposal, as shown in the below diagram. The proposed method has three stages, which use different devices, as shown in Fig. 2.

On the first stage, we perform the preprocessing, aiming to identify the relationship between the data obtained via the barcode's reader and the sequencing of production. At this stage, the workstation verifies the code obtained from the barcode label comparing it with the list of VINs to be recorded. This validation analyzes the sequenced VINs not recorded (A) and the set of VINs already performed (B). What can be expressed by Eq. 1, that is, verifies if (x) is available (to be recorded—A) but not was performed yet (B):

$$\exists x \in A/A{-}B = \{x|x \in A \wedge x \notin B\} \tag{1}$$

The Fig. 3 presents a flowchart to sequence of operations proposed in step one, highlighted in blue.

On the second stage is performed the VIN recording, based on information obtained because of first stage. Figure 2 shows the devices used, while the green flow (Fig. 3) exhibits the operations performed.

At stage three, the conference is held via machine vision, using the resources as shown at Fig. 2. The process flow realized by machine vision can be seen in Fig. 3,

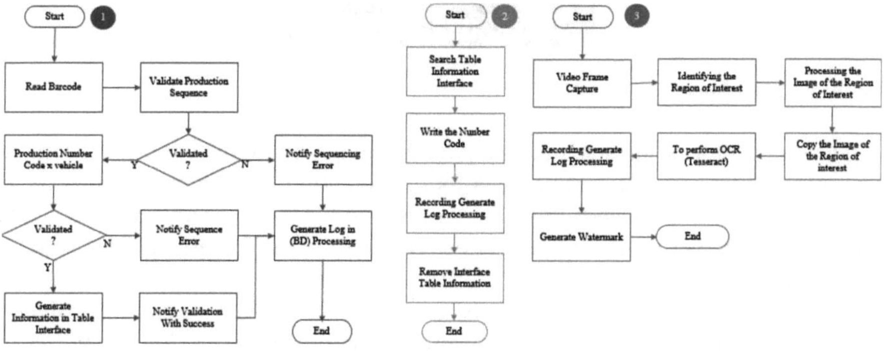

Fig. 3 Proposed flowchart

where occurs the process of image transcribing, where the image is converted to plaintext, using Optical Character Recognition (OCR).

The greatest effort happens in the third stage. For validation process, video frames are captured, after positioning the camera manually in the chassis, to ensure the best access angle and lighting for the considered region of interest. The camera captures the video and transmits him to a software, which uses the framework Emgu CV, which implements the functionality and features for capturing and manipulating frames from videos [6]. This software allows images to be captured as photos, discriminating the region of interest on the picture. We did a histogram inversion of captured gray levels image, by morphological operation.

We use the Niblack's binarization method to separate text and background. It works using a pixel wise threshold (T) by sliding a rectangular region over the greylevel image [7]. This algorithm is based on the local mean (m), and standard deviation (d) of gray values computed over a small location around each pixel. K is a parameter defined by the user [8]. The method's formula is given by Eq. (2):

$$\mathrm{T} = \mathrm{m} + \mathrm{k} * \mathrm{d} \tag{2}$$

The identification of the region of interest is defined using the concepts of erosion, dilation and growth of the region of interest. This process of expansion is used to fill and/or remove undesirable gaps in the frame, while the erosion is applied to eliminate irrelevant details, as noises, and create gaps in areas of unwanted connection [9].

The result of using the operations of erosion and dilation is the identification of the area that contains alphanumeric characters, so that, the system makes an outline around the code, which is performed using image segmentation based on growth in the region of interest (Fig. 4).

The structuring element applied over binary image, both in erosion and dilatation operations is a 3 × 3 matrix to reduce noises.

The coordinates of this area will be used for further processing contributing for a faster analysis, once that only the delimited area should be treated.

Fig. 4 Region of interest obtained

The image processing starts focusing on the region of interest, because the others parts of the image do not contain the VIN's information. So, the processing time of the algorithms is optimized, recognizing and working only with numbers in the chassis. Then is performed a contrast correction over the image, because was not applied diffused lighting.

Chassis commonly are black, so, the correction was also applied to distribute a gray-scale on the image, to assist the segmentation process of the information from the region of interest. To detect the edges of the image, we applied the Canny algorithm, implemented by Emgu CV [10]. During this phase, the Canny algorithm is applied twice, once for the grayscale and another for color scale, to relate the two results of the image, reaching a better location of the VIN.

Now the region of interest is forwarded to OCR for implementation of image analysis algorithms. The process of recognition of characters contained in the digital image is performed by Tesseract Engine, which performs the transcription of the text contained in the digital image as a plaintext.

After this, the plaintext is stored to be used in the comparison with the code originally entered for recording. To validates this method, were considered twenty photos for each chassis analyzed and twenty vehicles were submitted to test.

In the phase of comparison between codes released for recording and codes resulting from the automated inspection, to the cases where the sequence obtained in the result of OCR (inspection) is different from the code authorized for recording, a visual markup is inserted. This mark indicates that identified sequence does not match a valid code. For cases in which the chassis code obtained via OCR match the sequence of codes released, a visual marking is included in the picture to confirm the authenticity.

Figure 5 shows in the first image, labeled with a mark (X), meaning that does not match with de code released to be recorded. So, the operator sees that a mistake occurred, and the chassis cannot remain on the assembly line. On the other hand, the second image confirms that a VIN number was correctly recorded.

The result obtained in the OCR analysis process is recorded in the database, which can be accessed at any time. Also is possible to access the results obtained by this new process. Concern in the veracity of the information contained in the image, an authentication is included in the digital document. So that, the encoding, inserted in the images, ensures reliability and make easy future conferences [11].

Fig. 5 Bad and good recordings

Finally, the original video, and the images obtained via processing, containing the validation tags are stored together for future audits.

3 Results

The results were obtained after the operator positioned the camera focusing on the region where the VIN is registered. They were classified into two groups, the first represented by Table 1, that shows the performance (in time) of the algorithm in the processing of each frame.

Changing the resolution had direct impact on the processing time, according to Table 1, which was expected, due the largest number of pixels used. Thus, to identify the region of interest and determining the area important of the text, a greater number of mathematical operations are required to obtain a result.

It has been observed, in accordance with Table 2 that increasing the resolution of the camera, was not increased the amount of information relevant for the job. Consequently, was observed a greater amount of noise or information not necessary for the VIN validation process. This directly influences in the identification of the region of interest, apart from the increase of data to be analyzed.

The tests were also applied to the analysis of accuracy in detecting characters via OCR, varying only the resolution of the image inputted in Tesseract OCR Engine. The Tesseract OCR Engine considers as valid values, only images who the region of interest had identified correctly, as values shown in Table 2.

For all images transferred to Tesseract OCR Engine, the recognition was obtained successfully. With the region of interest known, the recognition of the characters in the images were validated without significant distortion, that could influence in the results.

Table 1 Results obtained by different camera resolutions

Analysis	Quantity	Resolution (pixels)	Mean time
Test—low resolution	20	480	~83
Test—high resolution	20	1080	~213

Table 2 Accuracy of post processing

Analysis	Resolution (pixels)	Examples	Success	Failure	(%)
Test—low resolution	480	20	16	4	80
Test—high resolution	1080	20	12	7	60

4 Conclusion

The solution developed for inspection of VIN, showed an alternative for assistance in validating the data recorded in the chassis. To be used in mass production, a device needs to be created in order to support the camera and maintain a standard position front of the VIN, which is always recorded in the same region. This device can contribute to avoiding any kind of obstruction of the region of interest or uncontrolled variations as the luminous flux on the VIN area. The article do not deepens over this issue because the characteristics of the environment are considered sufficient, as shown in tests.

In this experiment, for each test (low and high resolution) a total of 20 chassis were used for each chassis. The rate of success was 70 %, being necessary to improve the technique employed in the OCR process.

The restricted access in the industries prevents us to have sufficient information to make comparison or any other analyzes. The point is, we get a real problems and tried to develop a system that could mitigate it.

The developed proposal can be extended to many applications, supporting in routines of inspections and conference. The process presented can be customized for a range of applications, such as, traffic departments, vehicle inspection routines, allowing an easy identification of the VIN, recorded on the chassis of a vehicle.

For futures developments is possible to think in an application that uses mobile devices to validation.

References

1. Cerere V (2010) Estudo de Medidas Para a Melhoria da Identificação Veicular no Brasil. Dissertação (Mestrado Profissional em Engenharia Automotiva), Universidade de São Paulo
2. Bertagi V, Bertagi A, Bereza A, Bertagi C, Santos S (2012) Código Padronizado—Chassi. Fundação de Estudos Sociais Do Paraná, Curitiba
3. Pinheiro M (2000) Variabilidades dos Defeitos do Produto e Desempenho do Inspecionista. Universidade Federal de Minas Gerais—UFTM, Belo Horizonte
4. Shah P, Karamchandani S, Nadkar T, Gulechha N, Koli K, Lad K (2009) OCR-based Chassis-number recognition using artificial neural networks. Presented at the IEEE ICVES. pp 31–34
5. Rei J (2010) RFID Versus Código de Barras da Produção à Grande Distribuição. Faculdade da Engenharia da Universidade do Porto, Portugal. pp 30–31
6. Abubakar M (2014) Implementation of speed up robust feature for detection and tracking of inanimate objects. School of Electronics Engineering, Tianjin University of Technology and Education, Tianjin
7. Trived S, Gohil S, Bhatt H, Shah P (2014) Benchmarking Niblack's binarization algorithm using OpenCV library integrated in C++ and FPGA simulation. Institute of Technology, Nirma University, Ahmedabad
8. Feng M, Tan Y (2004) Contrast adaptive binarization of low quality document images. School of Electrical and Electronic Engineering, Nanyang Technological University, Singapore
9. Gonzalez W (2002) digital image processing. 2 edn. Pearson Education, Singapore

10. Zeilhofer P et al (2007) Técnicas de reconhecimento de formas para identificação de áreas de irrigação por imagens de satélite. In: Anais XIII Simpósio Brasileiro de Sensoriamento Remoto, Florianópolis, INPE
11. Rey C, Dugelay J (2002) A survey of watermarking algorithms for image authentication. Eurecom Institute, France

Concept Design Process for Robotic Devices: The Case of an Assistive Robot

H.A. Moreno Avalos, I.G. Carrera Calderón, S. Romero Hernández and V. Cruz Morales

Abstract Current demands for continuous innovation have made the product design and development process a branch of engineering itself. Several techniques have been proposed to address the problem of developing a new product in a systematic way [1, 2]. In this paper we use some of these techniques and adjust them to carry out the conceptual design of a robot for assistance and rehabilitation of elderly and disabled persons.

Keywords Concept design · Assistant robot

1 Introduction

The concept design process of a robot for assistance and rehabilitation is presented. The objective of this process is to determine the concept of robot's structure given the requirements of the application and users. The term concept refers to an approximate description of the technology, operating principles, and structure that the robot should have. The degree to which the robot meets the needs for which it was created depends importantly on the quality of the underlying concept.

The concept design process is performed once an opportunity for the application of a robotic system is found. In this phase the user's needs are identified, concepts for alternative systems are generated and evaluated. In the case of mechanical subsystem the concept generation includes the type synthesis of mechanisms. Once a concept is selected, the next phases are dimensional synthesis [3], detailed design, fabrication and testing of the system.

In a previous work [4], we found that the development of robotic systems for the disabled is an important assignment due to the expansion of the community with

H.A. Moreno Avalos (✉) · I.G. Carrera Calderón · S. Romero Hernández · V. Cruz Morales
Departamento de Ingeniería Industrial y Operaciones, Instituto Tecnológico Autónomo de México, Cd. de México, Mexico
e-mail: hector.moreno@itam.mx

M. Ceccarelli and E.E. Hernández Martinez (eds.), *Multibody Mechatronic Systems*, Mechanisms and Machine Science 25, DOI 10.1007/978-3-319-09858-6_28

disabilities caused by the increase in life expectancy. This means that the elderly and the institutions that take care of them will be the main consumers of these applications. From surveys conducted among care professionals and potential users it was found that the main issues that the robotic device must solve are related to assistance with gait, balance and lifting up from a sitting position.

In the following sections the user's needs list, the engineering specifications, the concept generation and selection of the robotic system are presented.

2 Needs Identification

For the identification of user's needs we used the data of surveys on disability in Spain [5, 6] as well as a research study performed in situ at the center for the rehabilitation of daily activities on the Hospital Infanta Sofia in Madrid, Spain [4]. The data was interpreted, organized and translated into a congruent set of customer needs; these are summarized in Table 1.

Based in different factors such as frequency, repeatability, customer fact statements and importance rating, five needs were identified as principal: (1) Support a person's weight, (2) Lift a person from a sitting position, (3) Balance a person while walking, (4) Autonomy for cruising within the house, (5) Transport objects from one place to the other within the house.

3 Engineering Specifications

The engineering specifications of the mechanisms are determined by performing a functional analysis. The functional analysis allows to clearly identifying the functions that the product is intended to accomplish [7, 8]. Having identified the functions, it is necessary to specify criteria to help their easy characterization. These criteria must be assessed objectively, in order to assign levels depending on the available resources, the product needs, or project's target. Finally, for the purpose of weighting these criteria, it is useful to define a degree of flexibility for each. For this

Table 1 Users Needs

1. Lift a person from a sitting position	9. Minimum number of drives
2. Balance a person while walking	10. Low dependence on sensors
3. Support a person's weight	11. Easy planning and localization algorithms
4. Help during rehab	12. No need to detect ground obstacles
5. Autonomy for cruising within the house	13. Not affecting the ground
6. Transport objects	14. Able to go up and down stairs
7. Al least 3 DOF	15. Allow the user to do other tasks with its upper limbs
8. Ability to go through doors	16. Use electricity

Table 2 Engineering specifications

Functions	Criteria	Measurable goals to reach	Flexibility	
			Class	Accepted limit
Help to assistance and rehabilitation	Velocity	1.2 m/s	n1	0.5 m/s < v < 1.5 m/s
	Load capacity	100 kg < *w* <150 kg	n0	
	Degrees of freedom	≥3	n1	
	Structural rigidity	Motion stable	n0	
	Work area	Move all around	n1	Reach important places
	Automation level	Full automatic	n1	From semi to full automatic
	Traction force	Capable to move a person of weight up to 150 kg	n0	
Allow Daily Life Activities, DLA	Motion area	Free of hands and legs	n1	
	Robot weight	≤15 kg	n2	According to the material and structure
Domestic use	Robot wide	70 cm	n1	±10 cm
	Robot height	100 cm > h > 200 cm	n1	According to the height of the roof
Satisfy customer expectation	Power and electric current	24 Vcc a 2.4 Amp. 24 h battery	n2	Battery > 12 h
	Degrees of freedom	≥3	n1	2 < d.o.f. < 4
	Structural rigidity	Kept fixed	n0	
	Automation level	Automatic	n1	From semi to full automatic

purpose 3 levels or flexibility classes are defined: n0 (for null flexibility), n1 (for relative level of flexibility), n2 (for a maximum level of flexibility). As a result of this procedure Table 2 presents the engineering specifications of the robot.

4 Generation of Concepts

When the functional analysis and the engineering specification of the mechanism have been defined the concept generation process begins. To perform this task, we use a 3-step method: (1) Separation of sub-problems, (2) Internal and external search, (3) Systematic exploration.

The first step divides the problem into smaller sub problems that satisfy the functions that we previously expressed. Three main sub-problems were identified. These are mainly related to the mechanical architecture of the device. Since the use of electric energy is a requirement, the search of alternatives to generate, store and control energy is reduced.

The mechanical sub problems found are:

- Translation system
- Lifting up system
- User's attachment point.

The main sources of information in the external search were: Professional care givers, disabled people, and previously developed products.

Although it was not possible to obtain technical solutions from the professional care givers and disabled people for the identified sub problems, we obtained information about the assistance that caregivers provide to the disabled. On the other hand, in [4] we studied the devices that help disabled in the displacement tasks, particularly those that allow transporting a person from one place to another. We included those devices on which the person is fully lifted and those which only aid the person while walking. These devices can be classified in: Walkers with handles, Walkers with harness and Ceiling hoists.

Robotic walkers with handles are mobile robots with wheels that allow users to swing with their hands. These walkers have navigation systems that can help users to go from one place to another without colliding with the environment. On the other hand there are the walkers with harness, for example the SAM and SAM Y platforms [4]. They hold the user from the waist. Naturally those devices provide more support than walkers with handles, but the space occupied by its structure is greater. The ceiling hoists are devices that allow transporting a person from one place to another. This action is accomplished by attachment to the roof rails. The rails support a carriage connected to the harness that supports the person.

Once the sub problems are defined and the internal and external search for possible solutions is finished, we perform a systematic exploration of the possible solutions. To manage the relative complexity of the having multiple possible solutions the methods named Solutions Classification Tree and Combination Table were used [1]. Solutions classification trees help to divide the possible solutions into separate categories. The combination table permits to generate different concepts given the different combinations of solutions of the sub problems.

The solutions classification trees for solving the sub problems *Translation system, Lifting up system* and *User's attachment point* are presented in Fig. 1a–c. These solutions trees allow visualizing all possible solutions, determining the most promising solutions and discarding those that clearly do not meet the requirements stated in the user's needs list and the engineering specification table.

In the solutions tree for the Translation system the branches of legs and caterpillar are discarded. Legs are dropped by the enormous complexity of achieving a stable system with this architecture. In addition, this system consists of various mechanical elements and actuators. Moreover, the caterpillar system is discarded because it is oriented to generate a large contact surface and it is generally used in outdoor environments. In the solutions tree of the lifting system the solution "chain" (which is used in ceiling hoists) is discarded, since this does not provide the necessary rigidity and control for the lifting task.

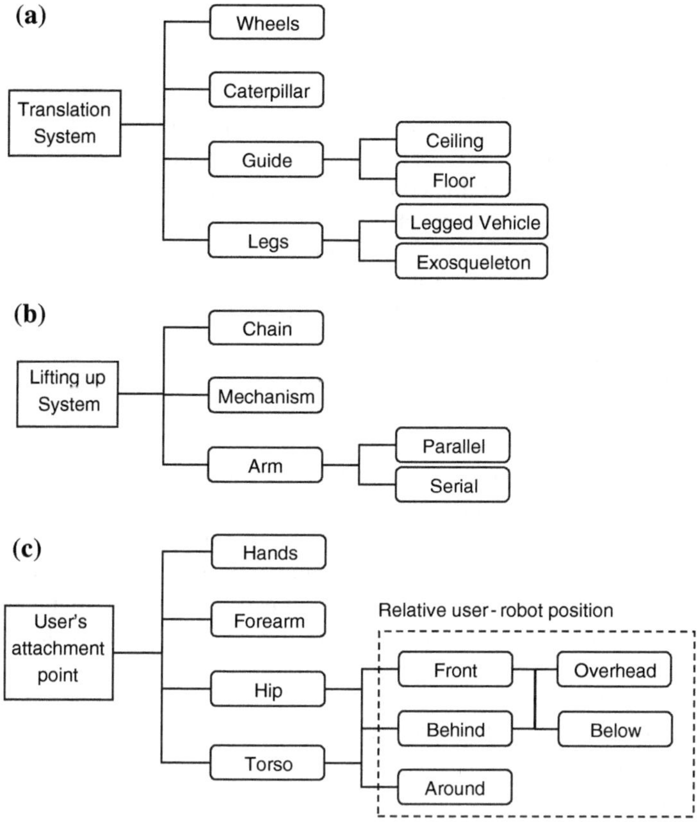

Fig. 1 Classification trees for the **a** Translation system, **b** Stand up system, and **c** User's attachment point

In the solutions tree for the user's attachment point the branches for hands and forearm were discarded, because there is a requirement to allow the person to perform other activities of daily living with the hands. In the case of the relative position of the robot structure with respect to the user, the option Around (Walkers with harness) was discarded because such structure clearly obstructs the execution of DLAs.

Based on the most promising branches of the classifications trees, a combination table of concepts was constructed. The table combines the possible solutions to the multiple sub problems considered in the functional decomposition. The multiple combinations of solutions were considered resulting in large universe of possible designs of robots. From this universe the design team selected 3 concepts to be compared with more detail on the concept selection phase.

4.1 Concept 1

Concept 1 is a robotic walker with harness, which itself has a wide and rigid base for the person to simultaneously support and assist in walking and balancing. The platform stores the energy and control. Concept 1 currently meets most of the important needs, and we use this concept as the reference product to compare and evaluate with the rest of the concepts. This concept has the advantage that if their navigation algorithm is accurate, it can move everywhere, however, has the disadvantage of the space occupied by the control and navigation equipment that it carries, i.e. if the equipment is in the front, it will be an obstacle for the DLA, otherwise if the equipment is put on the back, it limits the user in getting up from a sitting position. In the case of wheeled devices a Honda product named U3X personal mobility can be a suitable option, nevertheless the driving of such device requires a good level of dexterity of the user.

Below we present the concepts of robots for rehabilitation and assistance that emerged from the design team of this project. These robots are all mounted on the roof through guides and rails that allow translational movement. This feature gives the system stiffness because it will be firmly attached to the building by the guidelines. Moreover, the robot has predetermined roads hence reducing the need for planning algorithms, navigation systems for avoidance of obstacles that are needed in wheeled mobile robots (Fig. 2).

The movement through the different rooms is done by connecting the rails with exchangers or turntables that already exist commercially, so that it covers the largest number of spaces in the house.

4.2 Concept 2

The V robot is a parallel manipulator with 2 degrees of freedom. The kinematic chain consist of rotational-rotational-prismatic joints. The rail traction system is located on the ceiling and holds the entire structure. The parallel framework is supported directly by the traction system. Different types of tools can be set on the robot depending on the specific task to be performed.

4.3 Concept 3

The robot is composed of a U-shaped structural frame, which is connected to a carriage that moves through a rail on the ceiling, and is supported on the ground by a pair of omnidirectional wheels [9]. The movement of the carriage can be based on the well-known technology for transporting disabled people, the ceiling hoists. Furthermore within the frame structure there is a prismatic joint that in conjunction a carriage that moves through a rail can produce two-dimensional movements.

Fig. 2 Concepts for assistive robot: **a** Concept 2, **b** Concept 3 and **c** Concept 4

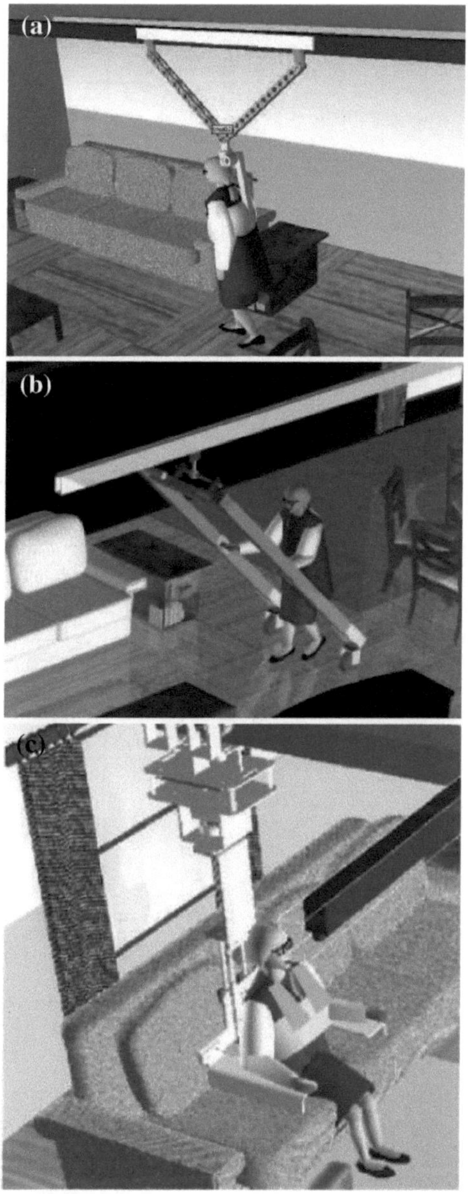

4.4 Concept 4

The robot has three links connected by three joints. These joints are: a prismatic joint (composed of a carriage and rails on the ceiling), a rotational joint on the vertical axis, and a prismatic joint that moves up and down the support system of

Table 3 Concept selection matrix

Selection criteria	Weight (%)	Concept 1 (Ref)		Concept 2		Concept 3		Concept 4	
		Grade	Weighted points	Grade	Weighted points	Grade	Weighted points	Grade	Weighted points
Support body weight	20	3	0.6	3	0.6	5	1	5	1
High structural rigidity	20	3	0.6	2	0.4	5	1	5	1
Balance assistance	15	3	0.45	3	0.6	5	0.75	5	0.75
Stand up assistance	15	3	0.45	1	0.15	5	0.75	5	0.75
Access to important places	10	2	0.16	5	0.5	2	0.2	5	0.5
Leave free arms	8	2	0.16	4	0.32	4	0.32	5	0.4
Simple navigation system	7	1	0.07	5	0.35	5	0.35	5	0.35
Objects transportation	4	2	0.08	4	0.16	5	0.2	4	0.16
Minimum # of mechanical elements	1	3	0.03	1	0.01	1	0.01	2	0.02
Total		–	2.6	–	3.09	–	4.58	–	4.93
Rank		4		3		2		1	

the user. This robot focuses on walking assistance tasks, as it allows balancing a person while walking. On the other hand, it may also assist a person when it performs other activities of daily living such as cooking, move objects etc. In addition, the design of this robot can perform the task of lifting, and once the person is standing it can continue with the walking phase.

5 Concept Selection

To make the selection of the concept, a comparative scale using a 5 point grading system was used. The points are awarded depending on the level of achievement of the concepts to the user's needs stated in Sect. 2. A concept selection matrix is prepared for the assessment of the previously presented designs. This matrix is composed of the selection criteria and concepts generated. Each criterion has a certain weight that defines its relative importance with respect to the others.

The selection criteria and weights for each criterion were determined based on the frequency, repeatability of customer needs. The grades given to each concept were given by design team through discussion and voting.

Each concept was evaluated with respect to the selection criteria in the selection matrix. In this table a total score is obtained for each concept (Table 3). The selected concept is the one that gets the highest score. The selection criteria used to determine measurable goals are chosen, the most important criteria are: support the weight of the person, assist in walking, balance and lifting.

From the ranking obtained using the concept selection matrix, concept 4 rises as the suitable candidate for further development. The detailed design of this robot is undergoing further studies and prototyping.

6 Conclusions

The concept design for an assistive robot has been presented. All along this research we have used a structured framework for product development.

Initial information to establish the user needs was obtained from care professionals, disabled users and health care information.

The needs were translated into a set of technical requirements using a functional decomposition technique; the variations allowed in the requirements were established using SADT: Structured Analysis and Design Technique.

A mixed methodology of function classification trees combined with a concept combination tables was used to explore and develop a series of robot concepts that could satisfy the technical requirements and functions.

Using a structured framework for product development, a multilinked/multi hinged concept was found to be the robot design that best suits the needs detected.

This selection was obtained by evaluating the different concepts against the set of user needs initially established.

Finally, the structured methodology employed allows for the development and design of a singular product (i.e. an assistive robot) by exploring a large set of possible designs and allowing for a costumer based product with high levels of user satisfaction.

References

1. Ulrich KT, Eppinger S (ed) (2011) Product design and development. McGraw-Hill, New York
2. Pahl G, Beitz W, Feldhusen J, Grote K (2007) Engineering design: a systematic approach. Springer, London
3. Puglisi LJ, Saltaren R, Moreno HA, Cárdenas P, Garcia C, Aracil R (2012) Dimensional synthesis of a spherical parallel manipulator based on the evaluation of global performance indexes. Robot Auton Syst 60(8):1037–1045
4. Carrera I, Moreno HA, Saltaren R, Perez C, Puglisi L, Garcia C (2011) ROAD—domestic assistant and rehabilitation robot. J Med Bio Eng Comput 49(10):1201–1211
5. Encuesta sobre discapacidades, deficiencias y estado de salud (1999). Instituto Nacional de Estadística. España
6. Encuesta de Discapacidad, autonomía personal y situaciones de dependencia, EDAD (2008). Instituto Nacional de Estadística. España
7. Moquet S, Wenger P, Pamanes A, Moreno HA, Franco G (2004) Aspectos del Diseño Conceptual de un Robot Ambulante Bípedo de 4 Grados de Libertad. Memorias de Congreso Mexicano de Robótica COMROB 2004. Torreón, México
8. Romero S, Romero O (2009) Product design optimization: an interdisciplinary approach. In: Tomovic M, Wang S (eds) Product realization: a comprehensive approach. Springer, New York
9. Carrera I, Campos A, Moreno H (2012) In: Proceedings of the Brazilian robotics symposium and Latin American robotics symposium (SBR-LARS), Fortaleza, Brazil

Design and Construction of a Submarine with Five Degrees of Freedom

Ortiz R. Floriberto, Barroeta Z. Carlos and J. Francisco Novoa C

Abstract One area that has seen an increase in recent years research either for academic research and industrial applications, is the use of submersible robots. There are two types of underwater vehicles: Autonomous Vehicles Non-Self (research submarines), which operate under the same principle. It is common to use these robots to perform tasks such as monitoring, inspection and maintenance of underwater structures in deep drainage, curtains in dams, groundwater, natural or artificial funds lakes, nuclear plants, oil platforms, etc. Besides in the field of education, a prototype of this kind, is able to show the basic principles of an underwater system, and is essential for monitoring and expansion of research from the lowest levels to the most advanced in the control area. The introduction provides an overview of the article, in the second part all the mathematical analysis for the underwater vehicle systems as well as some details of its physical construction involved, and some graphical results of the general behavior is concerned. The conclusions are finally future work.

Keywords Underwater · Mathematical model · Systems · Buoyancy · Propulsion

1 Introduction

A submarine is a ship or boat, using a system of weights and tanks are filled and emptied at will, has the ability to dive under the water and surf in these conditions or the surface itself by a variable buoyancy system.

O.R. Floriberto (✉) · B.Z. Carlos · J.F. Novoa C
Laboratorio de Control de I.C.E. E.S.I.M.E. Zacatenco I.P.N, Mexico, Mexico
e-mail: frodriguez@gmail.com

B.Z. Carlos
e-mail: cbarroet@hotmail.com

J.F. Novoa C
e-mail: jnovoa@ipn.mx

M. Ceccarelli and E.E. Hernández Martinez (eds.), *Multibody Mechatronic Systems*,
Mechanisms and Machine Science 25, DOI 10.1007/978-3-319-09858-6_29

The Underwater Research will feature an innovative system, we refer to the system that allows reposition some thrusters for more degrees of freedom and power more mobility, in addition to increasing the speed of forward or reverse.

For this system, the mathematical model will be analyzed to determine if it is controllable, observable and stable control is proposed and simulations that control will be performed. All this is necessary to find out if it really is appropriate to implement this new system.

Upon completion of the prototype drive mathematical model was obtained aft and propose this control speed.

By immersing an object in the water, a force on said object called hydrostatic pressure is exerted. To withstand the pressure and the helmet does not suffer ruptures, a structure with PVC coated fiberglass and resin, the easiness to mold, in addition to a strong, lightweight and waterproof structure was used. The pressure exerted on the hull is a function of depth and size thereof as well as the liquid in which it is immersed which in this case is water.

It was necessary to place and design all the mechanical transmissions, mechanical seals also were constructed to prevent water by arrows that go to the cross, and was resolved on how to place the foundations of the engines and their links to the arrows.

One of the problems faced was the dive. To accomplish this, water was introduced and to emerge, the water was removed and air was introduced into the ballast tanks. To solve this problem planned, introducing water into the ballast need compress air in a special tank and then expel pressure within the ballast water to drain, with this you can submerge and emerge the submarine.

For this system, the mathematical model was obtained from this model we analyzed whether this system is controllable, observable and stable, various types of control were proposed and simulations of these controls were performed. All this was necessary to know whether it was appropriate to implement this new system.

2 Physical and Mathematical Analysis

The fact that a submarine can float both surface and submerged, is due to the existence of two physical phenomena, which are listed under the names of **"ARQUIMEDES PRINCIPLE AND PRINCIPLE OF PASCAL"**.

2.1 Mathematical Model of the System Thrusters

The schematic model of an autonomous underwater with side flaps and the coordinate system shown in Fig. 1.

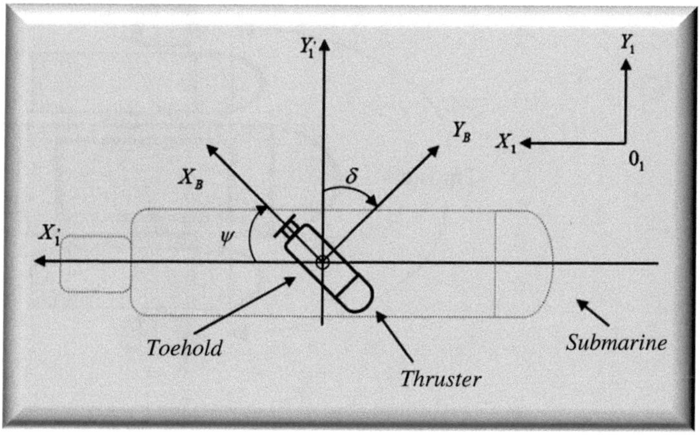

Fig. 1 Schematic model of the submarine

Here, $0_1X_1Y_1$ is the inertial coordinate system. The vehicle moves in the plane $X_1 - Y_1$. The variables X_BY_B axes are coordinated with the center of buoyancy at the source, i.e. $(X_B, Y_B) = 0$.

In the state space model of Eq. 1, ψ is the orientation angle, $\dot{\psi} = r$, v is the traveling speed, δ is the tilt angle from the vertical. Values $a_{21}a_{22}a_{23}a_{32}a_{33}$ are constant hydrodynamic vehicle [1—3].

$$x^{\cdot} = \begin{bmatrix} 0 & 1 & 0 \\ a_{21} & a_{22} & a_{23} \\ 0 & a_{32} & a_{33} \end{bmatrix} \begin{bmatrix} \psi \\ r \\ v \end{bmatrix} + \begin{bmatrix} 0 \\ b_2 \\ b_3 \end{bmatrix} \delta \tag{1}$$

In Fig. 2 you can better see the underwater thrusters, which are taken as independent autonomous submersible vehicles, this means that the previous mathematical model describes each of the thrusters, one by one. To apply this model is necessary to give the following initial conditions.

- Thrusters with respect to one another have the same position at all times.
- The rate of change of position of thrusters with respect to each other will be the same at all times.
- The speed of the two motors with thrusters will be the same at all times.

With these initial conditions, we can apply the above model to the submarine.

The system will control the depth of the submarine system, through its thrusters that can be defined by the following linearized system state space [2].

$$\dot{x} = \begin{bmatrix} 0 & 1 & 0 \\ -0.008 & -0.5 & 0.12 \\ 0 & 0.06 & -0.4 \end{bmatrix} \begin{bmatrix} \psi \\ r \\ v \end{bmatrix} + \begin{bmatrix} 0 \\ -0.1 \\ 0.8 \end{bmatrix} \delta(t) \tag{2}$$

Fig. 2 Thrusters

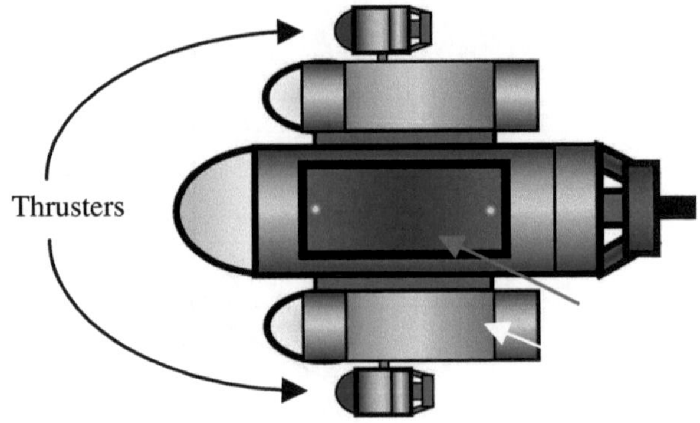

Here, ψ is the orientation angle.

$\dot{\psi}$ $=r$

v is the traveling speed

δ is the tilt angle relative to the vertical

Model of Eq. 2 can be obtained observability matrices (M_C) and controllability (M_O) which are shown in Eqs. 3 and 4 respectively.

$$Mc = \begin{bmatrix} 0 & -0.1 & 0.111 \\ -0.1 & 0.111 & -0.054 \\ 0.8 & -0.326 & 0.137 \end{bmatrix} \quad (3)$$

$$Mo = \begin{bmatrix} 0 & 0 & 1 \\ 0 & 0.06 & -0.4 \\ -0.0005 & 0.015 & 0.167 \end{bmatrix} \quad (4)$$

It can be seen that the matrices are of rank 3 which indicates that our system is controllable and observable.

On the other side of the Eq. 2 can also obtain the transfer function, using Eq. 5. And the transfer function of the submarine in Eq. 6.

$$\frac{Y(s)}{U(S)} = C(SI - A)^{-1}B + D \quad (5)$$

$$G(S) = \frac{0.8S^2 + 0.114S + 0.0064}{S^3 + 0.55S^2 + 0.0608S + 0.0032} \quad (6)$$

2.2 Submarine Systems

Figure 3 shows the submarine systems, these systems can be subdivided as follows:

1. Variable Buoyancy System.
    - Tyre Subsystem.
    - Hydraulic subsystem.
2. Propulsion system.
    - Horizontal propulsion subsystem.
    - Vertical propulsion subsystem.
3. Stabilization system.
4. Vision system.
    - Illumination subsystem.
5. Power system.
6. Control system.

The variable buoyancy system consists of subsystems such as pneumatic subsystem, which is shown in Fig. 7, control valves, temperature, pressure, electrical and mechanical subsystems, compressed air tank and hydraulic subsystem.

The propulsion system consists of a stern drive is a propeller located at the rear of the vehicle, side thrusters, a transmission shift position for thrusters, which gives the vehicle a lot of versatility and a rudder control direction [4].

The stabilization system aims to maintain the submarine horizontally or at an angle, for example, if you want shooting with the camera background where is submerged [5].

The vision system has a robot arm with two degrees of freedom and mounted thereon chamber, and a lighting subsystem that consists of a driver L293 and a

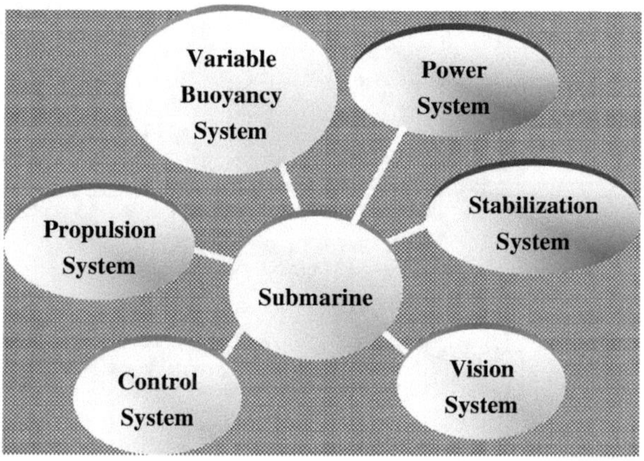

Fig. 3 The 6 submarine systems

timer circuit NE555, the output of the timer circuit was connected to two inputs L293B driver, in which 21 LEDs per output were connected [3].

The power system of the submarine is what gives us the power to all other systems. It consists of four batteries; two who are responsible for supplying power to all the actuators, while the other two are responsible for feeding all control circuits, this design was proposed for many problems of electromagnetic noise is not taken into control system [6].

3 Results

Now you have graphs of the responses of some systems, these graphs were taken on different initial conditions, trying to simulate the conditions of the field testing of the systems.

This chart was constructed from placing the submarine on the water in the test field (Pool), Figs. 4 and 5, where it remains floating partially submerged. An initial point is taken and through a distance aft impeller 19 m was shifted in 58 s. These data are covered with an error of +1 cm measurement errors.

Figure 6 shows the behavior of the fill weights for immersing the submarine.

Figure 7 shows the complete control system simulated in Simulink and the graph in Fig. 8 shows the simulation of the behavior of the system to a unit step input.

Figure 9 shows the behavior of the speed PID control in the field and can be seen to have a resemblance to the simulation. The x-axis values are multiplied by 10.

Finally in Fig. 10 shows the submarine submerged.

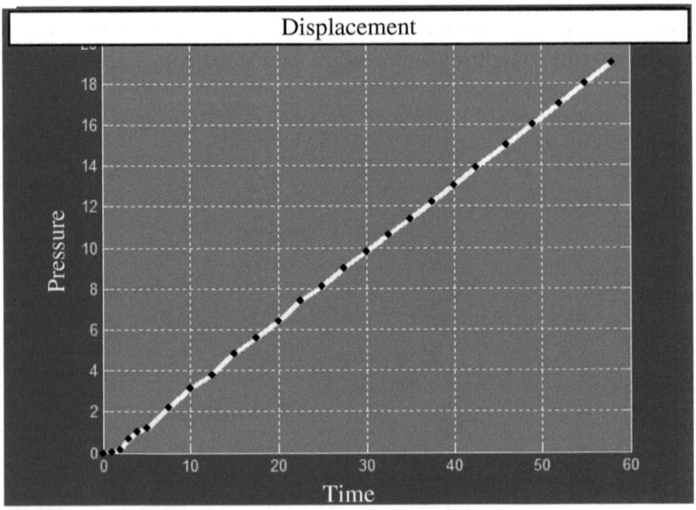

Fig. 4 Movement in the water

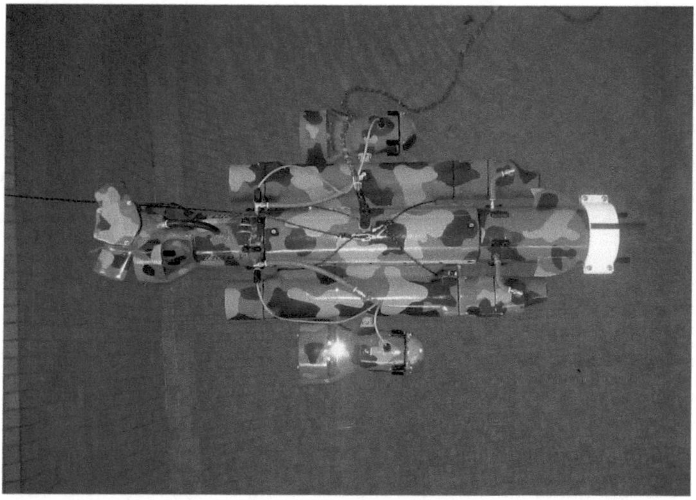

Fig. 5 Movement in the water

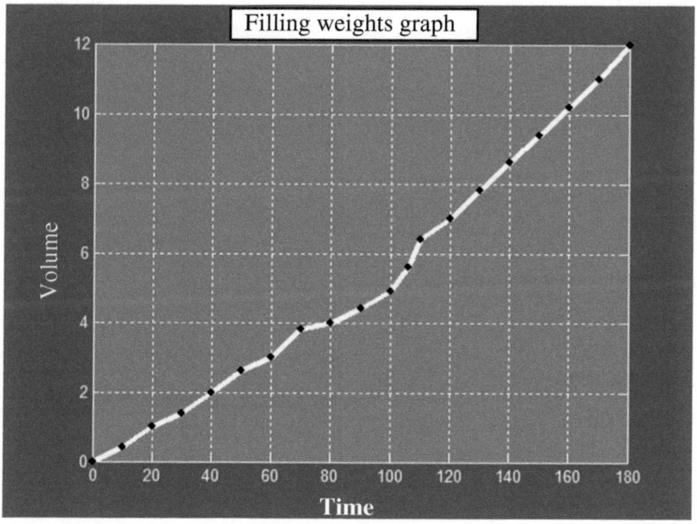

Fig. 6 Filling weights

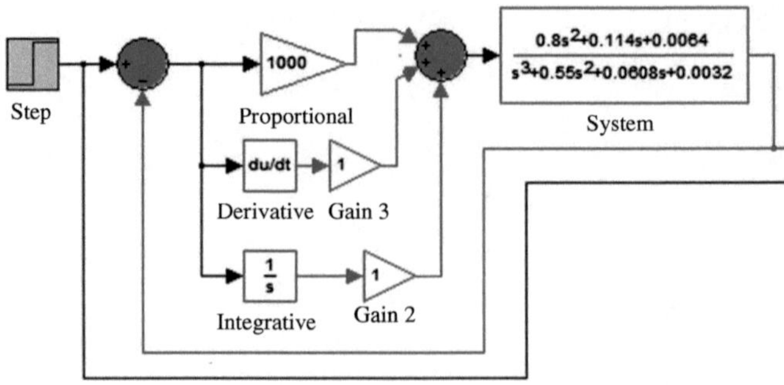

Fig. 7 Control system in simulink

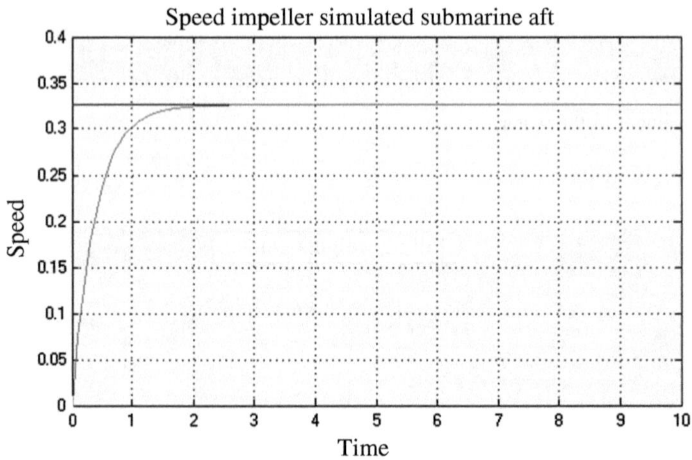

Fig. 8 Graph speed impeller simulated submarine aft

Fig. 9 Graph speed submarine with real stern drive

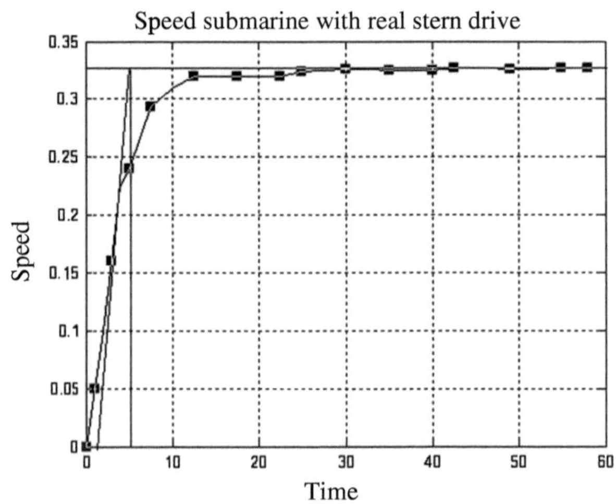

Fig. 10 Submerged submarine

4 Conclusions

Construction of submarine or submersible vehicles in Mexico, represent an excellent choice to provide technology for marine research, the planet is made up of 73 % water on its surface, this reflects the importance of marine research to find mineral resources natural for exploitation, besides knowing the important role played by oceans and aquifers with regulating the climate on our planet. In questions of power engineering repairs to all kinds of marine structures or systems that require it.

In this project, a prototype able to sail smoothly on the surface of water, power dives and the proper functioning of your vision system was achieved.

The problems were presented prototype, instability having being immersed in water, defined instability in regard to this prototype is not kept horizontal, consequently their operation is not immersed correct causing no you can prove all its dynamics.

In regard to all other systems, they operated at 100 % capacity with satisfactory results.

Therefore, this prototype provides an opportunity to repair the part of the storage of water which is responsible for its instability immersion in theory to fix the error the SIA will be able to show its dynamic. And the subsequent step will be to implement all their systems.

Another important conclusion is that by not implementing a prototype of this nature, working with times and speculations of their systems, which is why the prototype was not implemented due to economic and time reasons.

5 Future Work

Upon completion of this project is as future work:

1. Improve mathematical model.
2. Two. Improve the physical design of the vehicle.
3. Three. Improve the dynamics of buoyancy.
4. April. Improve propulsion system.

References

1. Creus AS (1998) Industrial instrumentation, Alfaomega, Barcelona (7th edn. in Spanish)
2. Nise NS (2002) Control system for engineering, CECSA, Mexico (3rd edn. in Spanish)
3. Boylestad R, Nashelsky L (2003) Theory of circuits and electronic devices, Pearson, Mexico (8th edn. in Spanish)
4. Tippens PE (2006) Physics: concepts and applications. McGraw-Hill, Mexico (3rd edn in Spanish)
5. Hewitt PG (2006) Conceptual physics, Pearson, Mexico (3rd edn. in Spanish)
6. Smits AJ (2003) Fluid mechanics—a physical introduction, Alfaomega, Barcelona (in Spanish)

Concurrent Structure-Control Design of Parallel Robots Using an Estimation of Distribution Algorithm

E. Chávez-Conde, S. Ivvan Valdez and Eusebio Hernández

Abstract In this paper, a structure-control design methodology for simultaneously optimizing both mechanical structure and control of parallel robots is proposed. It takes into count the dynamical model and the mechanical parameters for the optimization process. Thus, PID control and geometric variables are optimized in a simultaneously way. Through the concurrent procedure, an optimal combination of the robot structure and control gains is obtained. An estimation of distribution algorithm (EDA) is formulated and used as the search algorithm. The proposed methodology is verified through simulation experiments and applied to the design process of a parallelogram linkage system. The results obtained in experiments show the effectiveness of the proposed methodology. The presented approach is generic and can be applied to other mechanisms with similar structure.

Keywords Concurrent design · Parallelogram linkage · Optimization · Estimation of distribution algorithms

1 Introduction

In recent years, parallel mechanisms have attracted increasing attention from research community and industry. It is well known that compared with traditional serial mechanisms, parallel mechanisms have potential advantages in accuracy,

E. Chávez-Conde (✉)
Campus Loma Bonita, Universidad Del Papaloapan, Oaxaca, Mexico
e-mail: echavez@unpa.edu.mx

S. Ivvan Valdez
Centro de Investigación En Matematicas (CIMAT, a.C.), Guanajuato, Gto, Mexico
e-mail: ivvan@cimat.mx

E. Hernández
SEPI ESIME Ticomán Instituto Politécnico Nacional, Mexico, D.F, Mexico
e-mail: euhernandezm@ipn.mx

M. Ceccarelli and E.E. Hernández Martinez (eds.), *Multibody Mechatronic Systems*,
Mechanisms and Machine Science 25, DOI 10.1007/978-3-319-09858-6_30

dynamic and stiffness. Many architectures of robots using parallel mechanisms have been proposed and analyzed. However, the practical parallel structures show a significant difference between the expected and actual performance, due to passive joints that introduce geometric errors and deformations. In addition, most existing structures have limited workspace and singularities. This is a very demanding task for the design process of a parallel structure. Thus, the design and development of these kind of structures requires a multidisciplinary approach. The optimal design of these kind of structures, considered as mechatronic systems, can be stated as a traditional optimization problem. However, the optimization techniques to address these problems can be attracted by local minima, are sensitive to initial conditions or involve the computation of gradients in the objective function. In [1] is proposed a integrated structure/control design method, based on an iteration algorithm, for the development of a robotic system. The mechanical structure is modified iteratively by using the structure reinforcement and rapid prototyping techniques, coupled with the adjustment of the control parameters according to the mechanical structure updated. Recently, with the aim to circumvent these problems, optimization techniques based on heuristic methods have been used in the design process of different mechatronic systems. In [2], optimization techniques based on a heuristic evolutionary algorithm are selected for optimizing both the design parameters and non-linear PD controller gains of a two-link planar manipulator. In [3], a concurrent design process stated as a dynamic optimization problem is applied in a continuously variable transmission. In [4], a method based on a differential evolution algorithm is implemented to address the parametric reconfiguration in the optimal design of a five-bar parallel robot. Most of above design optimization problems are solved by widely used stochastic methods, which have been successful in approaching the problem, but lack of a consistent interpretation of its way of working from a mathematical viewpoint actually they are consistent since the artificial evolution point of view, additionally neither apriori information nor a posteriori information can be introduced or obtained, respectively, from these algorithms. We propose the use of Estimation of Distribution Algorithms (EDAs) [5] for addressing the design optimization problem. In this work we use an EDA based on the Boltzmann distribution, the Boltzmann Univariate Marginal Distribution Algorithm (BUMDA) [6]. It has been proven that conceptual EDAs based on the Boltzmann selection, such as the Boltzmann EDA (BEDA) [7], converge to the optimum [8]. The BUMDA uses the Boltzmann distribution through a Gaussian model, fitted according to the Kullback-Leibler divergence. The BUMDA aim is to preserve the desired characteristics of the Boltzmann distribution, while maintaining a low computational cost in the estimation and sampling steps.

In this paper the design of a parallel robot is carried out by this approach which integrates in a simultaneous way the structure design parameters of the parallel robot and the PID controller gains in order to optimize its mechanical design and to minimize position error for a specific path. Thus, both the structure design and the control design are simultaneously evaluated in order to obtain an appropriate system performance. The aim of this paper is to obtain the set of optimal mechanical and controller parameters in a concurrent way. The paper is organized as follows: Sect. 2

introduces the proposal of our concurrent design methodology. Section 3 describes the estimation of distribution algorithm based approach. Section 4 presents the design process for a Parallelogram Linkage System (PLS) as a case of study, while the optimization problem is depicted in Sect. 5. The results of experiments are showed and analysed in Sect. 6. Finally, the conclusions are presented in Sect. 7.

2 Concurrent Design Methodology

The concurrent design methodology proposed in this paper, considers the design process as a mapping from a requirement space to a structural space. In [9], suggests dividing the requirement space into two subspaces: real-time behaviours (e.g. desired path, desired tracking speeds), and nonreal-time behaviours (e.g. workspace, maximum payload). Following this division, the system parameters in the structural space can also be divided into two subspaces: real-time parameters (e.g. controller gains), and nonreal-time parameters (e.g. kinematics and dynamics parameters). This viewpoint has been adapted for the optimal integrated design of parallel robots proposed in this paper, to optimize the structural parameters and the controller gains simultaneously, satisfying the restrictions set imposed, by using an estimation distribution algorithm. It is recommended to include constraints for mass and force balance, or constraints for dynamic model simplification. The optimization process could consider the following criteria:

- **Design objectives**. (1) Minimizing position error for a given path. $E = \min \sum_{i=1}^{n} e_i(t),\ e_i(t) = q_i^* - q_i$. (2) Maximizing a manipulability measure (away from singularities). $S = \max w(\mathbf{q}),\ w(\mathbf{q}) = \sqrt{det(\mathbf{J}(q)\mathbf{J}^T(\mathbf{q}))}$.
- **Design constraints**. (1) Inequalities design constraints (kinematic and dynamic parameters, actuator power, workspace). (2) Equality design constraints (dynamic model simplification, desired path).
- **Design variables**. Design parameters vector (kinematic and dynamic parameters, controller gains), $\mathbf{X} \subset \mathbb{R}^n$.

The actual objective function depends on the particular problem. Thus, in the case of study we will provide the specific model for such application.

3 The Boltzmann Univariate Marginal Distribution Algorithm (BUMDA)

EDAs estimate and sample from a probability distribution. An EDA usually starts with an uniform initial sample (initial population), then the population is evaluated, the best individuals (samples) in the population are selected, the selected set is used to re-estimate the probability distribution, and so on. The BUMDA [6] is an EDA based on the Boltzmann distribution, the Boltzmann distribution is approximated by a Gaussian distribution because of its characteristics for estimating its parameters and sampling.

The BUMDA accomplish two desired characteristics in any optimization algorithm:

- It maintains a non-decreasing sequence of the expected value of the objective function. In order to obtain better samples than the generation before.
- It converges to the best solution found. This allows to refine the solution, and to determine when the algorithm rarely will improve the best solution known.

The convergence is achieved by applying a truncation selection method which increases the mean of the objective value of the population, such as explained in Fig. 1. We ensure that it is always at least one element in the selected set by preserving the elite individual.

The whole BUMDA is shown in Fig. 2. Notice that the BUMDA uses the truncation selection method to ensure an increasing average (mean estimator) of the objective function of the population. In addition, the objective function of the selected set is used to incorporate information about the fitness landscape into the Gaussian model. As can be seen it requires of two parameters: a population size and a minimum variance, in comparison with other similar optimization problems the number of parameters is quite small. As suggested in [6], the population size is chosen by setting an initial value, then increasing it until the algorithm performance does not change. The minimum variance allowed is of 5×10^{-5} and is related with the maximum error allowed.

Some of the BUMDA advantages are the following:

- The BUMDA converges to the best approximation to the optimum.
- The variance tends to 0 for a large number of generations.
- The BUMDA only needs **one** parameter (population size).
- The estimation of the search distribution parameters results in a fast automatic adaptation. The variance could be increased or decreased, according to the solutions in the selected set and their objective values, and the mean moves fast to the region where the best solutions are.

Truncation Selection Method

Consider a population of decreasingly sorted individuals (maximization case), such that x_1 are the decision variables of the individual with the maximum objective function in the population:

1. For the initial generation $t = 0$, let be $g(x_i, 0)$ for $i = 1..N$, the objective values of the initial population. Define: $\theta_0 = \min g(x_i, 0)$.
2. For $t > 0$, set: $\theta_t = \max\left(g(x_{N/2}, t), \min(g(x_i, t)|g(x_i, t) \geq \theta_{t-1})\right)$.
3. Truncate the population such that $g(x_s, t) \geq \theta_t$. Where x_s are all the individuals whose objective values are equal or greater than θ_t.

Fig. 1 Truncation method to ensure convergence in a population based algorithm

BUMDA

1. Give the parameter and stopping criterion:
   nsample ← Number of individuals to be sample.
   minvar ← minimum variance allowed.
2. Uniformly generate the initial population P_0, set $t = 0$.
3. While $v > minvar$ for all dimensions
   (a) $t \leftarrow t + 1$
   (b) Evaluate and truncate the population according algorithm in Figure 1.
   (c) Compute the approximation to μ and v (for all dimensions) by using the selected set (of size $nselec$), as follows:

   $$\mu \approx \frac{\sum_1^{nselec} x_i \bar{g}(x_i)}{\sum_1^{nselec} \bar{g}(x_i)}, \quad v \approx \frac{\sum_1^{nselec} \bar{g}(x_i)(x_i - \mu)^2}{1 + \sum_1^{nselec} \bar{g}(x_i)},$$

   where $\bar{g}(x_i) = g(x) - g(x_{nselec}) + 1$.
   Note: the individuals can be sorted to simplify the computation, and $g(x_{nselec})$ is the minimum (for maximization case) objective value of the selected individuals.
   (d) Generate $nsample - 1$ individuals from the new model $Q(x, t)$, and insert the elite individual.
4. Return the elite individual as the best approximation to the optimum.

Fig. 2 Pseudo-code for BUMDA

4 Case of Study: A Parallelogram Linkage System

The proposal is applied for the design process of a parallelogram robotic system, which can be considered as a parallel robot. Figure 3 shows a kinematic diagram of the parallelogram robot. The direct kinematic is given as,

$$\begin{bmatrix} x \\ y \end{bmatrix} = \begin{bmatrix} L_1 \cos \varphi_1 - L_4 \cos \varphi_2 \\ L_1 \sin \varphi_1 - L_4 \sin \varphi_2 \end{bmatrix} \tag{1}$$

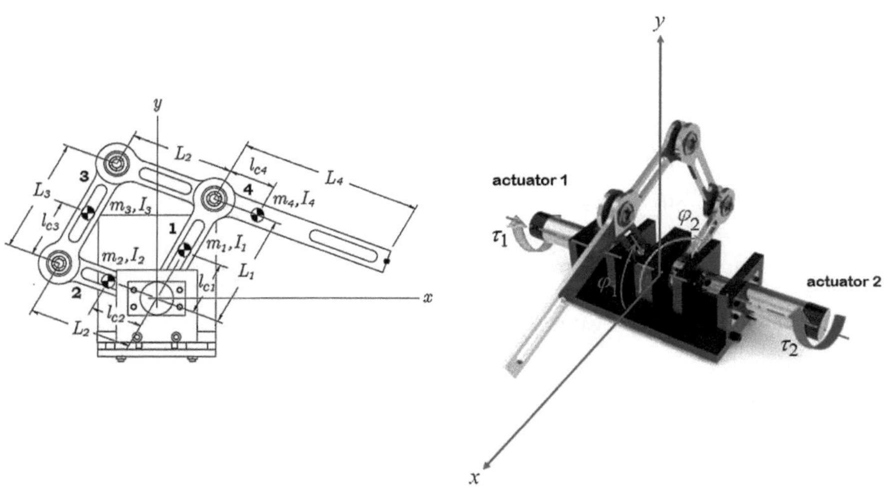

Fig. 3 A schematic diagram of the parallelogram linkage system

The solution of the inverse kinematic problem can be computed as,

$$\varphi_1 = \tan^{-1}\left(\frac{y}{x}\right) - \tan^{-1}\left(\frac{L_4 \sin\gamma}{L_1 + L_4 \cos\gamma}\right), \quad \gamma = \tan^{-1}\left(\frac{-\sqrt{1-d^2}}{d}\right) \tag{2}$$

$$\varphi_2 = \varphi_1 + (\gamma + \pi), \quad d = \cos\gamma = \frac{x^2 + y^2 - L_1^2 - L_4^2}{2L_1L_4} \tag{3}$$

where $L_1, L_2, L_3, L_2 + L_4$ are the link lenghts, $l_{c1}, l_{c2}, l_{c3}, l_{c4}$ are the locations of the mass centers for each link, φ_1, and φ_2 are the angular positions for the link 1 and link 2, respectively. In the same form, the differential kinematics can be computed as,

$$\dot{\mathbf{x}} = J\dot{\mathbf{q}} \tag{4}$$

where, $\dot{\mathbf{x}} = [\dot{x}\ \dot{y}]^T$, $\dot{\mathbf{q}} = [\dot{\varphi}_1\ \dot{\varphi}_2]^T$ and the Jacobian matrix is given by

$$J = \begin{bmatrix} -L_1 \sin\varphi_1 & L_4 \sin\varphi_2 \\ L_1 \cos\varphi_1 & -L_4 \cos\varphi_2 \end{bmatrix}$$

4.1 Dynamic Model of PLS and Its Controller

The dynamic model of the PLS without friction parameters has been reported in [10–12], which can be written as,

$$d_{11}\ddot{\varphi}_1 + d_{12}\ddot{\varphi}_2 + (m_3L_2l_{c3} - m_4L_1l_{c4})\sin(\varphi_2 - \varphi_1)\dot{\varphi}_2^2 + \phi_1(\varphi_1) = \tau_1, \tag{5}$$

$$d_{21}\ddot{\varphi}_1 + d_{22}\ddot{\varphi}_2 - (m_3L_2l_{c3} - m_4L_1l_{c4})\sin(\varphi_2 - \varphi_1)\dot{\varphi}_1^2 + \phi_2(\varphi_2) = \tau_2. \tag{6}$$

with

$$\begin{aligned}
d_{11} &= m_1l_{c1}^2 + m_3l_{c3}^2 + m_4L_1^2 + I_1 + I_3, \\
d_{12} &= d_{21} = (m_3L_2l_{c3} - m_4L_1l_{c4})\cos(\varphi_2 - \varphi_1), \\
d_{22} &= m_2l_{c2}^2 + m_3L_2^2 + m_4l_{c4}^2 + I_2 + I_4, \\
\phi_1(\varphi_1) &= g\cos\varphi_1(m_1l_{c1} + m_3l_{c3} + m_4L_1), \\
\phi_2(\varphi_2) &= g\cos\varphi_2(m_2l_{c2} + m_3L_2 + m_4l_{c4}).
\end{aligned}$$

Note that if,

$$m_3 L_2 l_{c3} = m_4 L_1 l_{c4}, \tag{7}$$

the inertial matrix is diagonal and constant. If this condition is satisfied, the dynamic model of the manipulator can be simplified and this is given by a set of decoupled equations without centrifugal terms as the next form,

$$d_{11}\ddot{\varphi}_1 + \phi_1(\varphi_1) = \tau_1, \tag{8}$$

$$d_{22}\ddot{\varphi}_2 + \phi_2(\varphi_2) = \tau_2. \tag{9}$$

Due to these considerations, the parallelogram linkage system has been adopted by industrial robots and can be considered very popular in this context. In fact, the angles φ_1 y φ_2 can be adjusted in independent form.

Considering that the gravity terms of the dynamic model have been eliminated in mechanical way by means of a force-balancing. It is possible to use a PD controller considering is not applied loads to parallelogram mechanism and it is friction free. In the other hand, if the gravity terms are not eliminated, a PID controller can be considered wich can be implemented for minimizing the steady state error that is generated by these terms.

$$\begin{aligned} u_i(t) &= k_p e_i(t) + k_i \int e_i(t)dt + k_d \dot{e}_i(t); \quad i = 1, 2. \\ e_i(t) &= \varphi_{id} - \varphi_i \end{aligned} \tag{10}$$

The closed-loop dynamics of thess controllers are in accordance with the Lyapunov stability theory, (see [11–13]), and have been widely applied in industrial robots.

5 Optimization Problem

As it has been mentioned before, the design process for both the geometric and controller parameters ought to be performed in a simultaneously way. Thus, the design process can be expressed as a optimization problem as follows:

5.1 Problem Definition

Design variables m_i, L_i, l_{ci} and as a consequence of these variables, I_i. The following constraints are considered:

$$L_{i,min} \leq L_i \leq L_{i,max} \tag{11}$$

$$u_{i,min} \leq u_i \leq u_{i,max} \tag{12}$$

$$m_{i,min} \leq m_i \leq m_{i,max} \tag{13}$$

$$L_1 = L_3 \tag{14}$$

$$m_3 L_2 l_{c3} = m_4 L_1 l_{c4} \tag{15}$$

The first three are the search limits, hence we only draw random values inside such limits. The last one is a model simplification considered in the simulation. The objective function is the following:

$$\begin{aligned} max\ f(m, L, l, I) = W - [\lambda(E(|\tau_1|) + E(|\tau_2|)) + E(|\int e_1(t)dt|) \\ + E(|\int e_2(t)dt|)]. \end{aligned} \tag{16}$$

where $E(x) = 1/T \cdot \int_0^T x dt$. And $\lambda = 1/10$ is a scaling factor. $W = 50$ is a translation parameter. Notice that the BUMDA is a maximization algorithm, hence the function to minimize is multiplied by -1 and translated in order to a maximization problem with a objective function that only returns positive values.

$$l_{c1} = \frac{1}{2} L_1 \tag{17}$$

$$l_{c2} = \frac{1}{2} L_2 \tag{18}$$

$$l_{c3} = \frac{1}{2} L_3 \tag{19}$$

$$l_{c4} = \frac{1}{2}(L_2 + L_4) \tag{20}$$

The optimization algorithms search for 9 optimal parameters, 3 of them are the lengths, notice that we only need 3 lengths because $L_1 = L_3$. And 6 parameters for the controls in Eq. 10.

6 Results

A test path is considered for optimization problem of parallelogram mechanism. The desired path is given by Eq. 21 (snail Pascal).

$$\begin{aligned} x(t) &= -0.15\cos(t)\sin(t) - 0.05\sin(t)\cos(3t) \\ y(t) &= 0.05\cos(t)\cos(3t) - 0.15\sin(t)\sin(3t) \end{aligned} \tag{21}$$

In the optimization process all the variables were considered in the interval $[0,\ 1]$, then the values are mapped to the corresponding intervals simply by scaling and translating the values. A population size of 73 was used and a maximum variance for stopping criterion of 2.5×10^{-9}. The execution of the algorithm is about 1 min 36 s. In a personal computer with Intel i7 processor and 6 GB of RAM, using OpenSuse Linux and GNU compiler (GCC) and *C* program. The number of evaluations of the objective function in a typical program execution is about 6,745. In Figs. 4 and 5 shows the error of φ_1 and φ_2, the τ_1 and τ_2 values, and desired path *versus* the delivered by the BUMDA.

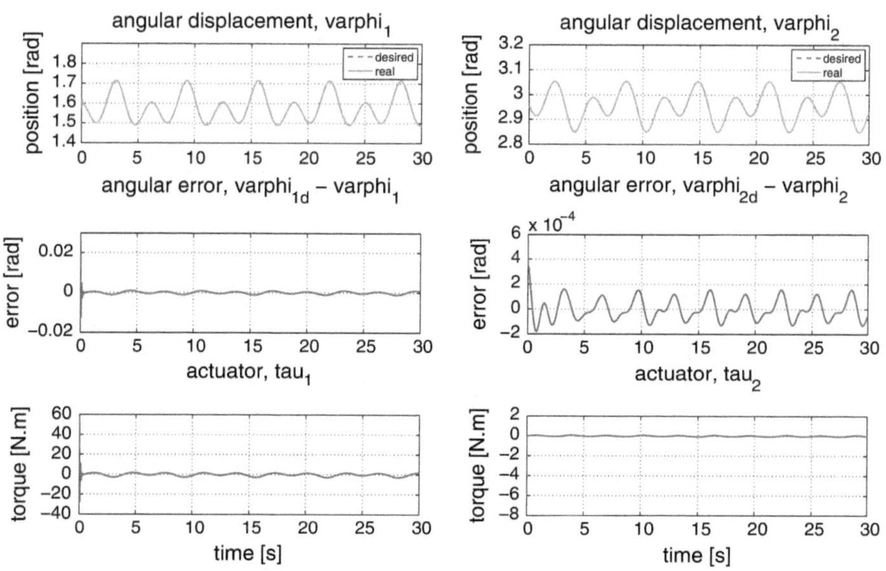

Fig. 4 Desired φ_{id} versus computed φ_i, angular error $\varphi_{id} - \varphi_i$, torque τ_1, τ_2

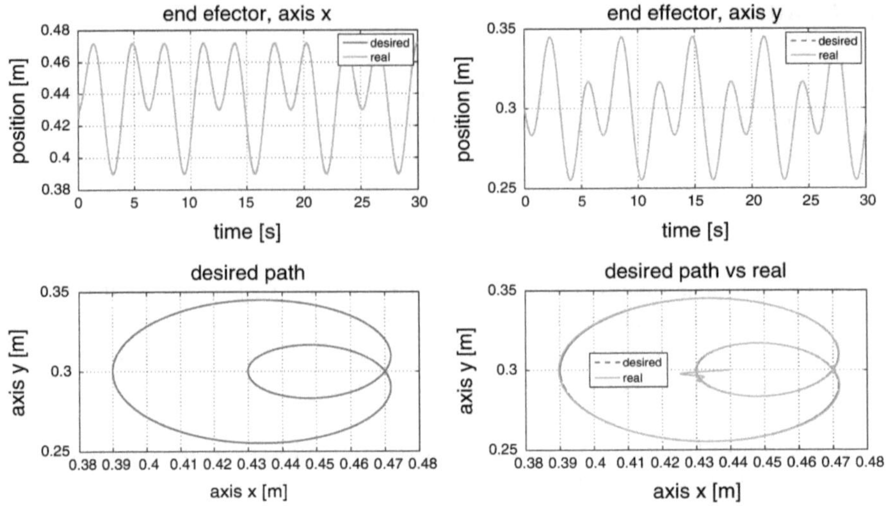

Fig. 5 Desired versus computed path, end effector

7 Conclusion

In this proposal, a structure-control design for a parallelogram linkage robot with trajectory tracking is presented. This is formulated as a concurrent process where both the kinematic and the dynamic models of the mechanical structure and the controller gains are considered for optimizing in a simultaneously way. An estimation of distribution algorithm was adopted for solving the optimization problem. This algorithm successfully found well performed solutions. One of the most remarkable characteristics of this algorithm is that practically it uses a unique parameter: the population size, because the stopping criterion is set according to the maximum allowed numerical error, which is a problem parameter. The population size is quite small for this kind of algorithms. The BUMDA performs quite well and we expect to tackle similar problems with this algorithm. Notice that the BUMDA uses a Gaussian model, which is a soft distribution, when this Gaussian model has a standard deviation less than ε, most of the samples from it will be in a radius of 3ε from the mean, hence if 3ε is small number the Gaussian distribution provides a way of looking for solutions in close neighbourhood, even this is not a prove of local optimality, it is an strong argument about our procedure, which samples intensively the neighbourhood of the best solution when the algorithm converges. In consequence we have a strong argument to say that very possibly the solutions delivered by the BUMDA are local optimums. In the other hand, when the algorithm is not in the convergence phase (when it has a large variance) it is performing a global exploration. Then it is possible that BUMDA finds the global optimum, but it is highly possible that at least it provides a local optimum.

In a subsequent work, other parallel mechanisms will be studied, in order to test the performance of the concurrent optimization methodology.

Acknowledgements The authors would like to thank CONACYT-Mexico under project CB-2011-01-169132 and SIP IPN (2013) for supporting part of this project. First author would like to thank financial support from PROMEP-SEP.

References

1. Pil A, Asada H (1996) Integrated structure-control design of mechatronics systems using a recursive experimental optimization method. IEEE/ASME Trans Mech 1(1):191 203
2. Ravichandran T, Wang D, Heppler G (2006) Simultaneous plant controller design optimization of a two-link planar manipulator. Mechatronics 16:233–242
3. Portilla-Flores EA, Mezura-Montes E, Alvarez-Gallegos J, Coello-Coello CA, Cruz-Villar CA (2007) Integration of structure and control using an evolutionary approach: an application to the optimal concurrent design of a CVT. Int J Numer Meth Eng 71(8):883–901
4. Portilla-Flores EA, Mezura-Montes E, Alvarez-Gallegos J, Coello-Coello CA, Cruz-Villar CA, Villarreal-Cervantes MG (2011) Parametric reconfiguration improvement in non-iterative concurrent mechatronic design using an evolutionary-based approach. Eng Appl Artif Intell 24:757–771
5. Mühlenbein H, Paaβ G (1996) From recombination of genes to the estimation of distributions I. binary parameters. In: Proceedings of the 4th international conference on parallel problem solving from nature, Springer, London, UK, pp 178–187
6. Ivvan Valdez S, Hernández A, Botello S (2013) A Boltzmann based estimation of distribution algorithm. Inf Sci 236(0):126–137
7. Mühlenbein H, Mahnig T, Rodriguez AO (1999) Schemata, distributions and graphical models in evolutionary optimization. J Heuristics 5(2):215–247
8. Zhang Q, Mühlenbein H (2004) On the convergence of a class of estimation of distribution algorithms. IEEE Trans Evol Comput 8(2):127–136
9. Li Q, Zhang WJ, Chen L (2001) Design for control- a concurrent engineering approach for mechatronic systems design. IEEE/ASME Trans Mech 6(2):161–169
10. Riving EI (1988) Mechanical design of robots. McGraw-Hill, New York
11. Spong MW, Vidyasagar M (1989) Robot dynamics and control. Wiley, New York
12. Sciavicco L, Siciliano B (2003) Modelling and control of robot manipulator. Springer, London
13. Murray RM, Li Z, Sastry SS (1994) A mathematical introduction to robotic manipulation. CRC Press, Boca Raton

Emulation of Mechanical Structures Through a Multi-agent Robot System: An Overview

J. González-Sierra and E. Aranda-Bricaire

Abstract This paper presents an overview of recent results on the use of a group of differential-drive mobile robots to emulate the motion of different types of mechanical structures. The type of mechanical structure is emulated by choosing suitable relative position vectors between the mobile robots. A control law is proposed to achieve such relative positions. It is shown that despite of the simplicity of the control law it is possible to emulate the mechanical structures changing only the corresponding relative position vectors and their time derivative. Numerical simulations and real-time experiments show the performance of the control law.

Keywords *N*–trailer · Mobile robots · Multi-agent systems · Formation tracking · Mechanical structure

1 Introduction

A multi-agent robot system (MARS) is integrated by a group of autonomous mobile robots, which are able to communicate, coordinate and interact among them to solve specific tasks [4]. One of the main research areas of MARS is Motion Coordination. Specifically the goal is to maintain a defined geometric pattern while the agents are following a desired trajectory as well as avoid collisions among them and the environment.

The particles of a rigid body maintain a fixed geometric relationship via a system of physical constraints. If any of the particles of the physical object is subject to a disturbance, it will be propagated to all the other particles. Based on the foregoing,

---

J. González-Sierra (✉) · E. Aranda-Bricaire
Electrical Engineering Department, Mechatronics Section, CINVESTAV,
A.P. 14-740, 07000 México, D.F, Mexico
e-mail: jamesgsjr@hotmail.com

E. Aranda-Bricaire
e-mail: earanda@cinvestav.mx

M. Ceccarelli and E.E. Hernández Martinez (eds.), *Multibody Mechatronic Systems*,
Mechanisms and Machine Science 25, DOI 10.1007/978-3-319-09858-6_31

it is desirable that a set of mobile robots that do not have any physical connections among them, behave as if they were physically linked. For the case of a MARS, it is possible to emulate the behaviour of rigid or articulated mechanical structures establishing time-varying position vectors. This type of problem has been addressed and solved in [8–10, 16, 17]. Paper [19] introduces the definition of virtual structures and proposes control strategies to force a group of robots to behave like particles in a rigid structure. Paper [8] proposes the use of time-varying formation vectors in order to emulate the behaviour of physical n–trailer systems through continuous- and discrete-time approaches. Reference [17] uses time-varying relative vectors to orientate a group of mobile robots to the heading angle of the leader vehicle.

This paper presents an overview of recent results obtained which allow the emulation of mechanical structures through a group of differential-drive mobile robots. The paper is organized as follows: Sect. 2 recalls the kinematic model of a differential-drive mobile robot. Section 3 describes the mechanical structures that can be emulated. The control law is presented in Sect. 4. Section 5 shows the performance of the control law through numerical simulations and real-time experiments. Finally, some conclusions are offered in Sect. 6.

2 Kinematic Model of Differential-Drive Mobile Robots

Denote by $\{R_1, \ldots, R_n\}$ a set of n differential-drive mobile robots moving in the plane. The kinematic model of each robot R_i, as shown in Fig. 1, is given by

$$\begin{bmatrix} \dot{x}_i \\ \dot{y}_i \\ \dot{\theta}_i \end{bmatrix} = \begin{bmatrix} \cos\theta_i & 0 \\ \sin\theta_i & 0 \\ 0 & 1 \end{bmatrix} \begin{bmatrix} v_i \\ w_i \end{bmatrix}, \quad i = 1, \ldots, n, \tag{1}$$

where (x_i, y_i) are the coordinates of the mid-point of the wheels axis, θ_i is the orientation of the robot with respect to the X-axis, v_i is the longitudinal velocity of the mid-point and w_i is the angular velocity of the robot. It is known [3] that the dynamic system (1) can not be stabilized by any continuous and time-invariant control law. To overcome this obstruction, in the literature it is common to study the kinematics of a point $\alpha_i = (p_i, q_i)^T = [x_i + \ell\cos\theta_i, y_i + \ell\sin\theta_i]^T$ outside the wheels axis [11, 14]. The kinematics of point α_i is given by:

$$\dot{\alpha}_i = A_i(\theta_i)\begin{bmatrix} v_i \\ w_i \end{bmatrix} = \begin{bmatrix} \cos\theta_i & -\ell\sin\theta_i \\ \sin\theta_i & \ell\cos\theta_i \end{bmatrix}\begin{bmatrix} v_i \\ w_i \end{bmatrix} \tag{2}$$

where $A_i(\theta_i)$ is the decoupling matrix for each robot R_i. The decoupling matrix is non-singular since $\det(A_i(\theta_i)) = \ell \neq 0$.

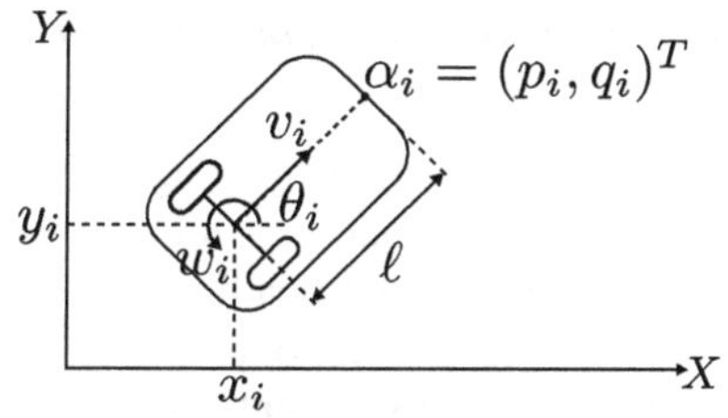

Fig. 1 Kinematic model of differential-drive mobile robots

3 Description of Mechanical Structures

3.1 Translational Motion of a Rigid Body

Figure 2a describes the motion in the plane of three agents. There is a leader agent which is responsible for follow up a predefined trajectory while the other agents, known as the followers, keep a certain position among them defined by constant relative position vectors. The front points of the agents form a geometric pattern, e. g., the vertices of a triangle. Since the positions among agents are defined by constant formation vectors, the geometric pattern that is emulated by the agents corresponds to a rigid body with translational motion only. Constant formation vectors are determined depending on the desired geometric pattern. In this case, we use direct open-chain formation graph with leader (Fig. 2b) [2, 5], where R_n is the leader and $\{R_1, \ldots, R_{n-1}\}$ are the followers. The desired relative positions α_i^* are establish as

$$\begin{aligned} \alpha_i^* &= \alpha_{i+1} + c_{(i+1)i}, \quad i = 1, \ldots, n-1 \\ \alpha_n^* &= m(t) \end{aligned} \tag{3}$$

where $c_{(i+1)i} = [c_{(i+1)i,x}, c_{(i+1)i,y}]^T \in \mathbb{R}^2$ is the desired relative position vector of agent R_i with respect to agent R_{i+1}, $m(t) = [x_d(t), y_d(t)]^T$ is the predefined marching trajectory which is supposed to be continuously differentiable.

Remark 1 Real-time experiments of translational motion can be found in [7, 11].

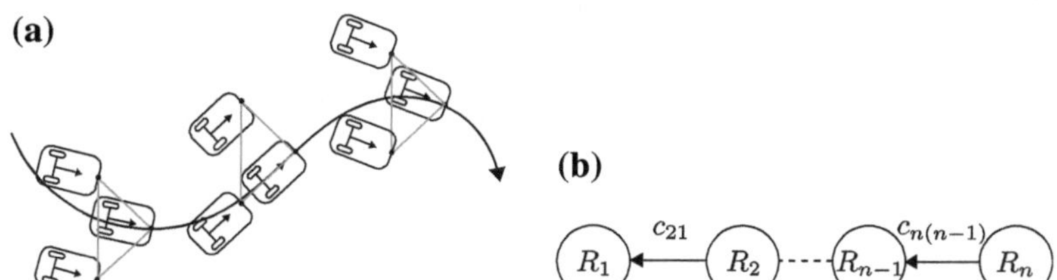

Fig. 2 **a** Translational motion of a mechanical structure. **b** Direct open-chain formation graph with leader

3.2 Translational and Rotational Motion of a Rigid Body

Fig. 3a shows the trajectory in the plane of three agents. While the leader agent is following a predefined trajectory, the followers are aligned in a geometric pattern and are oriented with respect to the leader heading angle by a rotation matrix and time-varying formation vectors. This corresponds to the emulation of a mechanical structure with rotational and translational motion. The rotation matrix $R(\theta_n) = \begin{bmatrix} \cos\theta_n & -\sin\theta_n \\ \sin\theta_n & \cos\theta_n \end{bmatrix}$, depends on heading angle of the leader and the static relative position vector C_{ni} which establishes the position between agent R_i and leader agent R_n. Figure 3b represents the leader centered formation graph where the desired relative positions are defined as

$$\alpha_i^* = \alpha_n + c_{ni}(t), \quad c_{ni}(t) = R(\theta_n)C_{ni}, \quad i = 1, \ldots, n-1, \tag{4}$$

where c_{ni} is the time-varying desired position vector. The desired velocity is

$$\dot{c}_{ni}(t) = \dot{R}(\theta_n)C_{ni}, \quad \dot{R}(\theta_n) = \dot{\theta}_n \begin{bmatrix} -\sin\theta_n & -\cos\theta_n \\ \cos\theta_n & -\sin\theta_n \end{bmatrix}.$$

This terms will be used later to define the control law.

3.3 The N–Trailer Configuration

Figure 4a presents the so-called standard n–trailer. This kind of articulated mechanical structures are composed of a certain number of trailers pulled by a tractor. The trailers are linked through mechanical joints. In the case when the mechanical joints coincide with the mid-point of the wheels axis of the trailer ahead, the n–trailer is called standard [12, 13, 18]. When the joints are placed off the

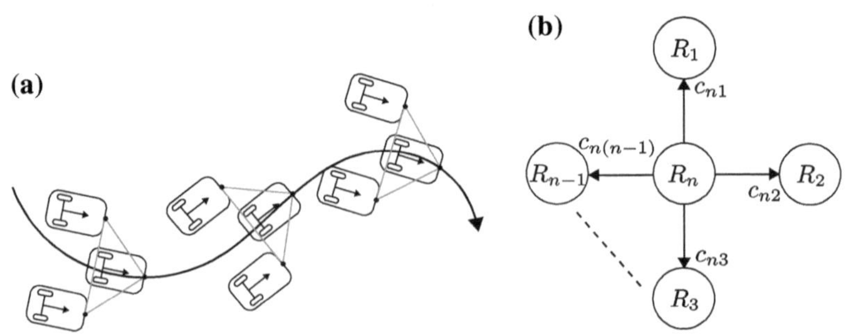

Fig. 3 **a** Translational and rotational motion of a mechanical structure. **b** Leader centered formation graph

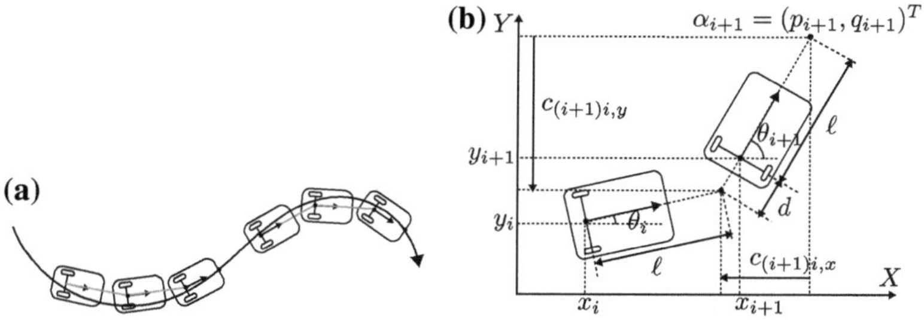

Fig. 4 **a** Standard *n*–trailer configuration. **b** Time-varying position vector for *n*–trailer configuration

mid-point of the wheels axis, the *n*–trailer is called generalized [1, 15]. The standard and generalized *n*–trailers typically exhibit a phenomenon called off-tracking, more specifically, oversteer or understeer, respectively. These phenomena occur when the tractor is following a trajectory with a certain radius of curvature and each intermediate vehicle deviates from the path of the leading one. Again the direct open-chain formation graph with leader (Fig. 2b) is used to emulate the behaviour of the $n-$trailer. The desired relative position vector α_i^* is similar to (3). The difference resides in the fact that the formation vector $c_{(i+1)i}(t) = [c_{(i+1)i_x}(t), c_{(i+1)i_y}(t)]^T \in \mathbb{R}^2$ is time-varying. Figure 4b shows the generalized $n-$trailer configuration. The desired relative position of robot R_i with respect to robot R_{i+1} is given by

$$\alpha_i^* = \alpha_{i+1} + c_{(i+1)i}(t), \quad c_{(i+1)i}(t) = -\begin{bmatrix} (\ell + d)\cos\theta_{i+1} \\ (\ell + d)\sin\theta_{i+1} \end{bmatrix} \tag{5}$$

Finally,the desired velocity vector is defined as

$$\dot{\alpha}_i^* = \begin{bmatrix} \dot{p}_{i+1} \\ \dot{q}_{i+1} \end{bmatrix} + \begin{bmatrix} (\ell + d)w_{i+1}\sin\theta_{i+1} \\ -(\ell + d)w_{i+1}\cos\theta_{i+1} \end{bmatrix} \tag{6}$$

The standard $n-$trailer configuration is obtained from (5) by letting $d = 0$.

3.4 The N*–Trailer Configuration with a Kingpin Mechanism*

Figure 5a shows the standard *n*–trailer configuration in which a sliding kingpin mechanism is implemented in order to reduce or eliminate the off-tracking phenomenon. This mechanism displaces the joint between two consecutive vehicles along the wheels axis of the leading trailer in the direction opposite to the center of curvature presented by the path. This mechanism allows the trailers to reduce the off-tracking and, in some special cases, to follow exactly the same trajectory as the

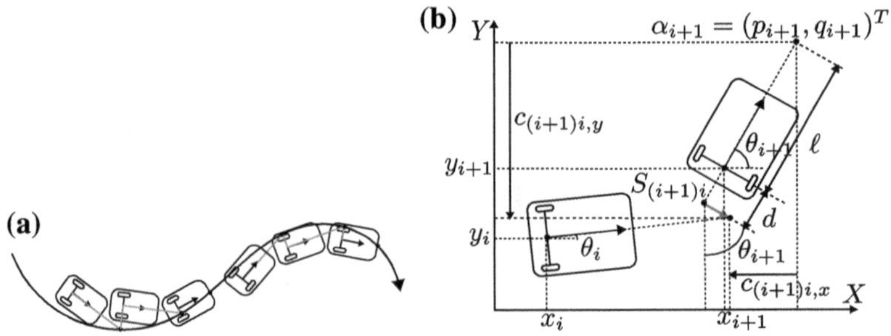

Fig. 5 **a** Standard n–trailer configuration with sliding kingpin mechanism. **b** Time-varying position vector for n–trailer with a virtual sliding kingpin point

tractor. Figure 5b shows two mobile robots where the virtual sliding kingpin position between them is located at a distance d from the middle point of the wheels axis of the agent that is moving ahead and at a distance $S_{(i+1)i}$ along the wheels axis of the same agent. The parameter $S_{(i+1)i}$ is the virtual sliding kingpin position which allows to reduce or eliminate the off-tracking, displacing each follower agent (trailer) with respect to the previous one. From Fig. 5b the desired position and desired velocity of agent R_i are given by

$$\alpha_i^* = \begin{bmatrix} p_{i+1} \\ q_{i+1} \end{bmatrix} - \begin{bmatrix} (\ell + d)\cos\theta_{i+1} - S_{(i+1)i}\sin\theta_{(i+1)} \\ (\ell + d)\sin\theta_{i+1} + S_{(i+1)i}\cos\theta_{(i+1)}, \end{bmatrix} \tag{7}$$

$$\begin{aligned} \dot{\alpha}_i^* = & \begin{bmatrix} \dot{p}_{i+1} \\ \dot{q}_{i+1} \end{bmatrix} + w_{i+1} \begin{bmatrix} (\ell + d)\sin\theta_{i+1} \\ -(\ell + d)\cos\theta_{i+1} \end{bmatrix} \\ & + \begin{bmatrix} \dot{S}_{(i+1)i}\sin\theta_{i+1} + S_{(i+1)i}w_{i+1}\cos\theta_{i+1} \\ -\dot{S}_{(i+1)i}\cos\theta_{i+1} + S_{(i+1)i}w_{i+1}\sin\theta_{i+1} \end{bmatrix} \end{aligned}$$

The virtual sliding kingpin position $S_{(i+1)i}$ is a function of the steering angle of the trailer moving ahead. For the standard $n-$trailer the time-varying virtual sliding kingpin and their velocity are given, for $i = 1, \ldots, n-1$, by

$$S_{(i+1)i} = \ell \left[\frac{1 - \cos(\theta_{i+1} - \theta_i)}{\sin(\theta_{i+1} - \theta_i)} \right], \quad \dot{S}_{(i+1)i} = \ell \left[\frac{w_{i+1} - w_i}{1 + \cos(\theta_{i+1} - \theta_i)} \right], \tag{8}$$

The expression (8) is ill-defined when $\theta_{i+1}(t_0) = \theta_i(t_0)$. This happens when instantaneously agents R_i and R_{i+1} are following the same straight line. However, the limit of this quantity is well defined as stated by the following:

Proposition 1 *González-Sierra and Aranda Bricaire* [9] *Let t_0 be a time instant such that $\theta_{i+1}(t_0) = \theta_i(t_0)$. Then $\lim\limits_{t \to t_0} S_{(i+1)i}(t) = 0$. Moreover $\dot{S}_{(i+1)i}(t_0) = 0$.*

For a generalized n–trailer the virtual sliding kingpin position and its time-derivative are given by

$$S_{(i+1)i} = \pm \frac{1}{w_{i+1}} \left(\sqrt{\dot{\alpha}_{i+1}^2 - d^2 w_{i+1}^2} - \sqrt{\dot{\alpha}_{i+1}^2 - \ell^2 w_{i+1}^2} \right) \tag{9}$$

$$\begin{aligned} \dot{S}_{(i+1)i} &= \pm g \dot{\alpha}_{i+1} \left(\frac{1}{\sqrt{\dot{\alpha}_{i+1}^2 - d^2 w_{i+1}^2}} - \frac{1}{\sqrt{\dot{\alpha}_{i+1}^2 - \ell^2 w_{i+1}^2}} \right) \\ g &= \frac{w_{i+1} \ddot{\alpha}_{i+1} - \dot{w}_{i+1} \dot{\alpha}_{i+1}}{w_{i+1}^2} \end{aligned} \tag{10}$$

Depending on the direction of rotation of the agents, the virtual sliding kingpin position has to be either positive or negative. More specifically, if the angular velocity of the agent R_{i+1} is positive we use the plus sign in (9)–(10) and if the angular velocity of the agent R_{i+1} is negative we use the negative sign.

Remark 2 Note that (9)–(10) is ill-defined when $w_{i+1}(t_0) = 0$. This occurs when the agent R_{i+1} pursues a path with a constant heading angle θ_{i+1}.

Proposition 2 *González-Sierra* [6] *Let* t_0 *a time instant such that* $w_{i+1}(t_0) = 0$ *and assume that* $\dot{w}_{i+1}(t_0) \neq 0$*, then* $\lim_{t \to t_0} S_{(i+1)i}(t) = 0$ *and* $\lim_{t \to t_0} \dot{S}_{(i+1)i}(t) = 0$.

4 Control Law

In this section we propose a control law which allows a group of differential-drive mobile robots to emulate the behaviour of the mechanical structures defined in the previous section. The control law is given by

$$\begin{aligned} \begin{bmatrix} v_i(t) \\ w_i(t) \end{bmatrix} &= A_i^{-1}(\theta_i) \left[\dot{\alpha}_i^* - k\left(\alpha_i - \alpha_i^*\right) \right], \quad i = 1, \ldots, n-1 \\ \begin{bmatrix} v_n(t) \\ w_n(t) \end{bmatrix} &= A_n^{-1}(\theta_n) [\dot{m}(t) - k_m(\alpha_n - m(t))] \end{aligned} \tag{11}$$

where k, k_m are positive design parameters, α_i^* and $\dot{\alpha}_i^*$ are the desired relative position and desired velocity vector of agent R_i, and $A_i^{-1}(\theta_i)$ is the inverse of the corresponding decoupling matrix. The unique control law (11) can be used to emulate all the mechanical structures defined in the previous section. The difference for each case study lies on the desired relative position vector α_i^* to be used. To emulate the translational motion we use (3). To emulate the translational and rotational motion we use (4). To emulate the behaviour of an $n-$trailer we use (5). To emulate the sliding kingpin mechanism we use (7).

Proposition 3 [8, 10, 16, 17] *Consider the system* (1) *and the control law* (11). *Suppose that* k, $k_m > 0$. *Then, in the closed-loop system* (1)–(11), *the follower robots converge to the desired formation, i.e.* $\lim_{t\to\infty}(\alpha_i(t) - \alpha_i^*(t)) = 0$, $i = 1, \ldots, (n-1)$ *whereas* R_n *converge to the marching trajectory i.e.* $\lim_{t\to\infty}(\alpha_n(t) - m(t)) = 0$.

Remark 3 Control law (11) for $i = 1, \ldots, n-1$ is a decentralized control. In this case the followers only known information about the agent that is moving ahead according to the formation graph that is being used.

5 Real-Time Experiments

Because of lack of space, we can not describe the experimental platform in detail. We just present in Fig. 6 the type of robots and cameras we are using for the real-time experiments. Each robot is furnished in their top with infrared markers which form a geometric pattern such that the centroid of this figure coincides with the middle point of the wheels axis of each robot. The cameras calculate the position of each marker. With this information, the software Motive calculates the position of each robot and its orientation. The control law is calculated in Visual C++ using Aria libraries which are also used to communicate with the robots. The protocol VRPN is used to share information between Motive and Visual C++. Finally, the velocities of each wheel are sent through Wi-Fi to the robots. The robots have an internal PID controller which ensures that the wheels reach the velocity needed for the trajectory tracking.

Figure 7 shows a real-time experiment of three agents emulating the translational an rotational motion of a mechanical structure using control law (11) and the desired relative position given by (4). The desired trajectory is a Gerono's Lemniscate defined by $[x_d, y_d]^T = [-0.6\sin(2\omega t), 1.5\cos(\omega t)]^T$ where $\omega = \frac{2\pi}{60}$. The initial conditions for the the follower agents are $[x_1, y_1, \theta_1]^T = [-0.68, 1.73, 0.023]^T$, $[x_2, y_2, \theta_2]^T = [-0.58, 1.24, 0.38]^T$ and for the leader are $[x_3, y_3, \theta_3]^T = [-0.16, 1.6, 0.013]^T$. The

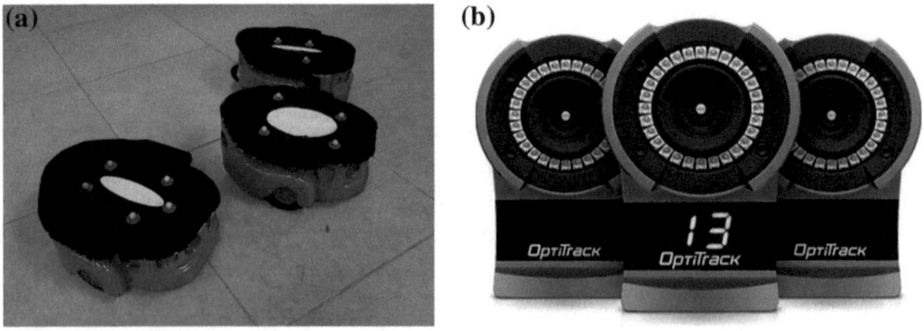

Fig. 6 **a** AmigoBot robots. **b** Flex13 Cameras

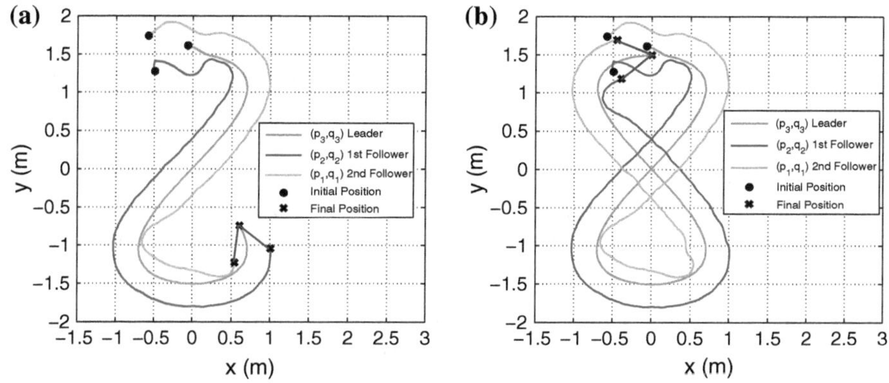

Fig. 7 Rotational and translational motion. **a** t = 40 s. **b** t = 60 s

distance from the mid-point of the wheels axis to the front point is $\ell = 0.1$ meters. The control parameters are $k = 2$, $k_m = 1$. The static position vector is given by $C_{31} = 0.5\left[-\cos\left(\frac{\pi}{6}\right), \sin\left(\frac{\pi}{6}\right)\right]^T$, $C_{32} = -0.5\left[\cos\left(\frac{\pi}{6}\right), \sin\left(\frac{\pi}{6}\right)\right]^T$. This choice corresponds to an equilateral triangle of 0.5 m per side. The agents are aligned in a triangle-type formation and they keep that formation until the end of the experiment.

Figure 8a displays a numerical simulation of the trajectories of three agents emulating the behaviour of a standard $n-$trailer using control law (11) and the desired relative positions given by (5). The desired trajectory is a Gerono's Lemniscate defined by $[x_d, y_d]^T = [-\sin(2\omega t), 1.5\cos(\omega t)]^T$ where $\omega = \frac{2\pi}{60}$. The initial conditions for the the follower agents are $[x_1, y_1, \theta_1]^T = [0.361, -0.61, 1.55]^T$, $[x_2, y_2, \theta_2]^T = [0.21, -0.02, 1.55]^T$ and for the leader are $[x_3, y_3, \theta_3]^T = [0.02,\ 0.58,\ 1.52]^T$, The control parameters are $k = 2$, $k_m = 1$. Due to the dimensions of the robots, the parameter $\ell = 0.6$ meters has been chosen so as to avoid collisions. It is interesting to

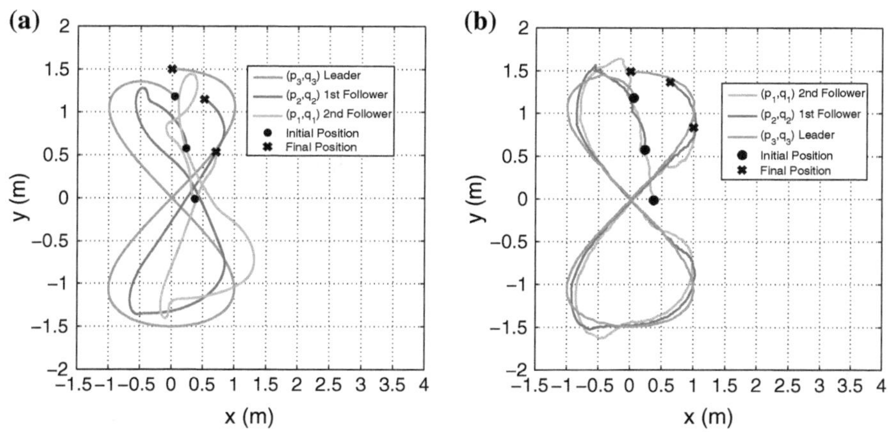

Fig. 8 **a** Standard $n-$trailer emulation. **b** Virtual kingpin emulation

note that off-tracking effects become apparent and follower agents lose track of the leader trajectory. This is due to two aspects. First, the agents reproduce the oversteer phenomenon present in the physical standard $n-$trailer. Second, the distances among the agents ($\ell = 0.6$ m) are relatively large.

Figure 8b shows a real-time experiment of the trajectories in the plane of three agents emulating the behaviour of a standard $n-$trailer using the control law (11) with the virtual sliding kingpin position defined in (8) and the desired relative position given by (7). The desired trajectory and the initial conditions are the same as in the above numerical simulation.

Remark 4 In Fig. 8b it can be seen that in the edges of the Gerono's Lemniscate there exists a slight deviation of the trajectory described by the agents followers from the trajectory of the leading vehicle. This is explained because the analysis to compute the position of the virtual sliding kingpin position is based on circumferences and, in these experiments, we attempt to follow a variable curvature trajectory. Even so, the off-tracking effects are significantly reduced allowing the follower agents a more accurate tracking of the leader trajectory.

6 Conclusions

In this paper we made an overview of recent results on the use of both constant and time-varying formation vectors to emulate the motion of mechanical structures through a group of differential-drive mobile robots. Despite the simplicity of the proposed control law, the only parameters that have to be changed from one structure to another are the relative position vectors. Numerical simulations and real-time experiments exhibit the good performance of the control law. Because of lack of space we have omitted the discrete-time analysis presented in [8]. As future work we propose to include artificial vector fields in the control law to avoid obstacles and collisions between the robots.

References

1. Altafini C (2001) Some properties of the general n–trailer. Int J Control 74(4):409–424
2. Belkhouche F, Belkhouche B (2005) Modelling and controlling a robotic convoy using guidance laws strategies. IEEE T Syst Man Cy B 35(4):813–825
3. Brockett RW (1983) Asymptotic stability and feedback stabilization. In: Millman RS, Brockett RW, Sussmann HJ (eds). Differential geometric control theory. Boston, Massachusetts, USA, pp 181–191
4. Cao Y, Fukunaga A, Kahng A (1997) Cooperative mobile robotics: antecedents and directions. Proc Auton Rob 4:226–234
5. Do K (2007) Formation tracking control of unicycle-type mobile robots with limited sensing ranges. IEEE Int Conf Rob Autom 16:527–538

6. González-Sierra J Emulation of mechanical structures through multi-agent systems, PhD. Thesis (In Spanish), CINVESTAV, Electric Engineering Department, Mechatronic Section, México, In preparation
7. González-Sierra J, Aranda-Bricaire E, Hernández-Martínez E (2013) Formation tracking with orientation convergence for groups of unicycles. Int J Adv Robot Syst 10(180):1–8
8. González-Sierra J, Aranda-Bricaire E, Hernández-Mendoza D, Santiaguillo-Salinas J (2013) Emulation of *n*–trailer systems through differentially driven multi-agent systems: Continuous- and discrete- time approaches. J Intell Robot Sys 1–18
9. González-Sierra J, Aranda Bricaire E (2013) Design of a virtual mechanism for trajectory tracking of convoys of mobile robots. In: Proceedings of 10th International Conference on Electrical Engineering, Computing Science and Automatic Control, pp 364–368
10. González-Sierra J, Santiaguillo-Salinas J, Aranda-Bricaire E (2013) Mechanical structures emulation through multi-agent systems (in spanish). In: Proceedings of National Congress of Automatic Control CNCA, pp. 348–353
11. Hernández-Martínez EG, Aranda-Bricaire E (2009) Marching control of unicycles based on the leader-followers scheme. In: Proceedings of 35th annual conference of IEEE of the industrial electronics, IECON '09, pp 2265–2270
12. Jean F (1996) The car with n trailers: characterisation of the singular configurations. Control Optimisation Calc Var 1:241–266
13. Laumond J (1993) Controllability of a multibody mobile robot. IEEE T Robot 9(6):755–763
14. Lawton JRT, Beard RW, Young.BJ (2003) A decentralized approach to formation maneuvers. IEEE T Robot, pp 933–941
15. Michalek M (2012) Application of the vfo method to set-point control for the n-trailer vehicle with off-axle hitching. Int J Control 85(5):502–521
16. Peñaloza Mendoza G, Hernández-Mendoza D, Aranda-Bricaire E (2011) Time-varying formation control for multi-agent systems applied to n-trailer configuration. In: 8th international conference on electrical engineering computing Science and automatic Control
17. Rendón-Benítez F, Santiaguillo-Salinas J, Aranda-Bricaire E, González-Sierra J (2012) Marching control with orientation to the leader heading angle for multi-agent systems (in spanish). In: Proceedings of 15th Latinoamerican congress of automatic control, Lime, Peru
18. Rouchon P, Fliess M, Levine J, Martin P (1993) Flatness, and motion planning: the car with n trailers. In: Proceedings of 2nd European control conference
19. KH Tan, Anthony-Lewis M (1996) Virtual structures for high-precision cooperative mobile robotic control. In: Proceedings of international conference on intelligent robots and systems, vol 1. pp 132–139

Design and Construction of a Nouvelle Vertical Axis Wind Turbine Experimental Platform

Lourdes García, David Lara, Azahel Treviño, Gerardo Romero, José G. Rivera and Esmeralda Lopez

Abstract This paper presents a methodology to design a nouvelle vertical axis wind turbine prototype, which can be used as mechatronic experimental platform for didactic use. Method engineering is used to plan the construction and assembly of the prototype, including the mechanical system design using computer aided design (CAD), which allows to develop drawings for each component of the entire system for visualization before its construction. Then, we can see if all parts fit in a suitable way. The results of this methodology are validated with the wind turbine prototype construction. Experimental test were conducted to obtain the best aerodynamical configuration of the blades. This platform can be used to develop control algorithms.

Keywords Vertical axis wind turbine · Prototype engineering · Mechatronic test platform

1 Introduction

Modern man need electricity for living, and its demand has grown. Then, to satisfy this need is a challenge, considering the reduction rate of fossil fuels as well the climate change due to greenhouse gas emissions. A solution to generate energy without using oils is the use of renewable energy sources, such as the wind power, which is abundant in almost planet surface [1]. Wind power has been used since the antiquity, for example, mechanisms powered by wind were used in Mesopotamia for irrigation and grinding. In the early seventeenth century, the wind was used with the purpose of moving mechanism for pumping water [2]. Since 1961, different

L. García · D. Lara (✉) · A. Treviño · G. Romero · J.G. Rivera · E. Lopez
Campus UAMRR, Departamento de Electronica, Universidad Autonoma de Tamaulipas, Carretera Reynosa San Fernando, SN Cruce Con Canal Rodhe, Col. Arco Iris, C.P. 88730 Reynosa Tamaulipas, Mexico
e-mail: dlara@uat.edu.mx

M. Ceccarelli and E.E. Hernández Martinez (eds.), *Multibody Mechatronic Systems*, Mechanisms and Machine Science 25, DOI 10.1007/978-3-319-09858-6_32

systems called wind turbines have been developed to produce electricity from wind energy. Wind turbines are designed and constructed based on knowledge of several disciplines and are operated with the most advanced technological development to be connected to any type of grid [3]. Today, in developed countries, many international organizations are interested on wind generation systems and they are making efforts, through research projects, to develop efficient systems. Latin America, including Mexico, generates 1 % of the total installed wind power worldwide, with funding from private or state projects [5]. Motivated by recent green energy polices in several countries, many universities and research centers conduct wind power generation research projects, such as the Autonomous University of Tamaulipas, in Mexico, whose objective is to develop small wind turbines. There is a vast literature concerning wind turbines, where each text is written with a specific goal in mind and the majority is focused on horizontal axis wind turbine (HAWT), due of its high efficiency, (see for example, [6] and [7]). Although the vertical axis wind turbines (VAWT) were the first types used to extract energy from wind, the majority of researchers of the modern era lost interest on these type because apparently they cannot be used for large scale electricity generation [4]. Then, the HAWT remained the focus of almost all the wind energy related research activity for the last few decades [8]. However, many research groups continue working on VAWT at a relatively smaller scale, developing various wind turbine configurations with different approaches for their analysis. For instance, the work presented on [9] is concerned with the study of a novel turbine design, that consists of several flapping blades hinged on a revolving drum, this turbine allows experimental test on a small model to provide some basic rules from which the movement of every flapping blade position could be determined. In [10] a VAWT formed of several asymmetric vertically-stacked stages was developed, then by computational fluid dynamics (CFD) analysis, the torque characteristics of the VAWT was determined. In [11], an experimental and computational study of the aerodynamics and performance of a small scale vertical axis wind turbine is presented. Using wind tunnel tests, the performance of the turbine was determined and using CFD dynamics a model was generated to establish its aerodynamics. In [12], a vertical axis H-rotor type wind turbine was designed and constructed to collect data to calculate the power coefficient, with measurement at different constant rotational speeds to observe the power coefficients dependence on tip speed ratio. In this paper we present a new configuration of vertical axis wind turbine rotor, (see Figs. 2 and 3 for better understanding), consisting of three dihedral shaped blades which are inclined with respect to the rotor disk with an elevation angle. In this way, the aerodynamic effects of lift and drag forces are used. This configuration takes into account the aerodynamic aspects in simplified form, to get nominal rotor angular velocity at relatively low wind speed. First of all, we focus on the design and construction of the wind turbine prototype, in order to obtain a combination of aerodynamic angles (elevation and dihedral) for the blades to maximize rotor speed in all wind speed range. As in previous work [13], using as main tools, methods engineering, computer aided design and prototype engineering, several sets of blades with different combinations were constructed and evaluated experimentally on the vertical axis wind turbine using different wind conditions, measuring the rotor speed in each case to collect data. The best combination of angles was obtained using MathLab to analyze experimental data. Then the rotor was instrumented to apply control algorithms for the VAWT.

2 Wind Turbine Aerodynamics Main Formules

Aerodynamics main formulaes are given in Table 1. They describe the forces acting on a wind turbine due an airflow, with density ρ. The airflow passes through the rotor, of diameter D, with an input and output velocities, V_1 and V_2, respectively, $V_2 < V_1$ and $V_2 = V_1(1 - 2a)$, where a is the induced velocity coefficient. The wind velocity at the rotor is given by V_r. The mechanical power P_m, is a function of the dissipated energy. In dimensionless form is the power coefficient C_p that indicates the amount per unit of useful energy extracted from the wind passing through the turbine. According to the Betz law $C_p = 16/27$, then, only up to 60 % of the wind kinetic energy passing by the rotor can be converted into mechanical energy [1]. Other important parameter is the torque Q transmitted by the rotor shaft to the generator system. This parameter can be represented by the coefficient C_q. The relation between the tangential rotor speed and the wind velocity is known as the Tip Speed Ratio (TSR) and is denoted by λ, where Ω is the rotor angular velocity. The power is given by the product of the angular velocity and the torque, this means that $C_q = C_p/\lambda$. The design of blades for wind turbines requires to analyze the action of wind on the surface of a blade element, which can be considered as a wing airfoil, (see Fig. 1). The chord c is the line connecting the leading edge with the trailing edge, and forms an angle α, (called angle of attack) with the relative direction of the airflow stream. The force acting on the blade of a wind turbine is the effect caused by the relative wind velocity V_{rel} on it [14]. This speed has as components the wind

Table 1 Wind turbine main formules

Variable	Formule
Rotor velocity	$V_r = V_1(1 - a) = (V_1 + V_2)/2$
Mechanical power	$P_m = \frac{1}{2}\rho\frac{\pi D^2}{4}V_1^3(1-a)(1-(1-2a)^2)$
Power coefficient	$C_p = \frac{P_m}{\frac{1}{2}\rho V_1^3\frac{\pi D^2}{4}} = 4a(1-a)^2$
Rotor force	$F = \frac{1}{2}\left(\rho\frac{\pi D^2}{4}\right)V_1^2(4a(1-a))$
Thrush coefficient	$C_T = \frac{F}{\frac{1}{2}\left(\rho\frac{\pi D^2}{4}\right)V_1^2} = 4a(1-a)$
Torque coefficient	$C_q = \frac{Q}{\frac{1}{2}\rho V_{turbina}^2\pi r^3}$
Tip speed ratio (TSR)	$\lambda = \Omega\frac{D/2}{V_w}$
Lift force	$L = \frac{\rho c V_{rel}^2 C_L}{2}$
Drag force	$D = \frac{\rho c V_{rel}^2 C_D}{2}$

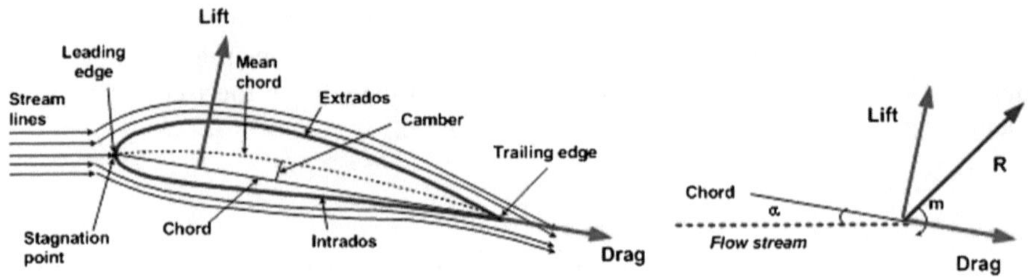

Fig. 1 Wing profile description and forces acting on it

speed and the rotational blade speed. The resultant force is the sum of the lift force, L and the drag force, D. The lift force, is produced by a blade element in the air, and acts from below upwards in a direction perpendicular to the relative wind, V_{rel}. In a wind turbine, the rotor generates the forward thrust, which produces the rotation. This force in dimensionless form is given by the lift coefficient C_L. The drag force D, acts in opposite direction to the movement of the blade element, and can be expressed as the drag coefficient, C_D. Both, the lift and drag coefficients, are functions of the angle of attack, which can be expressed as $\alpha = \gamma - \beta$, where γ is the angle between the local flow direction and the rotation plane, and β is the pitch angle of blade.

3 Methodology

3.1 Prototype Design Using Methods Engineering

Product prototype design is an important stage during the development of mechatronic systems and using methods engineering, it is separated into several phases to be followed systematically in order to obtain an optimal solution, preventing potential problems that could appear during the product life. As first stage, in the product lifecycle, the design arises from the fact to meet a need not covered or that can be improved according to the current or future needs. During the design, the principles of operation are considered, depending on product characteristics. The design determines the geometry, materials, and the construction process or implementation in mass production. Then, methods engineering technique, helps to reduce the production time during the prototype design, decreasing the cost per unit. This means, this technique subjects each operation of a particular part to a delicate analysis in order to eliminate any unnecessary operation, finding the fastest way to perform any operation. Methods engineering includes the equipment standardization, normalized procedures, work conditions and training for the operator to follow the precedent and not the posterior operation [16]. Moreover, using precise measurements, determines the amount of time to execute one task [15].

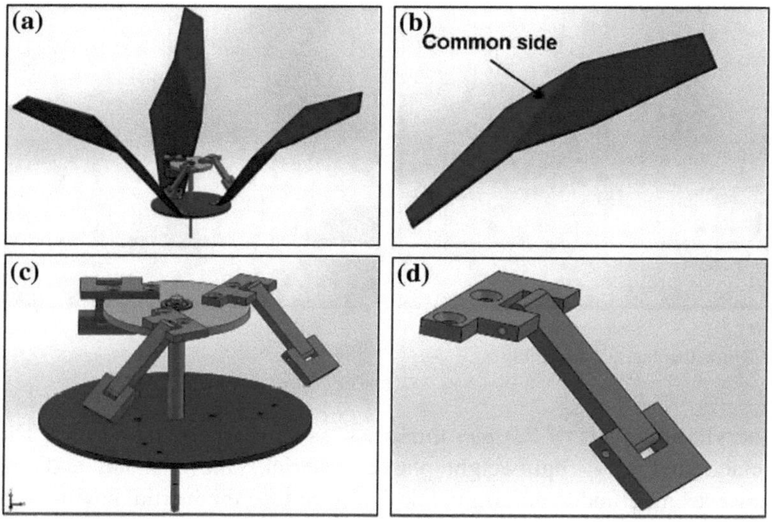

Fig. 2 Rotor main components CAD design

3.2 Prototype Aided Computer Design

Aided Computer Design software brings a better understanding of the structure and mechanical systems, visualizing in detail each part. For the rotor CAD design, after analyzing the aerodynamics of the blade with the shareware software Javafoil, a laminar wing profile geometry was proposed. Figure 2a shows the full assembly design for the rotor of the vertical axis wind turbine prototype, and as can be seen the rotor comprises three blades separated 120° each one. They are fabricated using two identical laminar stencils, (with five sides each one), as is shown in Fig. 2b, the shorter side is considered as the stencil base. The stencils are joined by a common edge, in mirror sense, forming a dihedral angle ϕ, resulting a cambered and twisted blade. The rotor consists of two parallel disks, aligned with respect to their center. A slim shaft passes through the centered ball bearing on each disk, having a free rotation. The upper disk has three articulated links to pull the blades, producing the inclination angle, as visualized in Fig. 2c. The lower disk has three hinges to attach each blade, allowing a free rotation to the blades. Then, pushing the shaft, the elevation angle can be modified (aerodynamic configuration of the rotor). Each link has three articulated blocks to join the upper disk with the blade, as depicted in Fig. 2d.

3.3 Prototype Assembly Brief Description

To build a mechatronic prototype, if precision and accuracy is required, then CNC type machines are recommended, using the appropriate materials for each part of the system. For the vertical axis wind turbine platform, the blades were builded

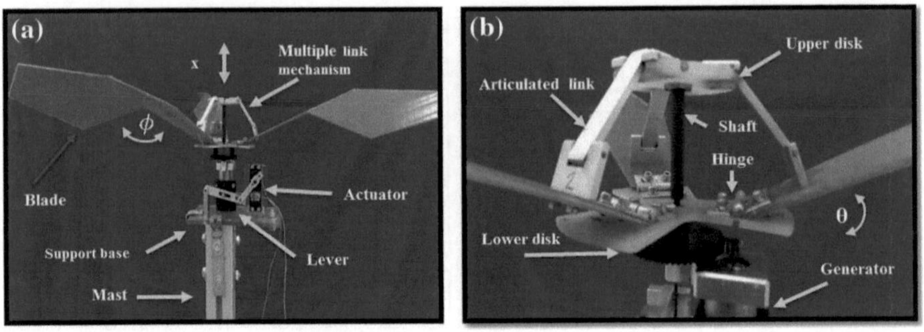

Fig. 3 Rotor mechanical assembly

using acrylic thin foils of 2.3 *mm* thickness, as can be seen in Fig. 3a. The disks were elaborated using lightweight plastic material. One disk has a diameter of 11 cm, it was modified from original design to reduce the inertia, having a shape of a three radius steering wheel, each radius separated 120° with respect to another, as can be appreciated in Fig. 3b. The blades are mounting in these radius by means of hinges. The second disk has a circular shape with a diameter of 7 cm, and is located on top of the rotor and acts as a multiple link. This mechanism consists of three articulated aluminium links, with 6.5 cm of length each one. Disks are aligned with respect to their center by means of a shaft that pushes the upper disk in the x direction (see Fig. 3a). This motion is transmitted to each blade to change their elevation angle. The carbon fiber shaft has a length of 11 cm and is pushed with small servo using an arm levers mechanism.

The rotor angular movement is transmitted to a small single phase permanent magnetic stator generator using a gear and pinion mechanism. The output voltage is proportional to the rotor speed.

4 Experimental Results

To carry out experimental tests in the absence of a wind tunnel, the test setup illustrated in Fig. 4 was implemented. The wind source was a multi-speed electric fan, and the tests were conducted in a closed room to avoid external disturbances.

A first experiment was conducted to obtain a combination of dihedral and elevation angles with the best performance at different wind speeds. Eleven sets of blades with different dihedral angles, were combined with six elevation angles. Then, each combination was submitted to ten wind speeds (0.5–5 m/s in steps of 0.5 m/s). The wind speed was modified: (a) changing the fan power and/or (b) moving the fan with respect wind turbine rotor. Wind speed and and rotor angular velocity were measured using an anemometer (EXTECH—AN100) and an optical tachometer (EXTECH—461920), respectively. Two of ten graphs of the rotor

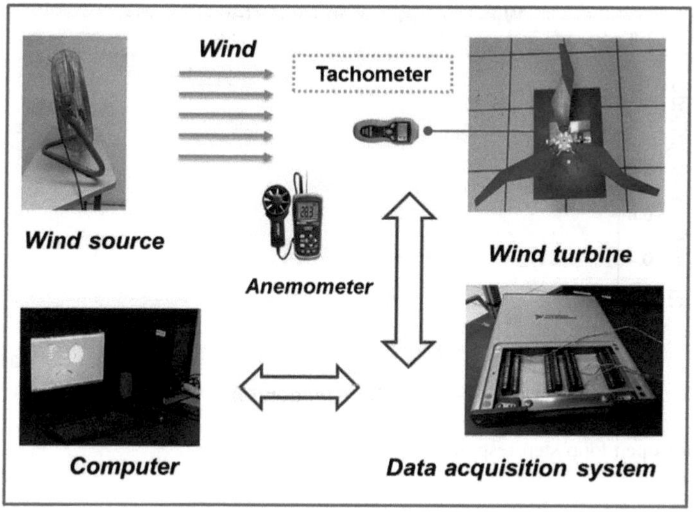

Fig. 4 Platform test setup

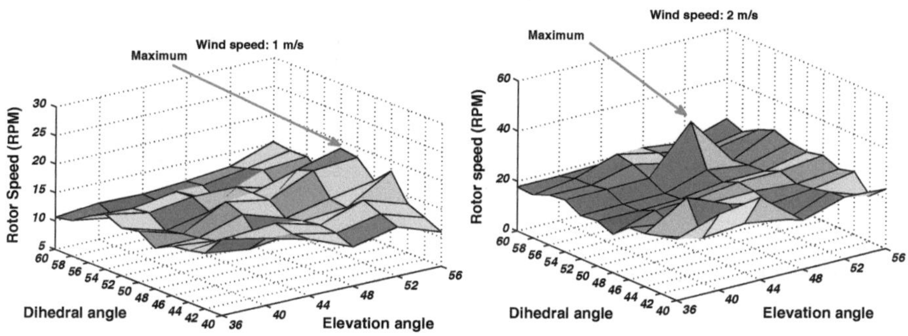

Fig. 5 Rotor response velocity combining elevation and dihedral angles at different wind speed

velocity response as function of the angles combination at different wind speeds are shown in Fig. 5.

The criteria to select the best angle combination is the maximal rotor speed and its linear response for a specific wind velocity. Therefore, the combination that better fits with the expected rotor speed behavior for a fixed wind speed is given by $\theta = 56°$ y $\phi = 48°$. The dihedral angle will be fixed to this value, and the elevation angle will serve to control the rotor speed, sending a signal to the servo. A second experiment consist on collecting data for the open loop step response of the wind turbine. This data is processed using MatLab system identification tool box to find parameters of a transfer function mathematical model, describing the platform dynamics. A unit step signal was applied to the servo to change its position from minimal to maximal, modifying the blades elevation angle and hence the rotor speed, proportionally, reaching its stationary state, as it is shown in Fig. 6.

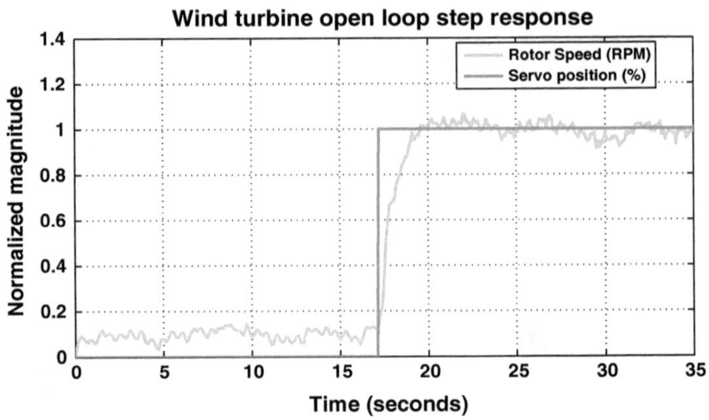

Fig. 6 Rotor open loop step response

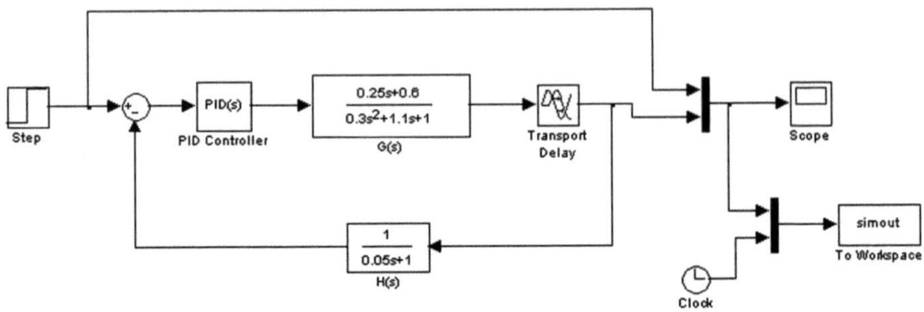

Fig. 7 Controller simulation

Once identified the system transfer function $G(s)$, it is possible to simulate the closed-loop system response implementing a control algorithm, as is shown in Fig. 7, where $H(s)$, represents the sensor dynamics, and the controller for this case is a PID. The gains for the controller can be obtained using different tuning methods for PID controllers, see for example [17]. In the practice the method given by Ziegler y Nichols described in [18] can be used as a first approach. Results were validated in the test platform in a Labview environment program.

5 Conclusions

A methodology to design and construct a nouvelle vertical axis wind turbine prototype was presented. This prototype shows can be used as experimental platform for didactic use, due that control algorithm can be implemented using graphical programming. Methods engineering helps us to plan the construction and assembly of the prototype. Also, the CAD brings the environment to develop all the parts and to visualize how they fit in the whole mechatronic system. The platform experimental data was used to identify the system dynamical model, represented by a transfer function, and in this way a control can be used to stabilize the system in close loop. This with the purpose of regulating the rotor velocity when there is a disturbance due to the wind variations.

Acknowledgments The authors wish to thank the Universidad Autonoma de Tamaulipas, and PROMEP-SEP for its support in this research project.

References

1. Anaya O et al (2009) Wind energy generation, modelling and control, 3rd edn. Wiley, Chichester
2. Hau E, Turbines W (2005) Fundamentals, technologies, application, economics, 2nd edn. Springer, Chichester
3. Manwell JF et al (2009) Wind energy explained, theory, design and application, 2nd edn. Wiley, Chichester
4. Pope K et al (2010) Energy and exergy efficiency comparison of horizontal and vertical axis wind turbines. Renewable Energy 35:2102–2113
5. Global Wind Energy Council, Global Wind Report: Annual Market Update 2012. http://www.gwec.net/wp-content/uploads/2012/06/Annual_report_2012_LowRes.pdf
6. Jamieson P (2011) Innovation in wind turbine design, 1st edn. Wiley, Chichester
7. Hemami A (2012) Wind turbine technology, 1st edn. Cengage Learning, USA
8. Mahmood M et al (2012) Vertical axis wind turbine, a review of various configurations and design techniques. Renew Sustain Energy Rev, Elsevier 16:1926–1939
9. Yang B, Lawnb C (2011) Fluid dynamic performance of a vertical axis turbine for tidal currents. Elsevier, Renewable Energy 36:3355–3366
10. McTavish S, Feszty D, Sankar T (2012) Steady and rotating computational fluid dynamics simulations of a novel vertical axis wind turbine for small-scale power generation. Elsevier, Renewable Energy 41:171–179
11. Howell R, Qin N, Edwards Jonathan, Durrani N (2010) Wind tunnel and numerical study of a small vertical axis wind turbine. Elsevier, Renewable Energy. 35:412–422
12. Kjellin J et al (2011) Power coefficient measurement on a 12 kW straight bladed vertical axis wind turbine. Renewable Energy 36:3050–3053
13. Trevino A et al (2012) Methods engineering for designing a didactic low cost small wind turbine generator. In: Proceedings of the IEEE-CCCA 2012 conference, Marseille France, pp 1261–1266, 6 Dec 2012
14. Bianchi FD, De Battista H, Mantz RJ (2010) Wind turbine control systems. Principles, modelling and gain scheduling design. Springer, London
15. Maynard HB (2004) Manual de la Ingeniería de la Producción Industrial. Quinta edición. McGraw Gill, México
16. Niebel BW, Freivalds A (2003) Methods, standards and work design, 11th edn. McGraw-Hill, USA
17. Astrom KJ, Hagglund T (1995) PID controllers: theory, design, and tuning. Instrument Society of America, USA, p 183
18. Smith CA, Corripio AB (2005) Principles and practices of automatic process control, 3rd edn. Wiley, Chichester

Motion Analysis of a Six-Legged Robot Using the Bennett's Linkage as Leg

J.C.M. Carvalho and T.R. Silvestre

Abstract Although the legged robots are considered as the most versatile type due to their major capability of moving on uneven terrain, the studied robots have not yet much mobility. In general this lack of mobility is related to the high number of degrees of freedom that makes it difficult to solve their mathematical model and applying it to the control system. This problem impose them a low operational velocity. Also the mobile robots which legs are based on parallel structures have not yet solved this problem. One alternative is to use simpler mechanisms, having small number of dof, for the legs. The solution analyzed in this paper uses the Bennett's linkage for a six legged robot. From the kinematic analysis of each leg, the robot motion is analyzed on two kinds of movement: the tripod type and the tetrapod type. In the present work the kinematic analysis of the robot behavior is analyzed in order to know, in advance, what can occur with the robot when it is walking. After the analysis some studies are proposed in such a way the robot can have a smooth movement, even considering its dynamics.

Keywords Mobile robot · Bennetts' linkage · Legged robot · Robot motion analysis

1 Introduction

In last year several researches has been developed about the mobile robots. In general the mobile robots can be classified as function of the environment in which they travel as "land robots" that can be wheeled, legged or move using tracks; the aerial robots named as "Unmanned Aerial Vehicle—UAV" and, underwater robots

J.C.M. Carvalho (✉) · T.R. Silvestre
University of Uberlandia, Uberlandia, Brazil
e-mail: jcmendes@mecanica.ufu.br

T.R. Silvestre
e-mail: tadeu_silvestre@hotmail.com

M. Ceccarelli and E.E. Hernández Martinez (eds.), *Multibody Mechatronic Systems*, Mechanisms and Machine Science 25, DOI 10.1007/978-3-319-09858-6_33

named as "Autonomous Underwater Vehicles—AUV". Initially, the main purpose of mobile robot was their capability of moving on the plants applying its ability where they were necessary or even substituting the human being in hazardous tasks. Today they are thinking for several applications like surveillance, inspections, agriculture and so on.

This paper is related to land mobile robots. One the most important mechanical characteristic of the mobile robot is related to its stability, which are in two forms: the static stability that is its ability to maintain a configuration from reaction forces; and dynamic stability, which is their ability to maintain a configuration from reaction and inertial forces. For this reason is that studies about legged robots show that their stability depends on both of its movement shape as the amount of legs. For a static equilibrium at least three legs are required, in such way the resulting reaction conditions over the ground cancel the weight of the robot. Although at least three legs are necessary to its equilibrium are used in general four or six legs. This is due to be easier to maintain the equilibrium of the robot during its movement: there is guaranteed to always have at least three legs in contact with the ground when it is walking.

The mobile robots still presents several problems to be solved. The wheeled and tracked robots are limited by the type of environment in which they travel, they can not move easily in soft soil or even rough terrain. However, in favorable environments, they can move at considerable speeds. Although several authors, like Raibert [1], say that the legged robots are the most versatile due to their major capability of moving on uneven terrain, the studied ones have not yet much mobility. In general this lack of mobility is related to the high number of degrees of freedom (dof) that makes it difficult to solve their mathematical model and applying it to the control system. This problem impose them a low operational velocity. Also the mobile robots which legs are based on parallel structures have not yet solved this problem. Examples whose can be seen on references shown the variety of studies about the legged robots [2—13].

One alternative is to use simpler mechanisms, having small number of dof, for the legs. The solution analyzed in this paper uses the Bennett's linkage for a six legged robot. From the kinematic analysis of each leg, the robot motion is analyzed on two kinds of movement: the tripod type and the tetrapod type.

In general the studies about six legged robots presents the dynamics model of its legs but don't the robot as an all. In the present work the kinematic analysis of the robot behavior is analyzed in order to know, in advance, what can occur with the robot when it is walking. After the analysis some studies are proposed in such a way the robot can have a smooth movement, even considering its dynamics.

2 The Bennett's Linkage

The Bennett's linkage is a spatial four-bar linkage that permits to describe spatial curves with only one dof, officially presented by Bennett in 1903 [14]. It is a 4-revolute 4-bar mechanism where the opposite links must have the same lengths and the same degree of skew or twist, Fig. 1a. Despite it was discovered few more than a century ago, this mechanism is still little investigated, and doesn't have many applications yet. It has been studied to apply on deployable structures formed by interconnected Bennett linkages like presented by Chen and You [15, 16] and Liu et al. [17].

The existence of the Bennett's linkage is conditioned to a geometric relation between the link lengths of its sides and the twist angles. This relation ensures that linkage moves with one degree-of-freedom and is given by $\sin(\theta_2)/\sin(\theta_3) = R/r = r_a$, where the angles θ_2 and θ_3 are the twist angles, and the dimensions R and r are the length of the bars [14, 18, 19], Fig. 1a.

The relation between the input angle θ_1 and the output one θ_4 can be obtained using the closing equation of the Bennett's linkage and homogeneous transformation matrices. For that, eight frames are defined, in the way that two consecutive frames always have common axes, given by: $y_i = y_{i+1}$ and $z_{i+1} = z_{i+2}$ for i = 1, 3, 5, 7, and the frames R_2 and R_3, R_4 and R_5, R_6 and R_7, R_8 and R_1 have, respectively, coincident origins. The y_i axes are placed along the bars of the mechanism. It were used auxiliary reference frames R_i' representing translations along the axes y_i, for i = 1, 3, 5, 7, as presented on Fig. 1a. For simplicity is presented only two axes of each frame.

By this way, we can define eight coordinate transformation matrices along the mechanism always in the following sequence: a translation along y_i, a rotation about y_i' and a rotation about z_i. The product of these eight matrices gives the closing equation of the Bennett's linkage:

$$I = T_{12} \cdot T_{23} \cdot T_{34} \cdot T_{45} \cdot T_{56} \cdot T_{67} \cdot T_{78} \cdot T_{81} \quad (1)$$

where I represents the identity matrix 4 × 4 and solving the Eq. 1 we have:

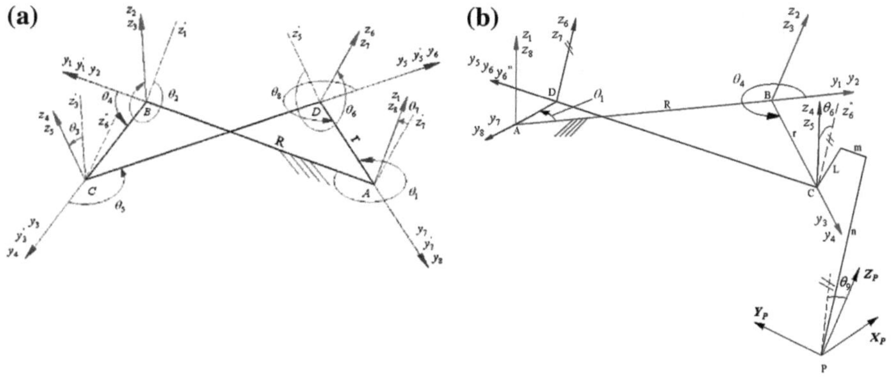

Fig. 1 **a** Reference frames attached to the Bennett's linkage. **b** The Bennett's linkage used as a leg

$$\sin(\theta_4) = \frac{s\theta_1 \cdot c\theta_2 - s\theta_1 \cdot c\theta_3}{1 - c\theta_2 \cdot c\theta_3 - c\theta_1 \cdot s\theta_2 \cdot s\theta_3} \quad \text{and}$$
$$\cos(\theta_4) = \frac{c\theta_1 - c\theta_1 \cdot c\theta_2 \cdot c\theta_3 - s\theta_2 \cdot s\theta_3}{1 - c\theta_2 \cdot c\theta_3 - c\theta_1 \cdot s\theta_2 \cdot s\theta_3} \tag{2}$$

where $s\theta_i$ and $c\theta_i$ means $\sin(\theta_i)$ and $\cos(\theta_i)$, respectively. The detail for transformation matrix and to obtain Eq. (2) can be seen in [19].

3 The Bennett's Linkage as Leg of the Hexapod

In order to obtain a path whose profile is similar to a robotic gait, that is considered as an almost semi-elliptical curve composed by a linear segment and a curve, the mobile connecting rod (*CD*) was extended as shown in Fig. 1b. The kinematic model that defines the foot trajectory, given by point P, and with the aim of systematizing a procedure of searching for an adequate configuration for this extension, it was modeled according to the three coordinate directions of the frame R_6'', which is attached to the mobile connecting rod and, the m length in the y_6'' direction (defined by the mobile connecting rod), the n length in the z_6'' direction and the L length in the x_6'' direction. The reference frame R_P, attached at point P, is rotated by θ_9 related to y_6'' axes and the auxiliary frame R_6'' has its origin coincident to R_5 one and rotate by θ_6 related to y_6.

To reorient the mechanism to set the quasi-linear segment of the trajectory and the direction of the robot's displacement in a coincident direction and put the path quasi-linear segment on the ground it was adopted three successive rotations of the mechanism defined by α, β and γ about z, x and y respectively. So, we can write:

$$T_P = T_\alpha \cdot T_\beta \cdot T_\gamma \cdot T_{18} \cdot T_{87} \cdot T_{76''} \cdot T_{6''P} = T_{\alpha\beta\gamma} \cdot T_{1P} \tag{3}$$

Then, one can to obtain the coordinates that define the trajectory of the robot foot, given by point P of the Bennett's linkage as:

$$\begin{aligned} X_P &= D_1 \cdot F_1 - s\alpha \cdot c\beta \cdot F_2 + D_2 \cdot (C_1 \cdot L - C_2 \cdot n) \\ Y_P &= E_1 \cdot F_1 + c\alpha \cdot c\beta \cdot F_2 + E_2 \cdot (C_1 \cdot L - C_2 \cdot n) \\ Z_P &= -c\beta \cdot s\gamma \cdot F_1 + s\beta \cdot F_2 + c\beta \cdot c\gamma \cdot (C_1 \cdot L - C_2 \cdot n) \end{aligned} \tag{4}$$

where:

$$C_1 = s\theta_3 \cdot c\theta_4 + c\theta_3 \cdot s\theta_4, \quad C_2 = -s\theta_3 \cdot s\theta_4 + c\theta_3 \cdot c\theta_4, \quad D_1 = c\alpha \cdot c\gamma - s\alpha \cdot s\beta \cdot s\gamma$$
$$D_2 = c\alpha \cdot s\gamma + s\alpha \cdot s\beta \cdot c\gamma, \quad E_1 = s\alpha \cdot c\gamma + c\alpha \cdot s\beta \cdot s\gamma, \quad E_2 = s\alpha \cdot s\gamma - c\alpha \cdot s\beta \cdot c\gamma$$
$$F_1 = A_1 \cdot L - s\theta_1 \cdot m - A_2 \cdot n - s\theta_1 \cdot (R + r), \quad F_2 = B_1 \cdot L + c\theta_1 \cdot m - B_2 \cdot n + c\theta_1 \cdot (R + r)$$

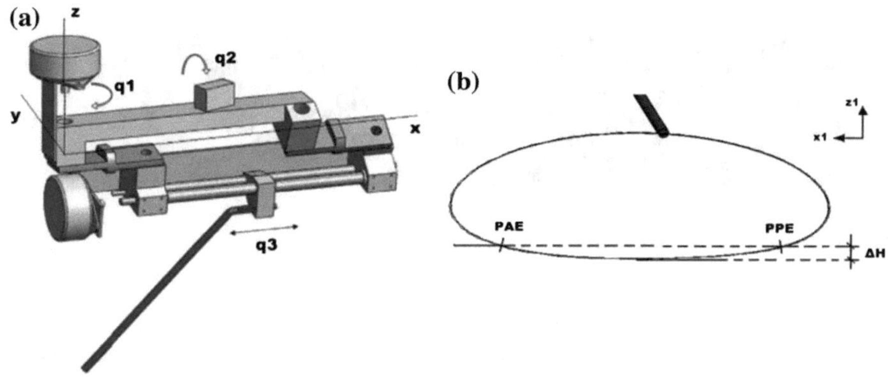

Fig. 2 **a** 3D-design of the leg. **b** The foot trajectory and definition of PPE and PAE

In order to obtain a good configuration for Bennett's linkage used as a robot leg, the following parameters must be analyzed: the twist angles θ_2 and θ_3 (which are related by the r_a parameter); the length links of the linkage R and r and then, its relation given by r_a; the three dimensions L, m and n of the extension rod and, the positioning angle of the robot foot, θ_9. For that, each parameter (θ_2, θ_9, L, m, n, r_a and R) was analyzed independently. The analysis had been made using the influence of each parameter on the coupler curve behavior which details can be seen in [19]. The data obtained for the leg are: $\theta_2 = 30°$; $\theta_9 = 0°$; $R = 200\,\text{mm}$; $r_a = 2.8$; $L = 2.1r$; $m = -30\,\text{mm}$; $n = 1.8r$, and the successive rotations $\alpha = -85.3°$; $\beta = -22.2°$ and $\gamma = -109.4°$. The projection of the obtained trajectory in a vertical plane is represented in Fig. 2b.

For transposition of obstacles two dof were included at each leg: one dof to uplift the leg, q_2, and another to change the foot pose, forward or backward, when exist an obstacle, q_3, as shown in Fig. 2a, where q_1 is the input motion of the Bennett's linkage, $q_1 = \theta_1$.

4 Gait Patterns

The gait pattern has been studied since the late nineteenth century by studying the horse trot. After that other animals had been your motion studied also like spider, centipedes, cockroaches, cats, camels, oxen and humans. These studies have been used as inspiration for analysis of legged mobile robots, principally the schemes and coordinating the movement of the legs which are named as "gait pattern". The gait pattern defines the sequence of steps of each leg with the duration of each phase, the stance phase and the swing phase. The gait patterns existing in nature vary greatly depending on the number of legs and size of each animal and there are different skills like walking, running, trotting, galloping, jumping, among other [2, 3, 5, 6].

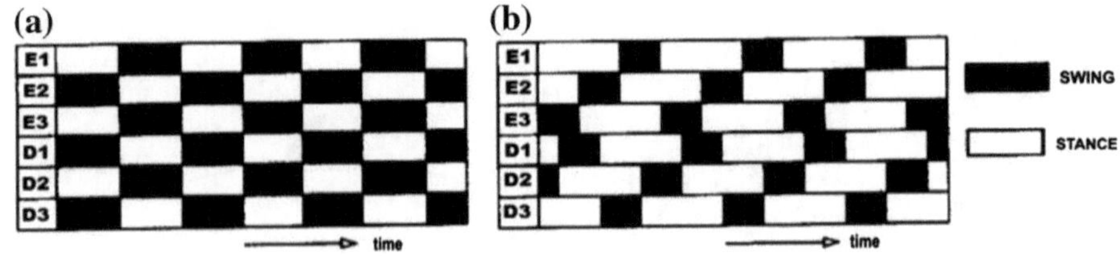

Fig. 3 Tripod and tetrapod gait

The design of hexapods considers that they are statically stable. This means that the projection of the center of gravity of the robot on the horizontal plane should remain within its supporting polygon. Two patterns are used: the *tripod* and the *tetrapod*. The tripod gait the legs move together in group of three alternating between the stance phase and the swing phase. For the tetrapod gait at least four legs are on the ground at the same time. In general the tripod gait can be observed at higher speeds motion, while at the low speeds tetrapod gait is used [2, 4, 6, 7, 9]. Figure 3 shows the tripod and tetrapod gaits where E1, E2 and E3 represents the left legs and D1, D2 and D3 the right ones.

For simulations it was defined from the foot trajectory the maximum forward pose (PAE) and the maximum backward pose (PPE). The PAE is the far pose in which the foot can be touch the ground and, for the forward motion this is the target position of the swing phase and the initial position of the stance phase. The PPE is the far position in which the foot can be touch the ground and, for the forward motion this is the target position of the stance phase and the initial position of the swing phase. As the foot trajectory at the stance phase is not really linear, it was defined the PAE and PPE for a $\Delta H = 5$ mm, as shown in Fig. 2b, where the $\boldsymbol{x1}$ direction defines the forward motion. For this consideration, the total length of a gait is about 138 mm. Since one motor cycle corresponds to one gait, it is possible to analyze each motion leg and then the robot motion.

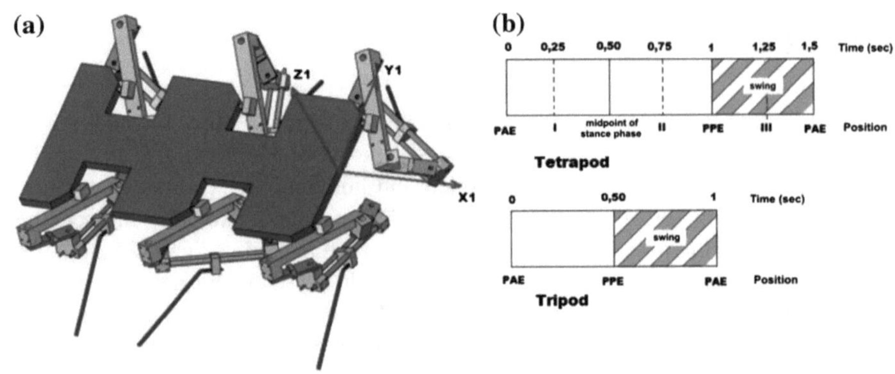

Fig. 4 **a** The 3D-design of the robot. **b** Motion cycle of each leg as function of the gait pattern

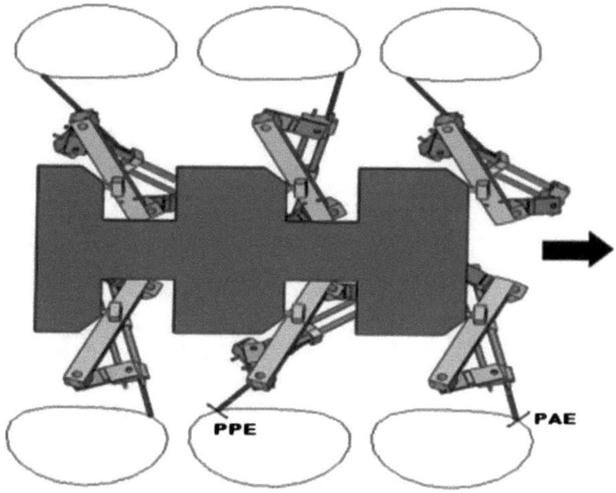

Fig. 5 Initial position of the robot legs for its motion

The robot body was designed in a symmetric distribution and in such a way that no collision between legs could occurs, in any configuration of them, as shown in Fig. 4a. Figure 5 shows the initial position of the robot's legs for its motion and the trajectory that each foot will perform identifying the PAE and PPE poses.

5 Simulations

For simulations each leg motion cycle is organizing in order to satisfy the gait pattern as represented on Fig. 3. The tetrapod gait the cycle can be divided in three regions where in two of them the leg is in the stance phase and the other ones the leg is in the swing phase. As Figs. 3b and 4b, in this case one side of the robot is out of phase of one and half region of the other side. For the tripod gait the cycle can be divided in only two equal regions where one the foot is in the stance phase and in the other one in the swing phase as can be seen in Fig. 4b.

For simulations it was used the Matlab and Cosmos Motion software, based on kinematic equations and considering a kinetic friction coefficient of 0.25 (dry rubber and steel) in such a way that the legs don't slip on the ground.

The stability motion conditions for the hexapod can be given by the vertical projection of the robot body into the front, rear and the sides. How smaller these values smoother is the robot motion. In the present study the stability can be done by the vertical variation of the center of mass (CM) related to the ground as between it and the center of mass of the legs. This analysis shows when the robot presents a kind of "balance" during its walking.

Figure 6a, b shows the vertical variation o the CM of the robot for tetrapod and tripod gaits for three different input motion of q_1. One can note that the difference between the two gaits is the variation amplitude. For tetrapod it is 1 mm for 5 rpm and approximately 2 mm for 25 rpm. On the other side, for tripod the amplitude was

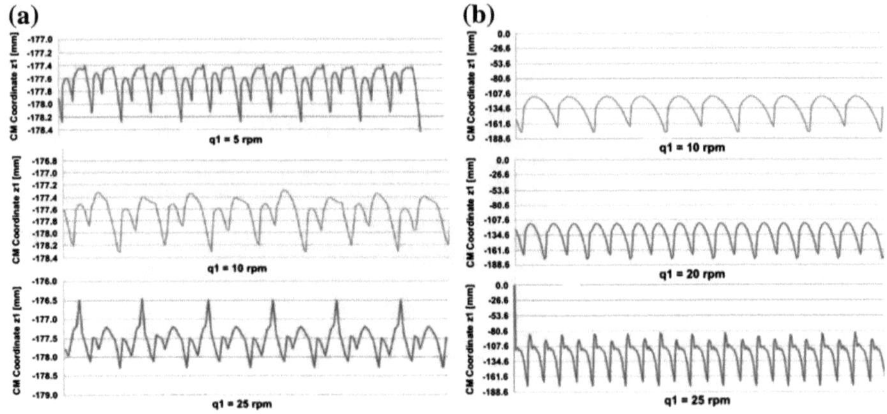

Fig. 6 Center of mass variation for different input motion q_1. **a** Tetrapod gait. **b** Tripod gait

approximately 5 mm for the three input velocities. This is due to the number of legs that contacts the ground at each time.

Although the tetrapod gait presents a smaller vertical amplitude when the robot is walking, its motion is not regular which can be seen using 3D simulations or plotting the six legs' behavior for a motion cycle. The behavior of the robot motion for tripod gait makes it "to jump" when considering its dynamics. On both gaits pattern when considering the robot dynamics, as it is, the robot will present a non uniform movement.

Simulations for analyzing the vertical and horizontal displacements of the point where each leg is attached to the structure had been made too. Similar results of those presented on Fig. 6 were obtained.

6 Conclusions

Mobile robots are increasingly efficient, with improvements in its motion systems due to, in part, to the study of new structures used in the legs.

In this paper an analysis of the stability of a six legged robot which use the Bennett's linkage as leg, based on its kinematics was presented. The advantage of using this mechanism for legs is the use of only one motor to obtain a spatial trajectory of the foot. For the studied leg, two other motors were included on each leg in order to transpose obstacles. But, in a usual motion without obstacles, only one motor is activated at each leg.

Two kinds of gaits were analyzed: the tetrapod and the tripod. Both gaits enable a stable robot motion. Although the tripod gait presents a more uniform motion, the amplitude of the vertical position of the robot CM is bigger. This behavior is justified by, at each instant there are three feet in contact with the ground and the other three at the swing phase, ensuring the motion uniformity but having a direct influence of the foot trajectory. The tetrapod gait presents more non uniform behavior but the vertical variation is smaller them the tripod gait.

From simulations it is clear that the stability is worst for bigger velocities.

The kinematic studies of the robot behavior are very important before to consider its dynamics. If it has important vertical variation of its CM, probably it will present misbehavior on its real motion, principally at high velocities when its dynamics must be considered. From obtained results, the robot will not have a smooth motion, both the tripod as the tetrapod gait.

In order to minimize this problem two kinds of analysis will be made: one is associating the input motion of two dof of the leg, given by q_1 and q_2, in order to obtain a linear stance phase. The second is to minimize the vertical displacement by reducing the stance phase, even if the robot speed is reduced too.

Acknowledgments The authors are thankful to CNPq, CAPES and FAPEMIG for the financing support of this research work.

References

1. Raibert MH (1986) Legged robots. Commun ACM 29(6):499–514
2. Pfeiffer F, Eltze J, Weidermann HJ (1995) Six-legged technical walking considering biological principles. Robot Auton Syst 14:223–232
3. Cruse H, Kindermanna T, Schumma M, Deanb J, Schmitza J (1998) Walknet—a biologically inspired network to control six legged walking. Neural Network 11:1435–1447
4. Altendorfer R, Moore N, Komsuoglu H, Buehler M, Brown HB Jr, McMordie D, Saranli U, Full R, Koditschek DK (2001) RHex: a biologically inspired hexapod runner. Auton Robots Neth 11:207–213
5. Malchano MD (2003) Biologically-plausible six-legged running: control and simulation. Master thesis, Massachusetts Institute of Technology, Cambridge
6. Erden MS, Leblebicioglu K (2005) Multi legged walking in robotics and dynamic gait pattern generation for a six-legged robot with reinforcement learning. In: Liu JX (ed) Mobile robots: new research. Nova Science Publishers, New York, pp 291–315
7. Alexandru G, Stroe I (2007) Stepping robot. Ann Oradea Univ Fascicle Manage Technol Eng VI(XVI):898–901
8. Collins CL (2007) Stiffness modeling and force distribution for the all-terrain hex-limbed extra-terrestrial explorer (ATHLETE). In: ASME 2007 international design engineering technical conferences and computers and information in engineering conference, 31st mechanisms and robotics conference, vol 8. Las Vegas, pp 781–789
9. Rosano-Matchain HL (2007) Decentralised compliant control for hexapod robots: a stick insect based walking model. Ph.D. thesis, Institute of Perception, Action and Behaviour, University of Edinburgh, Edinburgh (2007)
10. Wang Z-Y, Ding X-L, Rovetta A (2010) Analysis of typical locomotion of a symmetric hexapod robot. Robotica 28:893–907
11. Ordonez C, Gupta N, Collins EG Jr, Clark JE, Johnson AM (2012) Power modeling of the XRL hexapedal robot and its application to energy efficient motion planning. In: Proceedings of CLAWAR 2012
12. Bartsch S, Birnschein T, Römmermann M, Hilljegerdes J, Kühn D, Kirchner F (2012) Development of the six-legged walking and climbing robot space climber. J Field Robot (Special Issue on Space Robotics Part I) 29(3):506–532

13. Roy SS, Pratihar DK (2014) Kinematics, dynamics and power consumption analyses for turning motion of a six-legged robot. J Intell Robot Syst 74:663–688
14. Bennett GT (1903) A new mechanism. Engineering 76:777–778
15. Chen Y, You Z (2005) Mobile assemblies based on the Bennett linkage. Proc Real Soc A 461:1229–1245
16. Chen Y, You Z (2008) On mobile assemblies of Bennett linkages. Proc Real Soc A 464 (2093):1275–1293
17. Liu J, Yu Y, Huang Z, Huang X (2013) General order principle for multi-Bennett linkages. Chin J Mech Eng 26(2):275–281
18. Perez A, Mccarthy JM (2000) Dimensional synthesis of Bennett linkages. In: Proceedings of the ASME 2000 design engineering technical conference—DETC'00/MECH-14069, Baltimore
19. Oliveira AA, Carvalho JCM (2007) Modeling of the Bennett's linkage as leg of a mobile robot. In :12th IFToMM world congress, Besançon, pp 398–402

Implementation of Force and Position Controllers for a 3DOF Parallel Manipulator

J. Cazalilla, M. Vallés, A. Valera, V. Mata and M. Díaz-Rodríguez

Abstract The aim of this paper is to present the development of real-time controllers for a parallel manipulator (PM) of 3 degrees of freedom. The robot is able to generate one translation movement and two rotational movements (roll and pitch). Some applications for this type of manipulators can be found in driving simulation and biomechanics (rehabilitation of lower members, for example). An open control architecture has been implemented for this robot, allowing to implement and validate different dynamic control schemes for 3DOF PM. Thus, the developed robot can be used as a *test bench* to validate different control schemes. This article presents how several position and force control schemes have been implemented.

Keywords Parallel manipulator · Robot control · Force control · Mechatronics · Kinematics

1 Introduction

A parallel manipulator (PM) is a mobile platform connected to a fixed base through several kinematic chains. These manipulators have an end-effector connected to the mobile platform. PMs have some advantages over serial robots. For example, the

J. Cazalilla (✉) · M. Vallés · A. Valera · V. Mata
Universitat Politècnica de València, Valencia, Spain
e-mail: jcazalilla@ai2.upv.es

M. Vallés
e-mail: mvalles@ai2.upv.es

A. Valera
e-mail: giuprog@ai2.upv.es

V. Mata
e-mail: vmata@mcm.upv.es

M. Díaz-Rodríguez
Universidad de Los Andes, Bogota, Venezuela
e-mail: dmiguel@ula.ve

M. Ceccarelli and E.E. Hernández Martinez (eds.), *Multibody Mechatronic Systems*,
Mechanisms and Machine Science 25, DOI 10.1007/978-3-319-09858-6_34

load is distributed in parallel manipulators, being shared out between the joints that connect the mobile platform to the base. Thus, PMs have high stiffness, high load-carrying, high speed and accuracy. However, PMs have small work spaces and singularity problems. Furthermore, the forward kinematics and system dynamics solution, as well as the control of PMs are hard to develop as compared with serial robots.

Due to its advantages, parallel manipulators have varied applications. These robots are implemented for: motion simulators, tire testing machines, flight simulators, and medical applications. Different mechanical architectures can be found in works like [10, 14, 18, 19]. PMs Investigations initially focused on platforms with 6 degrees of freedom (DOF). However, for many applications, 6DOF weren't required. For example, the famous Delta Robot has 3 translational DOF (3T) [19]. This PM is suitable for tasks of pick-and-place [5]. The translational PM 3T is not only used in daily pick-and-place tasks, but are also used in medical applications such as cardiopulmonary resuscitation equipments [13], and as machine tools [15]. When translational and rotational movements are required, 3PRS [4] and 3RPS [12] architecture are proposed, where R, P and S notation means revolution, prismatic and spherical joint, respectively.

The aim of this paper is to present the developed robot, and how the real-time position and force controllers have been implemented, in order to perform one translational 1T (heave) and two rotary 2R movements (roll and pitch).

In order to implement and test dynamic control schemes for PM with 3-DOF, an open architecture has been developed. Schemes tested are position controllers based on inverse dynamics [16], using the model identified in [9].

Note that there are a variety of applications involving assembly tasks where one or more parts need to be handled, ensuring proper contact, or industrial operations such as milling or deburring [1]. Therefore, for such applications it is necessary to develop an accurate robot force control. In this paper, a force-position control for the parallel robot has been developed.

2 The Parallel Robot

The choice of a 3PRS architecture has been determined by the need to develop a low cost robot able to generate angular rotation in two axes, and a prismatic movement. Two different architectures were considered: 3-RPS and 3-PRS. The second one was the selected architecture after comparing the advantages and disadvantages of each proposal. For example, one advantage of the PRS architecture is that the actuators are located on the fixed base while the architecture 3-RPS, the actuators move with the joints of revolution.

Figure 1 shows the parallel robot developed. Its physical system consists of three legs connecting the moving platform to the base. Each leg consists of a direct drive ball screw (prismatic joints) and a coupler, besides the motor.

Fig. 1 The 3-PRS low-cost parallel manipulator

The motors in each leg are brushless DC servomotors equipped with power amplifiers. The actuators are Aerotech BMS465 AH brushless servomotors. Aerotech BA10 power amplifiers operate the motors. The control system was developed on an industrial PC. This fact presents several advantages: the first is that it is a completely open system, providing a powerful platform for programming high-level tasks. Therefore, applications such as trajectory generation or control strategies based on external sensing (machine vision and force sensors, etc.) can be developed. The second great advantage of the control system is its cost, because the software architecture is based on free and open software.

2.1 Kinematic Model

To establish the control of the robot, direct and inverse kinematic models have been developed.

Figure 2 shows a kinematic diagram of the robot. Nine generalized coordinates are used to model the robot (q_i, where $i = 1...9$).

Where the active coordinates q_1, q_6 and q_8 are associated with the actuated prismatic joints (P). In Fig. 2 can be seen that the length between the locations of the spherical joints P_i and P_j is constant and equal to l_m.

In the forward position problem, the position of the actuators is known (q_1, q_6 and q_8 are known). The Newton-Raphson (N-R) numerical method is chosen to solve this nonlinear problem. The method converges rather quickly when the initial guess is close to the desired solution [11].

Moreover, the inverse kinematics problem consists of finding the movement of the linear actuators from roll (γ), pitch (β) and height (z) of the platform.

The positions of the actuators can be obtained by the following expression [19]:

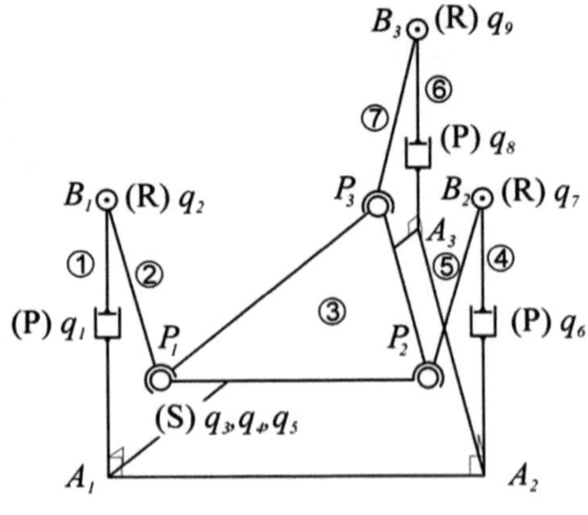

Fig. 2 Kinematic diagram of the 3-PRS parallel manipulator, type of joints and generalized coordinates

$$
\begin{aligned}
q_1 &= p_x^2 + p_y^2 + p_z^2 + 2h(p_x u_x + p_y u_y + p_z u_z) - 2g p_x - 2g h u_x + g^2 + h^2 \\
q_6 &= p_x^2 + p_y^2 + p_z^2 - h(p_x u_x + p_y u_y + p_z u_z) + \sqrt{3}h(p_x v_x + p_y v_y + p_z v_z) \\
&\quad + g\left(p_x - \sqrt{3}p_y\right) + gh\left(u_x - \sqrt{3}u_y\right)/2 + gh\left(v_x - \sqrt{3}v_y\right)/2 + g^2 + h^2 \\
q_8 &= p_x^2 + p_y^2 + p_z^2 - h(p_x u_x + p_y u_y + p_z u_z) - \sqrt{3}h(p_x v_x + p_y v_y + p_z v_z) \\
&\quad + g\left(p_x - \sqrt{3}p_y\right) - gh\left(u_x - \sqrt{3}u_y\right)/2 + gh\left(v_x - \sqrt{3}v_y\right)/2 + g^2 + h^2
\end{aligned}
\tag{1}
$$

In Eq. (1), $h = l_m/\sqrt{3}$, $g = l_b/\sqrt{3}$, $p_x = -hu_y$, $p_y = -h(u_x - v_y)$, $p_z = z$, l_b is the distance between A_iA_j, and u and v can be obtained from the rotation matrix of the moving platform.

2.2 *Jacobian of the Parallel Robot*

As is well known, the kinematic model of a robot seeks relationships between variables and joint position of the terminal end of the robot. This relationship does not take into account the forces or torques applied on the robot. However, it allows to know the relationship between the derivatives of the joint coordinates and the terminal end of the robot member.

The relationship between the velocities of the joint coordinates and the position and orientation of the robot is obtained through the Jacobian matrix.

To obtain the Jacobian, in Fig. 2 the following vector expression can be checked:

$$\vec{p} + \vec{b}_i = \vec{a}_i + q_1 \cdot \vec{r}_{1i} + l_r \cdot \vec{r}_{2i} \tag{2}$$

where $\vec{r}_{1i}$ and $\vec{r}_{2i}$ are unit vectors, l_r is the length of the coupler, $\vec{p}$ is the position vector of the terminal end, $\vec{b}_i$ is the position vector between the terminal end and the spherical joint i. Finally, $\vec{a}_i$ defines the position of the prismatic joint and the global coordinate axis.

Deriving Eq. (2) and multiplying $\vec{r}_{2i}$ times on both sides of the equation:

$$\vec{r}_{2i} \cdot \vec{v}_p + \vec{r}_{2i} \cdot \left(\vec{\omega}_p \times \vec{b}_i\right) = \vec{r}_{2i} \cdot \dot{q}_i \cdot \vec{r}_{1i}. \tag{3}$$

Applying the previous equation to each of the three legs of the robot:

$$\begin{bmatrix} \vec{r}_{21} \cdot \vec{r}_{11} & 0 & 0 \\ 0 & \vec{r}_{22} \cdot \vec{r}_{12} & 0 \\ 0 & 0 & \vec{r}_{23} \cdot \vec{r}_{13} \end{bmatrix} \cdot \begin{bmatrix} \dot{q}_1 \\ \dot{q}_6 \\ \dot{q}_8 \end{bmatrix} = J_q \cdot \dot{\vec{q}} \tag{4}$$

$$\begin{bmatrix} \vec{r}^{t}_{21} & \left(\vec{b}_1 \times \vec{r}_{21}\right)^t \\ \vec{r}^{t}_{22} & \left(\vec{b}_2 \times \vec{r}_{22}\right)^t \\ \vec{r}^{t}_{23} & \left(\vec{b}3 \times \vec{r}_{23}\right)^t \end{bmatrix} \cdot \begin{bmatrix} \dot{x} \\ \dot{y} \\ \dot{z} \\ \dot{\gamma} \\ \dot{\beta} \\ \dot{\alpha} \end{bmatrix} = J_x \cdot \dot{\vec{X}}. \tag{5}$$

In this way, it is verified that:

$$J_q \cdot \dot{\vec{q}} = J_x \cdot \dot{\vec{X}}. \tag{6}$$

However, it's necessary to remember that not all the variables of $\dot{\vec{X}}$ vector are independent. For this reason, it is necessary to find the following Jacobian:

$$\begin{bmatrix} \dot{x} \\ \dot{y} \\ \dot{z} \\ \dot{\gamma} \\ \dot{\beta} \\ \dot{\alpha} \end{bmatrix} = \underbrace{\begin{bmatrix} \frac{\delta x}{\delta z} & \frac{\delta x}{\delta \gamma} & \frac{\delta x}{\delta \beta} \\ \frac{\delta y}{\delta z} & \frac{\delta y}{\delta \gamma} & \frac{\delta y}{\delta \beta} \\ 1 & 0 & 0 \\ 0 & 1 & 0 \\ 0 & 0 & 1 \\ \frac{\delta \alpha}{\delta z} & \frac{\delta \alpha}{\delta \gamma} & \frac{\delta \alpha}{\delta \beta} \end{bmatrix}}_{J_r} \begin{bmatrix} \dot{z} \\ \dot{\gamma} \\ \dot{\beta} \end{bmatrix}. \tag{7}$$

Thus:

$$J_q \cdot \dot{\vec{q}} = J_x \cdot J_r \cdot \dot{\vec{X}}_c. \tag{8}$$

So, the relation between the coordinates of the joints and the terminal end is given by:

$$\dot{\vec{q}} = (J_q^{-1} \cdot J_x \cdot J_r) \cdot \dot{\vec{X}}_c. \tag{9}$$

2.3 Dynamic Model

In order to implement dynamic controllers the equation of motion can be described as follows,

$$M\left(\vec{q}, \vec{\Phi}\right) \cdot \ddot{\vec{q}} + \vec{C}\left(\vec{q}, \dot{\vec{q}}, \vec{\Phi}\right) \cdot \dot{\vec{q}} + \vec{G}\left(\vec{q}, \vec{\Phi}\right) = \vec{\tau}. \tag{10}$$

From Eq. (10) it can be seen that the system mass matrix M, the vectors corresponding to the centrifugal and Coriolis forces C, and the gravitational forces G depend on the dynamic parameters $\vec{\Phi}$ and the external generalized forces $\vec{\tau}$.

To identify the dynamic parameters, the model needs to be written in linear parameters [7],

$$\mathrm{K}\left(\vec{q}, \dot{\vec{q}}, \ddot{\vec{q}}\right) \cdot \vec{\Phi} = \vec{\tau}. \tag{11}$$

In Eq. (11), $\mathrm{K}\left(\vec{q}, \dot{\vec{q}}, \ddot{\vec{q}}\right)$ is the observation matrix corresponding to the set of generalized coordinates, velocities and accelerations. For this parallel robot, a complete and reduced model can be obtained and analyzed [8].

3 Development of the Simulated Parallel Robot

3.1 Position Control

When the dynamic problem is solved and the dynamic parameters are validated for the parallel robot, a real-time control can be addressed. In this work, various control strategies based on inverse dynamics have been implemented. This type of control is discussed in more detail in [3, 6, 17, 21]. The implemented control strategies are based on the motion of the robot (Eq. 10) and are given by the following general expression:

$$\vec{\tau}_c = M(\vec{q})\vec{a} + \vec{C}(\vec{q}, \dot{\vec{q}})\dot{\vec{q}} + \vec{G}(\vec{q}) \tag{12}$$

where τ_c is generalized torque control, $M(\vec{q})$, $\vec{C}(\vec{q}, \dot{\vec{q}})\dot{\vec{q}}$, $\vec{G}(\vec{q})$ are inertial matrix, Coriolis vector and system's gravitational forces, and $\vec{a}$ is the linear control action.

Table 1 Linear control action

Controller	$\vec{a}$
Point-to-point	$-K_d\dot{\vec{q}} - K_p\vec{e}$
Trajectory control	$\ddot{\vec{q}}_d - K_d\dot{\vec{e}} - K_p\vec{e}$
Trajectory control with integral action	$\ddot{\vec{q}}_d - K_d\dot{\vec{e}} - K_p\vec{e} - K_i\int_0^t \vec{e}(u)\,du$

Depending on the expression used in $\vec{a}$, different control strategies can be obtained, as seen in Table 1:

where $\vec{e} = \vec{q} - \vec{q}_d$, and K_d, K_p y K_i are the matrices (diagonal and positive definite) for derivative, proportional and integral gains.

The first controller is a point-to-point exponentially stable by an appropriate selection of the K_d and K_p. The second controller is similar to the first, but in this case a trajectory control is performed, in which the robot follows a trajectory $\vec{q}_d(t)$ and its successive derivatives $\dot{\vec{q}}_d(t)$ and, being speed and desired acceleration. The third one is a trajectory control in which is added an integral error term.

In order to validate the methodology proposed in this paper various Matlab/Simulink schemes have been developed. Figure 3 shows reference and real position of the robot joint q_1 for a point-to-point controller. A good response is obtained in following the trajectory. Joints q_6 and q_8 response is quite similar.

3.2 Force Control

Besides controlling position, more and more is being used force control in industrial robots. To establish the force control various types of control strategies can be chosen. The first type is based on the typical force control, and consists in following

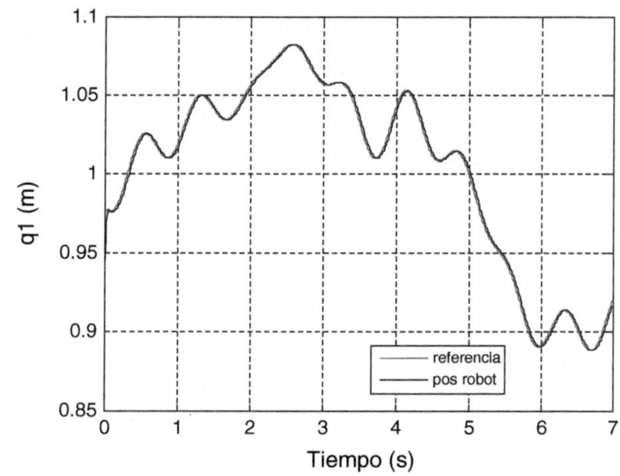

Fig. 3 Reference and q_1 position

a force reference. This control is restricted to a linear [20] control, like a PID force control. The effect of the three parameters of a PID controller is well known and has been extensively described in the literature as for example in [2].

However, PID force control presents some problems like: integral term can lead to system instability and the noise of the force sensor causes difficulty in derivative calculation. Therefore, in order to dampen the system, is common to use the speed ($\dot{x}$) since this is equivalent to the derivative of force. The resulting control law is given in Eq. (13).

$$\tau = F_{ref} + K_p(F_{ref} - f) - K_v\dot{x} \tag{13}$$

where τ is the control action, F_{ref} is the force reference, f is the force measured, and, K_p and K_v are proportional and derivative constants.

In order to simulate and analyze the force control, different Matlab/Simulink schemes have been developed. As mentioned above, due to the configuration of the parallel robot, it has 3 degrees of freedom: the height (z) of the platform and the orientation (γ and β). Thus, the robot has established a control force/position: force control in z axis, and orientation control of the robot platform.

4 Development of the Actual Parallel Robot

In order to implement the control architecture for the parallel robot, an industrial PC has been used. It is based on a high performance 4U Rackmount industrial system with 7 PCI slots and 7 ISA slots. It has a 2.5 GHz Intel® Pentium® Core 2 Quad/Duo processor and 4 GB SDRAM.

The industrial PC is equipped with 2 Advantech™ data acquisition cards: a PCI-1720 and a PCL-833. The first one has been used for supplying the control actions for each parallel robot actuator and the second one has been used for reading the measurements of the encoders.

The force control has been established with the ATI sensor Delta SI-330-30. This is a sensor with 6DOF able to measure forces and moments in the XYZ axes, made of silicon strain gauges, providing a measurement error nearly zero.

The programming language used to control the parallel robot has been C++ (using the middleware Orocos). The PC is equipped with Linux Ubuntu system, patched with Xenomai (a real-time kernel). Thus, real-time features are available.

Using this environment, different control algorithms have been implemented. Figure 4a shows the reference and the real position (for the joint q_1) of the parallel robot. As can be seen, a high accuracy is being obtained (the error is less than 0.1 mm).

Figure 4b shows the speed references of the active joints generated by the kinematic control, and the outputs obtained from the robot Jacobian. Their behavior is very good and fully corresponds to the results obtained in the simulation of the robot.

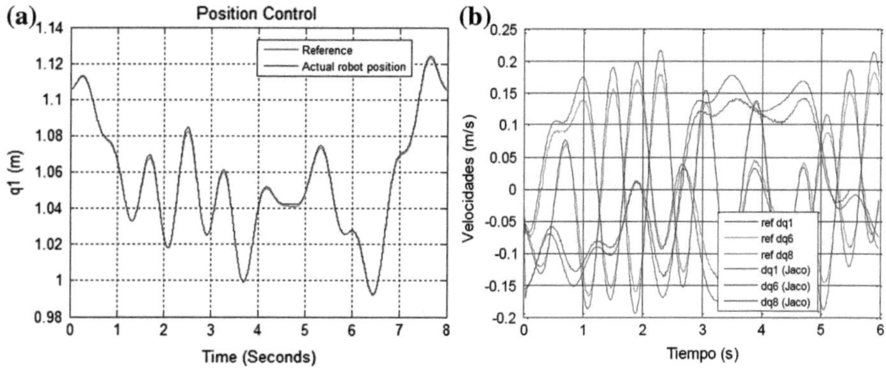

Fig. 4 Actual robot positions and velocities

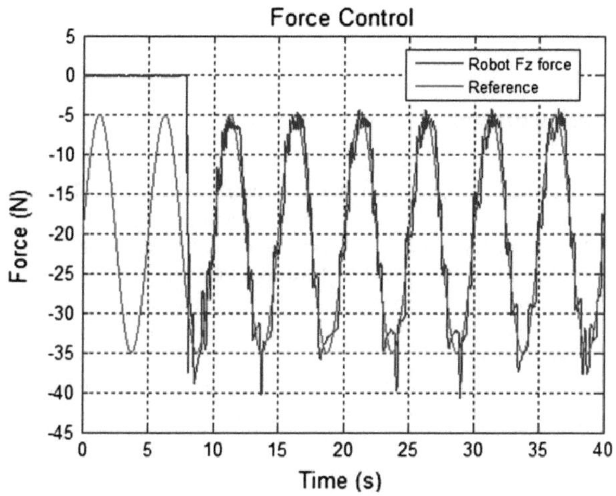

Fig. 5 Force control real robot

Finally, Fig. 5 shows the reference and the force applied by the parallel in Z axis. For the first 7 s (approximately) the force applied by the robot is 0 because during that time the platform is moving down and has no contact with the environment. From that point, the terminal element collides with the environment and the control force is established, using a sinusoidal force reference. As can be seen in the figure, the force applied by the robot follows the reference very accurately.

5 Conclusions

This paper has considered the position and force control for a parallel robot with 3DOF. In addition to design the robot, a control system (based on industrial PC) completely open and flexible has been developed. The PC is equipped with card

D/A converters, encoders reading and a force sensor. Programs and control applications have been developed with the programming language C++.

To establish control of the robot, direct and inverse kinematic model of the robot have been solved, as well as the dynamic model and the Jacobian.

For the development of position control, a model-based controller has been chosen. The response obtained with this position control both simulations and real robot provides a small position error.

Furthermore, in this work, a force controller has been implemented. A force sensor has been installed in the platform in order to measure forces and torques applied by the robot. The performance obtained in both simulation model and real robot is very good.

Acknowledgments The authors want to express their gratitude to the Plan Nacional de I + D (FEDER-CICYT) for the partial financing of this research under the projects DPI2010-20814-C02-01/02 and DPI2011-28507-C02-01. This research was also partially funded by the CDCHT-ULA Grant I-1286-11-02-B.

References

1. Afonso G, Pires JN, Estrela N (2007) Force control experiments for industrial applications: a test case using an industrial deburring example. Assem Autom J 26(2):148–156
2. Astrom KJ, Murray R (2008) Feedback systems. Princeton University Press, Princeton
3. Canudas C, Siciliano B, Basting G (1996) Theory of robot control. Springer, New York
4. Chablat D, Wenger P (2003) Architecture optimization of a 3-DOF translational parallel mechanism for machining applications, the Orthoglide. IEEE Trans Robot Autom 19(3): 403–410
5. Clavel R (1988) DELTA, a fast robot with parallel geometry. In: Proceedings of 18th international symposium on industrial robot, Lausanne, pp 91–100
6. Craig J (1989) Introduction to robotics: mechanics and control. Addison-Wesley, Reading
7. Díaz-Rodriguez M, Mata V, Valera A, Page A (2010) A methodology for dynamic parameters identification of 3-DOF parallel robots in terms of relevant parameters. Mech Mach Theory 45:1337–1356
8. Diaz-Rodriguez M, Valera A, Mata V, Valles M (2012) Model-based control of a 3-DOF parallel robot based on identified relevant parameters. IEEE/ASME Trans Mechatron. doi:10.1109/TMECH.2012.2212716
9. Farhat N (2006) Identificación de Parámetros Dinámicos en Sistemas de Cadena Cerrada. Aplicación a Robot Paralelos. Phd, Universidad Politécnica de Valencia
10. Gough VE, Whitehall SG (1962) Universal tire test machine. In: Proceedings of 9th international technical congress FISITA, pp 117–135
11. Jalón JG, Bayo E (1994) Kinematic and dynamic simulation of multibody systems: the real-time challenge. Springer, New York
12. Lee KM, Arjunan S (1991) A three degrees-of-freedom micromotion in parallel actuated manipulator. IEEE Trans Robot Autom 7(5):634–641
13. Li Y, Xu Q (2007) Design and development of a medical parallel robot for cardiopulmonary resuscitation. IEEE/ASME Trans Mechatron 12(3):265–273
14. Merlet JP (2000) Parallel robots. Kluwer, London
15. Pierrot F, Nabat V, Company O, Krut S, Poignet P (2009) Optimal design of a 4-dof parallel manipulator: From academia to industry. IEEE Trans Robot 25(2):213–224

16. Rosillo N, Valera A, Benimeli F, Mata V, Valero F (2011) Real-time solving of dynamic problem in industrial robots. Ind Robot 38(2):119–129
17. Spong M, Vidyasagar M (1989) Robot dynamics and control. Wiley, New York
18. Steward DA (1965) A platform with 6 degree of freedom. In: Proceedings of the institution of mechanical engineers, part 1, vol 15, pp 371–386
19. Tsai LW (1999) Robot analysis: the mechanics of serial and parallel manipulator. Wiley Interscience, New York
20. Volpe R, Khosla P (1993) A theoretical and experimental investigation of explicit force control strategies for robot manipulators. IEEE Trans Autom Control 38(11):1634–1650
21. Yoshikawa T (1990) Foundations of robotics: analysis and control. The MIT Press, Cambridge

Mechanical Reproduction of the Horse Movement from a Hippotherapy Cycle

C.S. López-Cajún, J.C. Jáuregui-Correa, C.A. González-Cruz and M. Rodríguez

Abstract In this paper, a methodology for obtaining the basic movements of a horse for children hippotherapy is presented. These were determined from actual measurements acquired from different body parts of a horse that influence the desired effects for therapy. Measuring results were analyzed to identify those movements that help the therapy. Accelerometers placed on several spots on the horse allowed recording measurements of the motion of an actual horse rider during the walking and trotting gaits of the horse. Based on those, the planar trajectory and main tilts of the motion of a horse rider were obtained. The main goal of this project is to reproduce the closest movement via a 4-DOF parallel mechanism.

Keywords Horse-therapy · 4-DOF-parallel mechanism · Mechanisms design

1 Introduction

Modern therapy equipment is intended to facilitate rehabilitating injure patients or motion illness. Clearly, the development of this type of equipment requires a deep understanding of muscle reaction to induced excitation. Indeed, one of the motion therapies that have proved good results is hippo therapy. This is a repeated activity that helps recovering motion capabilities for children with brain damage, accident shocks or any other disabilities. Horse riding provides tridimensional motions and impulses that stimulate muscle and body joints. It also helps patients by recovering the posture while the horse walks. As a matter of fact, when a patient feels insecure

C.S. López-Cajún (✉) · J.C. Jáuregui-Correa · C.A. González-Cruz · M. Rodríguez
Universidad Autónoma de Querétaro, Mexico, Mexico
e-mail: cajun@uaq.mx

J.C. Jáuregui-Correa
e-mail: jc.jauregui@uaq.mx

C.A. González-Cruz
e-mail: cgonzalez@uaq.mx

M. Ceccarelli and E.E. Hernández Martinez (eds.), *Multibody Mechatronic Systems*, Mechanisms and Machine Science 25, DOI 10.1007/978-3-319-09858-6_35

or fills an unstable position, he reacts impulsively correcting to the vertical position. This type of stimuli helps patient's coordination and, many times, the recovery of motion abilities [1]. Furthermore, the other effect that makes hippotherapy feasible for children is that the patients follow the horse motion trying to recover the natural posture [2]. In this way the upper body is forced to move in a similar way as a normal walking motion [3—5]. It is clear that riding a horse has a similar motion effect as human walking. Therefore, the spine and upper joints are stimulated while the patient rides a horse. Indeed, this is the main element that makes hippotherapy a good candidate for patient's recovery [6]. Several studies have demonstrated the effectiveness of hippotherapy. Indeed, researchers have reported how hippotherapy improved patient's walking skills, walking speed, and muscle strength [7—10]. Hippotherapy also provides long-term benefits and improvements on the overall health [11]. Unfortunately, this therapy require a large space and it cost limits the access to a large population. Therefore, it is desirable to have mechanic drivers capable of providing a similar therapy in a room. As there aren't statistically significant differences among the horse body movement parameters within one session of hippotherapy [12], the final goal of this project is the design a 4-DOF parallel mechanism that will reproduce, among horse movements, those that are effective in motion therapy. In general, horses have many movements, but there are few of them that are important helping patients with some motion illness. In this paper, the identification of horse movements is described. These movements will be reproduced via parallel mechanism. Moreover, the effective movements were identified measuring the acceleration on different body parts of a horse. These measurements were processed in order to identify the actual planar trajectory and the two main tilts. The mechanism do not replace at horse, this only will reproduce its movement.

2 Experimental Procedure

The first part of this research was the measurement of a horse riding. For this, six accelerometers were mounted on the horse back, namely, two at the hip, one at the loin, one at the whiter and two at the whither sides. Location of the above mentioned accelerometers is shown in Fig. 1. The recording data was done during the walking and trotting gaits of the horse.

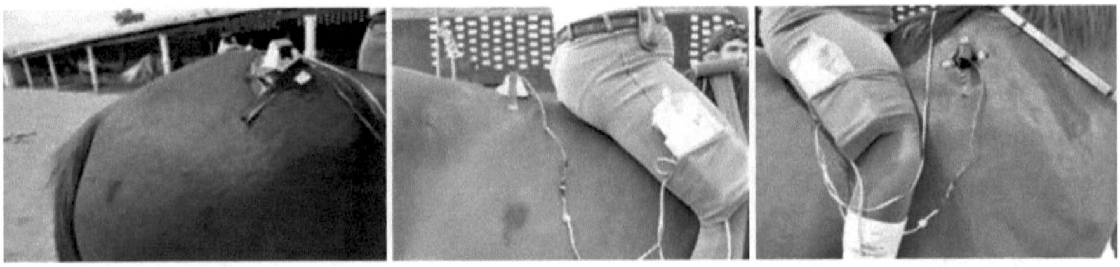

Fig. 1 Accelerometers location on the horse

The features of the accelerometers being used are described below. The model was the ADX-321 of Analog Devices. Its sensitivity was of 100 mV/g, whereas its reachable range was of ±18 g. The reconfigurable bandwidth was from 0.5 Hz to 2.27 kHz. The data acquisition was done using the NI-USB 6009 device of National Instruments. It was programmed from a LabVIEW graphic interface for getting 20 samples per second.

3 Analysis Techniques

The techniques to analyze the body parts movement of the horse were implemented in MATLAB. They are described below:

3.1 Analysis in the Frequency Domain

The recorded data of each sensing body part were analyzed in the frequency domain using the Fourier transform (FT). The FT decompose a time-domain signal into its constituent frequencies, it is given by

$$F(t) = \langle x, e^{i2\pi f_0} \rangle = \int_{-\infty}^{\infty} x(t) e^{-i2_0} dt \qquad (1)$$

where,

$x(t)$ input signal,

$e^{i2\pi f_0}$ exponential complex function,

f_0 fundamental frequency

3.2 Filtering Process

Infinite impulse response (IIR) band pass filters were implemented to accept the fundamental frequencies, including them harmonics and sub-harmonics, and reject all others. The design parameters of these were Chevyshev type II and order 10 and the band-pass width was fixed ±0.2 Hz around the interested frequency.

3.3 Phase Diagram

The stability of a signal can be analyzed using the phase diagram or phase plane. Its definition is based on the Hamilton's principle, where the energy conversion can be represented as

$$H(q,p) = \frac{p^2}{2m} + V(q) \tag{2}$$

q being the displacement and p is the linear momentum. From Eq. (2), it is possible to define a function of the form

$$\varphi_t = [q(t), p(t)] \tag{3}$$

If the function φ_t is continuous and smooth, then the data represents a stable behavior. Applying Eq. (3) to the original data it was possible to identify the smoothness of the horse ride.

4 Results

The movements of the horse were recorded during the walking and trotting gaits in a hippotherapy cycle. Figures 2 and 3 shows the recorded acceleration data for the left hip during the both gaits, respectively. From these the Fourier spectrum were obtained (Figs. 4 and 5) and the fundamental frequencies were found. From Fig. 4 we can see the fundamental frequency for walk gait is around 1.8 Hz, with harmonics and sub-harmonics of 0.9, 2.7, 3.6, 4.5 and 5.4 Hz. Figure 5 shows the fundamental frequency for trot gait is around 2.7 Hz with harmonics and sub-harmonics around 1.35, 4.1, 5.4, 8.1 Hz.

The identified frequencies were filtered from the corresponding original signals via pass-band filters. The results are shown in Figs. 6 and 7.

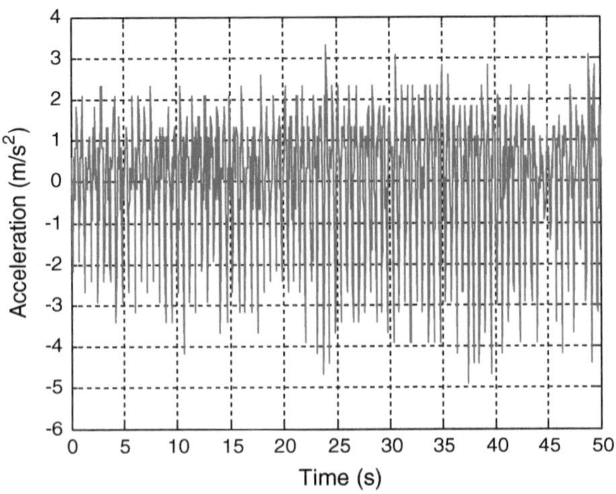

Fig. 2 Original signal of *left hip* during the walking gait

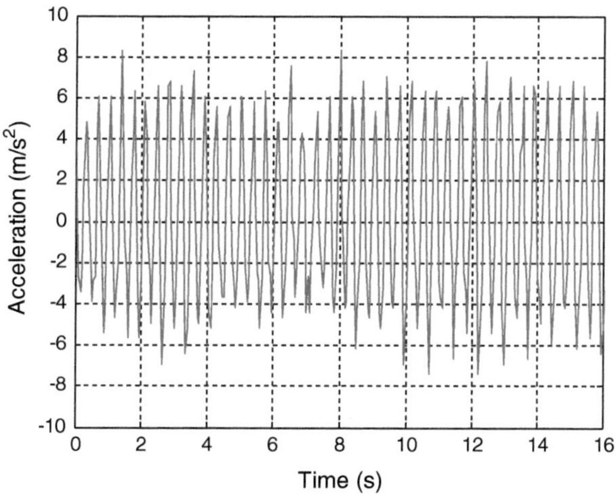

Fig. 3 Original signal of *left hip* during the trotting gait

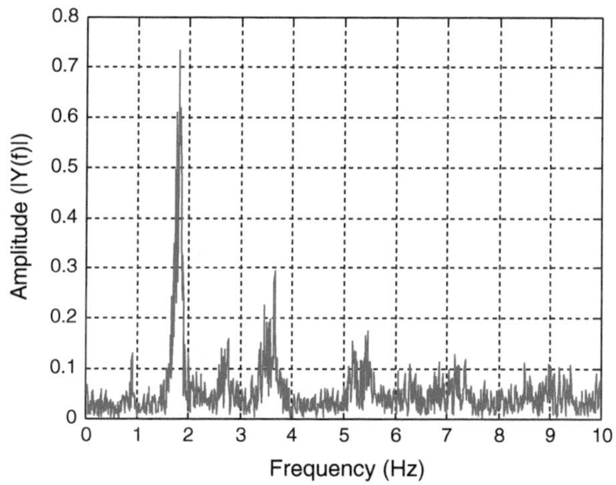

Fig. 4 Fourier spectrum from *left hip* during the walking gait

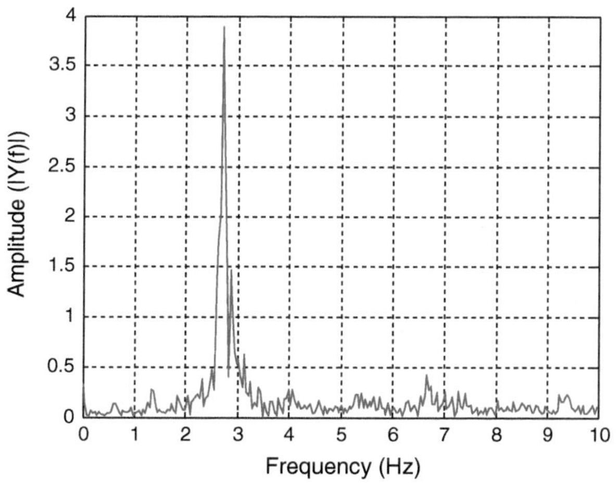

Fig. 5 Fourier spectrum from *left hip* during the trotting gait

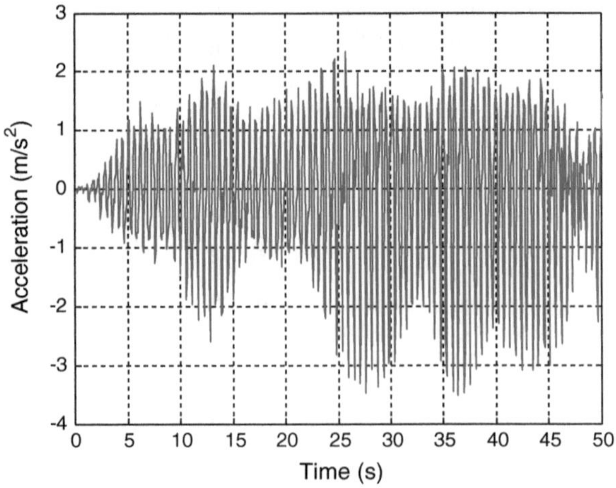

Fig. 6 *Left hip* signal from walking gait after filter processing

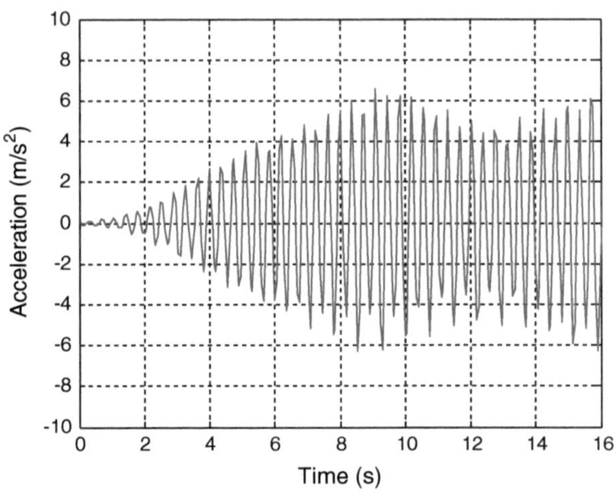

Fig. 7 *Left hip* signal from trotting gait after filter processing

After filtering process, the velocity and displacement were computed numerically and the phase plane were generated from these. The phase diagram for left and right hips during the walking gait are show in Figs. 8 and 9, respectively.

The phase diagrams from left and right hips are shown in Figs. 10 and 11, respectively, during the trotting gait test.

Finally the relative displacement between both articulations was obtained. And the orbit diagram was constructed. The results are show in Figs. 12 and 13.

From Figs. 12 and 13 the relation between the movement of both articulations can be estimated. Therefore, we calculated the angle between the hip articulations. The results for walking and trotting gaits are shown in Figs. 14 and 15, respectively. In these we can see the variation of the displacement in function of the angle and reproduce the movement of the horse in a mechanism of 4DOF. With this data, it is possible to construct a 4 DOF parallel robot capable of reproducing these motions.

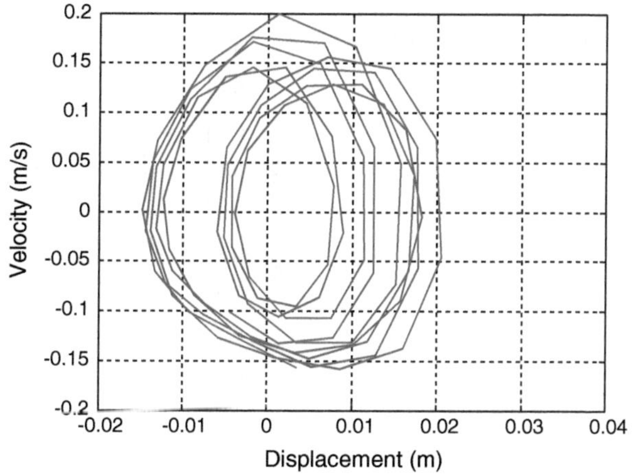

Fig. 8 Phase diagram of *left hip* from the walking gait

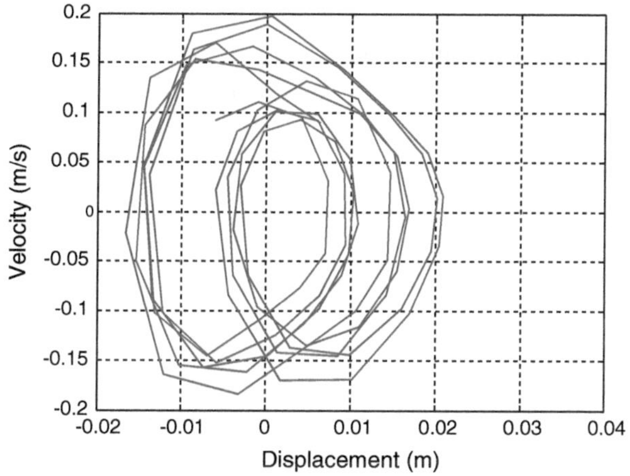

Fig. 9 Phase diagram of *right hip* from the walking gait

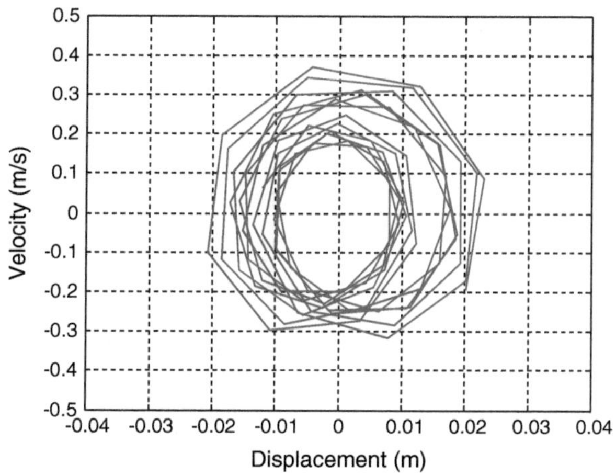

Fig. 10 Phase diagram of the *left hip* from the trotting gait

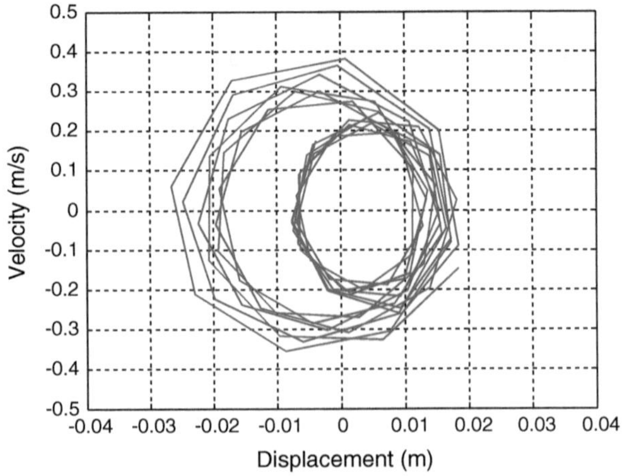

Fig. 11 Phase diagram of the *right hip* from the trotting gait

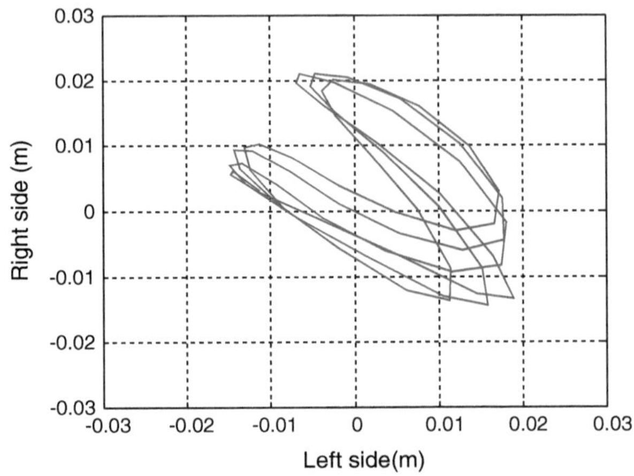

Fig. 12 Relative movement between the hip articulations from the walking gait

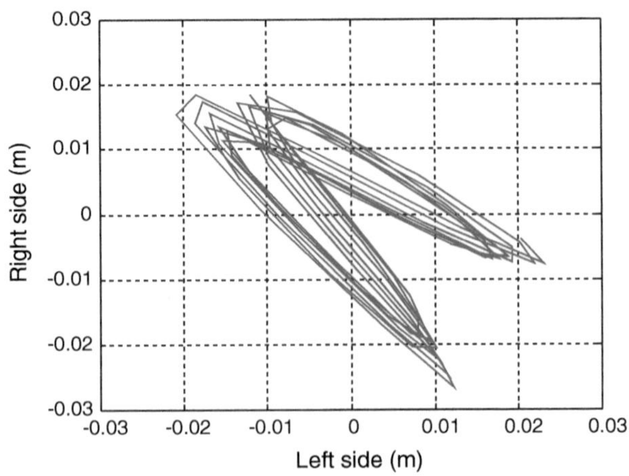

Fig. 13 Relative movement between the hip articulations from the trotting gait

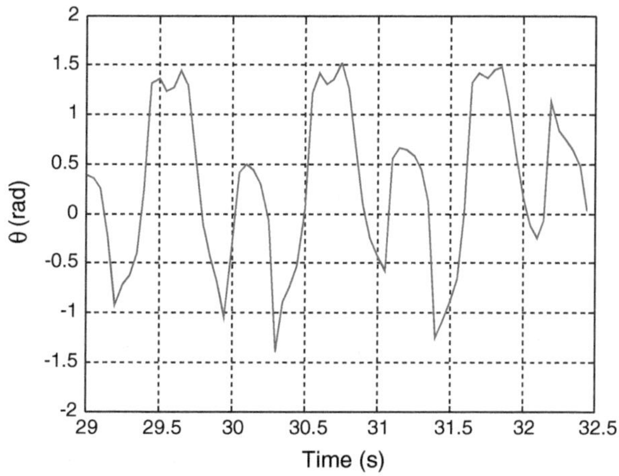

Fig. 14 Angular displacement between the *right and left hips* from the walking gait

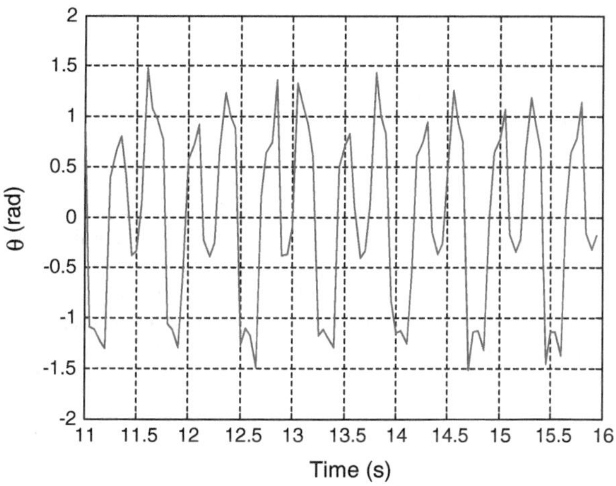

Fig. 15 Angular displacement between the *right and left hips* from the walking gait

5 Conclusions

An experimental methodology for determining some of horse motions used for hippo-therapy was presented. Based on psychological principles, those motions were identified from accelerations measurements. From these, via filtering, Fourier transforms, and numerical techniques the basic motions for hippo-therapy were determined. These motions basically are resumed by a planar trajectory and two tilts that help patient's therapy. Further implementation and design of a 4DOF parallel mechanism for reproducing such basic motions will be carried out.

The horse motion reproduced numerically will be used to program the parallel mechanism, and, in this way, it is possible to bring an equivalent therapy to a wider population.

Acknowledgments Ms. González-Cruz thanks CONACyT for her scholarship for pursuing her doctorate studies. Prof. López-Cajún also acknowledges CONACyT for its support during his sabbatical leave.

References

1. Januma M, Peham C, Dvorakova C, Elfmark M (2009) An assessment of the pressure distribution exerted by a rider on the back of a horse during hippotherapy. Hum Mov Sci 28:387–393
2. Tauffkirchen E (2000) Hippotherapie, neurophyisiologische Behandlungmit und aufdem Pferd, En Kinder Hippotherapie, pp 107–166
3. Bertotti D (1988) Effect of therapeutic horseback riding on posture in children with cerebral palsy. J Phys Ther 68(10):1505–1512
4. Casady RL, Nichols-Larsen D (2004) The effect of hippo-therapy on children with cerebral palsy. Pediatry Psychol Ther 16(3):165–172
5. Davis E, Davies B, Wolfe R, Raadsveld R, Heine B, Thomason P (2009) A randomized controlled trial of the impact of therapeutic horse riding on the quality of life, health, and function of children with cerebral palsy. Dev Med Child Neurol 51:111–119
6. Jeong-Yi K, Hyun JC, Young Lee J, Ha Y, Lee PK, Yun-Hee K (2011) Effects of hippotherapy on gait parameters in children with bilateral spastic cerebral palsy. Arch Psychol Med Rehabil 92:774–779
7. Exner G, Engelmann A, Lange K (1994) Basic principles and effects of hippo-therapy within the comprehensive treatment of paraplegic patients. Rehabilitation 33:39–41
8. Ionatamishvili, NI, Tsverava DM (Nov-Dic de 2002) Horseback riding therapy in development of motor skills in infantile cerebral palsy. Vopr Kurortol Fizioter Lech Fiz Kilt 6:45–47
9. Land G, Errington-Povalac E, Paul S (2001) The effects of therapeutic riding on sitting posture in individuals with disabilities. Occupational Therapy in Health Care 14(1):1–12
10. Shurtleff T, Standeven J, Engsberg J (2009) Changes in dynamic trunk/head stability and functional reach after hippo-therapy. Arch Phys Med Rehabil 90:1185–1195 (Ionatamishvili NI, 2002)
11. Shurtleff T (2010) American Hippotherapy Association. http://www.americanhippotherapy association.org/research/hippotherapy-research-articles/
12. Janura M, Svoboda Z, Dvorakova T, Cabell L, Elfmark M, Janurova E (2012) The variability if a horses' movement at walk in hippotherapy

Identification of a Cylindrical Robot Using Recurrent Neural Networks

Carlos Román Mariaca Gaspar, Juan Eduardo Velázquez-Velázquez and Julio César Tovar Rodríguez

Abstract Neural identification techniques are very useful for the problem of unknown dynamics and uncertainties during the development of a model that accurately represents the behaviour of a robot. In this paper we use the model of a Recurrent Trainable Neural Network (RTNN) for modelling a cylindrical robot. The RTNN proposal is a multilayer network local feedback into the single hidden layer, to approach the robot dynamics. The learning algorithm for this topology is the Backpropagation (BP) dynamic. The simulation results of the approximation obtained through RTNN showed a good convergence and accurate tracking.

Keywords Neural networks · Identification systems · Manipulator robot

1 Introduction

In general, robotic manipulators belong to the class of non-linear time variant systems. For these, manipulators drivers that are used to govern their behavior are advanced and robust type and necessarily require some knowledge about the dynamic behavior of the system under control [10].

In the case of robotic manipulators, most often it is difficult to obtain important parametric information, such as the matrix of inertia and mass center of the joints with sufficient accuracy. The effects of friction and inertia, for example, depend on the state of the system, coupled with the addition of fillers can affect the overall dynamic behavior. In this context, neural identification technique offers the possibility to learn

C.R.M. Gaspar (✉) · J.E. Velázquez-Velázquez · J.C.T. Rodríguez
Instituto Politécnico Nacional, Mexico, Mexico
e-mail: crmariaca@gmail.com

J.E. Velázquez-Velázquez
e-mail: jvelazquez@ctrl.cinvestav.mx

J.C.T. Rodríguez
e-mail: jctovar77@hotmail.com

M. Ceccarelli and E.E. Hernández Martinez (eds.), *Multibody Mechatronic Systems*,
Mechanisms and Machine Science 25, DOI 10.1007/978-3-319-09858-6_36

non-linear behavior through the presentation of patterns of input/output suitable to help overcome the difficulties mentioned above training. As reported in the literature, the results of several studies show that artificial neural networks (ANN) seem to be a very effective tool for the identification and control of nonlinear systems when you have partial or no information about them [8]. The ability to approximate a complex nonlinear model without prior knowledge of its dynamics makes the ANN model a very attractive alternative for modeling and control techniques. This property has been demonstrated by the universal approximation theorem [5].

The most widely used network architectures are Feedforward Neural Network (FFNN) and Recurrent Neural Network (RNN). In a FFNN signals are transmitted in one direction from the input layer and then through the hidden layers to the output layer and requires the application of global unit delay to achieve a nonlinear auto-regressive average model of the robot. RNN has local feedback connections in some of its hidden layers. This structure is a suitable alternative to FFNN when the task is to develop a model of a dynamic system. As mentioned above, the main advantage of the RNN in modeling is its ability to adaptively learn nonlinear functions whose forms of analysis are difficult to obtain, and the solutions are difficult to calculate. This is reflected in the widespread use of neural approaches to robotics problems; see [4, 6, 9]. In most of these works the drawback of using static NN higher order as they have a high complexity that makes difficult its application manifests. In [1—3], Baruch et al. defines the architecture of Recurrent Trainable Neural Network (RTNN) with backpropagation learning algorithm (BP) which simplified the problem of identification and control applications in different plants of biotechnological and mechanical nature. In the present work, this RTNN is used for nonparametric identification of a cylindrical robot modeled by a set of nonlinear differential equations of second order. The outline of the paper is as follows: In Sect. 2 an analytical model that describes the rigid body dynamics of a cylindrical robot manipulator is introduced. The description of the RTNN and learning algorithm is presented in Sect. 3. Simulation results of the identification of plants using RTNN are presented in Sect. 4. Finally, Sect. 5 draws together the conclusions of the document.

2 Analytical Model of a Cylindrical Robot and Problem Formulation

Through time several researchers have developed a variety of methods to derive the equations representing the dynamics of a robot manipulator. This section provides a general set of differential equations describing the time evolution of a cylindrical robot is presented. These are the Euler-Lagrange equations of motion.

The Lagrangian formulation is based on the generalized coordinates of strength and energy. If the manipulator moves freely in the working space then, the dynamic equation of motion of the robot is given in the form of the following matrix with n degrees of motion freedom:

$$M(q)\ddot{q} + C(q,\dot{q})\dot{q} + G(q) = \tau \tag{1}$$

where τ is the vector of dimension $n \times 1$ generalized torque applied to the joints q, $\dot{q}$ and $\ddot{q}$ are the vector of variables of the joints of the manipulator robot, its first and second derivatives, respectively. The first term $M(q)$ is the inertia matrix of dimension $n \times n$. The second term $C(q, \dot{q})$ is the matrix of centrifugal and Coriolis forces. The last term $G(q)$ is the vector that represents the action of gravitational forces.

The robot manipulator is a cylindrical three degrees of freedom, including two prismatic joints and a rotational joint. The common variables are the cylindrical coordinates of the end effector relative to the base. The cylindrical configuration is shown in Fig. 1.

In this case, the vector of generalized coordinates is $q = [q_1, q_2, q_3]^T$. Following the methodology of Euler-Lagrange following inertia matrix is obtained:

$$M(q) = \begin{bmatrix} I_{1zz} + I_{2yy} + I_{3yy} + m_3 q_3^2 & 0 & 0 \\ 0 & m_2 + m_3 & 0 \\ 0 & 0 & m_3 \end{bmatrix} \tag{2}$$

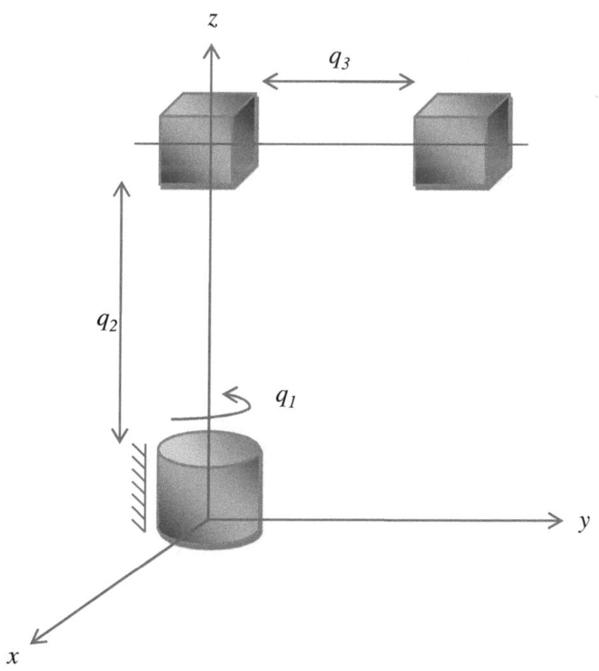

Fig. 1 Configuration of the cylindrical robot

The matrix of Coriolis forces is

$$C(q,\dot{q}) = \begin{bmatrix} m_3 q_3 \dot{q}_3 + f_{fric-1} & 0 & m_3 q_3 \dot{q}_1 \\ 0 & f_{fric-2} & 0 \\ -m_3 q_3 \dot{q}_1 & 0 & f_{fric-3} \end{bmatrix} \tag{3}$$

The vector of gravitational forces is $G(q) = [0,0,0]^T$ and vector torque is $\tau = [\tau_1, f_2 - (m_2 + m_3)g, f_3]^T$. Here q_1, q_2 and q_3 denote the angular position of the rotational joint and the displacement of the two prismatic joints respectively. For $i = 2$, 3, m_i denotes the mass of each link of the manipulator robot, located at the center of each link. $I_{1zz}, I_{2yy}, I_{3yy}$ are the inertia tensors of the link i around the mass center of the link i. τ_1, f_2 and f_3 denotes the torque and the vertical and horizontal forces, respectively. g is the acceleration due to gravity. f_{fric-i} denotes the frictional force on the board i. In fact, following the Newton-Euler method achieves the same dynamic model of the cylindrical robot [7].

Now, considering that $y = [q,\dot{q}]^T$ is the output of the manipulator dynamic system (1) when a generalized torque vector τ is applied to the cylindrical robot, then formulate the identification problem:

Suppose that the dynamic equation of motion (1) for cylindrical robot manipulator is unknown. The identification problem is to obtain a model that allows us to infer how the cylindrical manipulator robot responds to other inputs still unknown by approximating the output path of the manipulator system. That is, for an experiment of duration t_f, we want to determine a model with an output $\hat{y} = [q, \hat{\dot{q}}]^T$ that approximates the response used during the experiment, using only the observed data $\{\tau, y, t \in [0, t_f]\}$.

3 Description of the Topology and RTNN Learning

With the help of the diagrammatic method [11] and following Fig. 2, we present the topology of a RTNN described as vector-matrix as

$$\begin{aligned} \hat{x}(k+1) &= A\hat{x}(k) + B\hat{u}(k), \\ \hat{y}(k) &= F[v(k)], \\ v(k) &= Cz(k), \\ z(k) &= G[\hat{x}(k)]; \end{aligned} \tag{4}$$

where $F[\cdot]$ and $G[\cdot]$ are activation function vectors usually $\tanh(\cdot)$; $\hat{y}$, $\hat{x}$, $\hat{u}$ are the vectors of output, state and input RTNN with dimension p, N, $n+1$, respectively; here $\hat{u}^T = [\tau, u_0]$, where τ is the vector of generalized torque input and $u_0 = -1$ is an entry threshold. Here the output vector of the plant y is considered as the reference output for RTNN; A is a matrix of dimension $n \times n$ diagonal blocks,

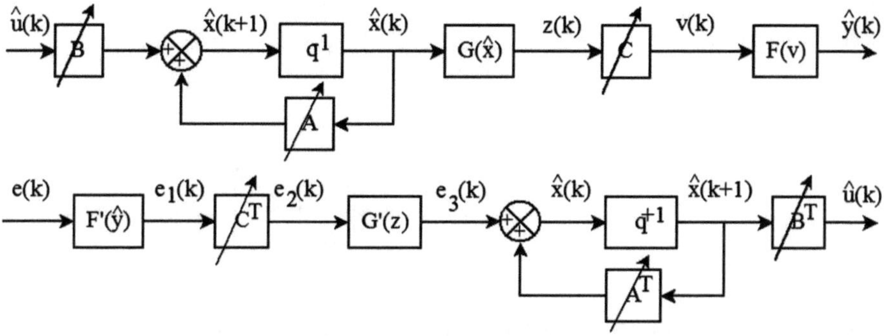

Fig. 2 Block diagram of the RTNN and attached version

defined as $A = block - diag(A_i)$; $B = [B_1, B_0]$ and $C = [C_1, C_0]$ are augmented matrices $N \times (m+1)$ and $L \times (N+1)$ respectively; B_0 and C_0 are threshold vectors of dimension $n \times 1$ and $l \times 1$ of hidden and output layers respectively.

The associated learning algorithm is described as:

$$\begin{aligned} w(k+1) &= w(k) + \eta \Delta w(k) + \alpha \Delta w_{ij}(k-1), \\ e(k) &= y(k) - \hat{y}(k), \end{aligned} \quad (5)$$

where w is a generalized weight matrix denoting each of the matrices (A, B, C) in the model RTNN to be updated; Δw is the correction of w, η and α are learning parameters; e is an error vector of appropriate dimensions, calculated from the attached model RTNN (Fig. 2). Correction of associated weights and the corresponding error vectors are:

$$\begin{aligned} e_1(k) &= F'[\hat{y}(k)]e(k), \\ \Delta C(k) &= e_1(k)z^T(k), \\ e_2(k) &= C^T(k)e_1(k), \\ e_3(k) &= G'[z(k)]e_2(k), \\ \Delta B(k) &= e_3(k)\hat{u}^T(k), \\ \Delta A(k) &= e_3(k)\hat{x}^T(k), \end{aligned} \quad (6)$$

where $F'[\hat{y}(k)] = [1 - \hat{y}^2(k)]$ and $G'[z(k)] = [1 - z^2(k)]$ are derived from the operation of activation functions; RTNN stability is ensured by the activation functions within the range $[-1, 1]$ bounds and by the local stability condition imposed on the weight matrix $|A_i| < 1$.

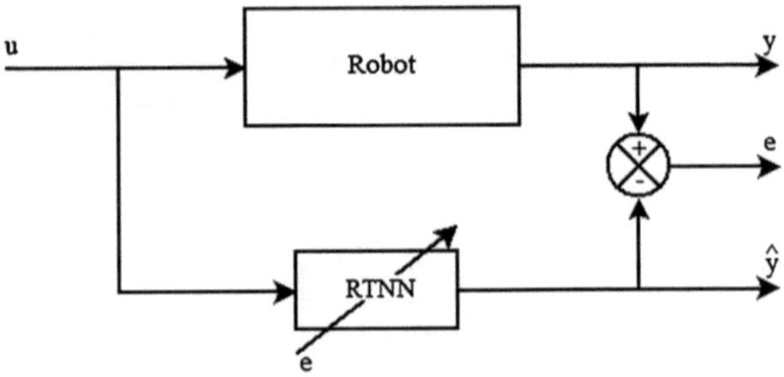

Fig. 3 Block diagram of model identification

4 Simulation Results for Identification of the Cylindrical Robot Using RTNN

Under the premise of identification problem, this section is devoted to obtain an RTNN to approximate the robot dynamics. Here we use the RTNN defined in (4), (5) and (6) set out in the previous section. The block diagram of the identification process is illustrated in Fig. 3.

Because of the necessity of experimental data of the cylindrical manipulator robot $\{\tau, y, t \in [0, t_f]\}$ we developed a computer simulation of the dynamics of the robot implemented on MATLAB® with the following parameter values: $I_{1zz} + I_{2yy} + I_{3yy} = 2.5\ (\mathrm{kg\,m^2})$, $g = -9.8\ (\mathrm{ms^{-2}})$, $m_2 = 5$ (kg), and $m_2 = 2.5$ (kg) taken from the literature [10]. It was also considered that $f_{fric-1} = 0.1\ (\mathrm{kg\,s^{-1}})$, $f_{fric-2} = 0.2\ (\mathrm{kg\,s^{-1}})$, $f_{fric-3} = 0.3\ (\mathrm{kg\,s^{-1}})$ and duration time $t_f = 50$ (s) with a sampling time of $t_f = 0.01$ (s). The generalized torque vector is taken as input $\tau = [\tau_1, f_2 - (m_2 + m_3)g, f_3]^T$, where

$$\begin{aligned} \tau_1 &= 5\sin\left(\frac{\pi}{4}t\right), \\ f_2 &= \frac{\pi}{3}\left(1 - \mathrm{e}^{-2t^3}\right) + \frac{\pi}{6}\left(1 - \mathrm{e}^{-2t^3}\right)\sin(3t), \\ f_3 &= \frac{\pi}{2}\left(1 - \mathrm{e}^{-2t^3}\right) + \frac{\pi}{3}\left(1 - \mathrm{e}^{-2t^3}\right)\sin(2t). \end{aligned} \tag{7}$$

It is worth to be mentioned that in order to obtain appropriate experimental data $\{\tau, y, t \in [0, t_f]\}$ we implement an action state feedback along the input vector τ. The data obtained are shown in Figs. 4 and 5.

Note that the range of activation functions $\tanh(\cdot)$ lies in the range $[-1, 1]$. Then, to approximate functions with a range outside such interval it is necessary to normalize these functions otherwise, the RTNN cannot approximate functions outside the range $[-1, 1]$. Thus, the exit path of the cylindrical robot is normalized.

Now, in order to illustrate the identification framework applied to the cylindrical robot, the nonlinear system (1)—(3) is considered as unknown. Thus, we develop a computer simulation using the identification strategy with a RTNN topology (4-10-6) (4 inputs, 10 neurons in the hidden layer, 6 outputs). The learning rate parameters

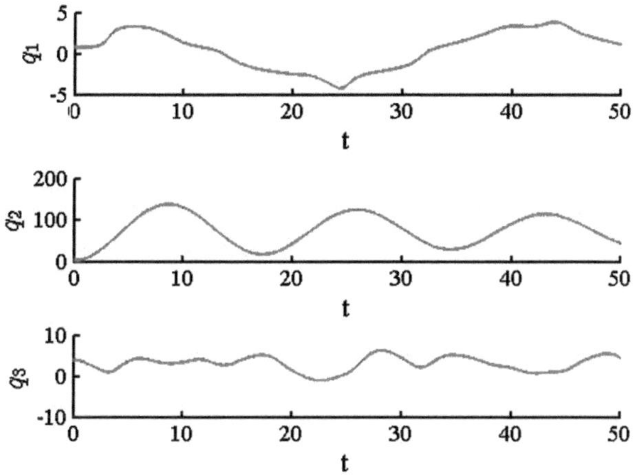

Fig. 4 Generalized position variables of joints of a robot manipulator

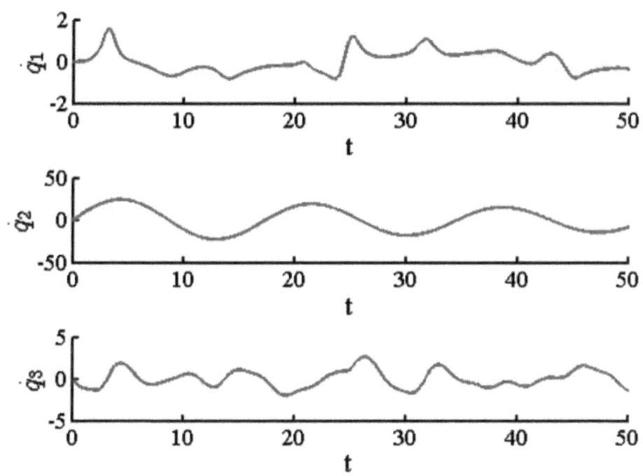

Fig. 5 Generalized variables of velocity joints for a robot manipulator

for the BP learning algorithm are $\alpha = 0.1$ and $\eta = 0.001$. Simulation results using RTNN identification are obtained online for a simulation time of 100 (s) with a step size of 0.01 (s). Identification entries are defined by (7) with a final time $t_f = 100$ (s).

It is noteworthy that the evolution of $\hat{q}$ was obtained and compared against corresponding reference signals q shown in Fig. 6.

The identification results for velocity $\hat{\dot{q}}$ and its comparison with the corresponding reference signal $\dot{q}$ are shown in Fig. 7.

The simulation results show the effectiveness of the proposed RTNN. This can be seen in the graph of the performance index given in terms of mean square error, where high accuracy is obtained convergence, see Fig. 8.

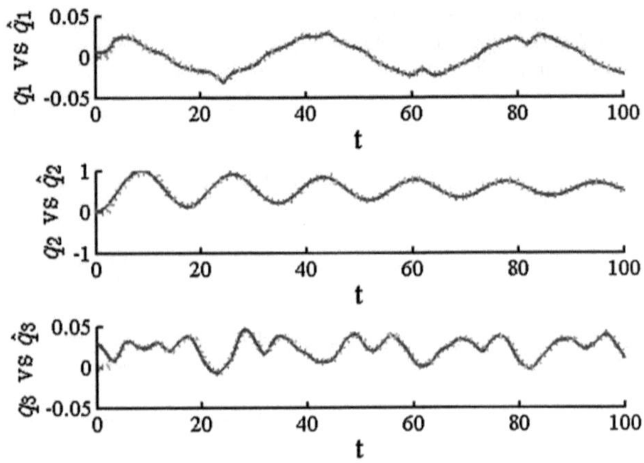

Fig. 6 Identification of the position of the robot joints: reference (*solid line*) and estimated (*dotted line*)

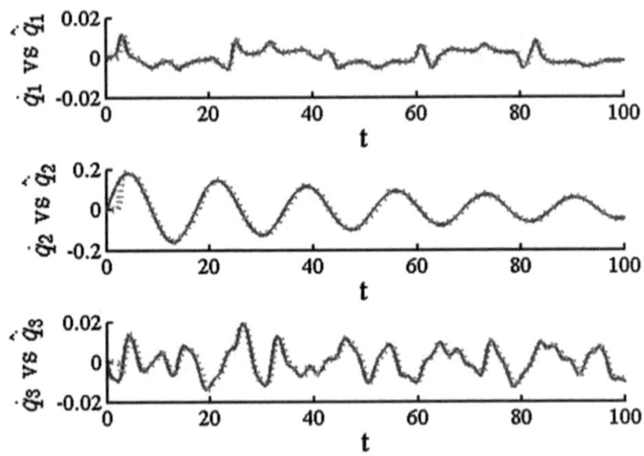

Fig. 7 Identification of the velocity of the robot joints: reference (*solid line*) and estimated (*dotted line*)

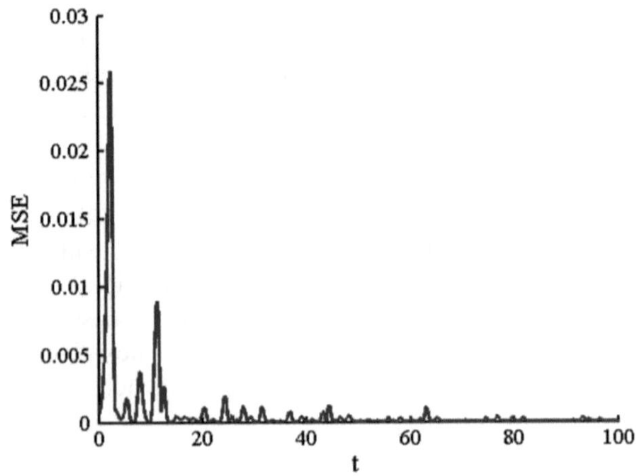

Fig. 8 MSE obtained using the proposed RTNN

5 Conclusions

In this paper the problem of identifying a cylindrical robot manipulator is solved using the RTNN. The use of neural networks allows obtaining an approximate model to infer how cylindrical robot manipulator responds to unknown inputs. The efficiency of the proposed algorithm is illustrated in the simulations where precise tracking and acceptable convergence is observed. Having a model that predicts the behavior of the robot when there is not a model described by a set of differential equations, will allow the development of controllers that use the benefits of RTNN to govern the behavior of the robot manipulator without necessarily having any knowledge of the dynamics of the system under control.

Acknowledgments The authors acknowledge the support provided to the *Secretaría de Investigación y Posgrado* of IPN (SIP-IPN), *Comisión de Operación y Fomento de Actividades Académicas* of IPN (COFAA-IPN), *Consejo Nacional de Ciencia y Tecnología* (CONACyT).

References

1. Baruch IS, Flores JM, Nava F, Ramirez IR, Nenkova B (2002) An advanced neural network topology and learning, applied for identification and control of a D.C. motor. In: Proceedings of 1st international IEEE symposium on intelligent systems, pp 289–295
2. Baruch IS, Barrera-Cortes J, Hernandez LA (2004) A fed-batch fermentation process identification and direct adaptive neural control with integral term. In: Monroy R, Arroyo-Figueroa G, Sucar LE, Sossa H (eds) MICAI 2004: advances in artificial intelligence, LNAI, vol 2972. Springer, Heidelberg, pp 764–773
3. Baruch IS, Georgieva P, Barrera-Cortes J, Feyo de Azevedo S (2005) Adaptive recurrent neural network control of biological wastewater treatment. Int J Intell Syst 20:173–194 (Wiley Periodicals, Inc. N. J.)
4. Ge SS, Lee TH, Harris CJ (1998) Adaptive neural network control of robotic manipulators. World Scientific Publishing Co., London
5. Haykin S (1999) Neural networks a comprehensive foundation. Prentice Hall, Upper Saddle River
6. Horne B, Jamshidi M, Vadiee N (1990) Neural networks in robotics: a survey. J Intell Robot Syst 3:51–66
7. Khosla P, Kanade T (1987) An algorithm to estimate manipulator dynamics parameters. Institute for Software Research, Irvine Paper 653
8. Narendra KS, Parthasarathy K (1990) Identification and control of dynamic systems using neural networks. IEEE Trans Neural Networks 1:4–27
9. Patino HD, Carelli R, Kuchen BR (2002) Neural networks for advanced control of robot manipulators. IEEE Trans Neural Networks 13:343–354
10. Spong MW, Vidyasagar M (1989) Robot dynamics and control. Wiley, New York
11. Wan E, Beaufays F (1996) Diagrammatic method for deriving and relating temporal neural networks algorithms. Neural Comput 8:182–201

New Design of Petal Type Deployable Space Mirror

V.I. Bujakas

Abstract A new kinematic design for a large petal-type space mirror package and deployment is considered. To check the approach computer and physical models of the multibody transformable structure were developed and studied. Results of simulation are presented in the paper.

Keywords Large space mirror · Kinematics of deployment · Computer and physical simulation

1 Introduction

Large solid deployable mirrors, intended for operation at mm, submm, IR and optical wavelengths are currently being developed within a number of programs and space projects [2–4, 8]. The classic design of petal type deployable reflector was proposed by Dornier Corporation during "FIRST" space project development ("FIRST"—far infrared space telescope) [7]. The similar design later was used for 10 m antenna of "Radioastron" space telescope and operates at cm wavelengths [1, 5]. Within this design a petal type reflector is a transformable structure and contains a central mirror and a set of petals. In folded state the petals are disposed in vertical position above the central mirror. In open state the petals and the central mirror form large parabolic reflector. Stages of petal type mirror deployment are presented in Figs. 1 and 2.

From kinematic point of view each petal is connected with central mirror by a cylindrical hinge and rotates around the axis of the hinge during deployment. It has been shown that there exist such directions of axes of the hinges under which synchronous rotation of the petals leads to the unfolding of the mirror without collision. Synchronous rotation of the petals is carried out by actuators, which are located on the back side of the central mirror.

V.I. Bujakas (✉)
P. N. Lebedev Physical Institute of Russian Academy of Science, Moscow, Russia
e-mail: bujakas@yandex.ru

M. Ceccarelli and E.E. Hernández Martinez (eds.), *Multibody Mechatronic Systems*, Mechanisms and Machine Science 25, DOI 10.1007/978-3-319-09858-6_37

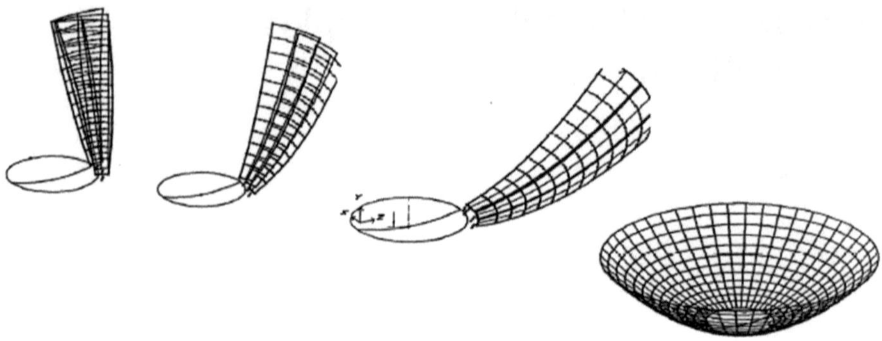

Fig. 1 The geometry of deployment

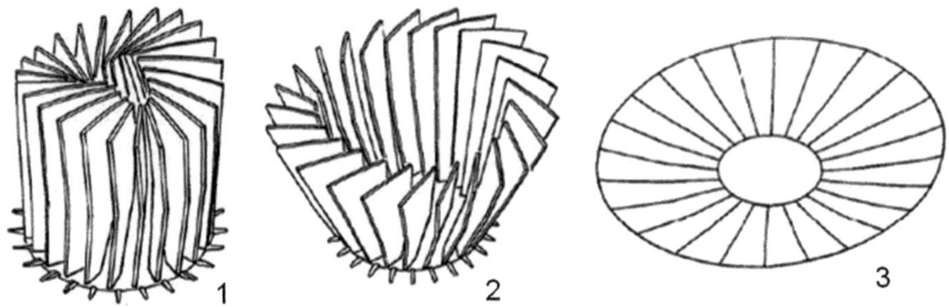

Fig. 2 Petal type mirror opening

However, there are drawbacks that limit the use of this kinematic scheme to create large deployable mirrors for short waves.

Main lacks of the design are as follows:

- after deployment the petals are not tied together at the outer rim of the mirror, therefore the rigidity of open reflector is not very high (the geometric rigidity of closed shell did not used),
- errors in the position of the axes of the cylindrical hinges and inaccuracies of petals opening distort the shape of the mirror surface, the value of which increases from the axis of rotation to the periphery of the mirror; however this part of mirror makes a decisive contribution to the effective area of the reflector and for short wave mirror must be made very precise.

Therefore attempts were made, unfortunately unsuccessful, to tie the adjacent petals together via locks along outer boundary of the mirror after deployment.

However drawbacks of ties after deployment are:

- uncontrolled disturbance of reflective surface arise and
- deployed mirror transforms in statically indeterminate, strained design, what is extremely undesirable for precision reflectors.

2 The New Kinematics for Mirror Deployment

To overcome these shortcomings another version of petal type mirror deployment kinematics was proposed and developed [6]. We propose to retain the link between adjacent petals during the whole deployment process. To implement this approach, it is necessary first of all to abandon the cylindrical hinges connecting the petals with central mirror, replacing them with links that provide a greater degree of freedom of the petals. We considered a structure in which each petal is connected to the central mirror by two cylindrical hinges (Fig. 3).

Moreover the vertex of the petal is connected to the edge of the adjacent petal with connections that introduce to the structure two kinematic constraints (Figs. 4 and 5). Miniature linear actuators are used for deployment. During the deployment the miniature linear actuator moves of the petal along the outer edge of the adjacent petal (Fig. 5).

The structure contains n petals and central mirror.

Let

N total number of solid elements in the structure

$$N = n + 1,$$

S_1 the number of kinematic constraints introduced by links between petals

$$S_1 = 2n,$$

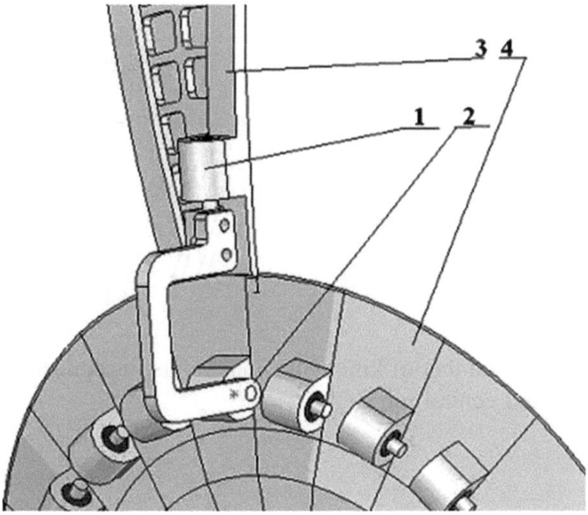

Fig. 3 Links between petal and central mirror. *1*, *2* cylindrical hinges, *3* petal, *4* central mirror

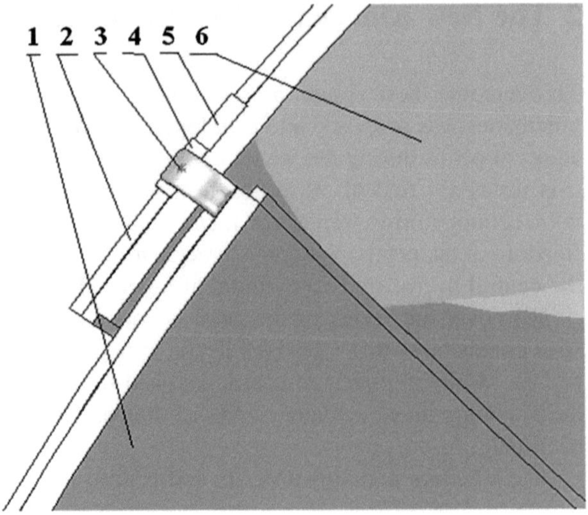

Fig. 4 Links between petals. *1*, *6* petals, *2* guide, *3* rod, *4* slider, *5* linear actuator

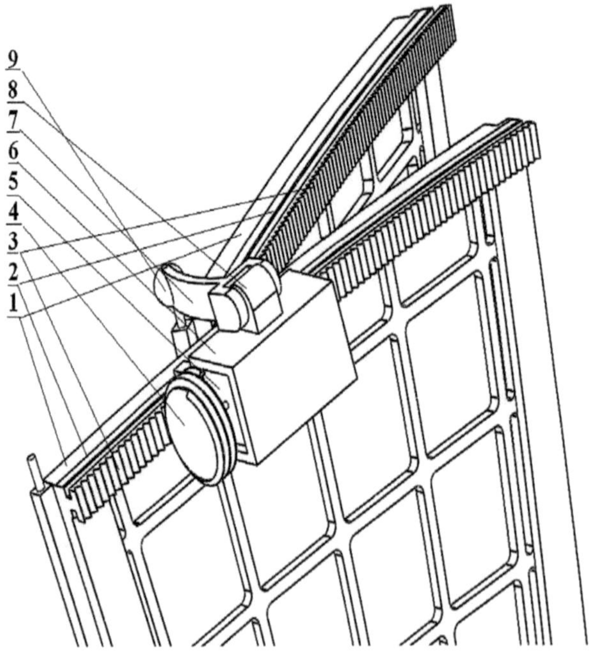

Fig. 5 Computer model of deployment mechanism. *1* petals, *2* guide, *3* worm rail, *4* screw, *5* motor, *6* holder of the motor (slider), *7* rod, *8* cylindrical hinge, *9* spherical hinge

S_2 the number of kinematic constraints introduced by the links between the petals and central mirror

$$S_2 = 4n,$$

S total number of kinematic constraints in the structure

$$S = S_1 + S_2 = 6n.$$

Thus Maxwell condition

$$S = 6N - 6$$

(necessary condition of statically determinability) in the structure is fulfilled. Solid Work model examination shows the structure remains during deployment statically determinate. Therefore the structure remains stress free during the deployment.

3 Solid Work Simulation

To check the approach the computer model within Solid Work package was made and studied. Some details of simulation are shown in Figs. 6 and 7.

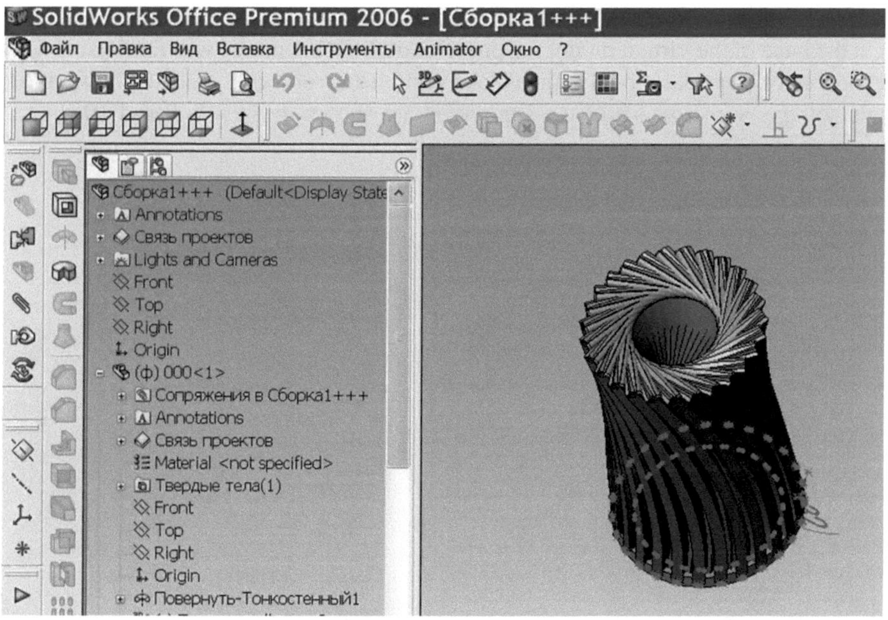

Fig. 6 Computer model of deployable reflector in closed state

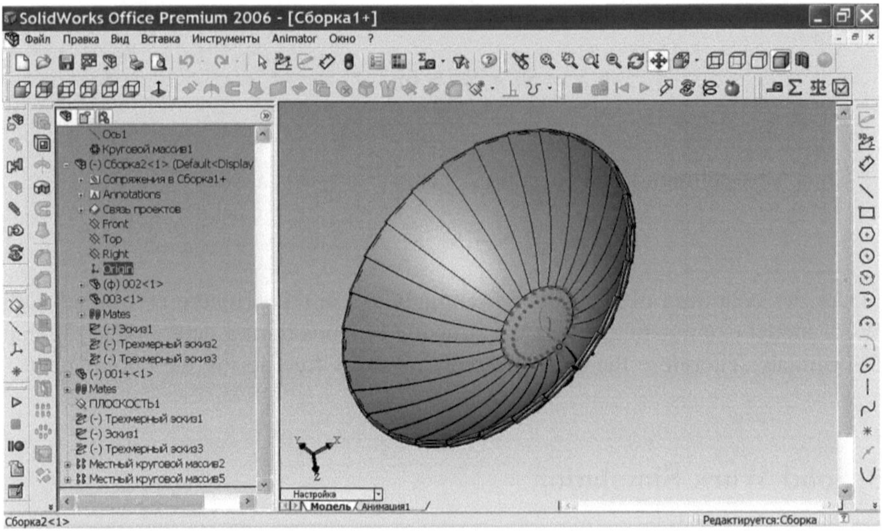

Fig. 7 Computer model of deployable reflector in open state

4 Physical Simulation

On the base of the virtual model the physical deployable model was fabricated. Main elements of physical model – petals and central mirror- were made by 3D printing technology. Some details of the simulation are shown in Figs. 8, 9 and 10, 11.

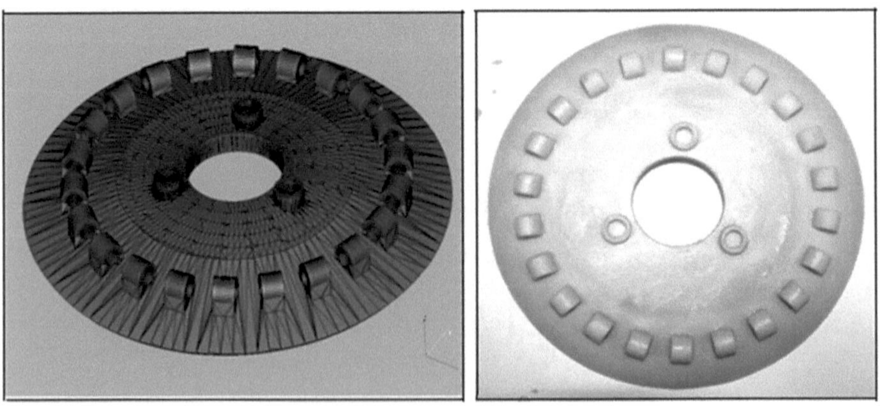

Fig. 8 Computer and physical models of central mirror

Fig. 9 Computer and physical models of petal

Fig. 10 Physical model of deployable reflector in closed and open state

Fig. 11 Physical model of the deployment mechanism, made using 3D printing technology. *1* petal, *2* guide, *3* worm rail printed together with petal as a whole, *4* screw, *5* motor, *6* holder of the motor (slider)

5 Conclusions

To increase surface accuracy and rigidity of large deployable space mirror a new kinematic scheme for petal-type mirror package and deployment was proposed and developed. Computer and physical models of deployable reflector were made and studied. Simulation confirmed the feasibility of the proposed technical solution. Results may be useful for advanced space projects development.

References

1. http://www.asc.rssi.ru/radioastron
2. http://www.asc.rssi.ru/millimetron
3. http://www.jwst.nasa.gov
4. http://safir.jpl.nasa.gov/whatIs.shtml
5. Arkhipov MY, Baryshev AM, Kardashev NS et al (2008) Deployable antennas for space radio telescope: radioastron and millimetron missions. In: Proceedings of the 30th ESA antenna workshop. Noordwijk, Netherlands
6. Bujakas VI (2013) Deployable antenna. Patent of Russian Federation №126199
7. Dornier, FIRST Technology study: multisurface control mechanism for a de- ployable antenna. Final report. Dornier Report RP-FA-D003, 1987
8. Kardashev NS, Bujakas VI et al (2007) The millimetron project. Radioastrono- mical Tools and Techniques, Cambridge

Corradino D'Ascanio and His Design of Vespa Scooter

M. Ceccarelli and G. Teoli

Abstract In this paper an overview is presented of the figure of Corradino D'Ascannio (1891–1981), who was inventor and designer of important mechanical designs. In particular, the paper discusses the successful design of the Vespa scooter by reporting an illustrated survey of the patent and a numerical analysis of a basic innovative performance.

Keywords History of machine design · History of designers · D'ascanio · Vespa scooter · Student paper

1 Introduction

Scooters have been developed from bicycles since their beginning, [1, 2]. The first motorcycle is the velocipede Michaux-Perreaux that was built in France in 1871 by using the first successful pedals bicycle on which an external combustion engine was installed. In the same year in the United States Roper steam velocipede was developed with double-cylinder engine. Later internal combustion engines were used. The first scooter was the Reitwagen design that was conceived and built by Daimler and Maybach in 1885, [3]. At the beginning of the1900s' motorcycles were designed with characteristics different from the previous ones as in the form of scooters. The first characteristic of a scooter is the platform whose design permits a driver to sit with legs in a more comfortable position for a more relaxed driving.

This paper is based on the work of Gennaro Teoli for the bachelor thesis of the first in mechanical engineering at University of Cassino in 2013.

M. Ceccarelli (✉) · G. Teoli
LARM, University of Cassino and South Latium, Cassino, Italy
e-mail: ceccarelli@unicas.it

G. Teoli
e-mail: teoli.gennaro@libero.it

M. Ceccarelli and E.E. Hernández Martinez (eds.), *Multibody Mechatronic Systems*, Mechanisms and Machine Science 25, DOI 10.1007/978-3-319-09858-6_38

A first scooter was the Auto-Fauteuil design that was made in France in 1902 by Gauthier. Auto-Fauteuil brings all features of a modern scooter such as small wheels, open frame, engine placed under the seat. Similar projects were carried out both in Europe and in America, but the one that most resembled modern scooter is the Unibus of 1920.

In this paper the history of VESPA scoter is presented by looking at his conceive and by analyzing basic performance that can explain the success with a mass production.

2 Biographical Notes of Corradino D'Ascanio

Corradino D'Ascanio, Fig. 1, was born 1 February 1891 in Popoli (Pescara), [4, 5]. He attended the Royal school of Engineering in Turin, where he got a degree in mechanical engineering in 1914.

His first engineering experience was in the military aviation during the First World War, when he manifested all his creativity by working out innovative solutions in many accidents. Then he started to work for Pomilio company and he went in the United States in 1917 to get experiences in aeronautic fields. In September 1919 he come back to Italy and he begun the study of vertical flight. In 1921 he married Paola Paolini, who gave him two sons, Giacomo in 1922 and Giorgio in 1927.

After long search and study on vertical flight in 1925 he founded a company together with baron Trojani for building a first helicopter DAT 1, which was improved later in DAT 2 design. Both the helicopters broke during fly tests. In 1929 he designed and built the DAT 3 version, Fig. 2. This helicopter was able to fly and it achieved several internationals primates. Nevertheless it was not a financial success and D'Ascanio lost lot of money in this project. In 1932 he patented a variable pitch propeller and he started a collaboration with Piaggio company for a

Fig. 1 A photo of Corradino D'Ascanio in 1955, [8]

Fig. 2 A flight test of PD 3 helicopter in the 1930s' [4]

screw design for propellers that were very successful. With the Piaggio company he developed the helicopter designs PD1, PD2 and PD3, although no one went into production.

After the Second World War Piaggio company converted the war production in a civil production. Thus, they set aside the helicopter design programs and proposed to D'Ascanio the design of a new motorcycle. D'Ascanio never designed a motorcycle before, but he recognized the traditional motorcycles uncomfortable and difficult to drive. Finally, in 1946 Piaggio started the production of Vespa scooter, Fig. 3, that was designed of D'Ascanio, [6–8].

D'Ascanio's notoriety grows enormously and in 1948 he was invited from the American Helicopter Society to take part to IV congress on helicopter technology in Philadelphia, where he was received like a pioneer of vertical flight. In 1951 he designed last helicopter PD4 for Piaggio company, but even PD4 design was not successful and never was produced.

Corradino D'Ascanio died on 5 August 1981 in Pisa and he is buried in Popoli.

The long life of D'Ascanio was characterized by his creativity as based on a great interest of technical knowledge. His great ingenuity is documented by a

Fig. 3 A photo of t he Piaggio Vespa scooter of the 1946 [6]

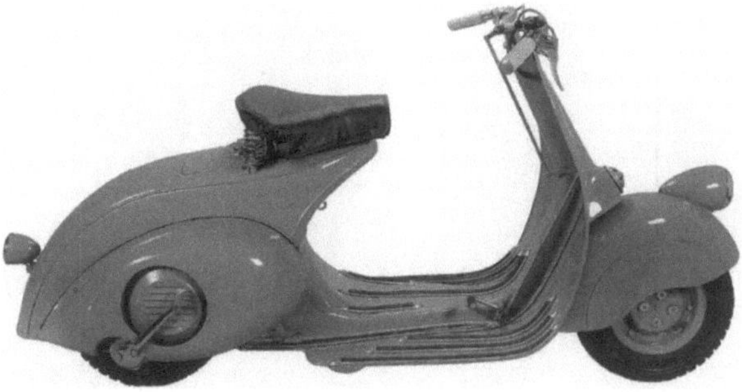

considerable production of new designs for several different systems, among which the Vespa scooter designs and helicopter designs are the most relevant.

The most significant patents by D'Ascanio can be listed as, [4, 5]:

1919 Oven with medium heat capacity for cooking bread and pastry.
1925 Electro-pneumatic machine for cataloging and searching of documents;
1925 Helicopter with two coaxial propellers with slow automatic lowering device.
1929 Helicopter with auto stability and controlled flight.
1932 Variable pitch propeller.
1935 DAT 2 Helicopter Piaggio-D'Ascanio 2.
1939 DAT 3 Helicopter Piaggio-D'Ascanio 3.
1946 scootet Vespa 98.
1948 Vespa 125 and Ape van design with three-wheels.
1951 Two three-blade contra-rotating rotors with synchronized tandem design.
1955 Motor bicycle frame with v-shaped beam and scooter Vespa 400.
1963 Two-seat dual-control glider for training helicopter pilots.

3 The Design of Vespa Scooter

On April 23, 1946 at Florence's Patent Office Piaggio & c. enterprise submitted the patent for a scooter model entitled "Motorcycle to rational complex of organs and elements combined frame with fenders and hood, covering the whole mechanical part", Fig. 4, [9].

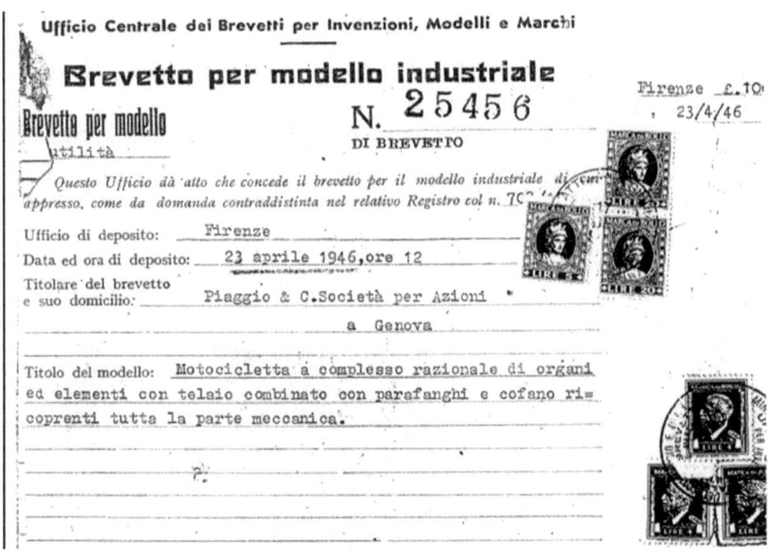
Ufficio Centrale dei Brevetti per Invenzioni, Modelli e Marchi

Brevetto per modello industriale

Firenze £.10

, 23/4/46

Brevetto per modello utilità

N. 25456 DI BREVETTO

Questo Ufficio dà atto che concede il brevetto per il modello industriale di cui appresso, come da domanda contraddistinta nel relativo Registro col n. 70...

Ufficio di deposito: Firenze

Data ed ora di deposito: 23 aprile 1946, ore 12

Titolare del brevetto e suo domicilio: Piaggio & C.Società per Azioni a Genova

Titolo del modello: Motocicletta a complesso razionale di organi ed elementi con telaio combinato con parafanghi e cofano ricoprenti tutta la parte meccanica.

Fig. 4 The patent of Vespa scooter in 1946 (Vespa patent 98)

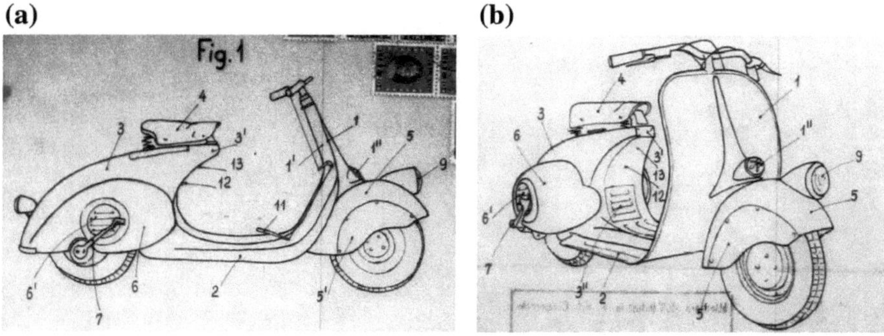

Fig. 5 The Vespa design by D'Ascanio from the patent: [6, 7, 9] **a** drawing of a side view; **b** drawing of a 3D view

From the description and drawings in the patent documents, Fig. 5, [9] one can understand how the whole project is aimed for a scooter design with comfortable riding, since a user will be seated with all driving units on front handlebars.

An additional new characteristic is the chassis, which is used to accommodate the user feet in a seated position, as well as to have the driver in a comfortable sitting position, with the peculiarity that it is also the structure connecting the front and back wheels. Another new feature of the Vespa is to have grouped all the mechanical parts in a single block that was placed under the saddle so that the engine was isolated and covered to avoid dirty to the driver. Even the transmission had new feature as working directly on the rear wheel without the presence of belt.

The design of "motor scooter 98 cc" (Vespa 98) was presented at Piaggio on August 10, 1945 with the drawing in Fig. 6, [6, 7].

The main features of the Vespa 98 can be recognized in:

- elastic suspensions with spiral steel spring for the front wheel and with rubber pads to the rear wheel and engine;
- single engine with two-stroke horizontal cylinder made of cast light alloy with 98 cm3 and stroke of 50 mm;
- direct transmission on the rear wheel via shift gears and coupling;
- 3-speed gear with clutch combined command on the left side of the handlebar;
- expansion brakes activated by hand on the right side of the handlebar for the front wheel, and by foot on the platform for the right rear wheel;
- wheels of 3.50 × 8 inches that are removable with nuts (like in cars);
- fuel tank of 5 liters capacity;
- maximum speed of 60 km/h;
- dimension as wheelbase of 1,170 mm, handlebar length of 700 mm, maximum length of 1,655 mm, maximum height of 860 mm, height of the saddle from the ground of 700 mm, total weight of 60 kg.

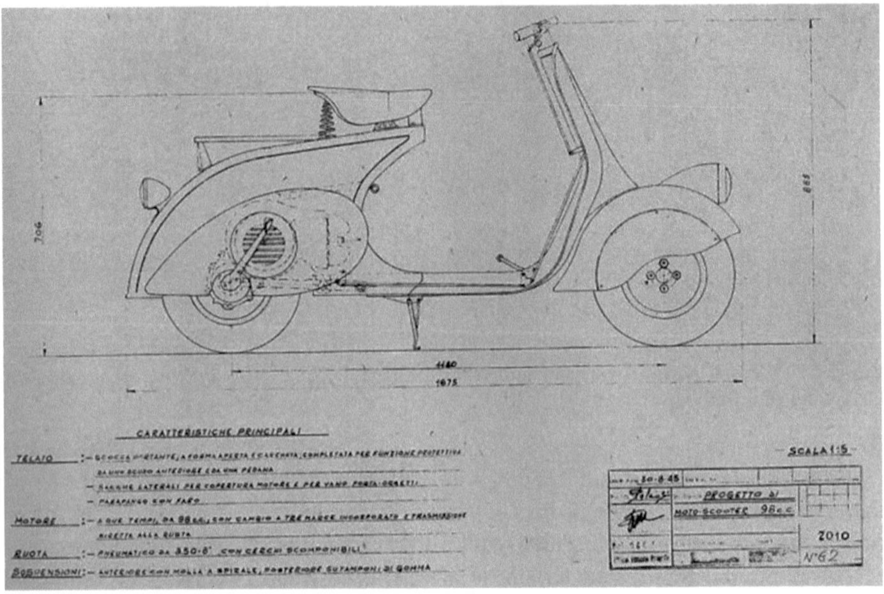

Fig. 6 Original design of motor scooter 98 cc of 1945 [9]

4 An Analysis of the Vespa Features

For a static analysis the Vespa model 125 of 1956 has been considered to compute front load Nsf and rear load Nsr with and without the driver, [10–12].

In the case of the Vespa without driver, the position of the centre of gravity G is influenced by the engine weight because it is the heaviest component, and G is located more near to the rear wheel. The force model can be summarized as the scheme in Fig. 7 with P = 843 N; a = 870 mm; b = 290 mm; p = 1,160 mm [Frisinghelli et al., 1998] and static equilibrium can be formulated as

$$-\mathrm{Pb} + \mathrm{Nsf}\ \mathrm{p} = 0 \tag{1}$$

Fig. 7 A model of Vespa scooter 125 for static analysis

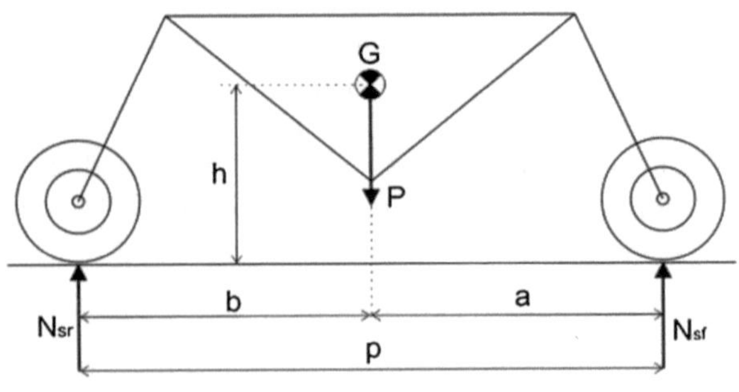

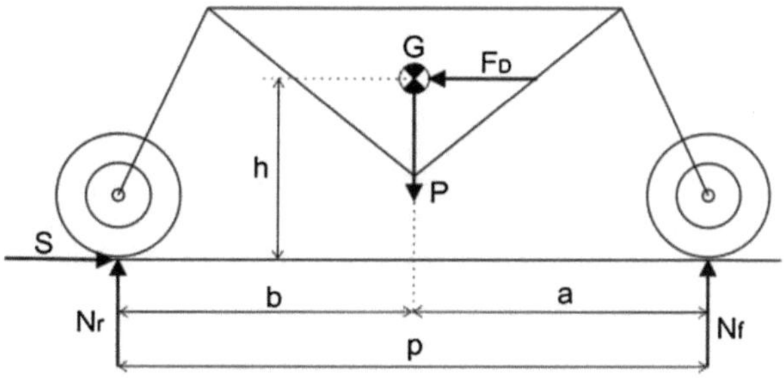

Fig. 8 Force acting on Vespa structure during straight motion

$$\mathrm{Nsr} - \mathrm{P} + \mathrm{Nsf} = 0 \tag{2}$$

By using Eqs. (1) and (2) in the case without driver, Nsf is computed as 211 and Nsr as 632 N. In case with 70 kg driver computations with P = 1529 N; b = 435; a = 725; p = 1160 mm give Nsf i as 573 and Nsr as 956 N.

Let's consider the case of Fig. 8 with the Vespa scooter moving in straight stationary motion with acting drag forces F_D given as

$$F_D = 0.5\,\rho\, C_D A\ v^2 \tag{3}$$

Assuming design characteristics for h = 750, a = 725, b = 435, p = 1160 mm, with a weight of **P** = 1529 N and A = 0.8 m^2; C_D = 1.8; ρ = 1.2 kg/m^3, A = 0.8 m^2; C_D = 1.8; ρ = 1.2 kg/m^3 [12], the drag force is computed form Eq. (3) as 167 N. Thus, from the equilibrium conditions given by

$$\mathrm{S} - \ = 0 \tag{4}$$

$$\mathrm{Nr} - \mathrm{P} + \mathrm{Nf}\ = 0 \tag{5}$$

$$\mathrm{Sh} - \mathrm{Nr}\,\mathrm{b} + \mathrm{Nf}\,\mathrm{a} = 0 \tag{6}$$

the dynamic loads on the front wheel and back wheel can be computed as $\mathbf{N_f}$ = 475 and $\mathbf{N_r}$ = 847 N, respectively.

5 An Analysis of the Front Suspension

The front suspension of the Vespa is represented in Fig. 9a with a functional scheme in Fig. 9b. The suspension consists of a coil spring and a double-acting shock absorber, Fig. 9a. In Fig. 9b k indicates the stiffness of the spring; c is the damping coefficient of shock absorber; Nf represents the dynamic load acting on the front wheel; and ε is the steering axis inclination. The values of these parameters can be assumed for: k = 150,000 N/m; c = 1,000 N s/m; Nf = 474 N; ε = 37 deg.

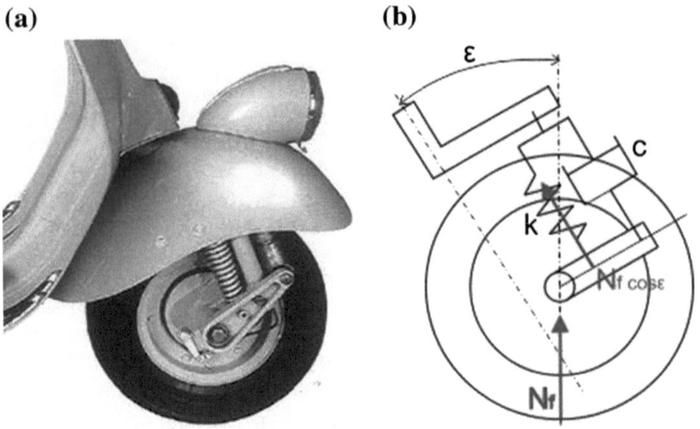

Fig. 9 Front suspension of Vespa scoter 125: **a** a photo; **b** a scheme for performance analysis

In order to facilitate the study the suspension is reduced to a equivalent suspension, that is represented in Fig. 10 as a vertical spring-damper unit between the two masses. The corresponding equivalent coefficient can be expressed as

$$k_f = k / \cos^2\varepsilon \tag{7}$$

$$c_f = c/\cos^2\varepsilon \tag{8}$$

The front suspension can be studied with the model with two degrees of freedom in Fig. 10b with motion of masses m_1 and m_2 as the vertical displacements z_1 and z_2, respectively. The model in Fig. 10b includes stiffness and damping coefficient of the tire kg and cg and the equation of motion can be expressed as

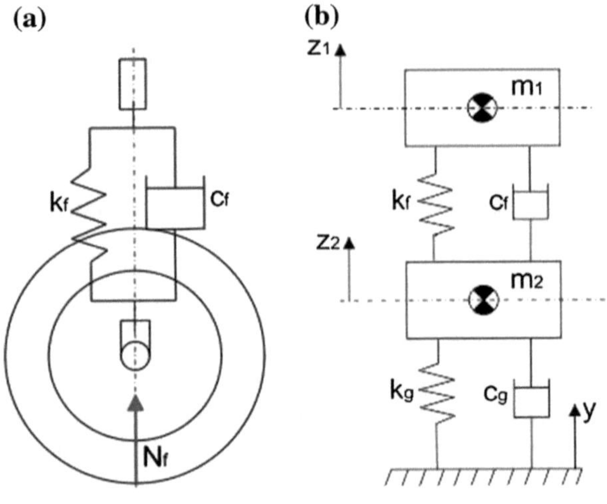

Fig. 10 A scheme of equivalent suspension for the model in Fig. 9: **a** mechanical model; **b** lumped parameter model

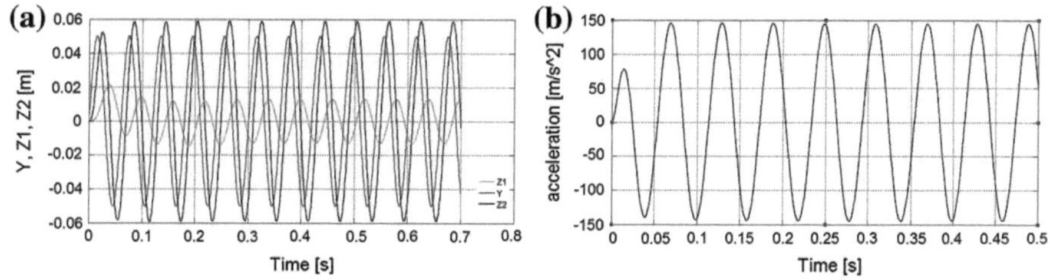

Fig. 11 Numerical results of performance evaluation for model in Fig. 10 when running at 20 km/h in a road with sinusoidal profile: **a** displacements of masses (in *blue* for road profile; in *yellow* for mass m_1 and in *red* for mass m_2); **b** acceleration for mass m_1 of a driver

$$m_1\ddot{z}_1 + c_f(\dot{z}_1 - \dot{z}_2) + k_f(z_1 - z_2) = 0 \tag{9}$$

$$m_2\ddot{z}_2 - c_f(\dot{z}_1 - \dot{z}_2) - k_f(z_1 - z_2) + c_g(\dot{z}_2 - \dot{y}) + k_g(z_2 - y) = 0 \tag{10}$$

A numerical characterization has been computed through a code in Simulink by using Eqs. (1) to (9) to evaluate basic performance characteristics that motivated the success of the Vespa design over many years and several model products. Results of numerical examples are reported in Figs. 11, 12 and 13 for three cases of road profiles and speeds. In particular Fig. 11 refers to the case of a road with a sinusoidal road profile with height of 5 cm in a wave length of 50 cm when the scooter runs and 20 km/h. The results in Fig. 11a show that the suspension is able to reduce the oscillation for the driver mass up to 1 cm. the proper output for driver comfort is also indicated in Fig. 11b by the mass acceleration that is computed within a range 0f +150 m/s^2.

In Fig. 12 results are reported as referring to the case of a scooter running at 30 km/h while passing over a hole of 7 cm high and 30 cm wide. Figure 12a shows computed results for mass displacements from which the driver mass suffers a maximum displacement of 5 cm that vanishes in 0.4 s. Even in this case the computed acceleration is limited and is vanished after 0.3 s, Fig. 12b.

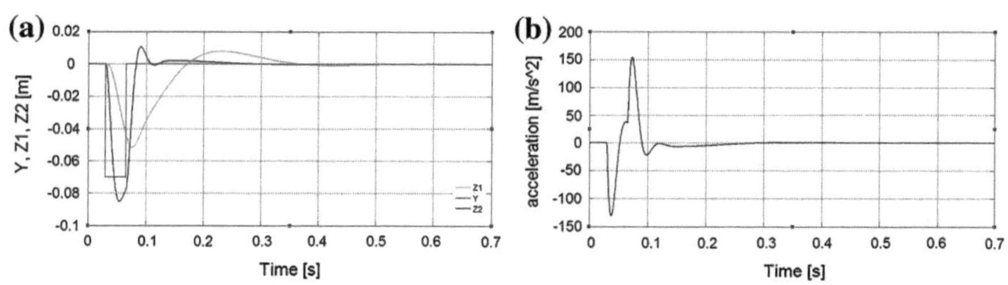

Fig. 12 Numerical results of performance evaluation for model in Fig. 10 when running at 30 km/h in a road hole of 7 cm high and 20 cm wide: **a** displacements of masses (in *blue* for road profile; in *yellow* for mass m_1 and in *red* for mass m_2); **b** acceleration for mass m_1 of a driver

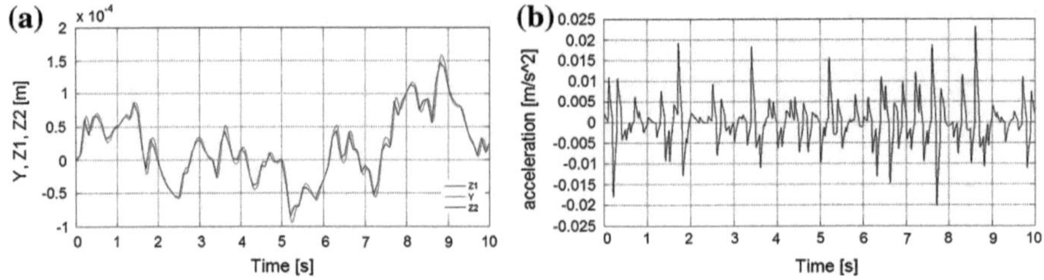

Fig. 13 Numerical results of performance evaluation for model in Fig. 10 when running at 50 km/h in a road with rough profile with maximum high of 15 μm: **a** displacements of masses (in *blue* for road profile; in *yellow* for mass m_1 and in *red* for mass m_2); **b** acceleration for mass m_1 of a driver

In Fig. 13 the numerical examples refer to a case of Vespa scooter running at 50 km/h on a road whose roof profile has a maximum rugosity of 15 μm. The mass response in Fig. 13a shows well adaptation of the suspension that gives a very limited acceleration in Fig. 13b.

6 Conclusions

The paper presents the successful design of Vespa scooter that was designed by Corradino D'Ascanio in the 1950 s'. The analysis has reported the original documents and has computed basic performance through suitable simplified models that explain the success of the Scooter. The computed performances through a fairly simple analysis show suitable behavior of the scooter with values that today are still successful for comfort and efficiency of that design. It is interesting to note that the design activity of Corradino D'Ascanio as in many fields of mechanical systems that has permitted him to conceive a brilliant solution for the Vespa scooter in terms of esthetics, comfort, and technical performance.

References

1. Histoire du scooter website. http://fr.questmachine.org/wiki/Histoire_du_scooter
2. Scooter's models website. http://www.scottermaniac.org/models.html
3. Reitwagen website. http://blog.mercedes-benz-passion.com/mercedes-benz-classic-november-1885-der-daimler-reitwagen-fohrt-von-connstatt-nach-unterturkheim/
4. Regional archive (1986) Corradino D'Ascanio from helicoptes to Vespa scooter. archives for Abbruzzo e Molise, Pescara (in Italian)
5. Fondazione Piaggio (2001) Corradino D'Ascanio Uomo Genio Mago Mito. Fondazione Piaggio, Pontedera
6. Frisinghelli L et al (1998) Vespa Tecnica N 1. CLD, Pontedera
7. Frisinghelli L et al (1999) Vespa Tecnica N 2. CLD, Pontedera

8. Mondini A (1995) Un'elica a due ruote. Nitri—Lischi, Pisa
9. Piaggio C (1946) Brevetto di motocicletta a complesso razionale di organi ed elementi con telaio combinato con parafanghi e cofano ricoprenti tutta la parte meccanica. Patenmt Office, Firenze (in Italian)
10. Cossalter V (1999) Kinematics and dynamics of motorcyles. edizione Progetto, Padova
11. Cai H, Liu H, Du P, Gao J, Chang H (2011) Suspension dynamics simulation based on simulink. 978-1-4244-8039-5/11/2011 IEEE
12. Drag coefficient of motorcycle website. www.engineeringtoolbox.com/drag-coefficient-d_627.html

Rigid Body Hyper-jerk Analysis Using Screw Theory

Jaime Gallardo-Alvarado and Mario A. Garcia-Murillo

Abstract Few decades ago screw theory appears to be an '*old fashioned*' mathematical tool confined to solve simple rigid-body first-order kinematic analysis. Still in our days the validity of the acceleration equations in screw form of kinematic chains introduced by Rico and Duffy in the 1990s has been bitterly disappointed by some kinematicians. This work deals with an application of the theory of screws not only in the rigid-body acceleration analysis but also in the jerk and hyper-jerk analyses of a six-degrees-of-freedom parallel manipulator. The Inverse/Forward kinematic equations of the robot are systematically obtained by resorting to reciprocal-screw theory. The contribution is a little proof that screw theory is a trusted and confident mathematical resource to investigate the rigid-body higher-order kinematic analyses. Numerical examples, which are verified with the aid of commercially available software, are provided in order to show the application of the method.

Keywords Jerk · Jounce · Parallel manipulator · Screw theory · Kinematics · Student paper

1 Introduction

The jerk is the time rate of change acceleration whereas the hyper-jerk is the time rate of change jerk, sometimes called *jounce* or *snap*. The jerk analysis dates back to the beginning of the 1970s and has been used to investigate the radii of torsion of spatial curves [1, 2], to control the position of planar mechanisms [3], and as a comprehensive resource to understand human movements [4–7]. In reality, examples of applications of the jerk yield a long list covering topics like the

J. Gallardo-Alvarado
Instituto Tecnologico de Celaya, Av. Tecnologico y A. Garcia Cubas, 38010 Celaya, GTO, Mexico

M.A. Garcia-Murillo (✉)
CICATA Queretaro-Instituto Politecnico Nacional, Mexico, Mexico
e-mail: mgarciam1114@alumno.ipn.mx

M. Ceccarelli and E.E. Hernández Martinez (eds.), *Multibody Mechatronic Systems*, Mechanisms and Machine Science 25, DOI 10.1007/978-3-319-09858-6_39

complete characterization of singularities in closed kinematic chains, the synthesis of spherical and spatial linkages, the computation of impact forces, the optimization of trajectories in the neighborhood of singular regions and so far. Furthermore, it is noticeable the quantity of remarkable contributions reported in the literature devoted to improve the performance of machine tools by controlling the jerk [8, 9]. Although the performance of machine tools is really improved by confining the jerk, smoother velocity functions may be obtained if the hyper-jerk is confined too. On the other hand, it is said that singularities in closed kinematic chains can be fully characterized based on the hyper-jerk analysis. Without doubt the higher-order analysis has a predominant role in new applications as well in the improving of existing processes. In that way, screw theory is a trusted and confident mathematical tool that can be used systematically in the higher-order kinematic analyses of robot manipulators.

The acceleration analysis of kinematic chains by means of screw theory dates back to 1996 when Rico and Duffy [10] published a pioneering contribution entitled "An application of screw algebra to the acceleration analysis of serial chains" [10]. By that time most kinematicians had been assumed that the theory of screws will be confined to the so-called first-order kinematic analysis due to the difficulty to express in screw form all the required expressions to achieve the acceleration analysis. This approach allowed to express in pure screw form the accelerator of Sugimoto [11, 12], named the reduced acceleration state of a rigid body. Furthermore, with the new mathematical tool, it was possible to extend the theory of screws not only to the acceleration analysis, but also to the jerk and hyper-jerk analyses [13, 14] with particular applications in parallel manipulators [15–18]. However, it is noteworthy that even today there is some skepticism about the validity of the acceleration equations in screw form introduced in [10]. As a little effort to reinforce the validity not only of the equations of acceleration but also the correctness of the equations of jerk and hyper-jerk, in this work the theory of screws is applied to the velocity, acceleration, jerk and hyper jerk analyses of a parallel manipulator.

The rest of the contribution is organized as follows. In Sect. 2 a brief summary of concepts applied in the contribution dealing with screw theory and high-order analyses of kinematic chains is provided. Next, an application example that consists of solving the velocity, acceleration, jerk and hyper-jerk of a parallel manipulator. Furthermore, the numerical results of the case study are verified with the aid of commercially available software. Finally, some conclusions are given at the end of the contribution.

2 Screw Theory and Higher-Order Analyses of Kinematic Chains

The kinematic state of a rigid body may be defined as a six-dimensional vector which results from the inseparable union of two three-dimensional vectors. The first vector, an angular entity, is called the primal part while the second vector is called

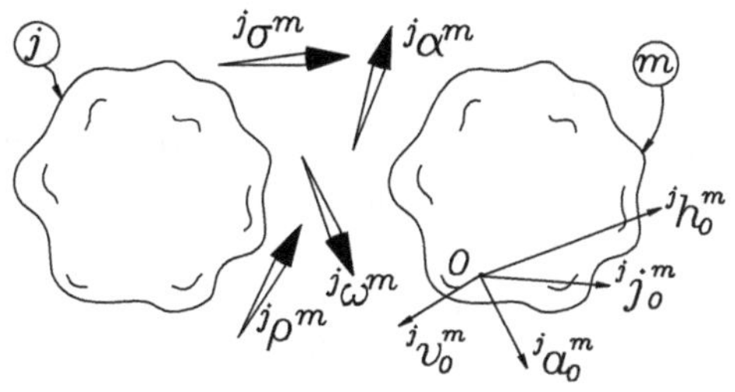

Fig. 1 The motion of body m as observed from body j

the dual part. The primal part is a property of the rigid bodies, in other words it is the same for all the points of the rigid body. On the other hand, the dual part is related with a particular point embedded to the body.

With reference to Fig. 1, let O be an arbitrary point embedded to body m. The twist about a screw of body m with respect to body j [19], known as the velocity state, is a six-dimensional vector ${}^j\mathbf{V}_O^m$ defined as

$$ {}^j\mathbf{V}_O^m \equiv \begin{bmatrix} {}^j\boldsymbol{\omega}^m \\ {}^j\boldsymbol{v}_O^m \end{bmatrix} \tag{1} $$

where ${}^j\boldsymbol{\omega}^m$ is the angular velocity vector of body m as observed from body j while ${}^j\boldsymbol{v}_O^m$ is the velocity vector of a point O embedded to body m. Point O is called the reference pole. Clearly, unlike the dual part, the primal part ${}^j\boldsymbol{\omega}^m$ does not depend on point O.

The reduced acceleration state of body m with respect to body j, the six-dimensional vector ${}^j\boldsymbol{A}_O^m$ may be obtained as the time derivative, via a helicoidal field, of the velocity state ${}^j\boldsymbol{V}_O^m$ and it may be expressed as

$$ {}^j\boldsymbol{A}_O^m = \begin{bmatrix} {}^j\boldsymbol{\alpha}^m \\ {}^j\boldsymbol{a}_O^m - {}^j\boldsymbol{\omega}^m \times {}^j\boldsymbol{v}_O^m \end{bmatrix} \tag{2} $$

where ${}^j\boldsymbol{\alpha}^m \equiv \frac{{}^jd}{dt}({}^j\boldsymbol{\omega}^m)$ is the angular acceleration vector of body m as measured from body j while ${}^j\boldsymbol{a}_O^m \equiv \frac{{}^jd}{dt}({}^j\boldsymbol{v}_O^m)$ is the acceleration of point O.

Taking into account that the jerk is the change of acceleration, e.g. ${}^j\boldsymbol{\rho}^m \equiv \frac{{}^jd}{dt}({}^j\boldsymbol{\alpha}^m)$ is the angular vector jerk of body m whereas ${}^j\boldsymbol{j}_O^m \equiv \frac{{}^jd}{dt}({}^j\boldsymbol{a}_O^m)$ is the jerk of point O attached to body m, where both vectors are measured from body j, a robust definition of the reduced jerk state, or jerkor for brevity, was obtained by Rico et al. [13] as follows

$$ {}^j\boldsymbol{J}_O^m = \begin{bmatrix} {}^j\boldsymbol{\rho}^m \\ {}^j\boldsymbol{j}_O^m - 2\,{}^j\boldsymbol{\alpha}^m \times {}^j\boldsymbol{v}_O^m - {}^j\boldsymbol{\omega}^m \times {}^j\boldsymbol{a}_O^m \end{bmatrix} \tag{3} $$

According with the methodology introduced by Rico et al. [13], there are no limits for the order of the desired analysis when the theory of helicoidal vector fields is employed. In that concern, the reduced hyper-jerk state of a rigid body, the six-dimensional vector ${}^j\boldsymbol{H}_O^m$, is given by

$$ {}^j\boldsymbol{H}_O^m = \begin{bmatrix} {}^j\boldsymbol{\sigma}^m \\ {}^j\boldsymbol{h}_O^m - 3\,{}^j\boldsymbol{\rho}^m \times {}^j\boldsymbol{v}_O^m - 3\,{}^j\boldsymbol{\alpha}^m \times {}^j\boldsymbol{a}_O^m - {}^j\boldsymbol{\omega}^m \times {}^j\boldsymbol{j}_O^m \end{bmatrix} \tag{4} $$

where ${}^j\boldsymbol{\sigma}^m \equiv \frac{{}^jd}{dt}({}^j\boldsymbol{\rho}^m)$ and ${}^j\boldsymbol{h}_O^m \equiv \frac{{}^jd}{dt}({}^j\boldsymbol{j}_O^m)$ are the angular and linear hyper-jerks of body m as observed from body j considering point O as the reference pole, for a detailed explanation of Eq. (4) the reader is referred to [14].

In what follows the kinematic states of rigid body are presented in screw form. '*A screw is a straight line with which a definite linear magnitude termed the pitch is associated*' [19]. Furthermore, a screw $\$ = (\hat{s}, \boldsymbol{s}_O)$ can be considered as a six-dimensional vector composed of a primal part and a dual part, where the primal part $\hat{s}$ is a unit vector along the screw axis, while the dual part $s_O = h\hat{s} + \hat{s} \times \boldsymbol{r}_{O/P}$, where $\boldsymbol{r}_{O/P}$ is a vector pointed from a point P on the screw axis to point O, is the moment produced by $\hat{s}$ about a selected point O, namely the pole, which is instantaneously coincident with a point of a reference frame, usually its origin. Any lower kinematic pair can be represented either by a screw or a group of screws.

Infinitesimal screws and kinematic states form inseparable kinematic entities. In serial manipulators, the velocity state of the end-effector, body m, with respect to the base link, body 0, may be written as a linear combination of the joint-rate velocities of the open kinematic chain [20], as follows

$$ {}_0\omega_1\,{}^0\$^1 + {}_1\omega_2\,{}^1\$^2 + \ldots + {}_{m-1}\omega_{m-1}\,{}^{m-1}\$^m = {}^0\boldsymbol{V}_O^m \tag{5} $$

while the corresponding accelerator can be written in screw form as [10]:

$$ {}_0\alpha_1\,{}^0\$^1 + {}_1\alpha_2\,{}^1\$^2 + \ldots + {}_{m-1}\alpha_{m-1}\,{}^{m-1}\$^m + \mathcal{A} = {}^0\boldsymbol{A}_O^m \tag{6} $$

where $\mathcal{A}$ is the *Lie screw of acceleration* which is computed as

$$ {}^i\mathcal{A}^m = \sum_{i=0}^{m-2}\left[{}_i\omega_{i+1}^i\,\$^{i+1} \quad \sum_{k=i+1}^{m-1} {}_k\omega_{k+1}^k\,\$^{k+1} \right]. $$

Furthermore, Rico et al. [13] found that the jerkor in screw form is given by

$$ {}_0\rho_1\,{}^0\$^1 + {}_1\rho_2\,{}^1\$^2 + \ldots + {}_{m-1}\rho_{m-1}\,{}^{m-1}\$^m + \mathcal{J} = {}^0\boldsymbol{J}_O^m \tag{7} $$

where $\mathcal{J}$ is the Lie screw of jerk which is computed as

$$\begin{aligned}\mathcal{J} =& 2\sum_{j=0}^{m-2}\left[{}_j\omega_{j+1}^{j}\$^{j+1} \quad \sum_{k=j+1}^{m-1} {}_k\alpha_{k+1}^{k}\$^{k+1} + {}^{k+1}\mathcal{A}^m\right]\\ &+\sum_{j=0}^{m-2}\left[{}_j\alpha_{j+1}^{j}\$^{j+1} \quad \sum_{k=j+1}^{m-1} {}_k\omega_{k+1}^{k}\$^{k+1}\right]\\ &+\sum_{j=0}^{m-2}\left[{}_j\omega_{j+1}^{j}\$^{j+1} \quad \left[\sum_{k=j+1}^{m-1} {}_k\omega_{k+1}^{k}\$^{k+1} \quad {}_k\omega_{k+1}^{k}\$^{k+1}\right]\right].\end{aligned} \tag{8}$$

Finally, the reduced hyper-jerk state in screw form [14] is given by

$${}_j\sigma_{j+1}^{j}\$^{j+1} +_{j+1} \sigma_{j+2}^{j+1}\$^{j+2} + \ldots +_{m-1} \sigma_{m}^{m-1}\$^{m} + \mathcal{H} = {}^j\mathbf{H}_O^m \tag{9}$$

where $\mathcal{H}$ is called the *Lie screw of hyper-jerk*, which is given by

$$\begin{aligned}\mathcal{H} =& 3\sum_{j=0}^{m-2}\left[{}_j\omega_{j+1}^{j}\$^{j+1} \quad \sum_{k=j+1}^{m-1} {}_{k+1}\rho_{k+2}^{k+1}\$^{k+2}\right]\\ &+3\sum_{j=0}^{m-2}\left[{}_j\alpha_{j+1}^{j}\$^{j+1} \quad \sum_{k=j+1}^{m-1} {}_k\alpha_{k+1}^{k}\$^{k+1} +^{j+1} \mathcal{A}^m\right]\\ &+3\sum_{j=0}^{m-2}\left[{}_j\omega_{j+1}^{j}\$^{j+1} \quad \left[{}_j\omega_{j+1}^{j}\$^{j+1} \quad \sum_{k=j+1}^{m-1} {}_k\alpha_{k+1}^{k}\$^{k+1} +^{j+1} \mathcal{A}^m\right]\right]\\ &+2\sum_{j=0}^{m-2}\left[{}_j\alpha_{j+1}^{j}\$^{j+1} \quad \left[{}_j\omega_{j+1}^{j}\$^{j+1} \quad \sum_{k=j+1}^{m-1} {}_k\omega_{k+1}^{k}\$^{k+1}\right]\right]\\ &+\sum_{j=0}^{m-2}\left[{}_j\rho_{j+1}^{j}\$^{j+1} \quad \sum_{k=j+1}^{m-1} {}_{k+1}\omega_{k+2}^{k+1}\$^{k+2}\right]\\ &+\sum_{j=0}^{m-2}\left[{}_j\omega_{j+1}^{j}\$^{j+1} \quad \left[{}_j\alpha_{j+1}^{j}\$^{j+1} \quad \sum_{k=j+1}^{m-1} {}_k\omega_{k+1}^{k}\$^{k+1}\right]\right]\\ &+\sum_{j=0}^{m-2}\left[{}_j\omega_{j+1}^{j}\$^{j+1} \left[{}_j\omega_{j+1}^{j}\$^{j+1} \quad \left[{}_j\omega_{j+1}^{j}\$^{j+1} \sum_{k=j+1}^{m-1} {}_k\omega_{k+1}^{k}\$^{k+1}\right]\right]\right]\end{aligned} \tag{10}$$

It is straightforward to demonstrate that expressions (5), (6) and (7) can be extended to parallel manipulators without significative effort, the inclusion of subscripts and superscripts makes the difference between distinct limbs.

3 Application: Hyper-jerk Analysis of a Six-Legged Parallel Manipulator

In order to show the application of the method, in this section the hyper-jerk analysis of a non-redundant spatial parallel manipulator is approached by means of the theory of screws.

3.1 Description of the Parallel Manipulator

The chosen parallel manipulator for the analysis is a 6DOF mechanism, see Fig. 2, consists of a moving platform (m) and a fixed platform (0), connected to each other by means of three RRP*S-type and three R*RRS-type kinematic chains where upper asterisks denote active kinematic pairs while R, P, and S stand for Revolute, Prismatic and Spherical joints, respectively.

Let XYZ be a reference frame attached to the fixed platform. The axes of the revolute joints connecting the limbs to the fixed platform are along the Y-axis whereas the orientations of the axes of the remaining revolute joints are located in the XZ-plane. Furthermore, the axes of the revolute joints in the same limb containing an active revolute joint have the same orientation $\hat{\boldsymbol{u}}_i$, in the remainder of the contribution unless otherwise $i = 1, 2, 3$. The moving platform is a regular hexagon inscribed in a circle of radius r whose corners are points B_i and $\underline{B}_i$ denoting the centers of the spherical joints, which are located, respectively, by vectors $\boldsymbol{B}_i$ and $\underline{\boldsymbol{B}}_i$. On the other hand, points A_i and $\underline{A}_i$, located by vectors $\boldsymbol{A}_i$ and $\underline{\boldsymbol{A}}_i$, assume to be the nominal positions of the lower passive revolute joints whereas a point D_i, located by vector $\boldsymbol{D}_i$, is the instantaneous position of the revolute joint connecting the arm and forearm of a RRRS-type limb. Finally, the generalized coordinates are specified by q_i and $\underline{q}_i$.

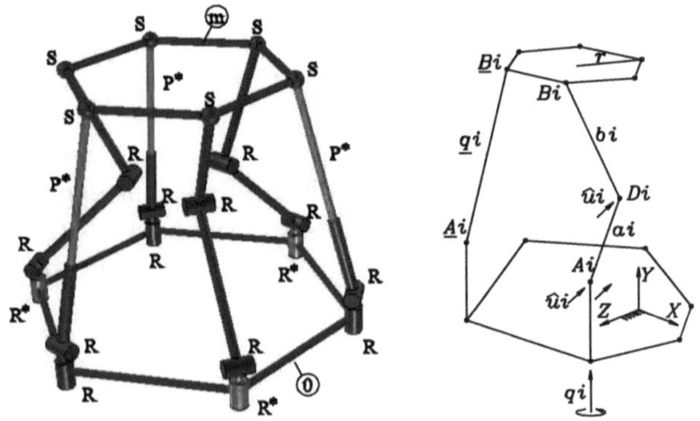

Fig. 2 The parallel manipulator under study and its geometric scheme

3.2 Velocity and Acceleration Analyses

Let $\boldsymbol{V}_C = [\boldsymbol{\omega}, \quad \boldsymbol{v}_C]^{\mathrm{T}}$ the velocity state of the moving platform as measured from the fixed platform, see Eq. (1). Furthermore, $\boldsymbol{V}_C$ can be written in screw form through the RRP*S-type limbs as follows

$$ {}_0\underline{\omega}_1^i \; {}^0\underline{\$}_i^1 + {}_1\underline{\omega}_2^i \; {}^1\underline{\$}_i^2 + \underline{\dot{q}}_i \; {}^2\underline{\$}_i^3 + {}_3\underline{\omega}_4^i \; {}^3\underline{\$}_i^4 + {}_4\underline{\omega}_5^i \; {}^4\underline{\$}_i^5 + {}_5\underline{\omega}_6^i \; {}^5\underline{\$}_i^6 = \boldsymbol{V}_C \qquad (11)$$

where the modeling of the screws is depicted in Fig. 3.

Taking into account that the screw ${}^5\underline{\$}_i^6$ is reciprocal to all the screws, excluding the screw ${}^2\underline{\$}_i^3$ associated to the active prismatic joint, in the same limb and, after reducing terms in Eq. (11) one obtains

$$\left\{ {}^5\underline{\$}_i^6; \boldsymbol{V}_C \right\} = \underline{\dot{q}} \qquad (12)$$

On the other hand, $\boldsymbol{V}_C$ can be written in screw form through the R*RRS-type limbs as follows

$$\dot{q}_i \; {}^0\$_i^1 + {}_1\,\omega_2^i \; {}^1\$_i^2 + {}_2\,\omega_3^i \; {}^2\$_i^3 + {}_3\,\omega_4^i \; {}^3\$_i^4 + {}_4\,\omega_5^i \; {}^4\$_i^5 + {}_5\,\omega_6^i \; {}^5\$_i^6 = \mathbf{V}_C \qquad (13)$$

where the screw ${}^4\$_i^5$ is reciprocal to all the screws in the same limb, excepting the screw ${}^0\$_i^1$. Hence, after reducing terms in Eq. (12) we have

$$\{ {}^4\$_i^5; \boldsymbol{V}_C \} = \dot{q}_i \{ {}^4\$_i^5; {}^0\$_i^1 \} \qquad (14)$$

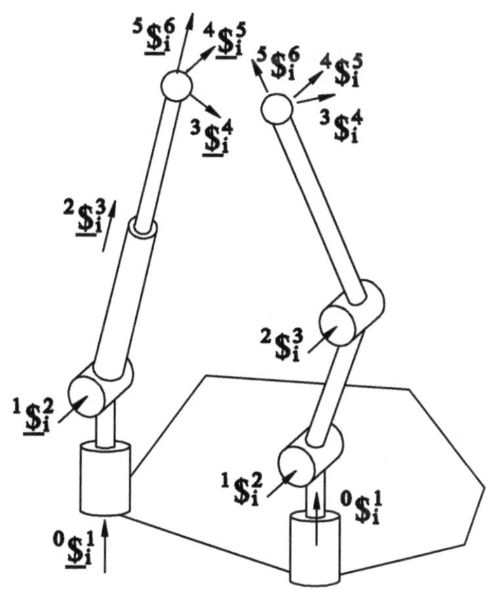

Fig. 3 Infinitesimal screws of two distinct limbs

Casting into a matrix-vector form Eqs. (12) and (14) the Inverse/Forward equation of velocity of the robot is obtained as

$$\boldsymbol{G}\boldsymbol{V}_C = \mathbf{H}\dot{\underline{\mathbf{q}}} \tag{15}$$

where $\dot{\mathbf{q}} = [\underline{q}_1\ \underline{q}_2\ \underline{q}_3\ \dot{q}_1\ \dot{q}_2\ \dot{q}_3]^T$ is the *first-order driver matrix* of the parallel manipulator, $\mathbf{G} = \mathbf{J}^T\,\boldsymbol{\Delta}$ is the first order coefficient matrix of the robot in which $\mathbf{J} = [{}^5\underline{\$}_1^6\ {}^5\underline{\$}_2^6\ {}^5\underline{\$}_3^6\ {}^4\$_1^5\ {}^4\$_2^5\ {}^4\$_3^5]$ is the active screw-coordinate Jacobian matrix of the hexapod and $\boldsymbol{\Delta}$ is a 6 × 6 operator of polarity, defined according to the identity matrix $\mathbf{I}$ and the zero matrix 0 as $\boldsymbol{\Delta} = \begin{bmatrix} \mathbf{0} & \mathbf{I} \\ \mathbf{I} & \mathbf{0} \end{bmatrix}$. Finally, $\mathbf{H}$ is a diagonal matrix given by

$$\mathbf{H} = \mathrm{diag}\left[1 \quad 1 \quad 1 \quad \{{}^4\$_1^5;{}^0\$_1^1\} \quad \{{}^4\$_2^5;{}^0\$_2^1\} \quad \{{}^4\$_3^5;{}^0\$_3^1\}\right] \tag{16}$$

The inverse velocity analysis consists of determining the generalized joint-velocity rates, matrix $\dot{\mathbf{q}}$, given the velocity state $\boldsymbol{V}_C$ while the forward velocity analysis consists of computing the six-dimensional vector $\boldsymbol{V}_C$ given matrix $\dot{\mathbf{q}}$.

Following the trend of the velocity analysis, the Inverse/Forward equation of acceleration of the robot results in

$$\mathbf{G}\boldsymbol{A}_C = \mathbf{H}\ddot{\mathbf{q}} + \mathbb{A} \tag{17}$$

where $\ddot{\mathbf{q}} = \left[\ddot{\underline{q}}_1\quad \ddot{\underline{q}}_2\quad \ddot{\underline{q}}_3\quad \ddot{q}_1\quad \ddot{q}_2\quad \ddot{q}_3\right]^T$ is the *second-order driver matrix* of the manipulator and $\mathbb{A} = \left[\left\{\underline{\mathcal{A}}_1;{}^5\underline{\$}_1^6\right\}\left\{\underline{\mathcal{A}}_2;{}^5\underline{\$}_2^6\right\}\left\{\underline{\mathcal{A}}_3;{}^5\underline{\$}_3^6\right\}\{\mathcal{A}_1;{}^4\$_1^5\}\{\mathcal{A}_2;{}^4\$_2^5\}\right.$ $\left.\{\mathcal{A}_3;{}^4\$_3^5\}\right]^T$ is the *complementary matrix of acceleration*. It is worth to say that terms concerned with the Coriolis acceleration are self-contained in matrix $\mathbb{A}$.

Given matrix $\ddot{\mathbf{q}}$, the forward acceleration analysis consists of finding the corresponding reduced acceleration state $\boldsymbol{A}_C$. On the other hand, given $\boldsymbol{A}_C$ the inverse acceleration analysis of the robot consists of determining the corresponding generalized joint-acceleration rates contained in matrix $\ddot{\mathbf{q}}$.

3.3 Jerk Analysis

The reduced jerk state of the moving platform with respect to the fixed platform $\boldsymbol{J}_C$, see Eq. (3), may be written in screw form through the connector chains of the robot manipulator as follows

$$ {}_0\underline{\rho}_1^i\,{}^0\underline{\$}_i^1 +_1 \underline{\rho}_2^i\,{}^1\underline{\$}_i^2 + \dddot{\underline{q}}_i\,{}^2\underline{\$}_i^3 +_3 \underline{\rho}_4^i\,{}^3\underline{\$}_i^4 +_4 \underline{\rho}_5^i\,{}^4\underline{\$}_i^5 +_5 \underline{\rho}_6^i\,{}^5\underline{\$}_i^6 + \underline{\mathcal{J}}_i = \boldsymbol{J}_C \tag{18}$$

and

$$ \dddot{q}_i\,{}^0\$_i^1 +_1 \rho_2^i\,{}^1\$_i^2 +_2 \rho_3^i\,{}^2\$_i^3 +_3 \rho_4^i\,{}^3\$_i^4 +_4 \rho_5^i\,{}^4\$_i^5 +_5 \rho_6^i\,{}^5\$_i^6 + \mathcal{J}_i = \boldsymbol{J}_C \tag{19}$$

where $\underline{\mathcal{J}}_i$ and $\mathcal{J}_i$ are the Lie screws of jerk of the limbs of the manipulator.

Following the trend of the velocity analysis, the Inverse/Forward equation of jerk of the robot results in

$$\mathbf{G}\boldsymbol{J}_C = \mathbf{H}\dddot{\mathbf{q}} + \mathbb{J} \tag{20}$$

where $\dddot{\mathbf{q}} = \begin{bmatrix} \dddot{\underline{q}}_1 & \dddot{\underline{q}}_2 & \dddot{\underline{q}}_3 & \dddot{q}_1 & \dddot{q}_2 & \dddot{q}_3 \end{bmatrix}^T$ is the *third-order driver matrix* of the parallel manipulator while

$$\mathbb{J} = \left[\left\{ \underline{\mathcal{J}}_1 ;^5 \underline{\$}_1^6 \right\} \left\{ \underline{\mathcal{J}}_2 ;^5 \underline{\$}_2^6 \right\} \left\{ \underline{\mathcal{J}}_3 ;^5 \underline{\$}_3^6 \right\} \{ \mathcal{J}_1 ;^4 \$_1^5 \} \{ \mathcal{J}_2 ;^4 \$_2^5 \} \{ \mathcal{J}_3 ;^4 \$_3^5 \} \right]^T$$

is the *complementary matrix of jerk*.

The inverse jerk analysis consists of computing matrix $\dddot{q}$ given the jerkor $\boldsymbol{J}_C$ while the forward jerk analysis consists of determining the six-dimensional vector $\boldsymbol{J}_C$ given the generalized joint-jerk rates contained in matrix $\dddot{q}$.

3.4 Hyper-jerk Analysis

Let $\boldsymbol{H}_C$ be the reduced hyper-jerk state, or hyper-jerkor, of the moving platform as observed from the fixed platform, see Eq. (4). Furthermore, $\boldsymbol{H}_C$ can be written in screw form through the limbs of the parallel manipulator as follows

$$ {}_0\underline{\sigma}_1^i\,{}^0\underline{\$}_i^1 +_1 \underline{\sigma}_2^i\,{}^1\underline{\$}_i^2 + \dddot{\underline{q}}_i{}^2_i\,{}^2\underline{\$}_i^3 +_3 \underline{\sigma}_4^i\,{}^3\underline{\$}_i^4 +_4 \underline{\sigma}_5^i\,{}^4\underline{\$}_i^5 +_5 \underline{\sigma}_6^i\,{}^5\underline{\$}_i^6 + \underline{\mathcal{H}}_i = \boldsymbol{H}_C \tag{21}$$

and

$$ \dddot{q}_i\,{}^0\$_i^1 +_1 \sigma_2^i\,{}^1\$_i^2 +_2 \sigma_3^i\,{}^2\$_i^3 +_3 \sigma_4^i\,{}^3\$_i^4 +_4 \sigma_5^i\,{}^4\$_i^5 +_5 \sigma_6^i\,{}^5\$_i^6 + \mathcal{H}_i = \boldsymbol{H}_C \tag{22}$$

where $\underline{\mathcal{H}}_i$ and $\mathcal{H}_i$ are the Lie screws of hyper-jerk of the limbs of the manipulator.

Following the trend of the velocity analysis, the Inverse/Forward equation of hyper-jerk of the parallel manipulator results in

$$\boldsymbol{GH}_C = \mathbf{H}\ddddot{q} + \mathbb{H} \tag{23}$$

where $\ddddot{\mathbf{q}} = \begin{bmatrix} \ddddot{\underline{q}}_1 & \ddddot{\underline{q}}_2 & \ddddot{\underline{q}}_3 & \ddddot{q}_1 & \ddddot{q}_2 & \ddddot{q}_3 \end{bmatrix}^T$ is the *fourth-order driver matrix* of the parallel manipulator while

$$\mathbb{H} = \left[\left\{\underline{\mathcal{H}}_1;^5 \underline{\$}_1^6\right\} \left\{\underline{\mathcal{H}}_2;^5 \underline{\$}_2^6\right\} \left\{\underline{\mathcal{H}}_3;^5 \underline{\$}_3^6\right\} \left\{\mathcal{H}_1;^4 \$_1^5\right\} \left\{\mathcal{H}_2;^4 \$_2^5\right\} \left\{\mathcal{H}_3;^4 \$_3^5\right\} \right]^T$$

is the *complementary matrix of hyper-jerk.*

The inverse hyper-jerk analysis consists of finding the joint-hyper-jerk rates of the robot contained in matrix $\ddddot{q}$ given the reduced hyper-jerk state $\boldsymbol{H}_C$ while the forward hyper-jerk analysis consists of computing the six-dimensional vector $\boldsymbol{H}_C$ for a given set of generalized joint-hyper-jerk rates.

3.5 Numerical Example

Using thorough the numerical example SI units, the arm and fore-arm lengths of the limbs containing actuated revolute joints are chosen as $a = 0.4472$ and $b = 0.5656$, respectively, while the radius of the moving platform is $r = 0.4$. The nominal positions of passive revolute joints are $\underline{A}_1 = (0.3, 0, -0.5196)$, $\underline{A}_2 = (-0.6, 0, 0.0)$, $\underline{A}_3 = (0.3, 0, 0.5196)$, $A_1 = (0.6, 0, 0)$, $A_2 = (-0.3, 0, -0.5196)$, $A_3 = (-0.3, 0, 0.5196)$. In the reference configuration of the parallel manipulator consider that $B_1 = (0.4, 0.8, 0)$, $B_2 = (-0.2, 0.8, -0.3464)$, $B_3 = (-0.2, 0.8, 0.3464)$, $\underline{B}_1 = (0.2, 0.8, -0.3464)$, $\underline{B}_2 = (-0.4, 0.8, 0)$, $\underline{B}_3 = (0.2, 0.8, 0.3464)$, $D_1 = (0.2, 0.4, 0)$, $D_2 = (-0.1, 0.4, -0.1732)$, $C = (0, 0.8, 0)$, and $D_3 = (-0.1, 0.4, 0.1732)$. Furthermore, the generalized coordinates are commanded to follow periodical functions given by $\underline{q}_1 = 0.8246 - 0.35 \sin(t) \cos(t)$, $\underline{q}_2 = 0.8246 + 0.1 \sin(t)$, $\underline{q}_3 = 0.8246 + 0.1 \sin(t)$, $q_1 = 0.5 \sin(t)$, $q_2 = \pi/6 + 0.25 \sin(t)$ and $q_3 = \pi/3 + 0.75 \sin(t)$, where the time t is given in the interval $0 < t < 2\pi$. As an intermediate step, the displacement analysis was carried-out by means of the method introduced in [21].

The method here reported to approach the kinematic analyses of the robot was translated into a Maple16© sheet. The time history of the hyper-jerk of the center C of the moving platform, by using the theory of screws is shown in Fig. 4.

In order to verify the results of the case study, computer simulations were generated with the aid of special software like ADAMS©. Taking into account that it is limited to the acceleration analysis, an hybrid algorithm was implemented in Maple16©: (i) adjust to spline curves, called SP functions, the results of the acceleration analysis obtained with ADAMS©. (ii) apply the first time derivative to the corresponding SP functions with the purpose to solve the jerk analysis (iii) apply the second time derivative to the corresponding SP functions in order to solve the hyper-jerk analysis. The results obtained are provided in Fig. 4.

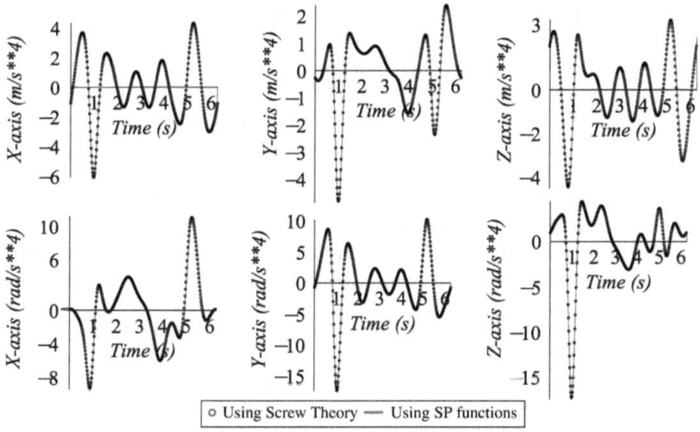

Fig. 4 Hyper-jerk of the center of the moving platform using two methods

Finally, it is worth to emphasize that the numerical results obtained via screw theory are in excellent agreement with those generated by using other strategies such as the application of commercially available software like ADAMS©. combined with an hybrid algorithm.

4 Conclusions

In this contribution the theory of screws is applied satisfactorily in the jerk and hyper-jerk analyses of a six-legged parallel manipulator equipped with a mixed scheme of actuation. Furthermore, the numerical results were verified by applying other methods such as the generation of computer simulations with the aid of commercially available software combined with an hybrid algorithm.

The material exposed in the contribution is a little proof that screw theory is a confident and trusted mathematical tool that can be used successfully in the higher-order kinematic analyses of robot manipulators.

Acknowledgments This work has been supported by CONACyT of Mexico

References

1. Suh CH (1971) Higher order analysis of spatial coupler curves. J Mech 6(1):81–95
2. Schot SH (1978) Jerk: the time rate of change acceleration. Am J Phys 46(11):1090–1094
3. Sparis PD, Mouroutsos SG (1984) A new matrix method for the kinematic analysis and motion simulation of planar mechanisms with lower pairs. Trans ASME: J Mech Transm Automation Des 106(4):429–436

4. Morasso P (1981) Spatial control arm movements. Exp Brain Res 42:223–227
5. Uno Y, Kawato M, Suzuki R (1989) Formation and control of optimal trajectory in human multijoint arm movements. Biol Cybern 61:89–101
6. Viviani P, Schneider R (1991) A developmental study of the relationship between geometry and kinematics in drawing movements. J Exp Psychol Human 17(1):198–218
7. Viviani P, Flash T (1995) Minimum-jerk, two-thirds power, law and isochrony: converging approaches to movement planning. J Exp Psychol 21(1):32–53
8. Schellekensa P, Rosiellea N, Vermeulena H, Vermeulena M, Wetzelsa S, Prila W (2008) Design for precision: current status and trends. CIRP Ann Manuf Tech 47(2):557–586
9. Zhang K, Gao X-S, Li H, Yuan C-M (2010) A greedy algorithm for feedrate planning of CNC machines along curved tool paths with jerk constraints. Math Mech Res Prepr KLMM 29:189–205
10. Rico JM, Duffy J (1996) An application of screw algebra to the acceleration analysis of serial chains. Mech Mach Theory 31(4):445–457
11. Sugimoto K (1987) Kinematic and dynamic analysis of parallel manipulators by means of motor algebra. Trans ASME J Mech Transm Autom Des 109(1):3–7
12. Sugimoto K (1989) Computational scheme for dynamic analysis of parallel manipulators. Trans ASME J Mech Transm Autom Des 111:29–33
13. Rico JM, Gallardo J, Duffy J (1999) Screw theory and higher order kinematic analysis of open serial and closed chains. Mech Mach Theory 34(4):559–586
14. Gallardo-Alvarado J (2013) Hyper-jerk analysis of robot manipulators. Intell Robot Syst. doi: 10.1007/s10846-013-9849-z
15. Gallardo-Alvarado J (2003) Jerk distribution of a 6-3 Gough-Stewart platform. Proc IMechE Part K: J Multi-body Dyn 217(1):77–84
16. Gallardo J, Lesso R, Rico JM, Alici G (2011) The kinematics of modular spatial hyper-redundant manipulators formed from RPS-type limbs. Robot Auton Syst 59(1):12–21
17. Gallardo-Alvarado J, Camarillo-Gómez KA (2011) Inverse jerk analysis of symmetric zero-torsion parallel manipulators. Robot Auton Syst 59(11):859–866
18. Gallardo-Alvarado J (2012) Jerk analysis of a six-degrees-of-freedom three-legged parallel manipulator. Robot Cim Int Manuf 28:220–226
19. Ball RS (1900) A treatise on the theory of screws. Cambridge University Press, Cambridge (reprint 1998)
20. Sugimoto K, Duffy J (1982) Application of linear algebra to screw systems. Mech Mach Theory 17(1):73–83
21. Gallardo-Alvarado J (2014) A simple method to solve the forward displacement analysis of the general six-legged parallel manipulator. Robot Cim Int Manuf 30:55–61

Joint Trajectory Optimization Using All Solutions of Inverse Kinematics of General 6-R Robots

U. Kuenzer and M.L. Husty

Abstract Based on a paper [1] that introduced an algorithm which allows to transform all solutions of the inverse kinematics of a general 6-R robot into continuous joint trajectories we present in this paper methods to use the continuous joint trajectories for path optimization. For this purpose the motion is discretized and at each instant the inverse kinematic is computed using a fast algorithm developed in [2, 3]. In the set of resulting joint angles continuous paths are determined and the resulting sets of points are interpolated with quintic splines. Different possibilities for trajectory optimization are discussed.

Keywords Inverse kinematics · General 6R-robot · Trajectory planning · Trajectory optimization · Student paper

1 Introduction

The inverse kinematics (IK) of general serial 6-R robots was in the 1980s of the 20th century considered to be one of the most challenging problems in robot kinematics. The first solution to this problem was given by Lee and Liang [4]. Many papers followed, the most popular being [5], a solution that is mostly cited in textbooks when the inverse problem is discussed. An overview of the existing literature can be found in the thesis [2, 3], where a new approach to the inverse problem was presented. This solution algorithm uses the Study parametrization of the Euclidean displacement group $SE(3)$ and needs much less equations than Raghavan's algorithm (eight compared to fourteen). Furthermore, the starting

U. Kuenzer (✉) · M.L. Husty
Institute for Basic Sciences in Engineering, AB Geometry and CAD, University Innsbruck, Innsbruck, Austria
e-mail: ulrich.kuenzer@student.uibk.ac.at

M.L. Husty
e-mail: manfred.husty@uibk.ac.at

M. Ceccarelli and E.E. Hernández Martinez (eds.), *Multibody Mechatronic Systems*, Mechanisms and Machine Science 25, DOI 10.1007/978-3-319-09858-6_40

equations are so simple that they can be formulated completely general, i.e. without specifying the Denavit-Hartenberg parameters, making the algorithm applicable to any thinkable robot architecture without reformulation. Study parametrization is an algebraic parametrization of SE(3), using eight parameters, that can be interpreted as homogeneous coordinates in a seven dimensional projective space P^7. To meet the dimensions of the Euclidean displacement group the coordinates have to fulfill a quadratic equation, corresponding to a quadric in P^7, the so called Study quadric. Detailed information on this parametrization and its use in kinematics can be found e.g. in [6].

Within a software developing project called Kinsoft this algorithm was implemented in C# and this package allows to compute all inverse kinematics solutions fast along a given motion trajectory at as many instances of the trajectory as specified. With motion trajectory (or equivalently motion curve) a curve on the Study quadric is meant. Note that a curve on the Study quadric corresponds to a one parameter motion in the Euclidean space. Mathematically the curve is represented by an eight dimensional vector function encoding position end orientation of the end effector (EE). At any instant the IK algorithm returns all solutions of the inverse kinematics. Having all solutions of the inverse kinematics at many instances the following problems arise:

1. How can the solutions for different joint angles be separated such that a continuous path for each joint angle and each of its different solutions can be generated?
2. How can the discrete solution set be transformed into curves that have some desired properties, like stable behaviour at the end points for at least three orders of differentiability?
3. How can the data be used to distinguish the different solutions in order to find out which joint trajectory path is optimal according to some optimizing criteria.

The first two items are discussed in [1]. The results of this paper are recalled briefly in Sect. 2. In Sect. 3 the optimization strategies are presented. Section 4 shows some examples where one can clearly see that the presented optimization algorithm allows to choose between different solutions of the inverse kinematics to optimize the desired behaviour of the manipulator.

2 Separating the Joint Paths

In this section results from [1] are recalled to make the optimization strategies of Sect. 3 understandable.

When the Denavit-Hartenberg parameters of a 6-R robot are specified and a desired trajectory of the EE in position and orientation is given then the Kinsoft program [7] will return a text file that graphically processed yields an output as shown in Fig. 1. On the axis of abscissae of this plot one can see that 1,000 points on the EE trajectory have been used to compute the inverse kinematics. On the axis

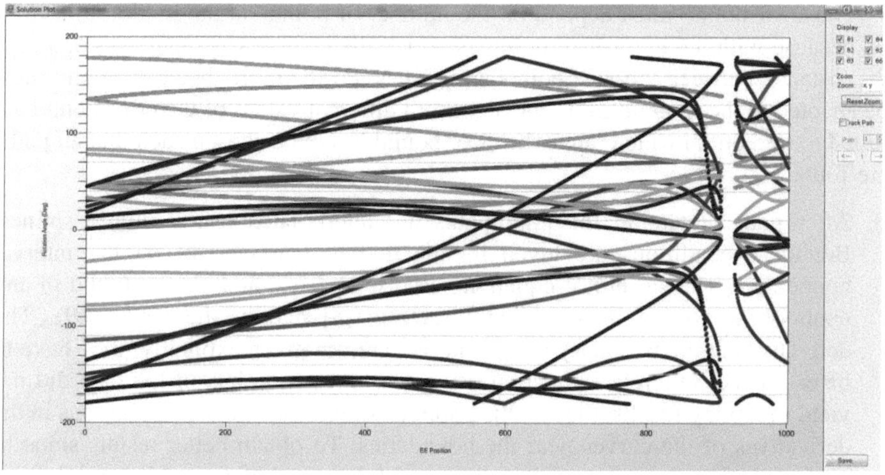

Fig. 1 Joint angle values resulting form inverse kinematics along an EE trajectory

of ordinates the corresponding joint angles are displayed in different colours in the range of $-\ 180°...180°$. Although it seems that there are continuous curves, the program can only return discrete sets of joint angles at every instant.

Remark The trajectory in the example of Fig. 1 had been chosen such that the manipulator hits its boundary which can be seen clearly because of the gap in all joint trajectories. The path generation algorithm follows the following steps:

1. Data import from the Kinsoft output and data storing in special arrays.
2. Separating the paths using the distance function.

Definition 1 Let $a, b \in [-\ 180^{\circ}, 180]^6$ be two vectors. Then the difference between the two sets of angles is defined as

$$\|a-b\|_p^W := \left\| \begin{pmatrix} min(|a_1-b_1|, 360-(|a_1-b_1|)) \\ min(|a_2-b_2|, 360-(|a_2-b_2|)) \\ min(|a_3-b_3|, 360-(|a_3-b_3|)) \\ min(|a_4-b_4|, 360-(|a_4-b_4|)) \\ min(|a_5-b_5|, 360-(|a_5-b_5|)) \\ min(|a_6-b_6|, 360-(|a_6-b_6|)) \end{pmatrix} \right\|_p$$

where on the right side any p-Norm from $\mathbb{R}^6$ can be used.

An upper bound of solutions is given by the number of solutions n_1 in the first point of the motion curve. In order to find a continuation of the first solution of the first point θ_1^1, the differences $\|\theta_1^1-\theta_i^2\|_2^W, i=1,\ldots,n_2$ are computed. If the minimum of these differences is smaller than an error bound ϵ, then the solution is added to the path.

The error bound must depend on the number of points on the motion trajectory m, because with the number of points in which the inverse kinematics is computed the distance between the data points diminishes. Furthermore the error bound must be adapted to the used norm. In the algorithm different values of the error bound are tested to find out in which range the error bound has to be chosen such that all paths are found.

3. After path separation the joint paths are interpolated using quintic splines. Boundary conditions are given by the first two derivatives on the interval boundaries. The system of equations used is classical and can be found in any textbook on spline interpolation (see. e.g. [8] and in the thesis [9]). The derivatives at the boundary of the interval are not given explicitly, they have to be estimated. Discrete estimation using the first two points of the data did not yield satisfying results, the resulting splines showed unwanted oscillations in the derivatives of the curves near the boundaries. To obtain better results smoothness of the curves was used. The data which have to be interpolated result from the solution of a polynomial system of equations. If the solutions are not complex then they have to describe a smooth C^∞ curve in $\mathbb{R}^6$. Using this fact the derivatives in the boundary point were approximated. To do this the first six and the last six points of the data were used to construct a quintic polynomial. The derivatives of those two polynomials were used as boundary conditions for the interpolation of the path splines.

Figure 2 shows the result of the three steps explained above for a classical example from the literature (example nr.7 from Wampler-Morgan [10], abbreviated CW7 in the following). The Denavit-Hartenberg parameters of the manipulator are given in Table 1.

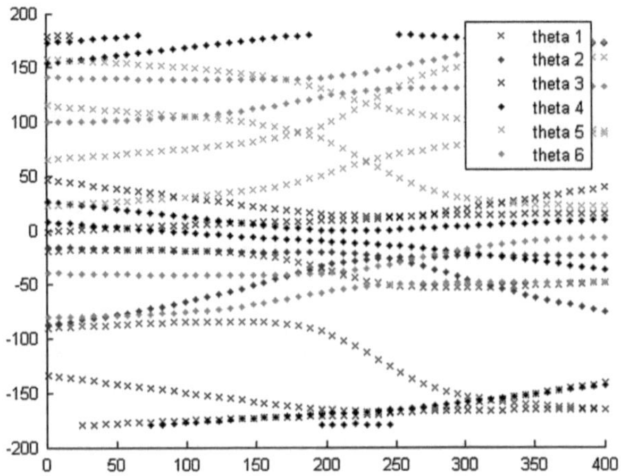

Fig. 2 Joint angles of the four continuous paths in joint space of 6-R robot performing a continuous EE motion

Table 1 DH-parameter of example CW7

a	$\frac{3}{10}$	1	0	$\frac{3}{2}$	0	0
d	0	0	$\frac{2}{10}$	0	0	0
α	$\frac{\pi}{2}$	$\frac{\pi}{18}$	$\frac{\pi}{2}$	$\frac{\pi}{18}$	$\frac{\pi}{2}$	$\frac{\pi}{18}$

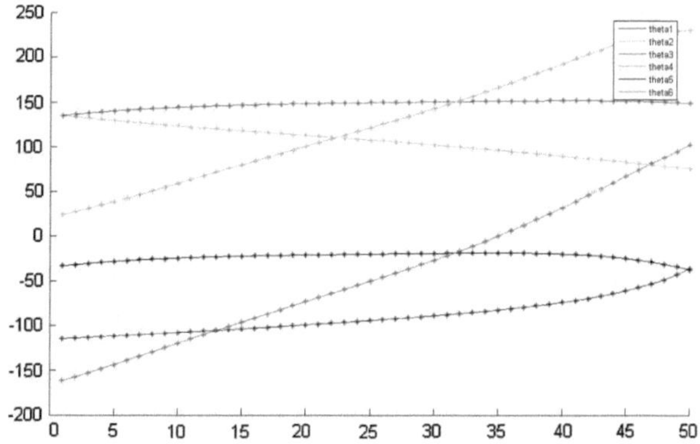

Fig. 3 Interpolated joint angle paths of one solution of a 6-R robot performing a continuous EE motion

The motion which has been performed is given by the Euclidean motion matrix $\mathbf{B}_C$:

$$\mathbf{B}_C = \begin{pmatrix} 1 & 0 & 0 & 0 \\ -1.140175 - t & -0.760117 & -0.641689 & 0.102262 \\ 0.133333 - t & 0.133333 & 0.991071 & 0 \\ \frac{t}{2} & -0.635959 & 0.766965 & 0.085558 \end{pmatrix}. \qquad (1)$$

Note that European notation is used in $\mathbf{B}_C$, writing the translation part of the motion in the first column and the homogenizing coordinate at first place. In Fig. 2 the upper boundary and the lower boundary have to be identified, and some angles (e.g. θ_5) seem to have four paths and others only two (e.g. θ_1). This is due to the fact that in these angles two solutions coincide. Figure 3 shows a typical result of this procedure where one of the possible continuous solutions of the joint trajectories of a robot motion has been plotted after performing the quintic spline interpolation.

3 Joint Trajectory Optimization

After the first three steps of the algorithm quintic splines of the joint trajectories are given. Because of the fact that the inverse kinematics yields more than one solution more than one continuous solution in the joint trajectories will exist. Therefore one has the possibility to choose among the solutions according to some optimization criteria. As joint angles are computed in the inverse kinematics, the chosen criteria should be linked to the joint angles along a given motion trajectory. In a first choice (type 1 optimization) we will minimize the overall change in the joint angles and in a second choice (type 2 optimization) we will ask for the solution which has a minimum change in joint angular velocities for all joints.

To perform this task a function $f : [t_1, t_2] \mapsto \mathbb{R}$ is considered, which determines the joint angles and which is at least twice continuously differentiable. Then $f'(t) = \frac{df}{dt}$ determines the change of joint angles or the joint velocity. The change of the velocity $f''(t) = \frac{d^2f}{dt^2}$ yields the joint acceleration. The overall joint angle changes are given by the integral $\int_{t_1}^{t_2} f'(t)dt$. This integral has the disadvantage that positive and negative velocities cancel therefore we have taken the absolute values of the functions $\int_{t_1}^{t_2} |f'(t)|dt$ and $\int_{t_1}^{t_2} |f''(t)|dt$. Integration of the absolute value of the velocity function yields information about the total length of the joint trajectory. The shorter the trajectory the less movement in the joint will be. Analogous consideration gives information that a smaller value for the integral $\int_{t_1}^{t_2} |f''(t)|dt$ yields a smaller change of joint velocities in the interval $[t_1, t_2]$.

The interpolation curves are polynomials of degree five. Therefore it is possible to determine the integral exactly with help of quadrature rules. Because the curves are continuous and are piecewise polynomial, integration can be done separately for each piece of the curve. For computation of the integrals Gauss quadrature rule was used

$$\mathcal{I}_n := \sum_{k=1}^{n} \sigma_k f(\lambda_k) \text{for } f \in [t_1, t_2], \tag{2}$$

with the nodes $\lambda_1, \ldots, \lambda_n \in \mathbb{R}$ and the weights $\sigma_1, \ldots, \sigma_n$.

4 Examples

4.1 Example 1

In a first example the Denavit-Hartenberg parameters of the CW7 example in Table 1 were slightly modified as in (Table 2). The motion performed is the motion of Eq. (1) in the interval $[-\frac{1}{2}, \frac{1}{2}]$.

Table 2 DH-parameter of Example 1

a	$\frac{3}{10}$	1	0	$\frac{3}{2}$	0	0
d	0	0	$\frac{2}{10}$	0	0	0
α	$\frac{\pi}{2}$	$\frac{(\pi+0.5)}{18}$	$\frac{\pi}{2}$	$\frac{\pi}{18}$	$\frac{(\pi-0.5)}{2}$	$\frac{\pi}{18}$

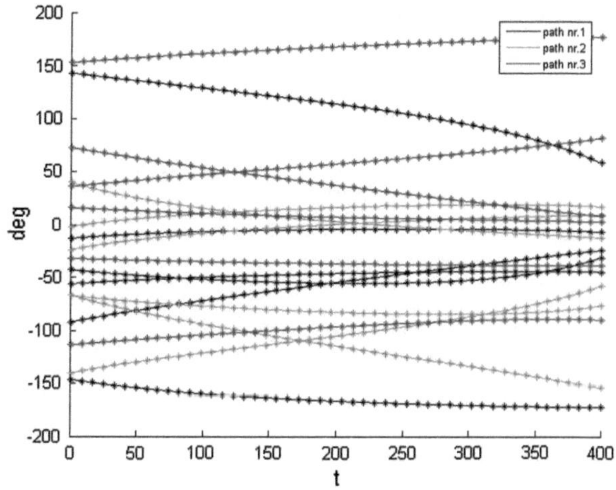

Fig. 4 Interpolated joint trajectories of Example 1

Table 3 Result of the optimization

Optimization type	Path Nr.1	Path Nr.2	path Nr.3
1. Type	39.9009	50.5817	30.0131
2. Type	0.2598	0.3153	0.0824

This example has three continuous paths (Fig. 4). The results of the optimization are shown in Table 3. The average of the six joint trajectory lengths is the smallest in the green path 3. Therefore this path is optimal when smallest overall change of joint parameters is desired. It can also be seen that the same joint trajectories are optimum when smallest overall change of joint velocities is desired. Once this result is obtained it is of course enough to put the manipulator in the starting values of path 3 and then it will stay on this path, because there is no singularity on this path. A boundary singularity would be seen in such a plot when a path ends, i.e. when two solutions meet in a point having a vertical tangent. This behaviour can be seen in Fig. 1 where the motion trajectory was chosen such that the manipulator hits its boundary and therefore no continuous path is possible. There is a gap in all joint trajectories.

4.2 Example 2

In this example the DH-parameters of CW7 are changed according to Table 4. The motion performed is the same as in Eq. (1) in the interval $t \in [-\frac{1}{4}, \frac{1}{4}]$. This example is especially interesting, because 14 continuous joint paths exist. The given task can be performed with 14 different configurations of the manipulator. It should be obvious that these 14 paths are not equivalent The 14 continuous joint trajectories are shown in Figs. 5 and 6.

Table 4 DH-parameter of Example 2

a	$\frac{3}{10}$	$\frac{11}{10}$	0	$\frac{3}{2}$	0	0
d	0	0	$\frac{25}{100}$	0	0	0
α	$\frac{\pi}{2}$	$\frac{(\pi+0.2)}{18}$	$\frac{(\pi+0.1)}{2}$	$\frac{\pi}{18}$	$\frac{(\pi-0.3)}{2}$	$\frac{\pi}{18}$

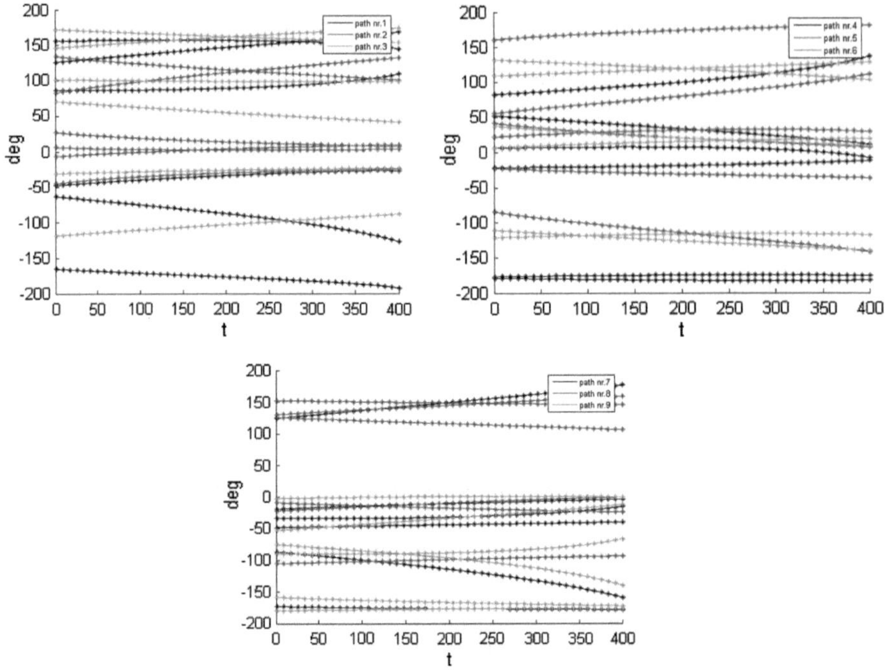

Fig. 5 Interpolated joint trajectories 1–9 of Example 2

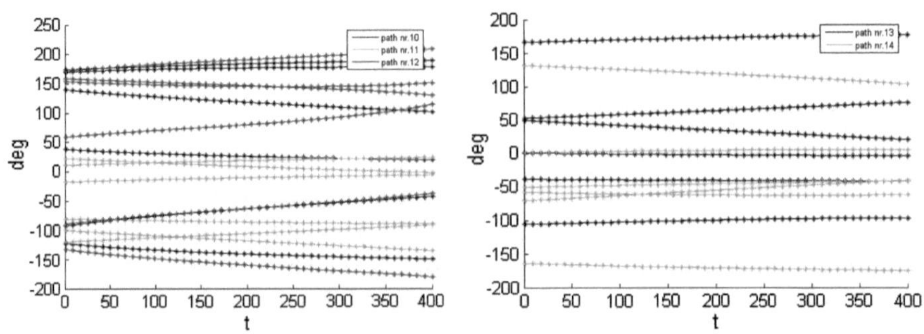

Fig. 6 Interpolated joint trajectories 10–14 of Example 2

Table 5 Results of the optimization Example 2

Optimization type	Path Nr.1	Path Nr.2	Path Nr.3	Path Nr.4	Path Nr.5	Path Nr.6	Path Nr.7
1. Type	31.7504	24.1101	19.4176	21.5736	32.5666	20.5959	28.8555
2. Type	0.1897	0.0846	0.0213	0.1273	0.1165	0.0542	0.0954
	Path Nr.8	Path Nr.9	Path Nr.10	Path Nr.11	Path Nr.12	Path Nr.13	Path Nr.14
1. Type	17.2200	25.4676	26.2039	20.1783	39.7655	13.5023	14.7125
2. Type	0.0157	0.1321	0.0984	0.0251	0.2140	0.0163	0.0280

In the result of the optimization one can see big differences in the different trajectories (Table 5). Concerning the average joint motion change trajectory 13 is optimal whereas the path 12 is the worst path. One can see this also in the plot of Fig. 6 on the right side. Trajectories of path 13 (black) have almost horizontal curves, whereas the trajectories of path 12 have much more vertical deviation. Concerning the overall velocity change averaged over all joints (type 2) one can see that path 8 does slightly better than path 13.

Remark In a real robot joint limits would have to be taken into account. This has not been done in this example because the joint limits of the Wampler example are not specified. But it would be no additional effort to plot these limits and make a decision if the given motion can be performed without hitting a joint limit.

Summarizing the results of both optimization algorithms one can state clearly that it matters which starting configuration is chosen from the different possibilities offered by the different solutions of the inverse kinematics. The better choice of a starting can reduce the cycle time of a desired task considerably.

5 Conclusion

The inverse kinematics of a general 6-R manipulator yields up to 16 solutions for the joint angles when the end effector pose is given. Using the fast algorithm developed in [2, 3] the inverse kinematics can be computed along a given end effector motion efficiently. The algorithm presented in this paper detects in the set of joint angles continuous paths and the joint trajectories of these paths are interpolated with quintic splines. Having polynomial curves for joint trajectories, velocities and acceleration a optimization procedures were discussed to decide which of the possible solutions is optimal according to a given optimization criterion like e.g. minimum overall change of joint motion or minimum average overall change of joint velocities. This work can help to decide in which starting pose the manipulator should be brought to perform a given task in a shorter time or with overall smaller change of joint velocities.

6 Acknowledgments

The research presented in this paper was supported by the Translational research project KineControl of the Standortagentur Tirol, Austria. M. Husty acknowledges the support of FWF project P 23831-N13, Algebraic Methods in Collision Detection and Path Planning.

References

1. Kuenzer U, Husty M (2014) Joint trajectory generation using all solutions of inverse kinematics of general 6-R robots. In: Proceedings of ARW (Austrian Robotics Workshop), 2014. accepted for publication and presentation
2. Pfurner M (2006) Analysis of spatial serial manipulators using kinematic mapping. PhD thesis, University of Innsbruck, October 2006
3. Husty M, Pfurner M, Schröcker H-P (2007) A new and efficient algorithm for the inverse kinematics of a general serial 6R manipulator. Mech Mach Theory 42(1):66–81
4. Lee H, Liang C (1988) Displacement analysis of the general 7-link 7R mechanism. Mech Mach Theory 23(3):219–226
5. Raghavan M, Roth B (1990) Inverse kinematics of the general 6R manipulator and related linkages. Trans ASME J Mech Des 115:228–235
6. Husty M, Schröcker H (2012) 21st century kinematics, ch. Kinematics Algebraic Geometry, pp 85–107. Springer, Berlin
7. Pfurner M, Husty ML (2010) Implementation of a new and efficient algorithm for the inverse kinematics of serial 6R chains. In: Pisla D, Husty ML, Corves B, Ceccarelli M (eds.) New trends in mechanism science, pp 91–98, Springer, New York
8. Späth H (1973) Spline-algorithmen zur Konstruktion glatter Kurven und Flächen. Oldenburg, München
9. Kuenzer U (2014) Bestimmung von glatten und optimierten Kurven im Raum der Gelenksvariablen eines allgemeinen 6-R Manipulators aus den Daten der inversen Kinematik. Master's thesis, Universität Innsbruck, 2014
10. Wampler C, Morgan A (1991) Solving the 6R inverse position problem using a generic-case solution methodology. Mech Mach Theory 26(1):91–106

Mill Setup Manual Aided by Augmented Reality

F. Suárez-Warden, E. González Mendívil, H. Ramírez, L.E. Garza Nájera and G. Pantoja

Abstract As a result of globalization, the degree of complexity in manufacturing systems and the existence of very elaborate processes are increasing. At the same time there is a contradiction, while high quality is demanded to the operators, reduction in setup and operation time is also required. It is crucial to deploy a particular interest in matters of efficiency that drives us to incorporate emerging technologies in these processes. One of them, The Augmented Reality (AR), is known as the combination of virtual and real images without completely replacing the real environment which is interactive and registered in 3D. This work expands the psychomotor and cognitive ability of an operator of a pipe manufacturing machine during the alignment process by incorporating AR in the correspondent manual.

Keywords Mill · Augmented reality · Setup manual · Cognitive ability · Complex process · Student paper

---

With the student contribution of H. Ramírez, L.E. Garza Nájera and G. Pantoja.

F. Suárez-Warden (✉)
Mechatronics and Automation Department, Tecnológico de Monterrey, Campus Monterrey, Monterrey, México
e-mail: fersuarezw@gmail.com

E. González Mendívil
Mechanical Engineering Department, Tecnológico de Monterrey, Campus Monterrey, Monterrey, México
e-mail: egm@itesm.mx

H. Ramírez · L.E. Garza Nájera · G. Pantoja
Center for Innovation in Design and Technology (CIDYT), Tecnológico de Monterrey, Campus Monterrey, Monterrey, México
e-mail: hramirez86@gmail.com

L.E. Garza Nájera
e-mail: lgarzan15@hotmail.com

G. Pantoja
e-mail: gaabriel.pantoja.432@gmail.com

M. Ceccarelli and E.E. Hernández Martinez (eds.), *Multibody Mechatronic Systems*, Mechanisms and Machine Science 25, DOI 10.1007/978-3-319-09858-6_41

1 Introduction

Augmented Reality technology permits the users interact with the real world and virtual animations at the same time on a screen with the main objective of supporting and enhancing activities and cognitive processes of users [4, 8].

Augmented Reality can be a conductor tool for technicians during maintenance and training. This technology reduces the working time and probability of accomplishing error because it has a guide with clear instructions and interactive animatronics for users.

The scope of this project is creating a prototype of augmented reality and interface for the operator who works in a productive line with complex assembly processes.

This enforcement has been effectuated for any mobile device with tactile screen and Android™ operative system. The motivation is to improve current methods for training by adding critical information which is not available at workstation and difficult to include in work manuals.

Vision field of the worker is upgraded or augmented via 3D models which encompass specific content according to alignment process for steel plates in the production of pipes for oil industry. The case study for this project is the change in diameter mill production. This operation consists of several tasks of which the most critical are identified and those where workers are statistically more likely to make mistakes, so more attention and thorough maintenance is required. As shown in Fig. 1, a mobile device is used with the augmented reality system that displays the 3D models and animations on the training manual, explaining the process of change in diameter. In the following sections, a description of the problem and the entire application will be described in detail.

Fig. 1 Tablet with application of AR

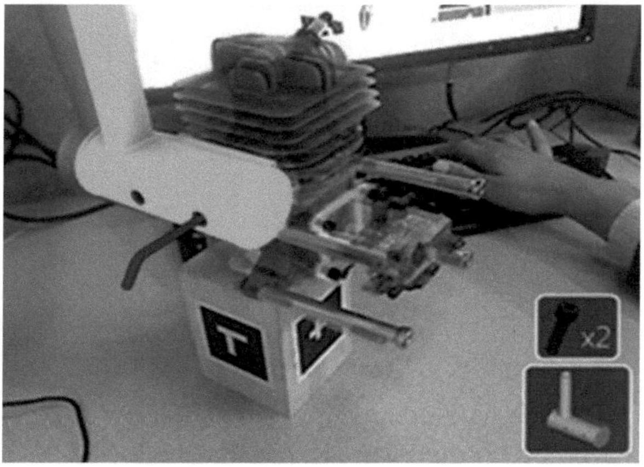

Fig. 2 Augmented pieces of a RF motor

2 Theoretical Framework

Augmented Reality (AR) seeks to superimpose additional information to our physical environment (Fig. 2) by using models, images, audio, vibration and text to search the user to have more critical information from the world around him in real time, unlike virtual reality where the user is immersed in a world made completely by computer.

Two definitions that are used to describe what is actually commonly raised in the literature, are the following: one of the author Azuma [2] "Augmented reality combines real and virtual elements, is interactive in real time and is registered in 3D", and other of Milgram [6] where he describes as a continuum from the real environment to a pure virtual environment, in between there are Augmented Reality (closer to the real environment) and Augmented Virtuality (is closer to the virtual environment).

3 Methodology

Normally the technique that takes to create an application of augmented reality is to first analyze the process to measure only the parts that you want to model, to show then virtually over the actual image and then apply to them textures and animations, such as Castro [3] and Mercado [5] explain. Already with the modeled part, a bookstore of AR with another graphics library is used to display the piece with the image that the camera is capturing and if there are changes to be made iterate the same steps above again (Fig. 3 with elements which has been mentioned by these Mexican authors).

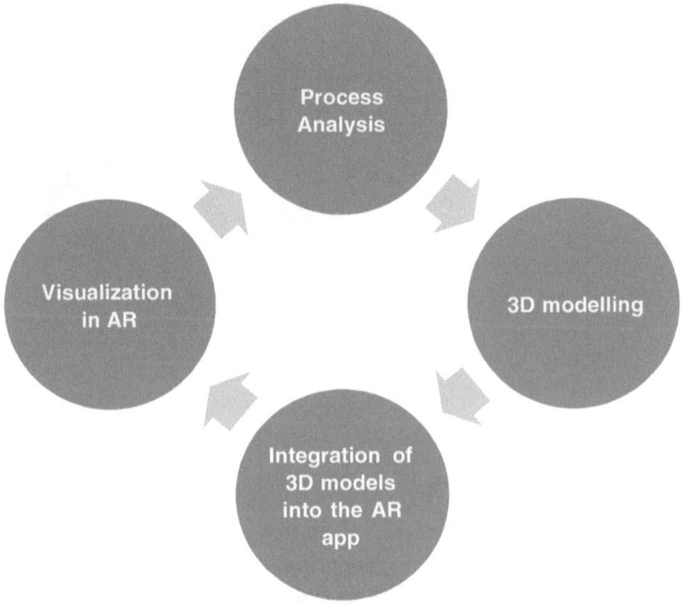

Fig. 3 Methodology for creating an application of AR

The main tasks to be performed for the application to work properly augmented reality are as follows, according to Wagner [7]:

(1) Starting the graphic user interface (GUI)
(2) Tracking the target image
(3) Estimation of scale and position
(4) Rendering

The overhead of the application is the segment that controls how the models will interact with the user and the tracking of the position. This section includes scripts with codes (usually written in C # or C + +) that govern the functions of the project. To read the patterns of the target image properly, you need to have a device that has a high resolution camera. The renders are used to create a more realistic animation but influence the user details to include.

This application can be used in manufacturing areas where workers must be trained in a specific function of assembly or maintenance of machines. It gives them the ability to work more efficiently, saving time and reducing the possibility of failure by a bug that could cause loss of money or production stoppages.

The first step in creating a content of AR is the modeling of the system or components that plans be supported by this technology. The modeling is performed in software capable of generating 3D geometries; you should always try to make an exact copy of the virtual system or physical component with the greatest amount of detail. Stage 3D modeling can be performed on any design software (CAD) aided by computer. Examples of this type of parametric software are: SIEMENS NX8 Unigraphics, PTC Creo Elements /Pro, Dassault Systemes CATIA V5, among others. Caution is advised to choose the format in which the file generated in 3D CAD is saved as it must then be processed into software to generate render so it seeks to ensure compatibility. Another important step in the modeling stage is to get the physical parts of the system or components that are going to work and perform correct classification to have good tracking parts through future stages of the process of forming AR application.

Once you have identified the components of interest, the modeler proceed with the capture of information. As shown in Fig. 4, physical measurements of the geometric characteristics (length, diameter, thickness, etc.) are taken and diagrams for the system or components are made. These steps of data collection should be properly planned and executed with the use of measuring devices such as micrometers, vernier, rules or any instrument to minimize errors in computer generated 3D models.

Fig. 4 Catching information about 3D geometries

After the models were created in CAD software, is required to modify some parameters using Autodesk 3Ds MAX or Maya to give them a specific look and feel. The skin and the color are applied individually to each section of the models. The panel color is used to apply the correct tone and intensity, depending on the characteristics of the real model. The second step is to apply the shadows where required. It is recommended to test the models in different positions and sizes to ensure that no further changes are required in the surface parameters.

You need to create animations with render models that resemble to the procedures as much as possible to avoid confusion on the end user. During the animation, the user should consider the position and movement of each of the pieces to successfully complete the installation or removal process (as applicable).

4 Problem Description and Work

Castles change is a process that is performed to adjust the diameters of the tubes produced. The mill has a number of stages in which the plates, introduced in the machine, are bending to finally form the tubes. The castles which allow you to change the position of the rollers that are responsible for bending the plates at different stages, so it is extremely important to run well change them so that the final product has the dimensions required by the customer.

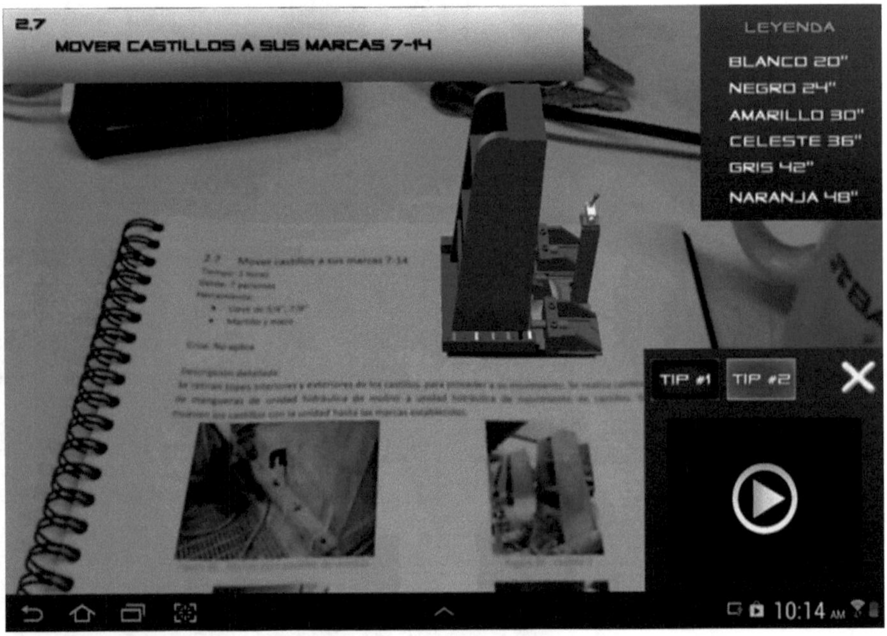

Fig. 5 AR Application: moving castles

The setting of castles is a process that takes about four times a year, due to the different orders received by the company. In the field was identified that had long in making the change to the new diameter was needed and there were no manual, and as new technicians who were joining the company found it difficult to be prevented when preparation will take place (setup) is also needed to streamline this process, since during the settlement of the castle should stand the machine, so a manual with parts became involved, and the steps required for positioning or replacement of castles, which allowed training technicians to see details without requiring that the process was in progress and provide guidance to those who already knew, but needed to consult specific details, such as dimensions and parts that required.

Once this manual was made, certain elements were very difficult to explain, so that was identified with the help of augmented reality that could make 3D models of the machine and to encourage them to provide additional information (Fig. 5).

5 Results

Once the performance of the manual was completed with markers for augmented reality we proceeded to organize tests to determine if this technology really brought an improvement in the process of adjustment of the castles. Figure 6 shows the

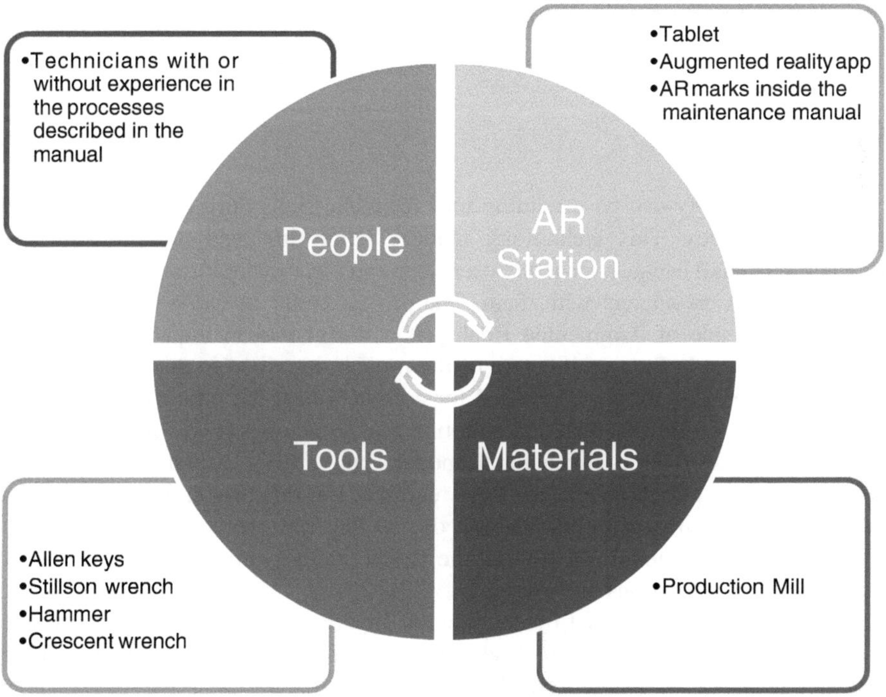

Fig. 6 Structure of the tests

structure of the tests to be carried out, in a scheme similar to that used by Mercado [5]. This diagram, presented with four phases, allows us to distinguish the subjects participating in the tests and the materials and tools that will be provided in their realization in the AR station.

We saw an improvement in time, with the help of the modified manual as a week working 24 h a day, was made, able to reduce considerably time. And also with the help of the most critical steps in positions of AR content, operators could better understand the process.

We also identified that as more operators using the augmented reality application better understand the benefit of this technology, we were making changes to certain contents of AR to be more explicit (see Fig. 3). For example in the process of positioning castles bender plates (mill), several crucial changes contained in one of their models as well as their animation were made. Also videos and explanatory text on each of these steps were added to give the user best practices in training and operation.

5.1 Disclaimer About Data to Analyze

Few technical data are included due to industrial secret reasons and/or lack of information about antique or used machines. This is the first stage of a work in progress. The paper presents production tasks with a mechatronic emphasis.

6 Conclusions

Augmented Reality can be a guiding tool for technicians during maintenance and training processes. This technology reduces labor time and the probability of making a mistake because leads the users through clear instructions and animations with the ability to interact with them. The project could be carried out effectively generate evidence of Augmented Reality is a useful tool in improving manufacturing processes. In the specific case of the project described in this paper process, markedly decreased the time required for completion of the process.

At the time of the first tests it was found that sometimes it was not enough display an animated three-dimensional model, but a brief explanation given by someone with experience in the process required. For this reason, it was decided to add a video explaining the application on, so that the operator may observe the three-dimensional model shown on the tablet along with a video that briefly explained the instructions to follow.

Finally, note the potential of Augmented Reality technology not only in manufacturing processes, but also in other types of activities where the transmission of knowledge is needed to successfully complete a task.

Acknowledgments The authors want to recognize the support of The Research Group in Intelligent Machines from ITESM (Tecnológico de Monterrey) and SEMARA project in collaboration with Polytechnic University of Valencia (Spain) and AEROMEXICO, with funding by Consejo Nacional de Ciencia y Tecnología (grant #116413 *Sistema de Entrenamiento para Mantenimiento Aeronáutico Mediante Realidad Aumentada*).

References

1. Alcázar JL (2010) Sistemas de transferencia experta con realidad aumentada en función de la complejidad de la tarea a desarrollar: aplicación a los procesos de mantenimiento aeronáutico. Instituto Tecnologico y de Estudios Superiores de Monterrey
2. Azuma RT (1997) A survey of augmented reality. Presence: teleoperator and virtual environments, vol 6(4). MIT Press, Cambridge, pp 355–385
3. Castro C (2012) A methodological framework for augmenter reality technological applications in industrial field. Instituto Tecnologico y de Estudios Superiores de Monterrey
4. Crecenzio FD (2011) Augmented reality for aircraft maintenance training and operations support. IEEE Comput Graphics Appl 31:96–101
5. Mercado E (2010) Improving training and maintenance operations in areonautic related process using augmented reality. Instituto Tecnologico y de Estudios Superiores de Monterrey
6. Milgram P, Takemura H, Utsumi A, Kishino F (1994) Augmented reality: a class of displays on the reality-virtuality continuum. Telemanipulator and Telepresence Technologies, pp 282–292
7. Wagner D (2009) Making augmented reality practical on mobile phones. IEEE Comput Soc 29 (3):12–15
8. Zanetti N. et al (2004) Assistance to maintenance in industrial process using an augmented reality system. Ind Technol 2:848–852

Dimensional Synthesis of a Planar Parallel Manipulator Applied to Upper Limb Rehabilitation

Ileana P. Corona-Acosta and Eduardo Castillo-Castaneda

Abstract Upper limb rehabilitation exercise programs have shown to be useful in reducing pain, improving mobility and functionality. This article presents dimensional design of a novel mechanism for rehabilitation based on joint trajectories extracted from common exercises for stroke recovery. A computer simulation was done in order to prove the performance of the rehabilitation trajectories by the mechanism.

Keywords Robotics rehabilitation · Dimensional synthesis · Trajectory tracking · PPM · Student paper

1 Introduction

The development of a mechanism to assist upper limb rehabilitation process appears by the need of having specialized equipment to automate, systematize and control the therapy process achieved at rehabilitation centers.

The protocol followed to rehabilitate upper limb is a combination of many techniques [8], one of them consists on providing movement to the arm (passive rehabilitation), this exercises [3] are made manually by a physiotherapist or with the use of some therapy devices.

Mechanotherapy [1] is a rehabilitation technique which uses mechanical apparatus to provide movement to certain body parts. Those apparatus are widely used in therapy sessions because of its geometric precision, ergonomic design, variety of movements, and so on.

I.P. Corona-Acosta (✉) · E. Castillo-Castaneda
IPN-CICATA Unidad Querétaro, 76090 Querétaro, Mexico
e-mail: icoronaa1300@alumno.ipn.mx

E. Castillo-Castaneda
e-mail: ecastilloca@ipn.mx

M. Ceccarelli and E.E. Hernández Martinez (eds.), *Multibody Mechatronic Systems*,
Mechanisms and Machine Science 25, DOI 10.1007/978-3-319-09858-6_42

Fig. 1 Arm skate used at CRIQ

Arm skate shown in Fig. 1 is a mechanotherapy device for range of motion exercises, commonly used by physiotherapists at *Centro de Rehabilitación Integral de Querétaro* (CRIQ) for treating patients with chronic hemiplegic.

There are many exercises that are performed with therapeutic skateboads, we selected the most frequently used in rehabilitation sessions. Figure 2 shows the 5 chosen exercises that were analyzed with image processing techniques.

2 Methodology for Trajectory Extraction

In this work, a procedure is proposed to obtain joint angles of shoulder, elbow and wrist while performing rehabilitation exercises. Four color markers were placed on the nearest point for each rotation axis of human limbs, the exercises were performed on the transverse plane of anatomical movement taking into account just rotational movement in each joint, for this reason just circular markers were needed to obtain such planar motion.

The method used to extract these trajectories is inexpensive and simple, a video of each exercise was taken from a parallel plane and decomposed into individual frames. A threshold was obtained from RGB value of the markers in order to implement color based segmentation. Once image filtering was done, found objects were labeled to finally get centroid coordinates of each marker in Cartesian space

To get a description of trajectories we need dynamic behavior of angles $(\theta_1, \theta_2, \theta_3)$ formed by limbs as shown in Fig. 3.

An angular description of these trajectories allows to characterize each movement taking into account the variability of anthropometric dimensions. This angular approach of rehabilitation exercises allows to adapt specific trajectory to the dimensions of any patient. With this, we have a good description of the movement performed by human arm during therapy sessions with 5 different exercises.

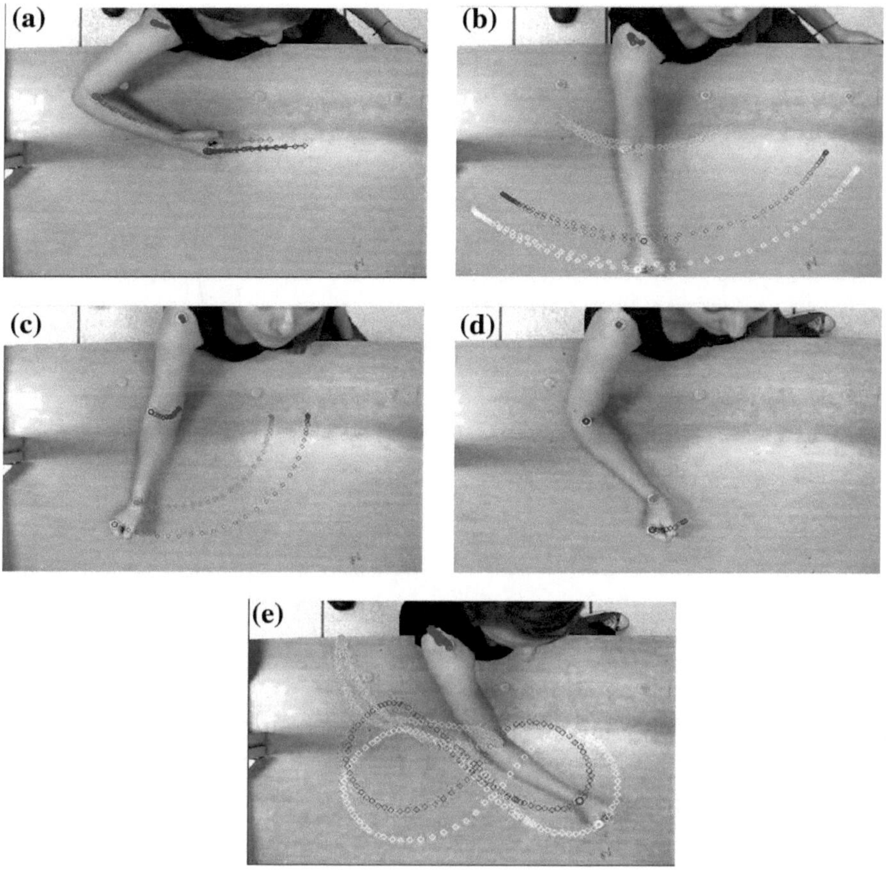

Fig. 2 Upper arm rehabilitation exercises: **a** shoulder adduction, **b** shoulder horizontal flexion, **c** elbow flexion, **d** wrist flexo-extension and **e** "8-figure" path

3 Estimation of Required Rehabilitation Workspace

In order to compute the workspace in cartesian coordinates required to effectively perform rehabilitation exercises, its necessary to transform from articular to cartesian description by using direct kinematics of human arm.

Direct geometric model of human arm (refer to Fig. 3) is calculated using Denavit-Hartenberg parameters as follows: (Table 1)

Position of knuckles (x, y) shown in Fig. 3, is calculated from matrix transformation $T = T_0^1\, T_1^2\, T_2^3$:

$$x = a\,\cos\theta_1 + c(f_1\cos\theta_3 - f_2\sin\theta_3) + bf_1 \tag{1}$$

Fig. 3 Human Arm basic diagram

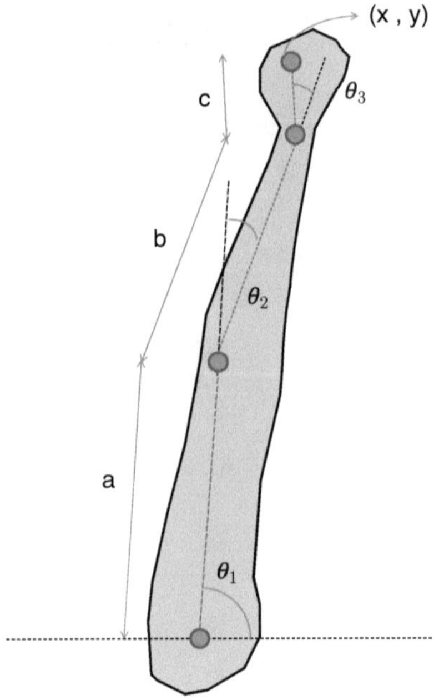

Table 1 Denavit and Hartenberg parameters of human arm

Link	θ	d	a	α
1	θ_1	0	a	0
2	θ_2	0	b	0
3	θ_3	0	c	0

$$y = a\,\sin\theta_1 + c(f_2\cos\theta_3 - f_1\sin\theta_3) + bf_2 \tag{2}$$

Where:

$$f_1 = \cos\theta_1\cos\theta_2 - \sin\theta_1\sin\theta_2$$
$$f_2 = \cos\theta_1\sin\theta_2 + \cos\theta_1\sin\theta_2$$

Direct geometric model of human arm requires as input data the length of each limb a, b, c, to obtain this values a review of anthropometric tables [9] was done. Maximum and minimum anthropometric measures of Mexican population is used to estimate required workspace region. After analyzing this data we computed lengths a, b, c (shown in Fig. 3) as follows:

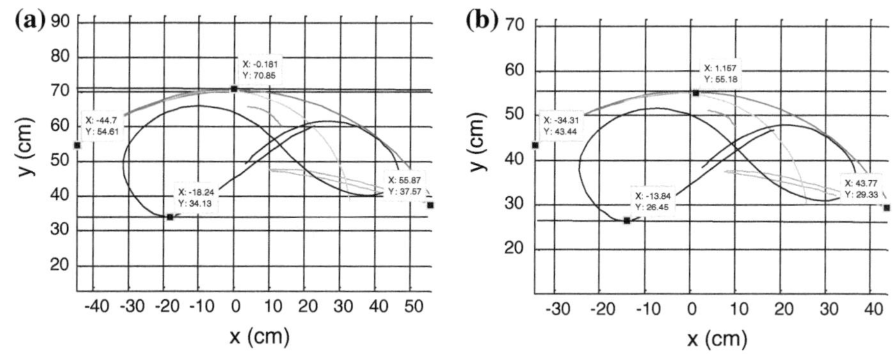

Fig. 4 Minimum and maximum exercise trajectories

Table 2 Boundary values of minimal and maximal exercise trajectories

Trajectories (cm)	X_{min}	X_{max}	Y_{min}	Y_{max}
Minimum	−34.31	43.77	26.45	55.18
Maximum	−44.7	55.87	34.13	70.85

Dimension	Min (cm)	Max (cm)
a	29	37
b	20	23
c	7	11.5

From above values, we can obtain exercise trajectories in cartesian space for minimum and maximum limb lengths of Mexican population (Fig. 4). An estimation of boundaries is made and we have that: (Table 2)

For convenience, by using these boundaries we approximate the workspace to a rectangular shape, which will be applied to estimate proper dimensions of our mechanism. This boundary was extracted taking the shoulder as a reference point (0,0). This means we can move this workspace in the Cartesian plane by changing the reference point.

4 Synthesis of the Mechanism

Since exercises are performed in a plane, we need a 3 d.o.f. mechanism, two for position and one for orientation; then a Planar Parallel Manipulator (PPM) can be used for this application. Positioning of mobile platform is needed in order to cover the workspace for analyzed trajectories and its orientation allows working the wrist at therapy sessions.

Fig. 5 3RRR planar parallel manipulator

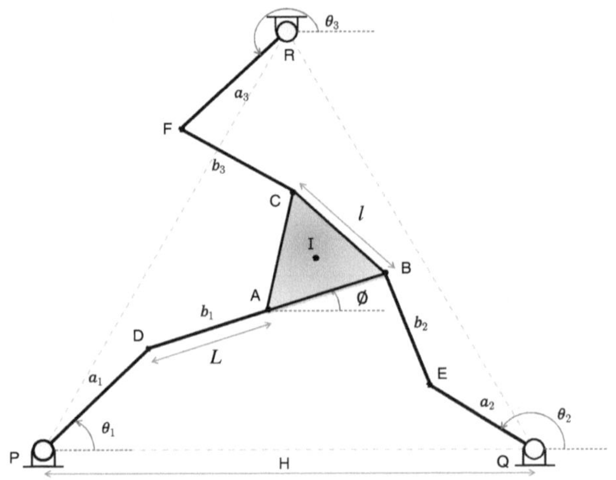

Nowadays, PPMs are widely used due to many inherent characteristics over the serial manipulators [6, 7], such as high stiffness, high payload capability, low moving inertia, and so on. Therefore, in this paper we propose to use a 3RRR planar parallel manipulator, which has been widely studied by many authors [2, 4, 5]. As shown in Fig. 5, the mechanism consists on a mobile platform (A, B, C) attached to a fixed platform (P, Q, R) by three kinematic chains. Each chain is formed by 2 links and 3 revolute joints; where its end-effector (A, B, C) and also its base (P, Q, R) are an equilateral triangle.

In this context, we need to know the length of the mobile platform (l), the length of fixed platform (H) and the length of each link (L), we consider that all links have the same length.

To cover the required workspace L can vary from $Y_{\min}$ to $Y_{\max}$; to avoid collisions between links a_i of mechanism l should be greater than L; finally by simple inspection we know that H should be greater than L. The following set of conditions are assumed:

1. $Y_{\min} \leq L \leq Y_{\max}$
2. $L \leq H \leq 3L$
3. $L < l < 2L$

At this point, we have a finite space of solutions for l, L and H lenghts. Additionally, we can define the mechanism by 4 main characteristic parameters (see Fig. 5): radius of mobile platform (R_1), radius of fixed platform (R_2), length of links a_i (R_3) and length of links b_i(R_4).

The relationship between geometric dimensions (l, H, L) and characteristic parameters (R_1, R_2, R_3, R_4) is:

$$R_1 = l/\sqrt{3}, \quad R_2 = H/\sqrt{3}, \quad R_3, R_4 = L \tag{3}$$

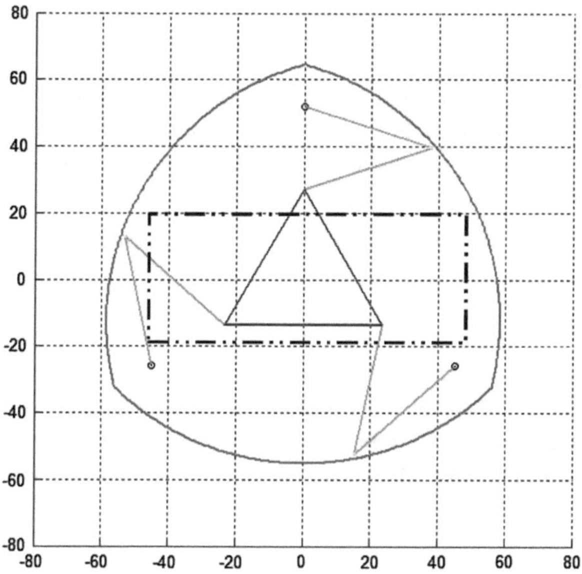

Fig. 6 Workspace of 3RRR planar parallel manipulator with estimated dimensions in gray line and required workspace for rehabilitation in blue dotted line

We used the *physical model of the solution space* technique explained by [5, 6] and used by [2], which claims that if a non-dimensional manipulator is optimal, the corresponding dimensional manipulator will be optimal as well. The dimensional parameters R_1, R_2, R_3, R_4 can be expressed as non-dimensional parameters r_1, r_2, r_3, r_4. Let:

$$D = \sum_{i=1}^{4} \frac{R_i}{4} \tag{4}$$

$$r_i = \frac{R_i}{D}(i = 1, 2, 3, 4) \tag{5}$$

Furthermore, in [5] authors show that the workspace volume can reach its maximum value if r_1 and r_2 are specified and $r_3 = r_4$, thus:

$$\frac{r}{2} = r_3 = r_4 \tag{6}$$

Besides, [2] studied motion/force transmition analysis of the planar 3-RRR parallel manipulator, based on the LTI (Local Condition Index) and introduced two global design indices: GTW (Good Transmition Workspace) and GTI (Global Transmition index). Using design space tool they found the relationship between the shapes of GTW and its geometric parameters, and conclude that architectures with the following characteristics of parameters, can be considered as candidates of optimal manipulator in terms of shape:

$$r_1 + r_2 < 2, \quad r_1 < \frac{r}{2}, \quad r_2 > \frac{r}{2} \tag{7}$$

With above considerations and with the use of inverse kinematics calculated as mention in [10], we found a set of possible non-dimensional parameters that satisfy our requirements. Finally we chose the combination with the smallest value of dimensional parameters, as follows:

$$L = 41, \quad H = 90, \quad l = 44 \quad \text{(cm)}$$

5 Results

The final workspace of the mechanism is shown in Fig. 6, as we can see the required workspace is inside the workspace of mechanism. To verify this, a simulation of the trajectories was done, taking into account minimum and maximum lengths of the anthropometric measurements (Fig. 7). Resulting exercises are shown in the Figs. 8, 9 and 10.

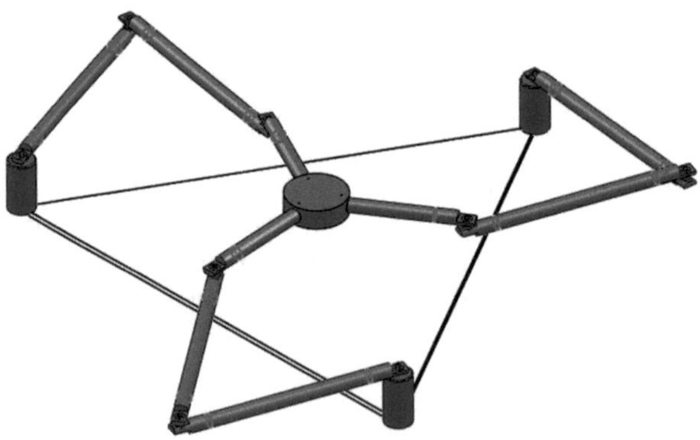

Fig. 7 CAD of mechanism with calculated dimensions

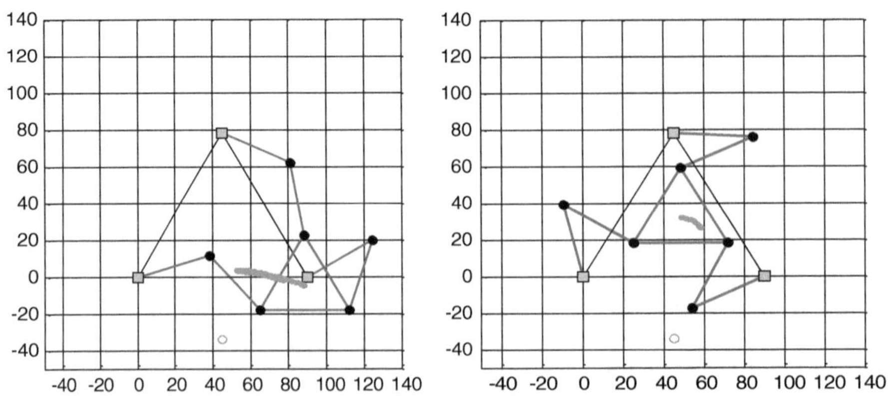

Fig. 8 Shoulder adduction and wrist flexo-extension exercises performed by mechanism

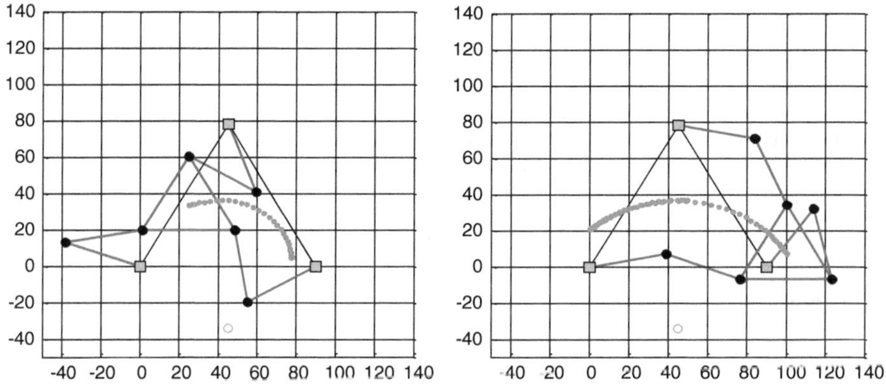

Fig. 9 Elbow flexion and shoulder horizontal flexion exercises performed by mechanism

Fig. 10 "8-figure" exercise performed by mechanism

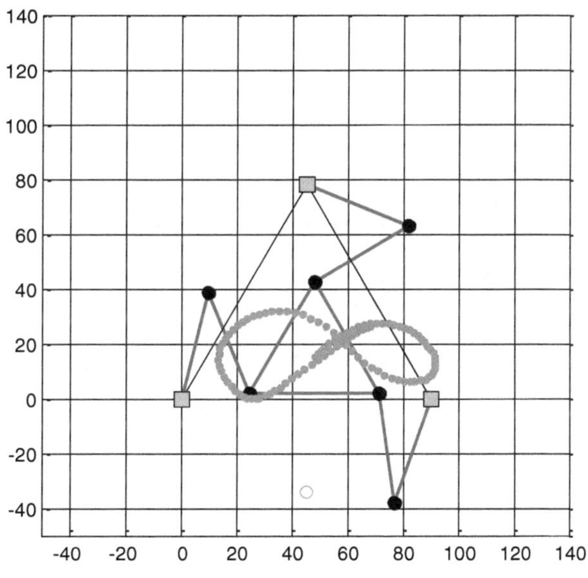

6 Conclusions

Some rehabilitation exercises done with arm skate at therapy sessions were analyzed, their angular description was extracted and a required workspace for this purpose was generated. Based on this calculation, we developed a way to estimate the design parameters of a mechanism which satisfies: degrees of freedom needed for trajectories, required workspace and no interference between legs of manipulator for rehabilitation tasks. The resulting mechanism was simulated in order to test if it was able to perform the exercises trajectories with the estimated geometric dimensions. As satisfactory results were obtained in the simulation, we proposed the design shown in Fig. 7. Future work includes the construction of mechanism based on this design and its control considering mentioned trajectories.

Acknowledgements This work is financially supported by IPN-SIP. The authors also like to acknowledge the help of CRIQ physiotherapists for their advice and discussion about rehabilitation protocols.

References

1. Arcas M, Galvez D, Leon C, Paniagua S, Pellicer M (2004) Manual de Fi-sioterapia. Modulo i Ebook. MAD-Eduforma
9. Ávila R, Prado L, González E (2001) Dimensiones antropométricas de población latinoamericana. Centro de Investigaciones en Ergonomia
2. Chen Y, Liu XJ, Chen X (2013) Dimension optimization of a planar 3-RRR parallel manipulator considering motion and force transmissibility. In: IEEE international conference on mechatronics and automation, pp 670–675
3. García MTF, Pérez FG, Cantalapiedra JA, Manzanero MAP, Pérez CE Programas de ejercicios. http://www.sermef-ejercicios.org/
4. Gosselin C, Angeles J (1988) The optimum kinematic design of a planar three-degree-of-freedom parallel manipulator. J Mech Transm Autom Des 110(1):35–41
5. Liu X, Wang J, Gao F (2000) Performance atlases of the workspace for planar 3-DOF parallel manipulators. Robotica 18:563–568
6. Liu XJ, Wang J (2007) A new methodology for optimal kinematic design of parallel mechanisms. Mech Mach Theory 42:1210–1224
7. Patel YD (2013) Workspace and singularity analysis of 3-RRR planar parallel manipulator. In: Proceedings of the 1st international and 16th national conference on machines and mechanisms, pp 1071–1077
8. Quesnot A, Chanussot JC (2010) Rehabilitacion Del Miembro Superior. Ed. Médica Panamericana
10. Tsai LW (1999) Robot analysis: the mechanics of serial and parallel manipulators. Wiley, New York

Workspace Analysis of a Delta-Like Robot Using an Alternative Approach

A. Gutiérrez-Preciado, M.A. González-Palacios and L.A. Aguilera-Cortés

Abstract This paper treats the workspace analysis of a delta-like robot using a different placement of the actuators to improve the overall reachable workspace. Instead of having the actuators separated 120° between each other, it is proposed to provide different separations to create a greatly improved workspace suitable for custom applications. The general analysis of inverse and direct kinematics is explained and implemented in a software created to analyse this type of robot. Finally a comparison to prove the advantages of changing the angles between legs is shown.

Keywords Delta · Robot · Design · Workspace · ADEFID · Student paper

1 Introduction

In recent years, parallel mechanisms have caught the attention of researches and industries due to their high range of potential applications. From their use as manipulators to applications as machining tools, this type of robots has several advantages with respect to their serial counterparts. Higher accuracy, stiffness, speed and better dynamic behavior allow them to be used in several production lines. These advantages can be reflected in lower production costs for example in industries that work with packing products on high volume.

A. Gutiérrez-Preciado (✉) · M.A. González-Palacios · L.A. Aguilera-Cortés
University of Guanajuato, León, Mexico
e-mail: gutierrezp.alfredo@yahoo.com.mx

M.A. González-Palacios
e-mail: max@ugto.mx

L.A. Aguilera-Cortés
e-mail: aguilera@ugto.mx

M. Ceccarelli and E.E. Hernández Martinez (eds.), *Multibody Mechatronic Systems*,
Mechanisms and Machine Science 25, DOI 10.1007/978-3-319-09858-6_43

One of the main disadvantages that are present on parallel robots is the reachable workspace. This is because of the fact that all parallel robots are made out of two or more serial chains that join together at some point (usually a platform). In the case of the delta robot, there are three serial chains (RSS) that join at the end effector. Even though each chain or leg might have a big workspace, when all of them join together, the reachable workspace reduces to only the places that all of the legs can reach. This kind of robots has been extensively studied but there are still some special cases that must not be left behind to maximize the potential of parallel robots.

The Delta robot is one of the most famous parallel robots in the academic and industrial environment. This robot consists of three equal serial chains separated by 120° on the fixed base that get together on a common link with the same separation. The 120° separation forms an equilateral triangle that gives its popular name, the Delta robot proposed by Clavel in 1986. Each chain has a revolute joint after the fixed base, after it, a parallelogram mechanism takes place and connects at the end with the mobile platform as shown of Fig. 1.

The synthesis of this type of robot depends highly on the type of application or optimization it is given. Vast work has been made with respect of the optimization of this robot with several parameters as stiffness, balance, dexterity, payload, accuracy, manipulability among others [1].

Liu et al. [2] present a synthesis method for a similar delta robot but with linear actuators. This design is made with a concept they call as "maximum inscribed workspace" that consists of the set of maximum inscribed circles that fall inside the robots workspace. Laribi et al. [3] synthetize the traditional Delta robot using genetic algorithms. In this work, they prescribe a workspace that the robot must

Fig. 1 A delta robot

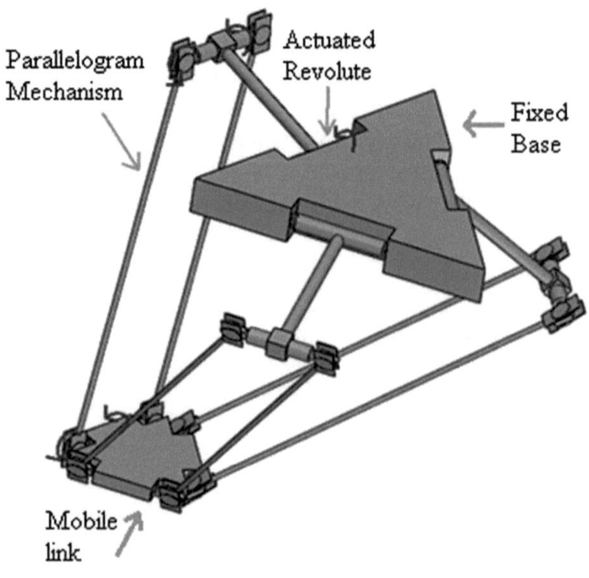

reach with the 8 corners of a cube and with the use of genetic algorithms they obtain an optimal design. Yun et al. [4] make a similar approach but using the golden section method. Other studies made by Courtellie et al. [5] focus on the design of the delta robot using the global stiffness of the manipulator. One more time, with the use of genetic algorithms they optimize the geometrical parameters to improve the static sensitivity for specialized applications in medicine.

Another great contribution comes from Karol Miller in 1998 when he introduced a new concept on the design of the delta robot. This design was called NUWAR and consists on changing the angles of the orientation of the motors acting on the base [6]. Miller concludes that when changing only the orientation of the motors, the workspace can be greatly increased.

In 2012, Mahdi and Mohsen [7] present the analysis and modeling of a delta robot variation where the first revolute of each leg is located on the corners of an isosceles triangle. If looked from the front, two legs form a planar five bar mechanism.

Until now, most delta robots are limited to perform pick and place tasks of small objects. Manufacture Industries prefer them for their capability to move really fast with high precision and repeatability. Actually, now that they are being widely accepted for use in industry, most of those that are being produced have relatively small workspace volumes. Even though the Delta robot has been looked in many different ways; little attention has been brought on the advantages or disadvantages that may cause the angle of separation between actuators.

Stamper et al. [8] made a more general study on the workspace of the delta robot taking into account the angle that the axis of each leg makes with the other two. In this work, they discretize a big area surrounding the workspace and then, they determine an objective function approximating the workspace volume with the number of points in the discretization that falls within the workspace of the robot using the inverse kinematics. They show that bringing the legs closer would result in an overall bigger workspace but would worsen the "Condition Index" that relates the velocities of the actuators to the velocity of the moving platform. With that, they preferred the symmetrical configuration of the robot and discarded the use of different angles between legs so. Even so, they did not give due attention on the advantages and disadvantages that can bring a change on the angles mentioned keeping the same physical parameters of the robot.

This work focuses only on the kinematic properties of the robot, the advantages and disadvantages of changing the angles between legs and the effect that these variables have on the workspace of the robot.

2 Kinematics of the Delta Robot

The geometrical solution of the kinematical equations of the Delta robot has been obtained and shown in several works so a short summary will be presented here for reference. For a deeper description of the calculations see Ref. [9].

2.1 Inverse Kinematics

The inverse kinematics must be calculated in order to check whether a point in the space falls inside the workspace or not. In this case, the workspace of each leg must be calculated, if a point can be reached by all three legs, then the Delta robot can reach it, if one of the legs fails on doing so, then that point is outside of the workspace. The skeleton of the delta robot can be drawn using only lines as shown on Fig. 2 to calculate the inverse kinematics.

Where A is the perpendicular distance from the center of the triangle to a side, L is de length of the first link, p is the length of the parallelogram links and B is the radius of the mobile platform. θ_j is the angle that makes the fixed base with the direction of the actuated link. α_j is the angle from a fixed line on the base axis to one of the legs that in almost all cases is 120° to form a Delta triangle. With the use of the previous variables, the following equations can be obtained:

$$\theta_j = 2\tan^{-1}\left(\frac{4LZ_j\sqrt{16L^2Z_j^2 - 4(M^2 - 4Q^2L^2)}}{2(M - 2QL)}\right) \tag{1}$$

where:

$$M = Z_j^2 + Y_j^2 - p^2 + L^2$$
$$Q = X_j - A + B$$

And X_j, Y_j, Z_j represent the coordinates of the end effector with respect of the fixed base and direction of each leg and can be found with:

$$X_j = X\cos(\alpha_j) - Y\sin(\alpha_j) \tag{2}$$

$$Y_j = X\sin(\alpha_j) + Y\cos(\alpha_j) \tag{3}$$

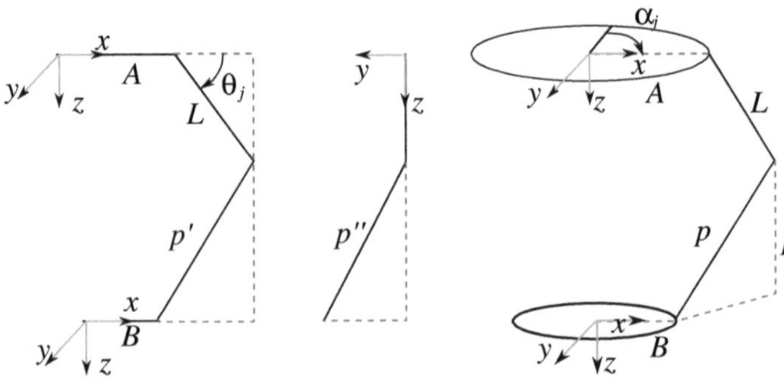

Fig. 2 Skeleton of the delta-like robot

$$Z_j = Z \tag{4}$$

Equation (1) can be used to obtain the inverse kinematics of each leg. When a point in the space can be reached by the three legs, the Delta robot will have a real solution for the inverse kinematics problem.

2.2 *Forward Kinematics*

The forward kinematics problem is usually more complex to find than the inverse kinematic, and the Delta robot is not an exception. Knowing the angles of each of the actuators, it can be predicted where the end effector will be. Its calculation can also be deeply understood checking Ref. [9]. Due to the limited space in the article and the length of the resulting equations, they will be omitted here.

3 Workspace Analysis

Each of the three legs in the Delta-like robot generates a Schönflies movement where the rotation occurs on parallel with the first revolute of each leg. Nevertheless, as a consequence of each leg being linked with the final platform, the extra rotation of each leg gets canceled and the overall movement of the platform is only translations in the space. So, supposing that none of the legs has that rotation at the end, the last link can be considered always parallel leaving an RSS mechanism.

With that, it can be inferred that the workspace of each leg would be the surfaces of a sphere whose center is located at the end of the first link (actuator), the radius of the sphere would be the length of the second link (parallelogram) and the volume workspace is formed by moving that sphere on the perimeter of a circle whose radius is the first link as it can be seen of Fig. 3.

It must not be left out the fact that the workspace of each leg behaves the same compared with a torus in the sense that when the radius of the torus tube increases, the torus turns from a ring torus, to a horn torus and finally a spindle torus (See Fig. 4). The ratio of the first and second links is the one that affects the shape of the torus. When their ratio is 1:1, the workspace would be a horn torus.

The resulting workspace of the Delta-Like robot is the intersection of the 3 tori which has a very irregular shape. Finding an algebraic equation for such an intersection turns out to be very complicated and out of this works scope. Using computational tools it is possible to draw a bounding surface that represents the real workspace of the robot. To do so the following algorithm is applied similar to what is done in [8].

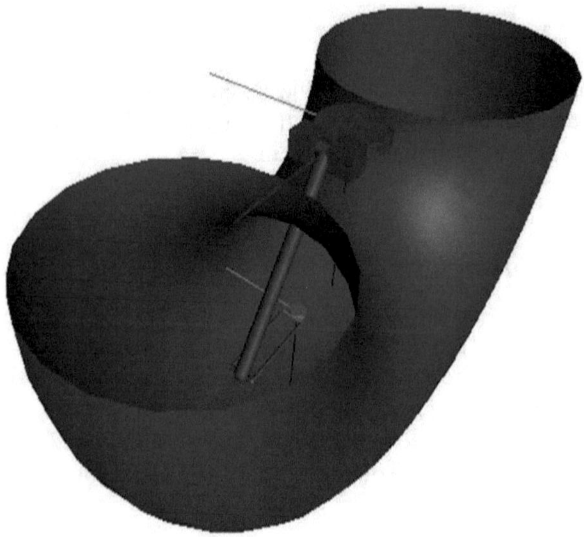

Fig. 3 Half of the workspace volume of a single leg of the delta-like robot

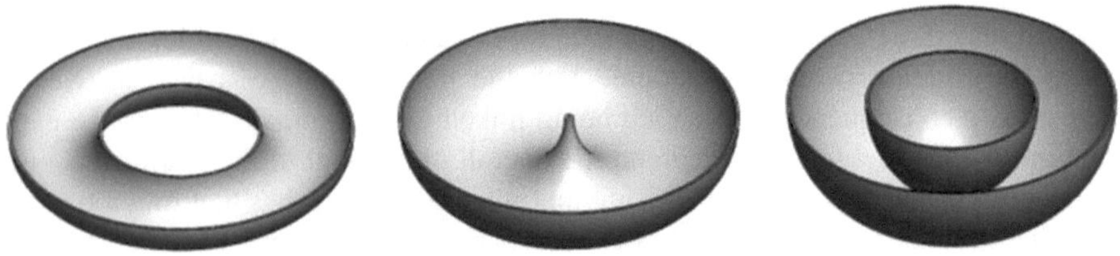

Fig. 4 From *left* to *right*, ring torus, horn torus and spindle torus

1. Find a space that surely will contain the workspace of the robot.
2. Divide the space in a 3 dimension matrix
3. Evaluate the inverse kinematics for each point for each leg. When all three legs can reach the desired point, save the location as reachable and otherwise unreachable. Repeat until all the space is covered.
4. Find the edges of the reachable points and join them in order to represent the workspace.

4 Software Developed to Study and Design the Delta-Like Robot

The software is coded inside Visual Studio with object oriented programming with C++, and using ADEFID [10] as a platform. ADEFID is very versatile and has been used to create several projects before treating the analysis of serial robots,

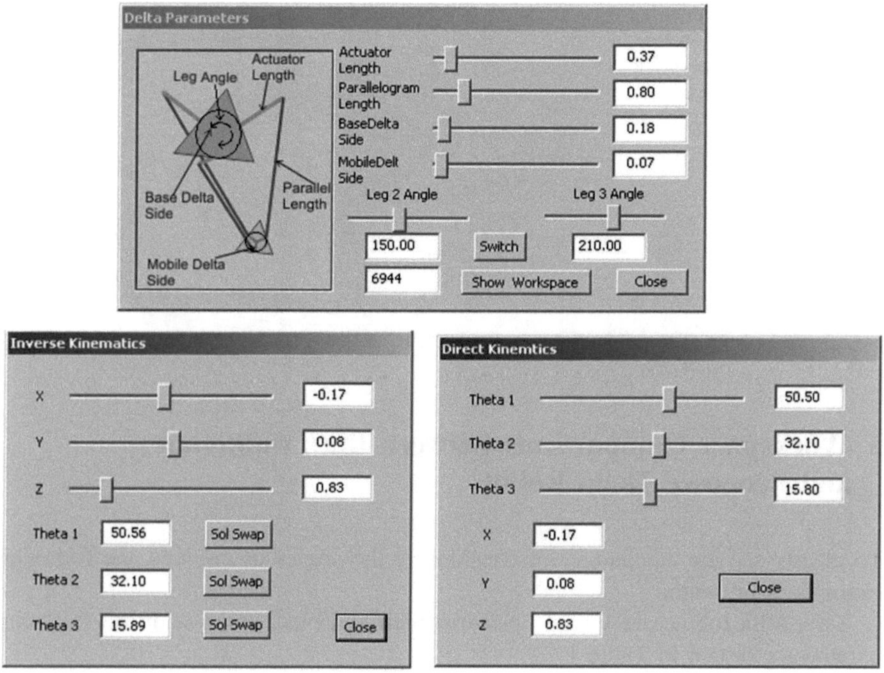

Fig. 5 Menus to handle the delta-like robot

vibrations, cams and gears, motor control among others. ADEFID contains several OpenGL functions to handle graphical environments that allowed a clean representation of what the delta robot can do.

The software consists of 3 interactive menus that allow a live interaction with the on-screen model of the delta robot. These menus can control the inverse and forward kinematics and design parameters of the Delta Robot as seen of Fig. 5.

The "Delta Parameters" menu allows the user to change the design variable of the robot. The "Inverse kinematics" menu is used to calculate the angles necessary to redraw the robot. It is possible to see the 8 different solutions of the inverse kinematics problem. The "Direct Kinematics" menu can be used to compute the final position of the robot by changing the angles of the legs. Figure 6 shows the resulting graphical representation of the delta robot with the parameters seen on Fig. 5 including the show workspace button.

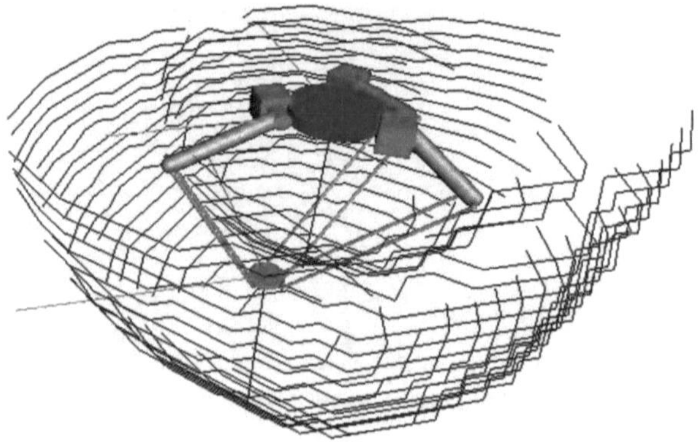

Fig. 6 Delta robot drawing with predefined parameters and showing the workspace boundaries with lines

5 Workspace Comparisons Between the Traditional and Proposed Delta Robot

To clearly see the advantages of considering the angles of the legs, the following example is presented.

Two delta robots are built to perform pick and place tasks with the following parameters shown in Table 1.

As it can be observed, both robots have the same links and the only change is the angular position of the legs. Using the programmed software, nine points that fall just next to the workspace boundaries for each robot are obtained to get a surface equation. All of those points were chosen in order to have surfaces that are reachable to the robots without getting in dangerous positions and that cover most of the usable workspace. Table 2 shows such points in the space.

As it can be seen in Table 2, changing only the angles of the legs, a slight improvement on the reachable workspace can be achieved. At a height of 31 cm, the robot can reach up to 38 cm (22 %) more of length in the x direction but 14 cm (9 %) less on the y direction. At a height of 65 cm, there is an increase of 23 cm (15 %) in the x direction and no change in the y direction. At a height of 1 m, there is only a slight increase of 5 cm (5 %) in the x direction. Figure 7 shows how the workspace of both robots would compare graphically.

Table 1 Parameters of the delta robot

Parameters	Standard delta size	Modified delta size
Actuator length	0.45 m	0.45 m
Parallelogram length	0.75 m	0.75 m
Fixed base radius	0.2 m	0.2 m
Platform radius	0.1 m	0.1 m
Angle leg 1	120°	160°
Angle leg 2	120°	200°

Table 2 Coordinates of workspace points

Standard delta robot boundary points (m)			Modified delta robot boundary points (m)		
x	*y*	*z*	*x*	*y*	*z*
−0.85	0.00	0.31	−1.04	0.00	0.31
0.83	0.00	0.31	1.02	0.00	0.31
0.00	0.79	0.31	0.00	0.72	0.31
−0.74	0.00	0.65	−0.89	0.00	0.65
0.79	0.00	0.65	0.87	0.00	0.65
0.00	0.71	0.65	0.00	0.71	0.65
−0.55	0.00	1.00	−0.55	0.00	1.00
0.49	0.00	1.00	0.54	0.00	1.00
0.00	−0.49	1.00	0.00	−0.49	1.00

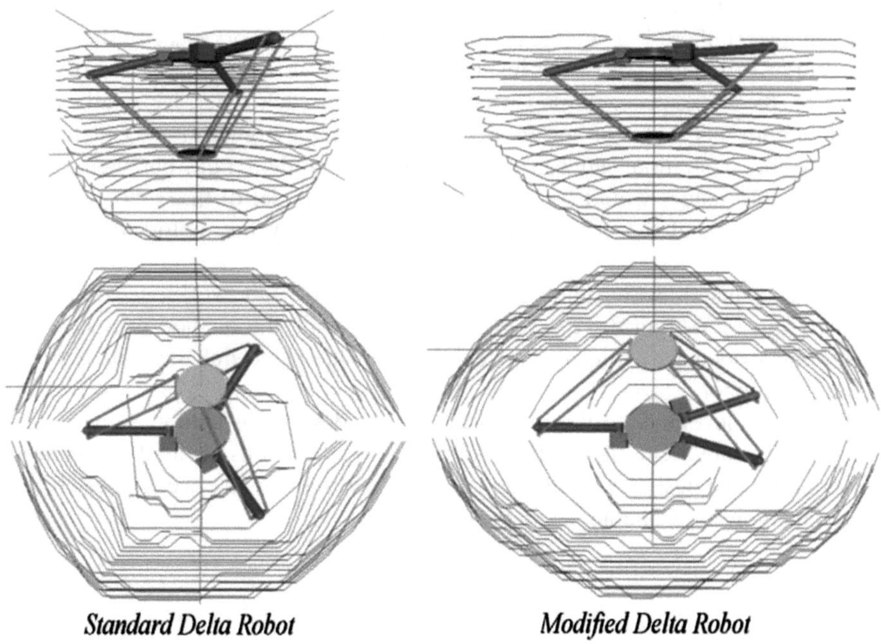

Fig. 7 Comparison between a standard delta robot and the modified one

The rate of increase of the workspace depends largely on the size of links. When the actuator link is small compared to the length of the parallel link, the change in workspace is minimum, in a contrary situation, the workspace increase is greater.

6 Conclusions

An alternative arrangement of the delta robot links is presented in this work. It can be seen that changing the angle that the legs make with each other can improve the workspace volume of the robot. An interactive software was specifically created to work and visualize the performance behavior of the robot when changing the design parameters. With the aid of this software, a brief graphical and numerical demonstration was made to show how the angle of the legs affects the size of the workspace. The result was a bigger workspace with the same sizes of links. In this case, the workspace of the modified Delta robot was relatively longer compared to the standard delta robot.

Even though there is a significant improvement on the workspace of the robot, a few design problems arise. For instance, when the robot tries to reach extreme points on the workspace, the legs may collide with each other, especially on the parallelogram links. This issue can be resolved having a bigger moving platform or reducing the separation between the rods of the parallelogram, but then, greater power on the motors would be needed. A dynamic model of this robot should be made to understand the change of the effort needed by the motors to move the platform.

Linear actuated delta robots can increase the overall workspace making it "higher" but by adding more construction parts, while the method presented here makes the workspace "longer" whereas keeping the same link sizes as the traditional delta robot.

Thanks to developed tool introduced here, the designer is able to define custom robot architecture, delta-like, with a workspace that best suits his application needs. For example, the designer can define link sizes and see how the workspace changes when changing only the angles between legs.

References

1. Zhao Y (2013) Dimensional synthesis of a three translational degrees of freedom parallel robot while considering kinematic anisotropic property. Robot Comput-Integr Manuf 29:169–179. (Elsevier)
2. Liu X-J, Wang J, Oh K-K, Kim J (2004) A new approach to the design of a delta robot with a desired workspace. J Intell Robot Syst 39:209–225. (Kluwer Academic Publishers, Netherlands)
3. Laribi MA, Romdhane L, Zeghoul S (2007) Analysis and dimensional synthesis of the delta robot for a prescribed workspace. Mech Mach Theory 42: 859–870. (Elsevier)

4. Yun Y, Wang L, Guan L (2004) Dimensional synthesis of a 3-DOF parallel manipulator. In: IEEE international conference on systems, man and cybernetics, pp 5317–5323
5. Courteillie E, Deblaise D, Maurine P (2009) Design optimization of a delta-like robot through global stiffness performance evaluation. In: IEEE/RSJ international conference on intelligent robots and systems. St. Louis USA, 11–15 Oct 2009
6. Miller K (1999) Synthesis of a manipulator of the new UWA robot. In: Department of mechanical and materials engineering, The University of Western Australia
7. Mahdi AA, Mohsen A (2012) Modeling and analysis of a novel 3-DOF spatial parallel robot. In: 19th international conference on mechatronics and machine vision in practice, pp 162–167
8. Stamper RE, Tsai L-W, Walsh GC (1997) Optimization of a three DOF translational platform for well-conditioned workspace. In: Proceedings of the 1997 IEEE international conference on robotics and automation. Albuquerque USA, pp 3250–3255
9. http://forums.trossenrobotics.com/tutorials/introduction-129/delta-robot-kinematics-3276/. Accessed 10 April 2014
10. Gonzalez-Palacios MA (2012) Advanced engineering platform for industrial development. J Appl Res Technol 10(3):309–326

Adaptive Low Cost Gravity Balanced Orthosis

Giuseppe Cannella, Dina S. Laila and Christopher T. Freeman

Abstract In this paper a novel design of a non-powered orthosis for stroke rehabilitation is reported. Designed for home based use, it is the first low-cost, passive design to incorporate an assistive level that can be adaptively varied within a closed-loop control scheme. This allows the device to be integrated with a dual robotic and electrical stimulation control scheme, to thereby enable full exploitation of the motor relearning principles which underpin both robotic therapy and Functional Electrical Stimulation (FES) based stroke rehabilitation. This embeds the potential for more effective treatment. The paper focuses on the mechanical design of the non-powered orthosis, providing detailed design, dynamic analysis and evaluation.

Keywords Student paper · Passive orthosis · Gravity balancing theory · FES · Stroke rehabilitation · Multibody dynamics

1 Introduction

Worldwide 12.6 million people live with moderate to severe disability following stroke, and this number is increasing. Due to the neuroplasticity of the brain, the function of the damaged cells can be transferred to surrounding areas of the motor cortex [3]. Hence impaired subjects can potentially regain lost motor function with the help of a tailored rehabilitation programme involving training of motor tasks,

G. Cannella (✉) · D.S. Laila · C.T. Freeman
Faculty of Engineering and the Environment, University of Southampton, Southampton, UK
e-mail: gc2g12@soton.ac.uk

D.S. Laila
e-mail: d.laila@soton.ac.uk

C.T. Freeman
e-mail: cf@ecs.soton.ac.uk

Electronics and Computer Science, University of Southampton, Southampton, UK

M. Ceccarelli and E.E. Hernández Martinez (eds.), *Multibody Mechatronic Systems*,
Mechanisms and Machine Science 25, DOI 10.1007/978-3-319-09858-6_44

with feedback (e.g. proprioceptive, haptic, visual) used to forge new motor connections. The therapeutic effectiveness of these therapies strongly depends on their frequency, intensity, and the amount of voluntary effort supplied by the patient. Unfortunately only 5 % of patients receiving conventional therapy, involving manual assistance supplied by a physiotherapist, recover useful movement [1].

In the field of non-conventional rehabilitation, new upper-limb technologies have the potential to provide intensive and motivating therapy. Moreover, there is evidence that non-conventional therapies offer more effective treatment than conventional ones [3, 9, 12]. Rehabilitation robotics is a leading non-conventional therapy and the employment of robots has shown to be very effective for patient recovery, but their design, control and use are complex, often expensive, hence they are not suitable as home-based devices. On the other hand non-powered orthoses can be considered low cost support devices. They are usually made of a mechanism combined with spring systems or counterweights that balance a mass on their extremity, for instance the user's arm, thus the patient can power the orthosis without carrying the weight of their own arm. The main drawback for these mechanical systems is that they are not as flexible and effective as rehabilitation robots, and they can be used only by patients with sufficient residual muscle strength to power the orthosis and perform the rehabilitation therapy. To overcome this drawback, a non-powered orthosis will be combined with Functional Electrical Stimulation (FES) [2, 10] in this research project. FES is another approach to therapy which facilitates movement by applying short electrical pulses to contract muscles in the same way as achieved by the unimpaired central nervous system. Furthermore, FES is often combined with a mechanical device to stabilise the system, especially during reaching tasks. Hence, using a biomechanical representation of the human arm within the FES control scheme, it is possible to adaptively adjust the assistance provided to each patient, encouraging maximum voluntary effort and therefore maximising effectiveness of treatment [4, 5].

This paper addresses the design of an orthosis that is suitable to combine with FES. The discussion focuses on the static gravity balancing analysis, the 3D CAD model and the preliminary dynamic analysis of the arm support.

2 Existing Prototypes and Gravity Balancing Theory

The most relevant non-powered orthoses to meet the aim of this research are the Wilmington Robotic Exoskeleton (WREX) [13] and Armon Orthosis [7, 8, 11]. These are both based on a novel mechanical design that makes use of spring systems to achieve gravity balancing.

WREX is an exoskeleton design-based body-powered orthosis that provides gravity support to allow patients to move their arm with very little effort [13]. This result was achieved by employing elastic bands rather than using counterweights, as in many other devices, giving rise to a lighter and more compact mechanism.

The main idea behind the Armon Orthosis was to design and manufacture a non-powered orthosis that could be mounted on a wheelchair and that could be used by different patients or also healthy users, for activities of daily living (ADLs) and upper limb training and rehabilitation [7, 8, 11]. The orthosis does not resemble an exoskeleton, as in the case of the WREX. Indeed, the joints of the mechanism do not have the function of emulating the anatomical human arm joints, but they are arranged to support the weight of the user's arm, when combined with a spring-cable-pulley system located at the lower extreme of the orthosis. The simplicity and efficiency of the Armon Orthosis is due to its mechanical design that is strongly based on gravity balancing theory (for more details refer to [6]).

Statically balanced systems are devices that stay in a static equilibrium state throughout their range of motion in the absence of friction. Mechanical devices designed to exploit this property present many advantages in terms of improved mechanical performance, including the ability to support the arm over a large workspace, while simultaneously presenting minimal inertial characteristics to the patient. This is critical in the design of a body worn orthosis, where it is important to reduce the effect of the weight of the supported limb, and the inertia of the mechanical system. A statically balanced system can be constructed by exploiting at least two conservative forces arranged in such a way as to provide stability. This result can be achieved by means of two potential energy storage devices, for instance springs and masses. Moreover these system components must be coupled in such a way that their energy characteristics, as a function of the d.o.f., add up to a constant value. For the purpose of this research it was decided to use the *zero-free length springs* model in the design of the balancing system, since they have the fundamental characteristic that the force is proportional to the actual length of the spring, rather than the sum of the initial length and the elongation as in normal springs. This means that the employment of this class of springs simplifies the analysis and the design of spring mechanisms, leading to linear mechanical models. It also simplifies the subsequent combined control scheme.

3 Mechanical Design

In this section the static gravity balancing analysis of the mechanism is developed to derive the most suitable mechanical solution for the adjustment system of the device. Then all the features and the mechanical subsystems of the preliminary 3D CAD model of the arm support are illustrated in detail.

3.1 Kinematic Chain and Gravity Balancing

The mechanism chosen for the aim of this research is shown in Fig. 1. The main reason for this choice is because this mechanism provides gravity balancing to the patient's arm acting only on a small area of the forearm, identified by the *Combined*

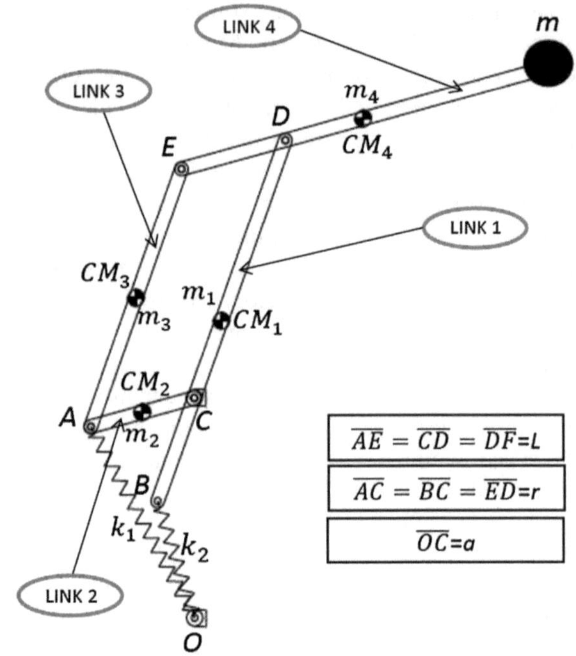

Fig. 1 Kinematic chain of the arm support

Center of Mass (CCM), making use of a parallelogram linkage and *zero-free length springs*. Moreover its hybrid kinematic chain keeps the favourable mechanical features of both parallel and serial linkages.

The model comprises a planar 2 d.o.f. kinematic chain made of a parallelogram linkage combined with two *zero-free length springs*, since each spring balances a single d.o.f. of the mechanism. The length of the links are L and r. Moreover r and a are also the geometrical parameters of the springs. The moving end of each of the two springs of stiffness k_1 and k_2 is connected to point A and B, respectively, whereas the fixed end is connected to the point O. The mass m of the user's arm is assumed to be attached to point F on the distal link, which corresponds to the CCM. Now it is possible to analyse the gravity balance behaviour of the mechanism, where the mass of the moving links is balanced by the springs [6]. The mass of each of the links is regarded as lumped at their center of mass. Focusing on the i-th spring, it is possible to decompose its force $k_i l_i$, where l_i is the spring elongation (which is equivalent to its length for *zero-free length springs*), into a component along the link length r, $k_i r$, on $\overline{BC}$ and the component along a, $k_i a$, on $\overline{OC}$.

For this gravity compensation analysis, it is possible to use the superposition principle. Considering Fig. 2a the first step is to fix spring k_1 and perform the moment equilibrium about pivot C. The only moment contribution of spring k_2 is that due to component $k_2 a$. The masses m and m_4 can be shifted to point D, and the mass m_3 parallel to link 2, from CM_3 to CM'_3, as shown by the arrows in Fig. 2a. This procedure is possible because after shifting the masses, the respective shape of

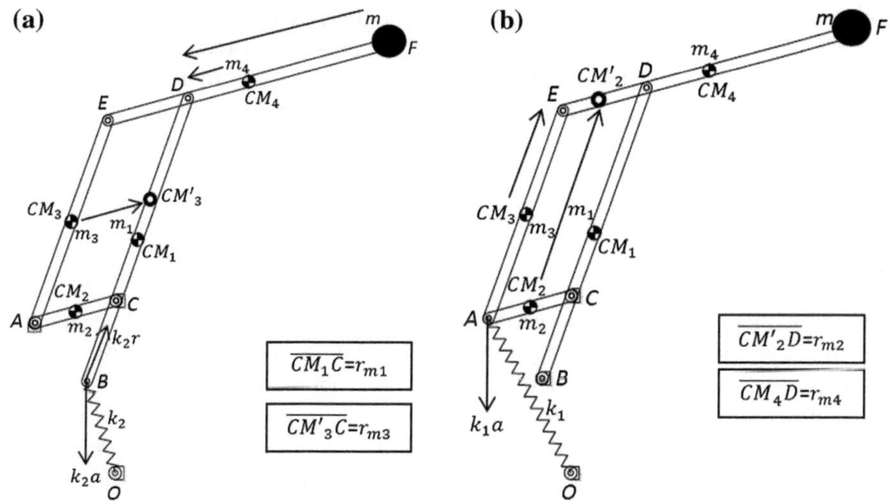

Fig. 2 Moment equilibrium analysis. **a** Moment equilibrium analysis when spring k_1 is fixed. **b** Moment equilibrium analysis when spring k_2 is fixed

their trajectories remains identical and so their potential energy differs only by a constant value. Hence the moment equilibrium equation about pivot C, when spring k_1 is fixed, can be written as

$$(m + m_4)gL + m_3gr_{m3} + m_1gr_{m1} = k_2ar, \tag{1}$$

where r_{m1} and r_{m3} are the respective distance of the centre of mass CM_1 and CM_3' from the pivot C. The second step is to release spring k_1 and fix spring k_2, in order to perform the moment equilibrium about the point D, as shown in Fig. 2b. The only moment contribution of spring k_1 is due to its component k_1a, while the masses m_3 and m_2 can be shifted to point E and CM_2' respectively, as shown in Fig. 2b, for the same reason as above. Hence the moment equilibrium equation is

$$mgL + m_4gr_{m4} - m_2gr_{m2} - m_3gr = k_1ar, \tag{2}$$

where r_{m2} and r_{m4} are the respective distance of the centre of mass CM_2' and CM_4 from point D. Both (1) and (2) are linear and can be combined in the linear system of equations

$$\begin{cases} (m + m_4)gL + m_3gr_{m3} + m_1gr_{m1} = k_2ar \\ mgL + m_4gr_{m4} - m_2gr_{m2} - m_3gr = k_1ar. \end{cases} \tag{3}$$

These contain all the geometrical and inertial elements of the mechanism and of the springs, that can be varied in order to adjust the gravity balancing behaviour of the device. Analysis reveals that the variation of the linkage parameters and of the spring stiffnesses is complex to achieve and it may require a heavy mechanism directly acting on the linkage. To avoid these cumbersome mechanical solutions, *a* was chosen as the adjustment parameter, and this leads to the design of an adjustment system that can be integrated within the lower part of the mechanism, without acting directly upon its links, but varying the position of the pivot *O*.

3.2 3D CAD Model

The most important issue to solve is the design of the adjustment system. Indeed, the arm support must be able to provide gravity balancing to a variety of different subjects, each of whom will have different anthropomorphic characteristics, and hence different arm weights. Moreover the overall design and manufacturing must be low cost. The 3D CAD model shown in Fig. 3 is drawn using SolidWorks and consists of a parallelogram linkage, with an interface connection for the user's arm, a balancing mechanism, an adjustment system and a frame. In total it has 5 d.o.f. divided as follows: 2 d.o.f. for the interface connection, 2 d.o.f. for the parallelogram linkage and 1 d.o.f. for the vertical axis of rotation of the mechanism (the 5th revolute joint in Fig. 3).

The interface connection between the user's arm and the support is to be made of a folded aluminium sheet with two revolute joints, one with a horizontal axis and another with the axis orthogonal to the sheet. The CCM of the user's arm is placed on it, so these 2 d.o.f. enable more forearm dexterity during use of the device. The interface connection will be later improved by covering it with molded foam and straps to ensure a comfortable, safe and stable contact with the user's forearm.

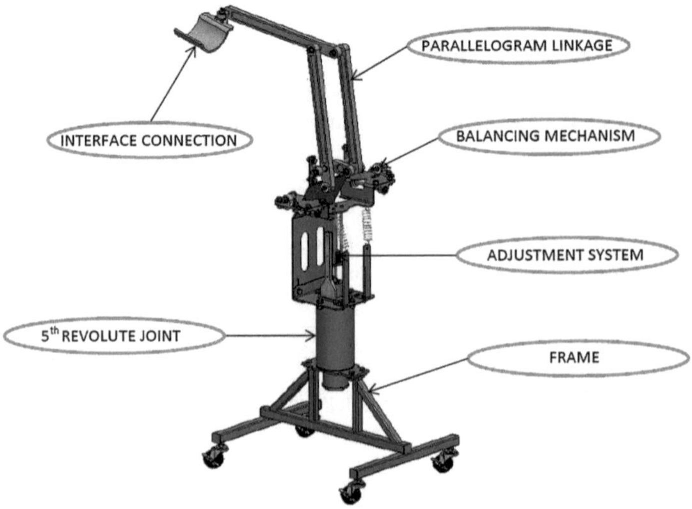

Fig. 3 3D CAD model showing subcomponent labels

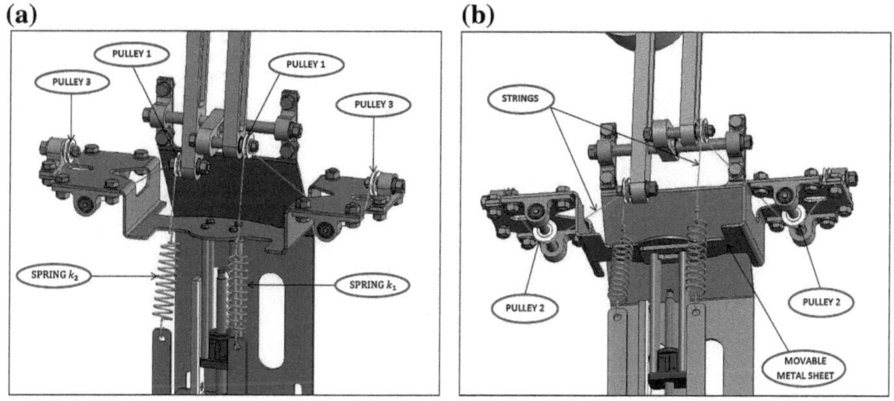

Fig. 4 Detailed views of the gravity balancing mechanism. **a** *Upper* detailed view. **b** *Lower* detailed view

Deep groove ball bearings were selected to be mounted in the joints of the device in order to reduce friction and so improve the mechanical performance of the overall system. Moreover, less friction translates into less effort for the patient when using the device during treatment.

The four links of the parallelogram mechanism are to be made of aluminium, with characteristic dimensions of L and r, as shown in Fig. 1. The length L was chosen as 320 mm in accordance with the design studies conducted for the Armon orthosis prototype [8, 11], while r was chosen as 50 mm in order to accommodate the balancing mechanism and the adjustment system. All the sheets and plates used for the arm support were selected to be made of steel.

The lower part of the linkage is connected to the springs via a *string-pulley* arrangement, shown in Fig. 4. This solution allows the gravity balancing system to have a linear mechanical behaviour, by using normal extension springs, rather than *zero-free length springs*, reducing the manufacturing cost of the device. Analysing in detail one of the two *string-pulley* arrangements (one for each d.o.f) in Fig. 4 it can be seen how one end of the string is connected to the upper hook of the spring, then wrapped around the first pulley placed at the bottom part of the linkage. The same string is wrapped around two more pulleys placed on the movable folded metal sheet. This sheet is movable since it is connected to the adjustment system, which varies the parameter a. The adjustment system, shown in Fig. 5, is made of a ball screw connected to a brushless motor, and a guide. This solution is designed so that it can be later integrated with a control system that allows the balancing of the device to be adjusted according to the value of the mass to support. At a later stage an advanced FES controller will be implemented to automatically vary the support given to the user's arm based on their rehabilitation progression. Ball screw systems were chosen for their high accuracy and precision in positioning, and lower friction.

The adjustment system allows variation of the parameter a within the range of 0 mm, when the mechanism is folded, to 50 mm, when $a = r$ and the system

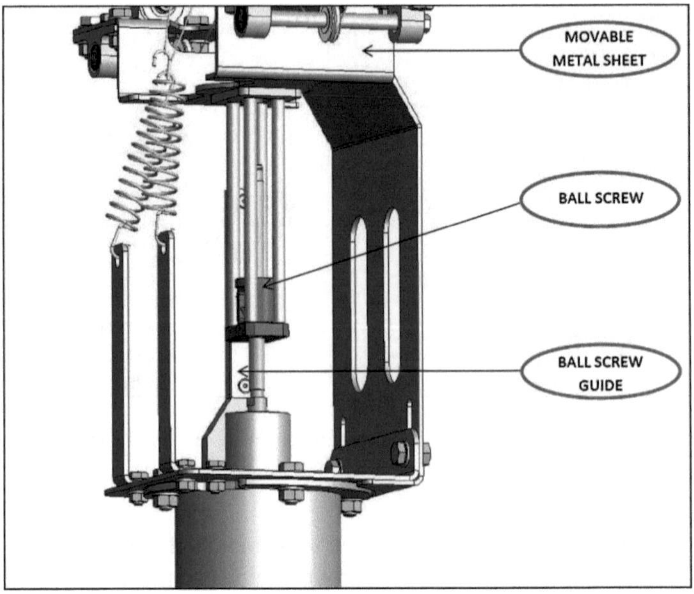

Fig. 5 Detailed view of the adjustment system

balances the maximum value of the mass. This maximum mass of 3 kg was chosen according to studies conducted for the design of the Armon orthosis [8, 11]. The *string-pulley* balancing system was designed to perform this adjustment avoiding interference with the links of the mechanism.

The lower hooks of both springs are connected to the plate, to which the fixed part of the ball screw system is attached. This specific plate is connected to a revolute joint with a vertical axis that allows the horizontal adduction of the user's shoulder. This means the whole mechanism is able to rotate around the vertical axis. This joint is bigger than the others since it has to carry all the weight of the mechanism, and fit the brushless motor in its upper flange, reducing the overall volume and the inertia about the vertical axis of the system.

The revolute joint is then connected to a frame, which comprises welded steel squares. The purpose of the frame is to carry the load of the whole mechanical system and to provide a stiff and steady support during its use. Moreover, 4 casters with brakes are installed at the bottom part of the frame, in order to make it easier to move and locate the orthosis in different environments or rooms.

4 Dynamic Analysis

In this section a preliminary dynamic analysis of the device is introduced to confirm appropriate performance. The model is based on the planar linkage of Fig. 1, where an extra d.o.f. has been added to describe the rotation around the vertical axis of the

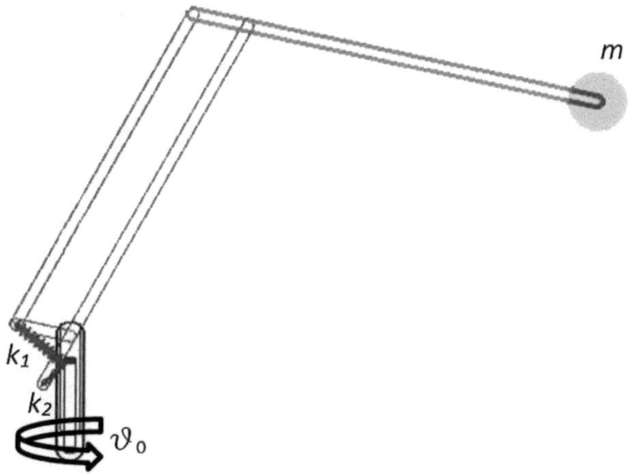

Fig. 6 ADAMS model for the dynamic analysis

overall parallelogram linkage. Its detailed view is shown in Fig. 6. The general dynamic equation for robotic systems, derived from the Lagrangian formulation is

$$\boldsymbol{B}(\boldsymbol{q})\ddot{\boldsymbol{q}} + \mathbf{C}(\boldsymbol{q}, \dot{\boldsymbol{q}})\dot{\boldsymbol{q}} + \boldsymbol{F}_v\dot{\boldsymbol{q}} + \boldsymbol{F}_s sgn(\dot{\boldsymbol{q}}) + \boldsymbol{g}(\boldsymbol{q}) = \boldsymbol{\tau} - \boldsymbol{J}^T(\boldsymbol{q})\boldsymbol{h}_e, \tag{4}$$

where, $\boldsymbol{q}, \dot{\boldsymbol{q}}$ and $\ddot{\boldsymbol{q}}$ are the $(n \times 1)$ generalized coordinate, velocity and acceleration vectors, respectively, $\boldsymbol{B}(\boldsymbol{q})$ is the $(n \times n)$ *inertia matrix*, $\boldsymbol{C}(\boldsymbol{q}, \dot{\boldsymbol{q}})$ is the $(n \times n)$ *Coriolis matrix*, $\boldsymbol{\tau}$ is the $(n \times 1)$ vector of the actuation torque at the actuated joints, $\boldsymbol{F}_v(\dot{\boldsymbol{q}})$ and $\boldsymbol{F}_s sgn(\dot{\boldsymbol{q}})$ are the $(n \times 1)$ vectors of the viscous and static friction torques, respectively, $\boldsymbol{h}_e$ is the $(n \times 1)$ vector of the contact forces and moments acting on the end-effector of the manipulator and $\boldsymbol{J}$ is the $(m \times n)$ *Jacobian matrix* of the mechanism.

In this specific case the number *(n)* of active joints of the mechanism is 3, and the generalized coordinate vector is $[\theta_0, \theta_{1'}, \theta_{1''}]$. The overall dynamic study of the system has been performed using ADAMS (MSC Software). The first step was to draw the mechanism, and add all the revolute joints and links with their inertial properties. These linkage mechanical features have been added in a code, based on the static gravity balance analysis, written using Matlab in order to obtain the value of the stiffness of both springs, $k_1 = 3869$ N/mm and $k_2 = 4454$ N/mm. Hence two springs with the previous specific stiffness have been added to the ADAMS model. Then, the problem was analysed for the case where joints are equipped with either bushings or bearings, which can be considered as corresponding to the case of either high or low friction, respectively. In the case of bushings, using a steel-brass contact model, the static friction coefficient is $\mu_s = 0.35$, and the dynamic friction coefficient is $\mu_d = 0.19$. For bearings, they are $\mu_s = 0.0024$ and $\mu_d = 0.0012$, respectively. The dynamic analysis has been made for three different values of mass *m*: 1, 2 and 3 kg, acting on the end-effector, while the whole mechanism is under the action of gravity.

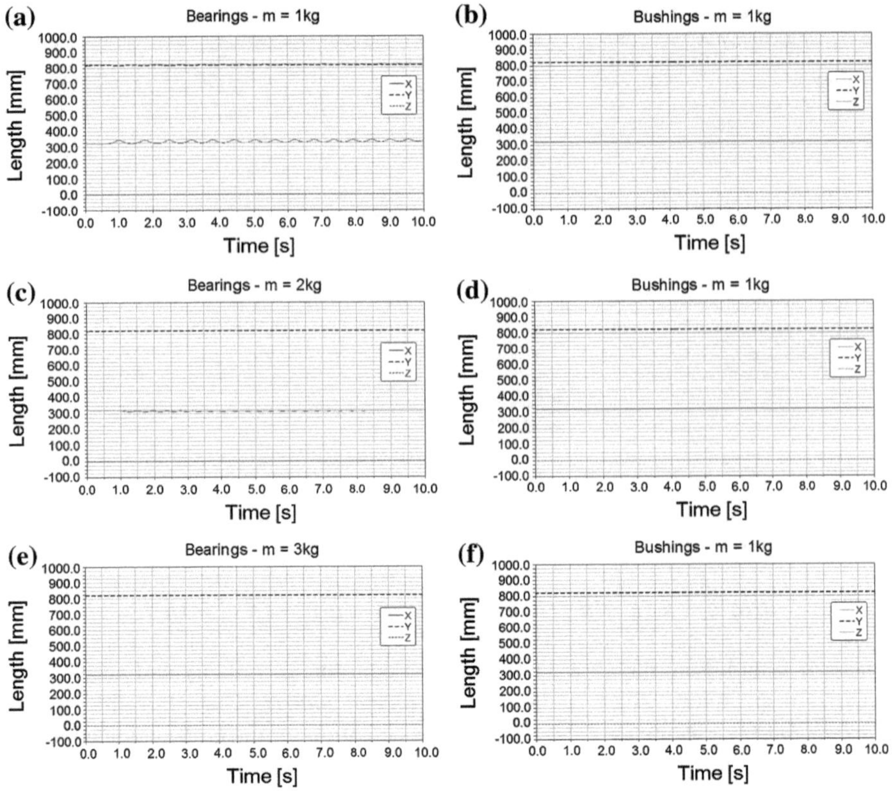

Fig. 7 Position of center of mass. **a** With bearings and $m = 1$ kg. **b** With bushings and $m = 1$ kg. **c** With bearings and $m = 2$ kg. **d** With bushings and $m = 2$ kg. **e** With bearings and $m = 3$ kg. **f** With bushings and $m = 3$ kg

The results in Fig. 7 show the position of the center of mass of the end effector during a simulation of 10 s, for the two friction cases. In the one with bearings the system tends to become less oscillatory in response to initial conditions when the mass is increased. The reason is that the stiffness of the springs are fixed parameters that are derived from the static gravity balance analysis, and the chosen values are the ones corresponding to $m = 3$ kg. Furthermore, the results in the plots correspond to a dynamic analysis that takes into account more detail than the static one, such as inertia effects. In the case with bushings, the system is less oscillatory for all values of m, but this means that the device will not be suitable for weaker patients, since they might not be able to perform a rehabilitation task due to the high friction in the joints of the mechanism. Thus the solution with bearings is chosen as the optimal one, but a controller must be added in order to vary the parameter a of the springs, so that it will adapt its dynamic behaviour to different masses, making it more stable and effortless to use by different patients.

5 Conclusion and Future Work

In this paper the preliminary 3D CAD model and dynamic analysis of the first low cost mechanical support, that can be combined with FES in a model-based control system, was discussed. Although the design of the device is based on static gravity balancing theory, the preliminary dynamic analysis shows an appropriate dynamic behaviour as well. Moreover the 3D CAD model shows how these mechanical properties can be achieved as a result of an innovative analysis of the arm support and in particular the balancing mechanism and the adjustment system.

Future work will involve further design analysis, in particular based on simulation with the 3D CAD model. Then the integration of FES with a specific controller will be investigated. Once an optimal solution is found, the first prototype will be built and tested on users.

References

1. Barreca S, Wolf SL, Fasoli S, Bohannon R (2003) Treatment interventions for the paretic upper limb of stroke survivors: a critical review. Neurorehabilitation Neural Repair 17:220–226
2. Brend O, Freeman CT, French M (2012) Application of multiple model adaptive control to upper limb stroke rehabilitation. In: IEEE multi-conference on systems and control. Dubrovnik, pp 69–74
3. Egglestone SR, Axelrod L, Nind T, Turk R, ans Burridge JAW, Fitzpatrick G, Mawson S, Robertson Z, Hughes A, Ng KH, Pearson W, Shublaq N, Probert-Smith P, Rickets I, Rodden T (2009) A design framework for a homebased stroke rehabilitation system: identifying the key components. In: Pervasive computing technologies for healthcare, London, pp 1–8
4. Freeman CT, Hughes A, Burridge JH, Chappell PH, Lewin PL, Rogers E (2009) Iterative learning control of fes applied to the upper extremity for rehabilitation. Control Eng Pract 17:368–381
5. Freeman CT, Rogers E, Hughes A, Burridge JH, Meadmore KL (2012) Itertive learning control in health care: Electrical stimulation and robotic-assisted upper-limb stroke rehabilitation. IEEE Control Syst Mag 32:18–43
6. Herder JL (2001) Energy-free systems: theory, conception, and design of statically balanced mechanisms. Ph.D. thesis, Delft University of Technology
7. Herder JL (2005) Development of a statically balanced arm support: Armon. In: 9th international conference on rehabilitation robotics, Chicago, pp 281–286
8. Herder JL, Vrijlandt N, Antonides T, Cloosterman M, Mastenbroek PL (2006) Principle and design of a mobile arm support for people with muscular weakness. J Rehabil Res Dev 43:591–604
9. Loureiro RCV, Harwin WS, Nagai K, Johnson M (2011) Advances in upper limb stroke rehabilitation: A technology push. Med Biol Eng Comput 49:1103–1118
10. Lynch CL, Popovic MR (2008) Functional electrical stimulation: closed-loop control of induced muscle contractions. IEEE Control Syst Mag 28:40–50
11. Mastenbroek B, de Haan E, van den Berg M, Herder JL (2007) Development of a mobile arm support (Armon): design evolution and preliminary user experience. In: Proceedings of IEEE 10th international conference on rehabilitation robotics, Noordwijk, pp 1114–1120

12. Mehrholz J, Platz T, Kugler J, Pohl M (2009) Electromechanical and robot-assisted arm training for improving arm function and activities of daily living after stroke. J Am Heart Assoc 40:392–393
13. Rahman T, Sample W, Jayakumar S, King MM, Wee JY, Seliktar R, Alexander M, Scavina M, Clark A (2006) Passive exoskeletons for assisting limb movement. J Rehabil Res Dev 43:583–590

Design and FEM Analysis of a Novel Humanoid Torso

Daniele Cafolla and Marco Ceccarelli

Abstract An effective trunk behaviour is one of the most important factors for a humanoid to accomplish any given task using stable operations. In this paper, a novel structure that can reproduce a human-like movement of the torso-pelvis part with a pure simple low-cost rotation mechanism easy to realize and control is proposed.

Keywords Humanoid robots · Design · FEM analysis · Student paper

1 Introduction

Humanoid robots have always inspired the imagination of robotics researchers as well as the general public. Up until 2000, the design and construction of humanoid robots was very expensive and limited to a few well-funded research labs and companies (for example, Honda Asimov, Fujitsu HOAP). Starting in about 2001 advances in material sciences, motors, batteries, sensors, and the continuing increase in processing power available to embedded systems developers has led to the development of many small humanoid robots, but they have also allowed the creation of the next generation of humanoid robots that are between 1.3 and 1.8 m tall. There has been increasing academic and commercial interest in humanoid robots for various applications since the recent significant advancements in robot control technology. An effective trunk attitude control is one of the most important factors for a humanoid robot to continue dynamically stable walking which is necessary to accomplish any given task.

D. Cafolla (✉) · M. Ceccarelli
LARM: Laboratory of Robotics and Mechatronics, DICEM, University of Cassino and South Latium, Cassino, Italy
e-mail: cafolla@unicas.it

M. Ceccarelli
e-mail: ceccarelli@unicas.it

M. Ceccarelli and E.E. Hernández Martinez (eds.), *Multibody Mechatronic Systems*, Mechanisms and Machine Science 25, DOI 10.1007/978-3-319-09858-6_45

Existing human-size biped robots can achieve stable and dynamic walking. Many of them have classical leg joint kinematics where all pitch, roll and yaw joints are respectively along longitudinal, lateral and vertical axes of the robot's body. They have 3-DOF at the hip, 1-DOF at the knee and 2-DOF at the ankle. Examples of such humanoids are robots P2, P3, and ASIMO from HONDA research group [1, 2]. HRP-2P, described in [3, 4], from the Japanese National Institute of Advanced Industrial Science and Technology (AIST) and Kawada Industries, Inc., H6 and H7, as mentioned in [5, 6], from the University of Tokyo, WABIAN from Waseda University [7, 8]. Finally KHR-2 from Korea Advanced Institute of Science and Technology (KAIST) [9].

All previously mentioned humanoids have classical joint axis orientation except Johnnie, as in [10], from the Technical University of Munich and NAO from the French Aldebaran Robotics company, [11]. The hip yaw joint axis of these two robots is inclined in the frontal plane.

2 The Attached Problem

Human body mechanism includes bones as rigid links, cartilage that links the joints, muscles and tendons actuation of each part of the body. It is impossible to replace all of this muscular-skeletal system by current mechanical components. Therefore, a primary goal of the mechanical design is to develop a robot that can imitate equivalent human motion. An effective trunk behaviour is one of the most important factors for humanoids dynamically stable walking in any given task. It is essential for a human-like robot to have the ability to move its trunk properly. Trunk motions are used mainly for rising a sitting position and walking. It is also important for the balance of the whole body. It is considered that a human usually use its trunk motions unconsciously in these cases.

In this last decade the interest on parallel manipulators has been growing since new applications have been recognized feasible for these devices, mainly because of their greater stiffness and payload capacity with respect to the open chain manipulators [12]. This interest has produced a deep knowledge of parallel structures, which can be considered as an emblematic example, and as further results some new designs [13].

A specific research line has been carried out in the Laboratory of Robotics in Cassino with the aim to design a mechanical parallel manipulator which may be easily used and may have a friendly structure. The last prototype called CaPaMan 2bis has been the result of various enhancements [14]. It has been implemented as a part of a hybrid robotic architecture for surgical tasks in the work as well as a trunk module in a humanoid robot design that is named as CALUMA (CAssino Low-cost hUMAnoid robot) [15]. Previous works were developed to analyse the CaPaMan 2bis prototype in terms of stiffness and basic [16, 17]. Since the solution of direct kinematic problem for CaPaMan 2bis can be complex, the system behaviour can be conveniently simulated and experimentally measured [18].

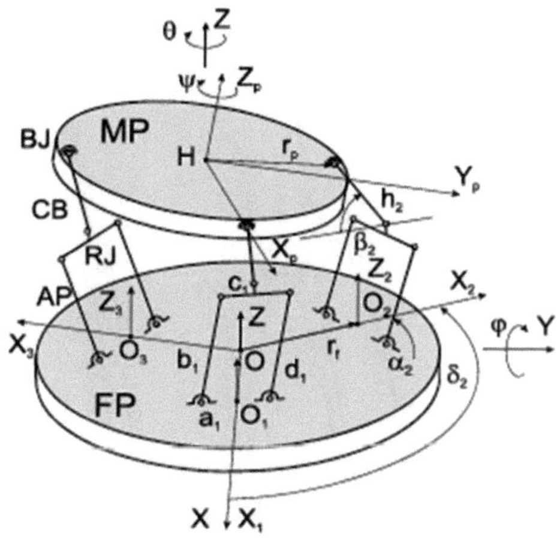

Fig. 1 A kinematic scheme of CaPaMan 2bis

A kinematic scheme of CaPaMan 2bis is shown in Fig. 1. It is composed of a movable platform (MP) connected to a fixed base (FP) through three leg mechanisms. Each leg mechanism is composed of an articulated parallelogram (AP) whose coupler carries a revolute joint (RJ), a connecting bar (CB) that transmits the motion from AP to MP through RJ, and a spherical joint (BJ), which is installed on MP at point J. Revolute joint RJ installed on the coupler of AP has the rotation axis coinciding with the parallelogram plane. Each leg mechanism is rotated $2\pi/3$ with respect to the neighbouring one so that the leg planes lie along two vertices of an equilateral triangle, giving symmetry properties to the mechanism. They can be identified for the k-th leg mechanism (k = 1, 2, 3) as: ak is the length of the frame link; bk is the length of the input crank; ck is the length of the coupler link; dk is the length of the follower crank and hk is the length of the connecting bar. The size of movable platform MP and fixed base FP are given by distances rp and rf, as shown in Fig. 1. Points H and O are the center points of MP and FP, respectively. Points Ok are the middle point of frame link ak, Jk are the connecting points between an inertial frame O-XYZ has been assumed to be fixed to base FP. A moving frame HXPYPZP has been attached to platform MP. O-XYZ has been fixed with Z-axis orthogonal to the FP plane and X axis as coincide with the line joining O to O1. Moving frame HXPYPZP has been fixed to platform MP with ZP orthogonal to the MP plane and XP axis as coincident to the line joining H to J and YP to give a Cartesian frame. Angle δk is the structure rotation angle between OX1 and OXk as well as between HJ1 and HJk. They are equal to δ1 = 0, δ2 = $2\pi/3$ and δ3 = $4\pi/3$, the kth leg mechanism and platform MP.

The attached problem is to choose a suitable structure that can reproduce a human-like movement of the torso with a pure simple low-cost rotation mechanism easy to realize and control.

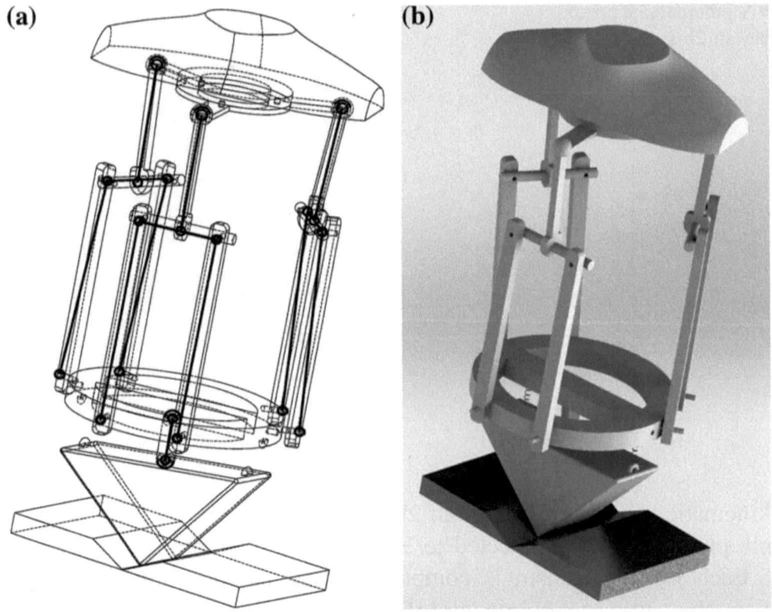

Fig. 2 Proposed torso design: **a** Kinematic Scheme of the proposed torso design, **b** CAD representation

3 Design and FEM Simulation

The CaPaMan 2bis structure has been chosen suitable kinematic structure due to his characteristics as part of the torso structure to simulate the human trunk behaviour. To make the prototype more human-like a pelvis structure has been added connecting the parallel structure and the latter with bar with two spherical joint at the extremities. A kinematic scheme of the proposed prototype is shown seen in Fig. 2a.

The trunk and the pelvis have been fixed together with two springs on the edges to give a smoother movement. The virtual prototype is shown in Fig. 2b.

A static FEM simulation has been carried out considering ABS material, whose properties are listed in Table 1.

In the simulations, forces acting on the human trunk and pelvis have been calculated. The forces acting on a shoulder, shown in Fig. 3. Considering if the weight of the arm is 33 N, the moment arm for the total arm segment is 30 cm, and the moment arm for the deltoid muscle (F_m) is 3 cm [19].

Thus assuming data in Table 2

$$\sum T_s = 0 \tag{1}$$

Table 1 Material Properties

Name	ABS generic
Model type	Linear elastic isotropic
Tensile strength	3e + 007 N/m^2
Elastic modulus	2e + 009 N/m^2
Poisson's ratio	0.394
Mass density	1,020 kg/m^3
Shear modulus	3.189e + 008 N/m^2

Fig. 3 A scheme for forces acting on a shoulder [19]

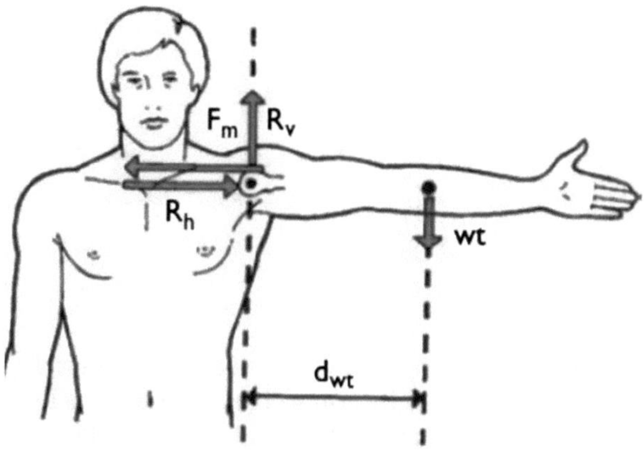

Table 2 Assumed Shoulder Data

Parameter	Value
wt	33 N
d_{wt}	30 cm
d_m	3 cm

$$\sum T_s = (F_m)(d_m) - (wt)(d_{wt}) \tag{2}$$

$$0 = (F_m)(3\text{ cm}) - (33\text{ N})(30\text{ cm}) \tag{3}$$

$$F_m = \frac{(133\text{ N})(30\text{ cm})}{3\text{ cm}} \tag{4}$$

$$F_m = 330\text{ N} \tag{5}$$

$$R_h = 330\text{ N} \tag{6}$$

The force that must be supplied by the deltoid to maintain the arm in position is $F_m = 330$ N and the magnitude of the horizontal component of the joint reaction force is $R_h = 330$ N [19].

Table 3 Assumed Pelvis Data

Parameter	Value (N)
wt	250
F_m	600

Fig. 4 Forces acting on Lower Extremity [19]

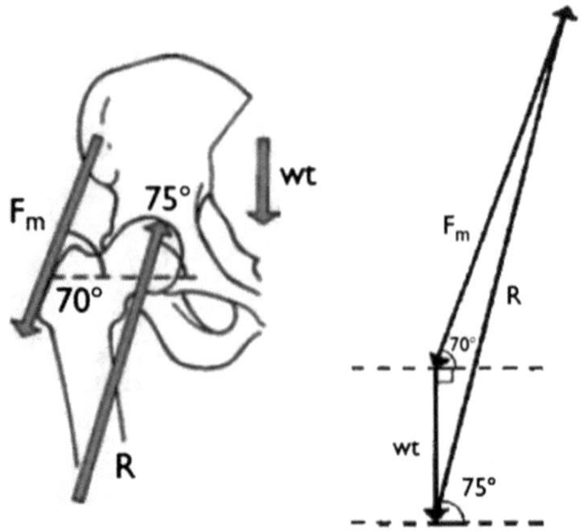

The torque at the shoulder created by the muscle force must equal the torque at the shoulder created by arm weight, yielding a net shoulder torque of zero. Another important force that has been calculated is the compression that acts on the hip during two-legged standing, given that the joint supports 250 N of body weight and the abductor muscles are producing 600 N of tension, as calculated in [19].

Thus assuming data in Table 3.

The law of cosines can be used with the same triangle, Fig. 4, to calculate the length of R.

Fig. 5 Forces acting on the Mobile Platform

Table 4 Mobile platform acting forces

Load Name	Load Details
Entities	2 faces
Values	0, 0, 330 N

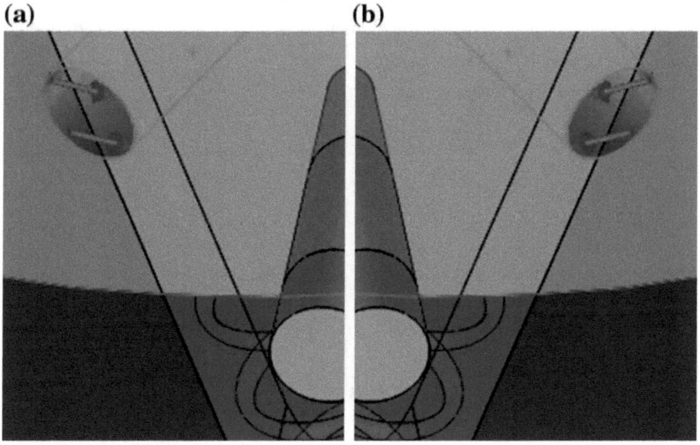

Fig. 6 Forces acting on Hip: **a** *right* side, **b** *left* side

Table 5 Pelvis acting forces

Load Name	Load Details
Entities	2 face
Values	0, 0, 840 N

Fig. 7 Prototype Model Mesh

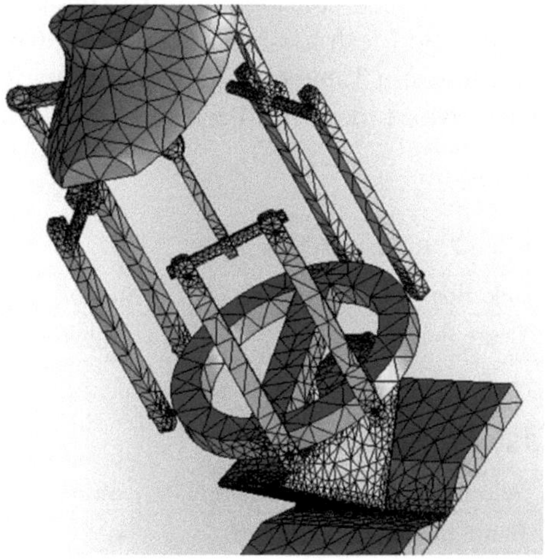

Table 6 Mesh Characteristics

Parameters	Value
Mesh type	Solid mesh
Mesher used	Curvature based mesh
Jacobian points	4 Points
Maximum element size	12.4068 mm
Minimum element size	0.620341 mm
Mesh quality	High
Total nodes	2,69,359
Total elements	1,59,537

$$R^2 = F_m^2 + wt^2 - 2(F_m)(wt)\cos 160° \tag{7}$$

$$R^2 = 600\ \text{N} + 250\ \text{N}^2 - 2(600\ \text{N})(250\ \text{N})\cos 160° \tag{8}$$

$$R = 839.3\ \text{N} \tag{9}$$

The compression that acts on the hip during two-legged standing is $R = 839.3$ N.

A Static FEM Analysis using SolidWorks environment has been chosen because human-like motion are done with velocities that are constant. Calculated forces has been put in the simulation. On the mobile platform act the forces acting on a shoulder, values of the force are listed in Table 2 and a graphic representation is shown in Fig. 5. On the mobile platform act the forces acting on a shoulder, values of the force are listed in Table 4 and a graphic representation is shown in Fig. 6. On pelvis structure act the forces acting on the hip, values of the force are listed in Table 5 and a graphic representation is shown in Fig. 7a (right part), Fig. 7b (left part). Furthermore, gravity perpendicular to the platform of the parallel mechanism force has been considered in the simulation.

A solid Mesh has been built, as shown in Fig. 7, Mesh parameter characteristics are shown in Table 6. A high quality mesh has been used to obtain better results. A curve based mesh has been used due to the chosen prototype shape.

Table 7 Reaction forces

Selection Set	Units	Sum X	Sum Y	Sum Z	Resultant
Entire model	N	−162.84	15.7239	1,879.03	1,886.14

Table 8 Reaction moments

Selection Set	Units	Sum X	Sum Y	Sum Z	Resultant
Entire model	N m	0	0	0	0

Table 9 FEM Analysis results

Type	Min	Max
Von mises stress	4.4494e–006 N/ m^2	7.80663e + 007 N/ m^2
	Node: 38,206	Node: 35,677
Equivalent strain	8.41642e–011	0.0199874
	Node: 21,525	Node: 24,853
Resultant displacement	0 mm	8.19112 mm
	Node: 5,692	Node: 43,074

(a)
(b)
(c)

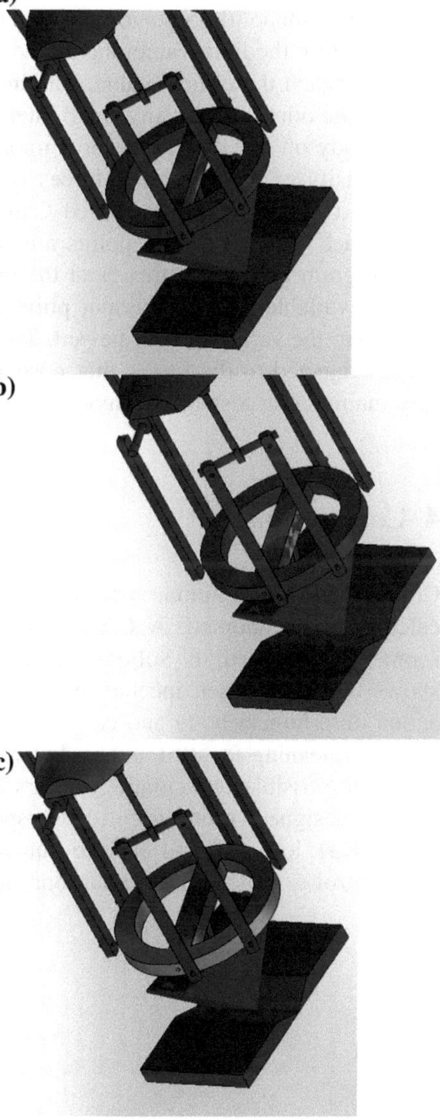

Fig. 8 FEM Analysis: **a** Von Mises Stress, **b** Equivalent Strain, **c** Resultant Displacement

The small element size allows size allow simulation to be more real and results has been used to check the human-like behaviour.

Table 7 shows the computed reaction forces. The material can afford these forces without any problem.

Table 8 shows the computed Reaction Moments. These results shows that the model has properly a static behaviour.

The computed result listed in Table 9 and shown in Fig. 8 are useful to understand if the chosen material is worth for the purpose.

According to Von Mises criterion, the yield strength of the material is reached when the strain energy distorting reaches a limit value, meaning distortional deformation energy of a body, the rate of the energy of elastic deformation due to variation in shape (deformation distorting). The results of the simulation show that we are before the limit value, the material shape remain the same, as the yield field does not reach the critical value. The internal forces in a body will vary from one point to the other. Across any small internal plane area, loads are exerted by the part of the body on one side of the area upon the part on the other side. Stress denotes the intensity of these internal forces (force per unit area). Stress results are first calculated at special points, called Gaussian points or Quadrature points, located inside each element. These points are selected to give optimal numerical results. The program calculates stresses at the nodes of each element by extrapolating the results available at the Gaussian points. According to Equivalent strain and displacement the variation is irrelevant. This results show that the proposed prototype when subjected to the forces that normally act on human body reacts in a proper way maintaining a stable behaviour.

4 Conclusions

Characteristics of a human torso including dimension and forces that act on it was calculated and adopted. A CAD model has been designed in SolidWorks environment and tested in SolidWorks Simulation FEM Analysis. The Simulation shows that the chosen mechanism behaviour suitably can afford common forces acting on a human body and can be used in operation modes such as: turning left and right, leaning forward and backward, and leaning left and right. This operation mode are possible by actuating motors in different combinations.

The designed prototype fully respects the design target; it is very light (10.38 Kg), low-cost and easy to manufacture (ABS), functional and easy to use and control as it has pure revolutional joints.

References

1. Hirai K, Hirose M, Haikawa Y, and Takenaka T (1988) The development of honda humanoid robot. In: Proceedings of the IEEE international conference on robotics and automation, pp 1321–1326
2. Sakagami Y et al (2002) The intelligent ASIMO: system overview and integration. In: Proceedings of the IEEE/RSJ international conference on intelligent robots and systems, pp 2478–2483
3. Kaneko K et al (2004) Humanoid robot HRP-2. In: Proceedings of the IEEE international conference on robotics and automation, pp 1803–1809
4. Fujiwara K et al (2003) The first human-size humanoid that can fall over safely and stand-up again. In: Proceedings of the IEEE/RSJ international conference on intelligent robots and systems, pp 1920–1926
5. Nishiwaki K et al (2000) Design and development of research platform for perception-action integration in humanoid robot H6. In: Proceedings of the IEEE/RSJ international conference on intelligent robots and systems, pp 1559–1564
6. Nishiwaki K, Kagami S, Kuniyoshi Y, Inaba M, Inoue H (2002) Online generation of humanoid walking motion based on a fast generation method of motion pattern that follows desired ZMP. In: Proceedings of the IEEE/RSJ international conference on intelligent robots and systems, pp 2684–2689
7. Ogura. Y, Aikawa H, Shimomura K, Kondo H, Morishima A (2006) Development of a new humanoid robot WABIAN-2. In: Proceedings of the IEEE international conference on robotics and automation, pp 76–81
8. Lim H, Kaneshima Y, Takanishi A (2002) Online walking pattern generation for biped humanoid with trunk. In: Proceedings of the IEEE international conference on robotics and automation, Washington, pp 3111–3116
9. Kim J et al (2005) System design and dynamic walking of humanoid robot KHR-2. In: Proceedings of the IEEE international conference on robotics and automation, pp 1443–1448
10. Lffler K, Gienger M, Pfeiffer F (2003) Sensor and control design of a dynamically stable biped robot. In: Proceedings of the IEEE international conference on robotics and automation, pp 484–490
11. Hugel V, Gouaillier D et al (2009) Mechatronic design of NAO humanoid. In: Proceedings of the 2009 IEEE international conference on robotics and automation, pp 2124–2129
12. Ceccarelli M (1997) A new 3 D.O.F. parallel spatial mechanism. IFToMM J Mech Mach Theory 32(8):895–902
13. Ceccarelli M (2004) Fundamentals of mechanics of robotic manipulation. Kluwer Academic Publishers, Dordrecht
14. Hernandez-Martinez EE, Conghui L, Carbone G, Ceccarelli M, Lopez-Cajun CS (2010) Experimental and numerical characterization of CaPaMan 2bis operation. J Appl Res Technol 8(1):101–119
15. Nava-Rodriguez NE, Carbone G, Ceccarelli M (2006) CaPaMan 2bis as trunk module in CALUMA (Cassino Low-Cost hUMAnoid Robot). In: 2nd IEEE international conference on robotics, automation and mechatronics, RAM 06, Bangkok, pp 347–352
16. Carbone G, Ceccarelli M (2006) A procedure for experimental evaluation of cartesian stiffness matrix of multibody robotic systems. In: 15th CISM-IFToMM symposium on robot design, dynamics and control, pp Paper Rom04–24
17. Aguirre G, Acevedo M, Carbone G, Ceccarelli M (2003) Kinematic and dynamic analyses of a 3 DOF parallel manipulator by symbolic formulations. In: Thematic conference on advances in computational multibody dynamics, ECCOMAS, Lisbon, pp Paper MB2003–010

18. Carbone G, Ottaviano E, Acevedo M, Ceccarelli M (2003) Validación experimental del modelo dinámico del manipulador parallelo capaman 2. In: 6th congreso iberoamericano de ingeniería mecánica, CIBEM 6, Coimbra, pp 869–874
19. Hall JS (2013) Basic biomechanics. LAVOISIER S.A.S, France

Nonlinear Full-Car Model for Optimal Dynamic Design of an Automotive Damper

Carlos A. Duchanoy, Carlos A. Cruz-Villar and Marco A. Moreno-Armendáriz

Abstract In this paper a nonlinear full-car model, comprising the dynamic behavior of the suspension system, which includes the body displacement, body acceleration, wheel displacement, tire deformation, suspension travel, suspension geometry, pitch and roll has been designed. The main improvement introduced to this model is that it considers the nonlinearities caused by the geometry of the suspension system and includes a detailed tire model. This is used by a dynamic optimization methodology in order to improve the passenger comfort and the vehicle safety, which are represented by the chassis displacement and the contact area of the tires, respectively. The optimization algorithm used to solve the problem at hand is a multi-objective artificial bee colony algorithm (MOABC). As result of the optimization a set of nondominated solutions is presented.

Keywords Ful-car model · Suspension system · Multi-objective optimization · Student paper

1 Introduction

Today it is not enough to build quality vehicles, it is necessary to build them at the lowest cost possible to be competitive in the automotive market. This has changed the means of production. In the past multiple prototypes were built to test new

C.A. Duchanoy (✉) · M.A. Moreno-Armendáriz
Centro de Investigación En Computación, Instituto Politécnico Nacional, Av.Juan de Dios Bátiz S/N, 07738 Mexico, D.F., Mexico
e-mail: duchduchanoy@gmail.com

M.A. Moreno-Armendáriz
e-mail: mam_armendariz@cic.ipn.mx

C.A. Cruz-Villar
Sección de Mecatrónica, CINVESTAV Av, Instituto Politécnico Nacional, 2508, Apdo Postal, 14740 Mexico, D.F., Mexico
e-mail: cacruz@cinvestav.mx

M. Ceccarelli and E.E. Hernández Martinez (eds.), *Multibody Mechatronic Systems*, Mechanisms and Machine Science 25, DOI 10.1007/978-3-319-09858-6_46

designs. The new trend is to make numerical simulations of the vehicle behavior to ensure the functionality of the systems before their construction. These simulations are made using computerized models of the automotive systems. For this reason, models of automotive systems have become more detailed and considered precisely all elements that are part of the vehicle.

In the particular case of the automotive suspension multiple models have been developed. The first suspension car models are linear approximations of a quarter-car [6], these models consider two degrees of freedom, the displacement of the vehicle body and the displacement of the tire. The second kind of models were linear half-car models [4], which integrates two quarter-car models and have four degrees of freedom, the displacement of each tire, the displacement of the vehicle body and one inclination angle. Subsequently, a third kind of model was proposed which is the linear full-car model [5]. The full-car model integrates four conventional linear quarter-car models, this model has seven degrees of freedom one for each tire, the displacement of the vehicle body, pitch and roll. All these models are linear approximations of the suspension system. A linear model captures the basic behavior of vehicle suspension with low computational cost [2], this is useful when the computer resources are limited. However, the suspension is a nonlinear system and when a linearized model of the suspension is used important information of the system is lost.

New computers have more resources, which allow the execution of complex models that accurately represent the suspension system. An example of this type of model is presented in [8]. Complex models allow more analysis as testing the influence of different parameters [7]. This motivated us to design a full-car model that considers the nonlinearities caused by the geometry of the suspension system. The geometry influences the behavior of the suspension system, as the geometry changes the direction of the forces depending on the suspension position. Another improvement we make was the addition of a detailed tire model which computes the deformation of the tire and from that deformation, the contact area and the pushing force of the tire are calculated. The model is implemented in Matlab Simulink® and is used to optimize the suspension system using an evolutive algorithm. We select this kind of algorithm because they have been useful for this kind of problems [3]. The main goal of the optimization is designing a passive damper which improves the performance of the suspension in two factors, the vehicle safety and comfort of the passengers.

2 Mathematical Model

As a test vehicle we select a Baja SAE off-road vehicle due our previous knowledge and because it is a competition benchmark vehicle, designed to test different automotive systems. In Fig. 1 the physical Baja SAE vehicle is shown. The vehicle has been drawn using computer assisted design (CAD) tools, in order to easily

(a) **(b)**

Fig. 1 Baja SAE. **a** Physical Car. **b** CAD

acquire the parameters required by the model, as the mass center, the body inertias, among others. Figure 1 shows the CAD model.

The proposed full-car model is expressed in Eqs. (1)–(14). We select as state variables the displacement of each unsprung mass ($z_t(fr)$, $z_t(fl)$, $z_t(rr)$, $z_t(rl)$), the velocity of each unsprung mass ($\dot{z}_t(fr)$, $\dot{z}_t(fl)$, $\dot{z}_t(rr)$, $\dot{z}_t(rl)$), the body displacement z_c, the body velocity $\dot{z}_c$, the roll displacement φ_r, the roll angular velocity $\dot{\varphi}_r$, the pitch displacement θ_p and the pitch angular velocity $\dot{\theta}_p$. The state variables are expressed in Eqs. (15)–(19). The inputs of the system are the displacement of the ground below each tire, this are expressed in Eqs. (20) and (21).

The mathematical model is used to measure the performance of each possible solution in two aspects the vehicle comfort and the vehicle safety. The comfort of the vehicle is represented by the chassis displacement x_9 and the safety is represented by the contact area A_{ct} of the tires Eqs. (44)–(47). The proposed model takes into account the following subsystems; front right (fr) and left (fl) suspensions, rear right (rr) and left (rl) suspensions, tires and vehicle body. The main difference between the proposed model and the conventional full car model is that the proposed model considers the geometry of the suspension systems. Both suspensions are a double wishbone system, this configuration generates a constant change in the angles at which the forces are transmitted. The geometry is expressed in Eqs. (24)–(35), where α represents the angle of the lower arm, β represents the angle of the upper arm and δ is the angle of the damper. The forces that interact in the car model are the force of the tires F_t Eqs. (52)–(55), the force of the rear arms F_b, shown in Eqs. (38) and (39), the force of the lower front arm F_{bi}, shown in Eqs. (36) and (37), and the force of the upper front arm F_{bs}, shown in Eqs. (22) and (23). and the force of the damper F_D Eqs. (40)–(43). The model of the body uses the forces of the suspension in the upward direction, these components are expressed in Eqs. (56)–(60). The suspension model needs the displacement of the anchor point of each damper, these are calculated by the body model using Eqs. (61)–(64). This mathematical model allows to relate the contact area of the tire with the forces transmitted from the ground to the suspension. This allows to observe how both objectives are affected by modifying the suspensions.

One design objective seek to ensure the tire contact with the ground at all times. In order to measure the contact area of the tires we propose a model capable of representing the behavior of the tire. This is achieved describing the contact area of the tire by Eqs. (44)–(47), which is calculated using a geometrical analysis of the wheel by estimating the tire deformation. The deformation is used to calculate the area of the tire that is in contact with the soil and the volume of the tire that is compressed, which are used later to calculate the thrust force of the tire that is transmitted to the suspension. The second objective, the passenger comfort, aims to reduce the cockpit vertical movement during the travel. This is obtained by solving the differential equations describing the vehicle behavior in function of the ground variations.

$$\dot{x}_1 = x_5 \tag{1}$$

$$\dot{x}_2 = x_6 \tag{2}$$

$$\dot{x}_3 = x_7 \tag{3}$$

$$\dot{x}_4 = x_8 \tag{4}$$

$$\dot{x}_5 = \frac{F_t(fr)}{M_t(f)} - \frac{F_{bi}(fr)\sin\alpha_f(fr)}{M_t(f)} - \frac{F_{bs}(fr)\sin(\frac{\pi}{2} - \beta_f(fr))}{M_t(f)} - g \tag{5}$$

$$\dot{x}_6 = \frac{F_t(fl)}{M_t(f)} - \frac{F_{bi}(fl)\sin\alpha_f(fl)}{M_t(f)} - \frac{F_{bs}(fl)\sin(\frac{\pi}{2} - \beta_f(fl))}{M_t(f)} - g \tag{6}$$

$$\dot{x}_7 = \frac{F_t(rr)}{M_t(r)} + \frac{F_b(rr)\sin(\alpha_r(rr))}{M_t(r)} - \frac{F_D(rr)\sin(\delta_r(rr))}{M_t(r)} - g \tag{7}$$

$$\dot{x}_8 = \frac{F_t(rl)}{M_t(r)} + \frac{F_b(rl)\sin(\alpha_r(rl))}{M_t(r)} - \frac{F_D(rl)\sin(\delta_r(rl))}{M_t(r)} - g \tag{8}$$

$$\dot{x}_9 = x_{10} \tag{9}$$

$$\begin{aligned}\dot{x}_{10} = {} & \frac{Fz_{bi}(fr) + Fz_{bs}(fr) + Fz_D(fr) + Fz_{bi}(fl) + Fz_{bs}(fl)}{M_c} \\ & + \frac{Fz_D(fl) + Fz_b(rr) + Fz_D(rr) + Fz_b(rl) + Fz_D(rl)}{M_c}\end{aligned} \tag{10}$$

$$\dot{x}_{11} = x_{12} \tag{11}$$

$$\dot{x}_{12} = \frac{d_{bi}(fP)(Fz_{bi}(fr) + Fz_{bi}(fl)) + d_{bs}(fP)(Fz_{bs}(fr) + Fz_{bs}(fl))}{J_p} - \frac{d_D(rP)(Fz_b(rr) + Fz_D(rr) + Fz_b(rl) + Fz_D(rl))}{J_p} + \frac{d_D(fP)(Fz_D(fl) + Fz_D(fr))}{J_p} \tag{12}$$

$$\dot{x}_{13} = x_{14} \tag{13}$$

$$\dot{x}_{14} = \frac{d_{bi}(fR)(Fz_{bi}(fl) - Fz_{bi}(fr)) + d_b(rR)(Fz_b(rl) - Fz_b(rr))}{J_r} + \frac{+d_{bs}(fR)(Fz_{bs}(fl) - Fz_{bs}(fr) + Fz_D(fl) - Fz_D(fr))}{J_r} + \frac{d_D(rR)(Fz_D(rl) - Fz_D(rr))}{J_r} \tag{14}$$

$$x_1 = z_t(fr) \quad x_2 = z_t(fl) \quad x_3 = z_t(rr) \tag{15}$$

$$x_4 = z_t(rl) \quad x_5 = \dot{z}_t(fr) \quad x_6 = \dot{z}_t(fl) \tag{16}$$

$$x_7 = \dot{z}_t(rr) \quad x_8 = \dot{z}_t(rl) \quad x_9 = z_c \tag{17}$$

$$x_{10} = \dot{z}_c \quad x_{11} = \theta_p \quad x_{12} = \dot{\theta}_p \tag{18}$$

$$x_{13} = \varphi_r \quad x_{14} = \dot{\varphi}_r \tag{19}$$

$$u_1 = Q(fr) \quad u_2 = Q(fl) \quad u_3 = Q(rr) \tag{20}$$

$$u_4 = Q(rl) \tag{21}$$

where:

$$F_{bs}(fr) = \frac{F_{bi}(fr)\cos\alpha_f(fr)}{\cos(\frac{\pi}{2} - \beta_f(fr))} \tag{22}$$

$$F_{bs}(fl) = \frac{F_{bi}(fl)\cos\alpha_f(fl)}{\cos(\frac{\pi}{2} - \beta_f(fl))} \tag{23}$$

$$\alpha_f(fr) = \arccos\left(\frac{(z_c(fr) + z_{co}(fr)) - (x_1 + z_{to}(fr)) - D_1}{D_2}\right) \tag{24}$$

$$\beta_f(fr) = \arccos\left(\frac{(z_c(fr) + z_{co}(fr)) - (x_1 + z_{to}(fr)) - D_3}{D_4}\right) \tag{25}$$

$$\delta_f(fr) = \arctan\left(\frac{D_5 + D_6\cos\left(\varepsilon_f(fr)\right)}{D_7\sin\left(\varepsilon_f(fr)\right) - D_8}\right) - \left(\frac{\pi}{2} - \varepsilon_f(fr)\right) \tag{26}$$

$$\alpha_f(fl) = \arccos\left(\frac{(z_c(fl) + z_{co}(fl)) - (x_2 + z_{to}(fl)) - D_1}{D_2}\right) \tag{27}$$

$$\beta_f(fl) = \arccos\left(\frac{(z_c(fl) + z_{co}(fl)) - (x_2 + z_{to}(fl)) - D_3}{D_4}\right) \tag{28}$$

$$\delta_f(fl) = \arctan\left(\frac{D_5 + D_6\cos\left(\varepsilon_f(fl)\right)}{D_7\sin\left(\varepsilon_f(fl)\right) - D_8}\right) - \left(\frac{\pi}{2} - \varepsilon_f(fl)\right) \tag{29}$$

$$\alpha_r(rr) = \arcsin\frac{(z_c(rr) + z_{co}(rr)) - (x_3 + z_{to}(rr)) - D_9}{D_{10}} \tag{30}$$

$$\beta_r(rr) = \alpha_r(rr) \tag{31}$$

$$\delta_r(rr) = \arctan\frac{(z_c(rr) + z_{co}(rr)) - (x_3 + z_{to}(rr)) - D_{11}}{D_{12}\cos\alpha_r(rr) - D_{13}} \tag{32}$$

$$\alpha_r(rl) = \arcsin\frac{(z_c(rl) + z_{co}(rl)) - (x_4 + z_{to}(rl)) - D_9}{D_{10}} \tag{33}$$

$$\beta_r(rl) = \alpha_r(rl) \tag{34}$$

$$\delta_r(rl) = \arctan\frac{(z_c(rl) + z_{co}(rl)) - (x_4 + z_{to}(rl)) - D_{11}}{D_{12}\cos\alpha_r(rl) - D_{13}} \tag{35}$$

$$\begin{aligned} F_{bi}(fr) = {} & \frac{(M_{bi}g)(d_c\cos\varepsilon_f(fr) - d_r\sin\varepsilon_f(fr))}{d_t} \\ & + \frac{F_D(fr)(d_a\sin\delta_f(fr) - d_r\cos\delta_f(fr))}{d_t} \end{aligned} \tag{36}$$

$$\begin{aligned} F_{bi}(fl) = {} & \frac{(M_{bi}g)(d_c\cos\varepsilon_f(fl) - d_r\sin\varepsilon_f(fl))}{d_t} \\ & + \frac{F_D(fl)(d_a\sin\delta_f(fl) - d_r\cos\delta_f(fl))}{d_t} \end{aligned} \tag{37}$$

$$F_b(rr) = \frac{F_D(rr)\cos(\delta_r(rr))}{\cos(\alpha_r(rr))} \tag{38}$$

$$F_b(rl) = \frac{F_D(rl)\cos(\delta_r(rl))}{\cos(\alpha_r(rl))} \tag{39}$$

$$F_D(fr) = K_d(fr)(z_c(fr) - x_1) + B_d(fr)(x_{10} - x_5) \tag{40}$$

$$F_D(fl) = K_d(fl)(z_c(fl) - x_2) + B_d(fl)(x_{10} - x_6) \tag{41}$$

$$F_D(rr) = K_d(rr)(z_c(rr) - x_3) + B_d(rr)(x_{10} - x_7) \tag{42}$$

$$F_D(rl) = K_d(rl)(z_c(rl) - x_4) + B_d(rl)(x_{10} - x_8) \tag{43}$$

$$A_{ct}(fr) = 2a_n R_t \sin(\arccos(\frac{R_t - (u_1 - x_1)}{R_t})) \tag{44}$$

$$A_{ct}(fl) = 2a_n R_t \sin(\arccos(\frac{R_t - (u_2 - x_2)}{R_t})) \tag{45}$$

$$A_{ct}(rr) = 2a_n R_t \sin(\arccos(\frac{R_t - (u_3 - x_3)}{R_t})) \tag{46}$$

$$A_{ct}(rl) = 2a_n R_t \sin(\arccos(\frac{R_t - (u_4 - x_4)}{R_t})) \tag{47}$$

$$P_c(fr) = \frac{V_t P_t}{V_t - a_n\left(R_t^2 \arcsin\left(\frac{A_{ct}(fr)}{2a_n R_t}\right) - \frac{A_{ct}(fr)(R_t - (u_1 - x_1))}{2a_n}\right)} \tag{48}$$

$$P_c(fl) = \frac{V_t P_t}{V_t - a_n\left(R_t^2 \arcsin\left(\frac{A_{ct}(fl)}{2a_n R_t}\right) - \frac{A_{cl}(fl)(R_t - (u_2 - x_2))}{2a_n}\right)} \tag{49}$$

$$P_c(rr) = \frac{V_t P_t}{V_t - a_n\left(R_t^2 \arcsin\left(\frac{A_{ct}(rr)}{2a_n R_t}\right) - \frac{A_{ct}(rr)(R_t - (u_3 - x_3))}{2a_n}\right)} \tag{50}$$

$$P_c(rl) = \frac{V_t P_t}{V_t - a_n\left(R_t^2 \arcsin\left(\frac{A_{ct}(rl)}{2a_n R_t}\right) - \frac{A_{ct}(rl)(R_t - (u_4 - x_4))}{2a_n}\right)} \tag{51}$$

$$F_t(fr) = P_c(fr) A_{ct}(fr) \tag{52}$$

$$F_t(fl) = P_c(fl) A_{ct}(fl) \tag{53}$$

$$F_t(rr) = P_c(rr) A_{ct}(rr) \tag{54}$$

$$F_t(rl) = P_c(rl) A_{ct}(rl) \tag{55}$$

$$Fz_{bi}(fr) = F_{bi}(fr) \cos \alpha_f(fr) Fz_{bs}(fr) = F_{bs}(fr) \cos \beta_f(fr) \tag{56}$$

$$Fz_D(fr) = F_D(fr)\sin\delta_f(fr)Fz_{bi}(fl) = F_{bi}(fl)\cos\alpha_f(fl) \tag{57}$$

$$Fz_{bs}(fl) = F_{bs}(fl)\cos\beta_f(fl)Fz_D(fl) = F_D(fl)\sin\delta_f(fl) \tag{58}$$

$$Fz_b(rr) = F_b(rr)\sin\alpha_f(rr)Fz_D(rr) = F_D(rr)\sin\delta_f(rr) \tag{59}$$

$$Fz_b(rl) = F_b(rl)\sin\alpha_f(rl)Fz_D(rl) = F_D(rl)\sin\delta_f(rl) \tag{60}$$

$$z_c(fr) = x_9 - d_D(fP)\sin(\theta_p) - d_{bs}(fR)\sin(\varphi_r) \tag{61}$$

$$z_c(fl) = x_9 - d_D(fP)\sin(\theta_p) + d_{bs}(fR)\sin(\varphi_r) \tag{62}$$

$$z_c(rr) = x_9 + d_D(rP)\sin(\theta_p) - d_D(rR)\sin(\varphi_r) \tag{63}$$

$$z_c(rl) = x_9 + d_D(rP)\sin(\theta_p) + d_D(rR)\sin(\varphi_r) \tag{64}$$

3 Optimization Problem

The suspension system has two main objectives: The first one, is to guarantee the safety of the passengers during the trip and the second one, is improving the comfort of the passengers. For the off-road vehicles, the comfort of the driver is particularly important because the characteristics of the road generate displacements in the cockpit that can be distracting to the driver and may cause an accident. So, we define the safety and comfort as follows: It should be understood by comfort C, a reduction in the amplitude of the cockpit displacement and as safety S, an increase of the contact area of the tires. The problem is formulated, as the minimization of the chassis displacement and the maximization of the contact area of all the tires. We propose as design variables the stiffness and damping coefficients of each damper. The problem is expressed in Eqs. (65)–(67). It is necessary to establish that the vehicle has an initial displacement, which is caused by the compression of the suspension due to the vehicle weight. This initial displacement is known as displacement of the suspension in steady state (Z_{cs}), which changes for each configuration of the damper, therefore, it is necessary to calculate its value for each configuration. Both objectives are measured during the car travel, for this reason the objectives are integrated over a time interval (t_n) in which the vehicle travels the track. The objectives are subject to Eqs. (15)–(64), which describe the mathematical model

$$\max_{\phi}\left[\int_{t=0}^{t_n} A_{ct}(fr) + A_{ct}(fl) + A_{ct}(rr) + A_{ct}(rl)\right] \tag{65}$$

$$\min_{\phi}\left[\int_{t=0}^{t_n} \|x_9 - Z_{cs}\|\right] \tag{66}$$

$$\phi = [K_d(fr), K_d(fl), K(rr), K_d(rl), B_d(fr), B_d(fl), B_d(rr), B_d(rl)] \quad (67)$$

4 Multi-Objective Artificial Bee Colony Algorithm

The MOABC [1] represents the foraging behavior of honey bees and the colony has three types of bees: Employees, onlookers and scouts. This algorithm assumes a fixed number of food sources, each food source represents a solution to the problem, these food sources have been found by employed bees and will be presented in the dance area. The onlooker bees wait on the dance area to choose a food source by its quality. When a food source could not be improved after some cycles, a scout bee will do a random search in order to find a new food source. The MOABCM is a Pareto based algorithm with an external archive used to store non-dominated solutions. This algorithm requires as parameters: Population size, Max-trial, maximum number of iterations. Population size is the number of food sources, employed and onlooker bees, Max-trial is the value that the algorithm uses to find out which food sources should be abandoned. The maximum number of iterations indicates when the algorithm must be stopped (end condition).

5 Numerical Simulations

The mathematical model is programmed in Matlab Simulink® in order to simulate the behavior of the car. The model was set up to fit the characteristics of the proposed car. This is achieved by setting the parameters shown in Table 1. The performance of the suspension system is tested on a track. The track selected for this experiment is conformed by a pair of cross-caps 20 cm tall, this track is shown

Table 1 Baja SAE phisical parameters

$M_t(f) = 7.98$ kg	$M_t(r) = 7$ kg	$M_c = 275.35$ kg
$J_p = 88.31$ kg	$J_r = 135.82$ kg	$d_{bi}(fP) = 0.9029$ m
$d_{bs}(fP) = 0.8237$ m	$d_D(fP) = 0.7910$ m	$d_D(rP) = 0.40879$ m
$d_{bi}(fR) = 0.2468$ m	$d_{bs}(fR) = 0.1694$ m	$d_b(rR) = 0.2468$ m
$d_D(rR) = 0.4217$ m	$z_{co}(fr) = z_{co}(fl) = 0.6857$ m	$z_{to}(fr) = z_{to}(fl) = 0.3046$ m
$z_{co}(rr) = z_{co}(rl) = 0.9586$ m	$z_{to}(rr) = z_{to}(rl) = 0.3046$ m	$D_1 = 0.2447$ m
$D_2 = 0.4622$ m	$D_3 = 0.2524$ m	$D_4 = 0.2982$ m
$D_5 = 0.3445$ m	$D_6 = 0.180cos(\varepsilon_f)$ m	$D_7 = .180sin(\varepsilon_f)$ m
$D_8 = 0.079$ m	$D_9 = 0.508$ m	$D_{10} = 0.4031$ m
$D_{11} = 0.080$ m	$D_{12} = 0.403$ m	$D_{13} = 0.2617$ m
$d_r = 0.0127$ m	$d_a = 0.180$ m	$d_t = 0.240$ m
$d_c = 0.180$ m	$a_n = 0.254$ m	$V_t = 0.05367$ m^3
$P_t = 48.26$ KPa	$R_t = 0.304$ m	$M_{bi} = 1.19$ kg
$\mu_r = 0.18$		

Fig. 2 Schematic diagram of the cross-caps terrain

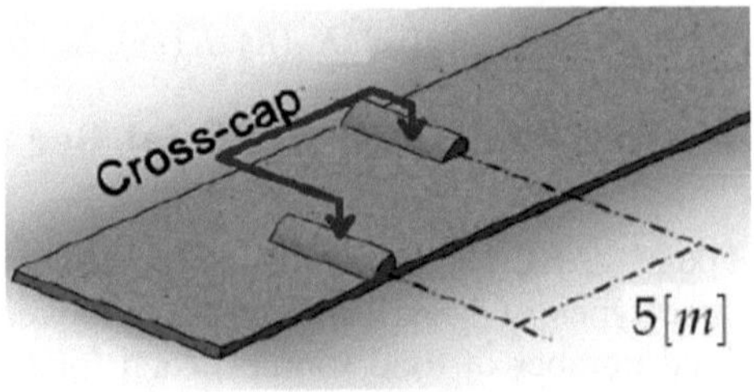

in Fig. 2. The vehicle travels trough the track in 10.75 s, which is the integration time t_n for the objectives. The first design objective is the minimization of the cockpit displacement which is related to the passenger comfort. The second objective is the maximization of the contact area of the tires which is related to the safe handling of the vehicle, as a consequence of the maximization of the contact area, the tires tend to keep contact with the ground.

Applying the algorithm described in Sect. 4 the optimization was realized. The method was performed using the following parameters for the MOABC: Population = 60, max-trial = 10 and max-Iterations = 40. The execution of the algorithm spent around of 26 h, with a total of 7,320 evaluations of the fitness function. The algorithm found 48 possible solutions for the problem which are graphically expressed in Fig. 3a.

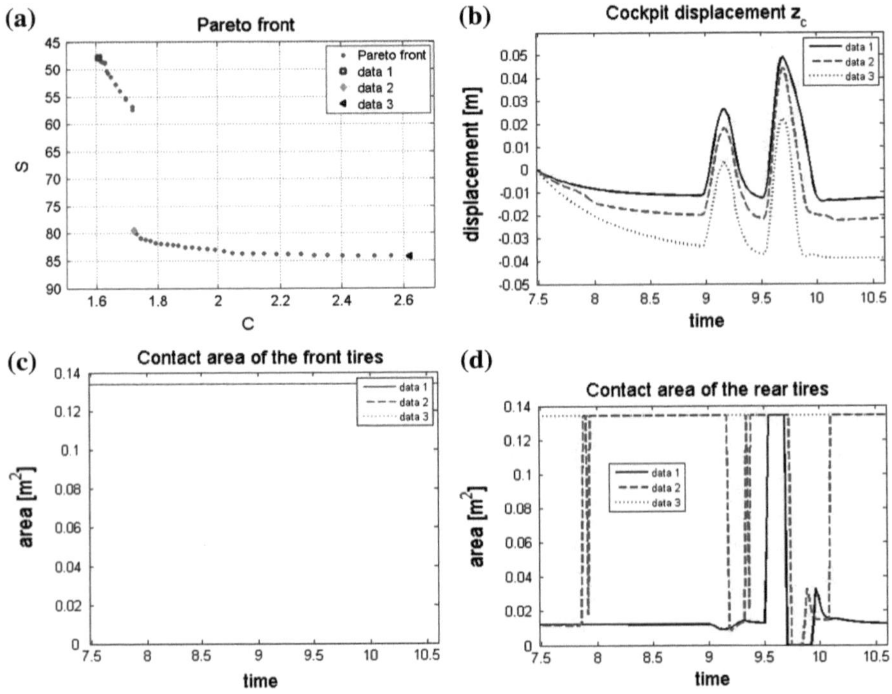

Fig. 3 Dynamic behaviour of the vehicle. **a** Pareto front. **b** Cockpit displacement. **c** Contact area of the front tire. **d** Contact area of the rear tire

To analyze the dynamic behavior of the vehicle, we select three possible solutions of the ones obtained by the optimization algorithm. We selected the solution with the best performance in the comfort *C* objective (data 1), the solution that is in the center of the Pareto front (data 2) and the solution with the best performance in the safety *S* objective (data 3). These solutions are shown in Fig. 3a.

From the selected solutions we compare: The cockpit displacement, the contact area of the front tires and the contact area of the rear tires, which are shown in Fig. 3.

The cockpit displacement is different in the three proposed solutions but the amplitude and duration of the changes in each signal are similar. These changes are the ones that affect the comfort of the vehicle, for these reason we decide that the three solutions have an acceptable comfort. On the other hand, the safety of the vehicle is determined by the contact area of the tires, the front tires have the same contact area during all the travel in the three solutions, but the rear tires behave different in each solutions. The first solution (data 1) has a reduced contact area of the tire and loses contact with the ground at 9.65 s, the second solution (data 2) has a bigger contact area than the first solution but it also loses contact with the ground at 9.68 s. Finally, only one solution (data 3) guarantees that the tires are always in contact with the ground all the time.

6 Conclusion and Future Work

In this paper we propose a nonlinear full-car model which includes the geometry of the suspensions systems and a geometric model of the tire. This model is used to optimize the suspension system in two objectives, the comfort of the passengers and the safe driving of the vehicle. The model proposed for the vehicle is able to represent more accurately its behavior, but it consumes a lot of time for each execution, thus delaying the optimization process. To solve this problem, we use a modified MOABC algorithm with a reduced number of evaluations of the fitness function. The algorithm obtains a set of nondominated solutions which are analyzed to determine its performance. The safe handling of the vehicle is critical, in order to ensure it, the steering system needs that the front tires are always in contact with the ground. The optimal design of the suspension system guarantees that the front tires kept the same contact area during the travel of the vehicle in the proposed track. This is shown in Fig. 3c. Because safety is more important than the passenger comfort, we suggest that the best solution is the one with the best performance in the safety objective. This solution guarantees that all the tires are always in contact with the ground as shown in Fig. 3c, d. As future work, we plan to test the optimal damper in different tracks to measure its performance.

The authors wish to thank the Government of Mexico (SNI, SIP-IPN, COFAA-IPN, PIFI-IPN and CONACyT, for providing necessary support to carry out this research work.

References

1. Akbari R, Hedayatzadeh R, Ziarati K, Hassanizadeh B (2011) A multi-objective artificial bee colony algorithm. Swarm Evol Comput 2:39–52
2. Darus R, Sam YM (2009) Modeling and control active suspension system for a full car model. In: 5th international colloquium on signal processing and its applications, vol 4(7), pp 13–18
3. Goga V, Klúčik M (2012) Optimization of vehicle suspension parameters with use of evolutionary computation. Procedia Eng 48:174–179
4. Hrovat D, Krtolica R (1990) Optimal active suspension control based on a half-car model. In: Proceeding of the 29th conference on decision and control, pp 2238–2243
5. Kruczek A, Stribrsky A (2004) A full-car model for active suspension—some practical aspects. In: Proceedings of the IEEE international conference on mechatronics, pp 41–45
6. Natsiavas S, Verros G, Papadimitriou C (2005) Design optimization of quarter-car models with passive and semi-active suspensions under random road exitation. J Vib Control 11(5):581–606
7. Tey JY, Ramli R, Kheng CW, Chong SY, Abidin MAZ (2013) Identification of vehicle suspension parameters by design optimization. Eng Optim (November) 1–18
8. Torres JL, Gimenez A, Lopez-Martinez J, Carbone G, Ceccarelli M (2013) Analysis of the dynamic behavior of an electric vehicle using an equivalent roll stiffness model. New Trends Mech Mach Sci 7:599–607

Design of a Parallel Mechanism for Knee Rehabilitation

B.D. Chaparro-Rico, E. Castillo-Castaneda and R. Maldonado-Echegoyen

Abstract This paper discusses the design of lower limb rehabilitation device based on a five-bar parallel mechanism. The proposed mechanism reproduces exercises for knee rehabilitation, wherein the end-effector of the five-bar mechanism guides patient's leg. Forward and inverse position kinematics of the mechanism was revised. Detailed mechanical design using CAD was performed and it was used for deformation analysis and construction of the mechanical structure.

Keywords Position kinematics · Five-bar mechanism · Knee rehabilitation · Deformation analysis · Student paper

1 Introduction

In the search for technological solutions to improve and speed up the results of lower and upper limb rehabilitation, researchers have worked around the development of robotic devices to assist therapies.

The University of Applied Sciences Rapperswil, Switzerland [1], proposes an automatic device that performs therapy exercises and record of progress. Others works propose alternatives to mimic the physiotherapist intervention as the "Knee rehabilitation using an intelligent robotic system" [2] that propose a rule based intelligent control and Therapeutic Exercise Machine (TEM) [3], which memorizes movements, guided by the therapist, and reproduces them afterwards. Active Knee Rehabilitation Orthotic Device (AKROD) [4] was designed for gait retraining in

B.D. Chaparro-Rico (✉) · E. Castillo-Castaneda · R. Maldonado-Echegoyen
Instituto Politecnico Nacional CICATA Unidad Queretaro, Queretaro, Mexico
e-mail: betsychaparro@hotmail.com

E. Castillo-Castaneda
e-mail: ecastilloca@ipn.mx

R. Maldonado-Echegoyen
e-mail: rodrigorobotics@hotmail.com

M. Ceccarelli and E.E. Hernández Martinez (eds.), *Multibody Mechatronic Systems*,
Mechanisms and Machine Science 25, DOI 10.1007/978-3-319-09858-6_47

stroke patients, correcting knee hyperextension during standing posture and gait. The patent described in [5] presents a device that increases the range of motion of the knee joint.

Recently, researchers have shown interest in introducing parallel mechanisms in the rehabilitation technology since they offer advantages compared to serial robots [6] in terms of accuracy, load capacity, stiffness and speed. An example is the auto tune exoskeleton for knee rehabilitation based on parallel structure 3-RRP of two degrees of freedom [7].

The novelty of the work presented in this paper is the application of a known parallel structure in the design of a mechatronic system for knee rehabilitation that offers advantages over other devices as, among others, the ability to reproduce multiple exercises in a single apparatus. Our device adds the advantages already mentioned regarding the serial mechanism. Furthermore it allows flexion and extension of the knee and hip, individually or in combination, while other devices only allow one way.

2 Concept of the Mechanism

According to literature review and discussions with physiotherapists from Centro de Rehabilitación Integral de Querétaro (CRIQ), a large number of exercises for knee rehabilitation take place on vertical plane [8–13]. The ankle joint angle with respect to tibia remains constant and hip and knee joints rotate depending on the exercise, as is shown in Fig. 1.

Fig. 1 Four common rehabilitation exercises

Fig. 2 Basic 5-bar mechanism

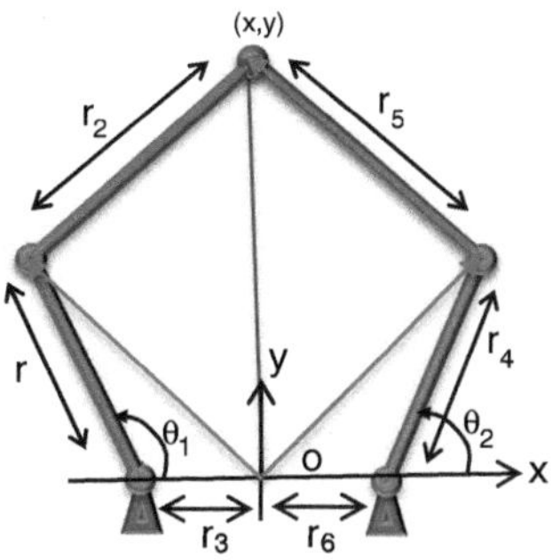

Then, two degrees of freedom are needed to perform most part of knee rehabilitation exercises. For this reason, we propose a 5-bar parallel mechanism to reproduce knee rehabilitation exercises. Figure 2 shows the basic 5-bar mechanism with 5 rotational joints, where the angles θ_1 and θ_2 are the joint variables, coordinates (x,y) correspond to end-effector position and *ri* with (*i* = *1*,..., *6*) are the link lengths.

3 Position Kinematics

3.1 Forward Kinematics

In Fig. 2, red lines are vectors describing the 5-bar mechanism; from scalar product operations one can obtain an implicit solution such as:

$$(x - r_1\cos\theta_1 + r_3)^2 + (y - r_1\sin\theta_1)^2 = r_2^2 \tag{1}$$

$$(x - r_4\cos\theta_2 - r_6)^2 + (y - r_4\sin\theta_2)^2 = r_5^2 \tag{2}$$

If lengths *ri* with (i = 1,..., 6) are known, equation's system (1) and (2) can be solved by symbolic computation using MATLAB.

3.2 Inverse Kinematics

Inverse kinematics [14] was solved also from Eqs. (1) and (2). To calculate θ_1 Eq. (1) is expanded as:

$$\theta_1 = 2\tan^{-1}\frac{-B_1 \pm \sqrt{B_1^2 - 4A_1C_1}}{2A_1} \tag{3}$$

With:

$$A_1 = r_1^2 + y^2 + (x + r_3)^2 - r_2^2 + 2r_1(x + r_3)$$
$$B_1 = -4yr_1$$
$$C_1 = r_1^2 + y^2 + (x + r_3)^2 - r_2^2 - 2r_1(x + r_3)$$

To calculate θ_2, Eq. (2) is expanded as:

$$\theta_2 = 2\tan^{-1}\frac{-B_2 \pm \sqrt{B_2^2 - 4A_2C_2}}{2A_2} \quad (4)$$

With:

$$A_2 = r_4^2 + y^2 + (x - r_6)^2 - r_5^2 + 2r_4(x - r_6)$$
$$B_2 = -4yr_4$$
$$C_2 = r_4^2 + y^2 + (x - r_6)^2 - r_5^2 - 2r_4(x - r_6)$$

Using inverse kinematic, the workspace of the 5-bar mechanism can be obtained. It was verified that workspace covers the space needed to reproduce four exercises for knee rehabilitation characterized previously (see Fig. 1).

4 Mechanical Design

4.1 Mechanical Structure

The mechanical design was performed using CAD software, the final design is shown in Fig. 3. It consists of a movable base supporting two articulated arms (5-bar mechanism), where the end-effector guides the patient's leg. The sizing of the mechanism is based on anthropomorphic dimensions. The devise can slide on a stretcher to be placed on the patient. In the detailed design, commercially available parts were used to facilitate the construction of the prototype.

The prototype was validated geometrically by CAD simulations using an articulated mannequin to simulate the patient's leg. The mannequin is placed on a stretcher and his leg is fixed on the end-effector at the level of the ankle joint. The mechanism does not act directly on the knee; the end-effector follows the path of the ankle in all the exercises of knee rehabilitation. Flexion and extension of the leg was performed and the four rehabilitation exercises were performed.

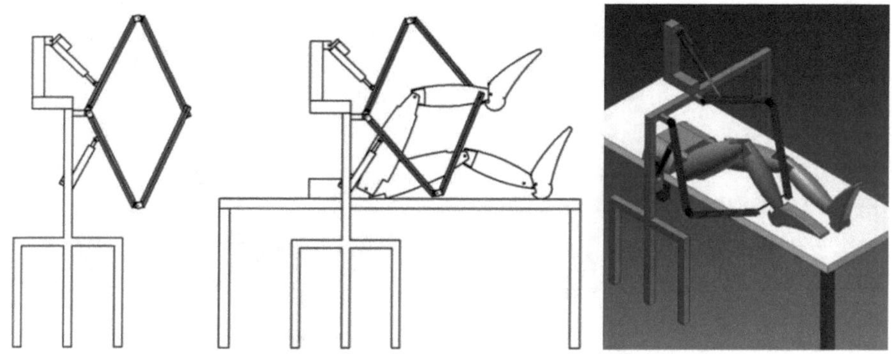

Fig. 3 Mechanical design using CAD

4.2 Deformation Analysis

Deformation analysis, using ANSYS software, was conducted to simulate the behavior of the mechanism during rehabilitation exercises. The maximum value of the mass of the leg, which corresponds to 15 kg is considered, the positions of the mechanism revised were those shown in Fig. 4.

The applied force at end-effector level was calculated by considering the leg, with length l, as a simple beam supporting a load P of 15 kg, where the hip and the end-effector represent the beam ends. The exerted force F depends on the angle θ between the hip and ankle joints; it can be computed from equilibrium of momenta from:

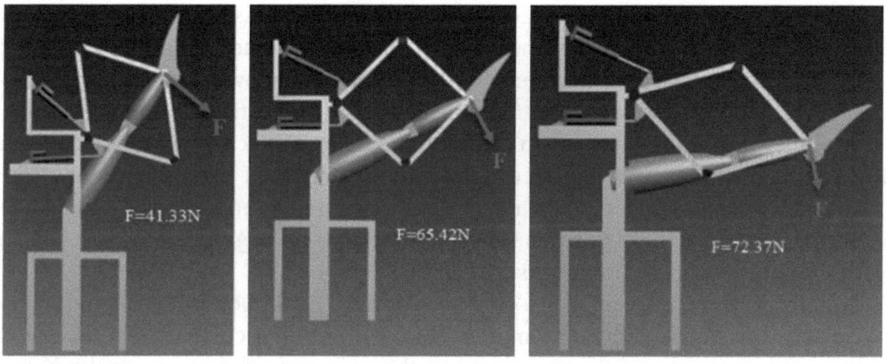

Fig. 4 Revised mechanism positions

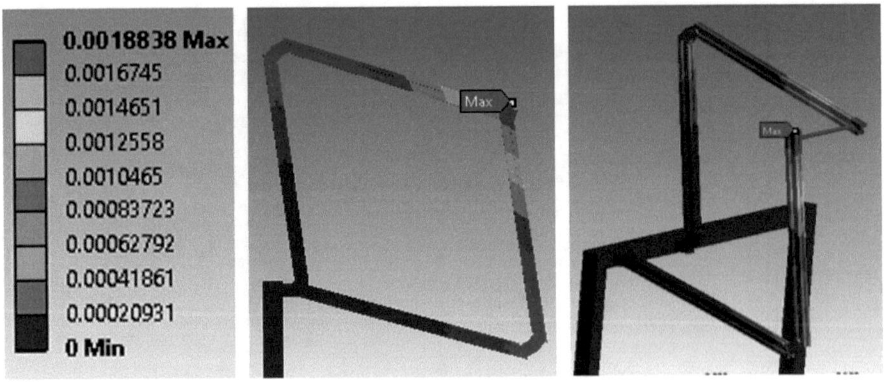

Fig. 5 Deformation analysis considering an applied force of 41.33 N

$$Pg\left(\frac{l}{2}\right)\cos\theta + Fl = 0 \tag{5}$$

The simulation result, considering an applied force of F = 41.33 N, is shown in Fig. 5. In that case, the maximum deformation is 1.88 mm. When the applied force is 65.42 N maximum deformation is 4.5 mm, and 5.2 mm for an applied force of 72.37 N. For all cases, maximum deformation takes place at the end-effector level.

5 Construction

The mechanism was built according to design specifications; mechanism mobility was tested manually by moving the end-effector at different positions on the vertical plane. Also a mannequin was used to verify that all exercises can be reproduced. The leg was attached to the end-effector mechanism and mobility was tested applying four exercises for knee rehabilitation. Figure 6 shows some movements of the rehabilitation exercises using the mechanism and an articulated mannequin.

Mobility test of the mechanism ensures that the mechanical structure is collision free and to verify that workspace allows reproduction of exercises. A two-axis controller is being developed for trajectory tracking in closed loop using two linear actuators and encoders as position sensors. In Fig. 7 a photograph of the mechanism with the motors and position sensors can be observed.

Motors current curves were obtained during reproduction of rehabilitation exercises. The Fig. 8 shows the motors (M1, M2) current curves when the mechanism reproduces one of the exercises and the trajectory of exercise on the X axis and Y-axis. The exercise was carried round back.

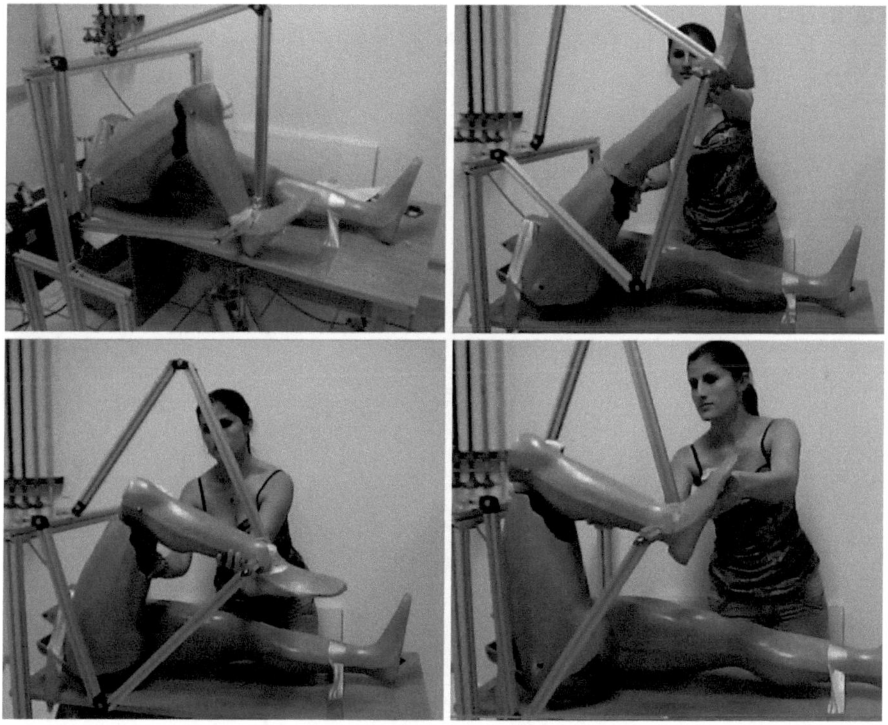

Fig. 6 Mechanism performing rehabilitation motions with a mannequin

Fig. 7 Actuated mechanism performing rehabilitation

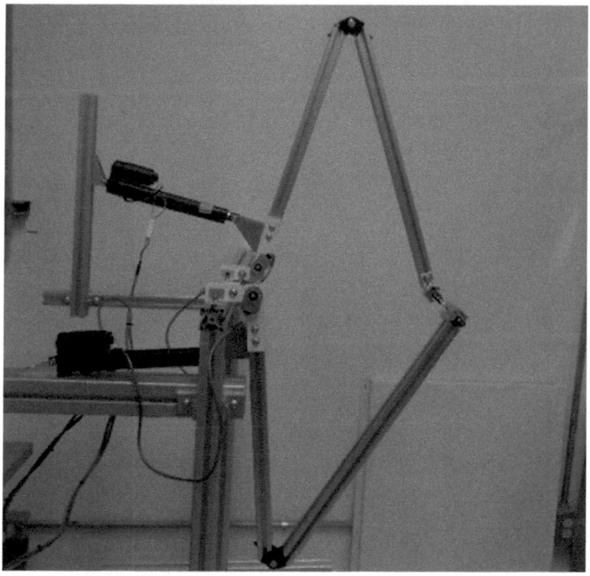

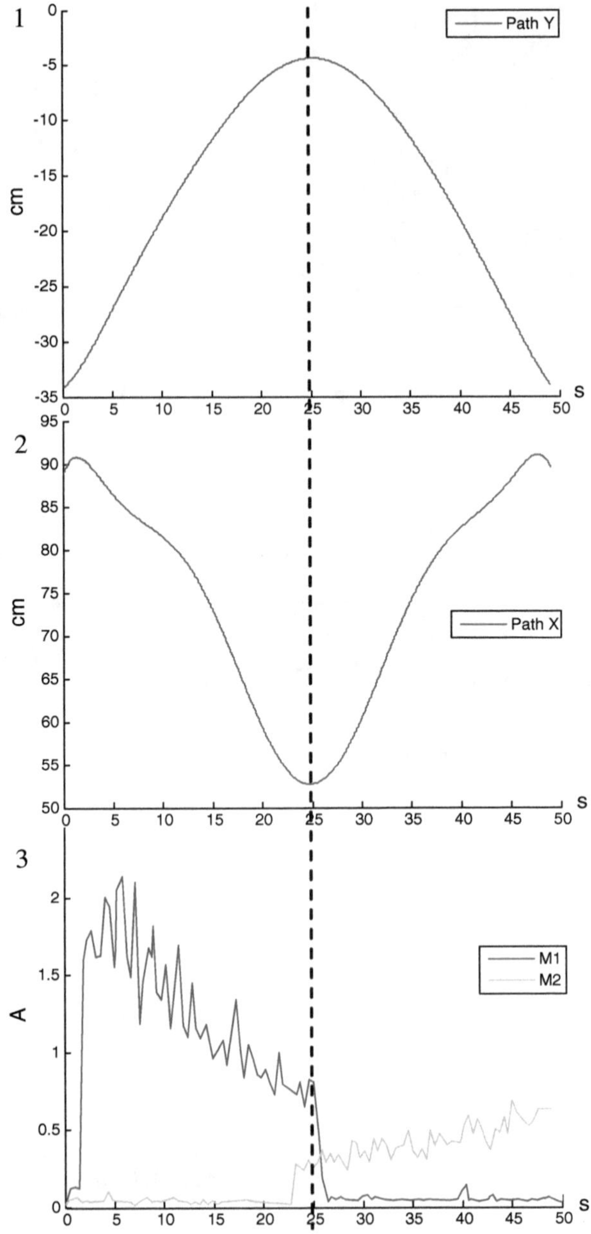

Fig. 8 (*1*) Exercise path in Y-axis, (*2*) exercise path in X-axis, (*3*) motors current curves

6 Conclusions

The mechanism for knee rehabilitation based on a parallel mechanism of 5-bars allows the reproduction of exercises for knee rehabilitation in the vertical plane.

The patient should be placed on a stretcher or chair, after the mechanism should slide along the patient and stretcher, finally the patient's leg is attached to the end-effector. Tests are being done to validate the prototype mobility and performance evaluation.

Acknowledgments Authors thank to CRIQ specialists for their help in the selection of exercises and to IPN-SIP for the financial support.

References

1. Koller-Hodac A, Leonardo D, Walpen S, Felder D (2011) Knee orthopaedic device how robotic technology can improve outcome in knee rehabilitation. In: IEEE international conference on rehabilitation robotics. ETH Zurich, Switzerland
2. Akdogan E, Arif Adli M (2011) The design and control of a therapeutic exercise robot for lower limb rehabilitation. © Elsevier. Mechatronics 21:509–522. (Received: 23 Feb 2010/ Accepted: 17 Jan 2011/Published online: 12 Feb 2011)
3. Sakaki T, Okada S, Okajima Y, Tanaka N (1999) TEM: therapeutic exercise machine for hipand knee joints of spastic patients. In: Sixth international conference on rehabilitation robotics ICORR'99, 1999
4. Weinberg B, Nikitczuk J et al (2007) Design, control and human testing of an active knee rehabilitation orthotic device. In: IEEE international conference on robotics and automation. Roma, Italy, 10–14 abril de 2007
5. John Weiner J (2010) Device and method for knee joint rehabilitation. Patent No US 7,695,416 B2. United. States Patent. Apr 13, 2010
6. Tsai L-W (1999) Robot analysis: the mechanics of serial and parallel manipulators. Wiley, New York, 505 p. ISBN 0-471-32593-7
7. Mehmet Alper E, Volkan P (2011) A self-adjusting knee exoskeleton for robot-assisted treatment of knee injuries. Faculty of engineering and natural sciences of Sabanci University, İstanbul, Turkey. In: IEEE international conference on intelligent robots and systems (IROS), IEEE/RSJ 2011. San Francisco, CA
8. Ejercicio después de reemplazo de rodilla (Knee Replacement Exercises) (2009) Ortho-Info. The American academy of orthopaedic surgeons, AAOS. April 2009. http://orthoinfo.aaos.org/topic.cfm?topic=A00494
9. Schoen DC (2000) Adult orthopaedic nursing, 1st edn, Lippincott, Philadelphia, p 345
10. Knee Arthroscopy Exercise Guide OrthoInfo (2000) The American academy of orthopaedic surgeons AAOS. http://orthoinfo.aaos.org/topic.cfm?topic=A00300
11. Knee Conditioning Program (2012) GuideOrthoInfo. The American academy of orthopaedic surgeons AAOS. http://orthoinfo.aaos.org/topic.cfm?topic=A00664#7. Accessed Oct 2012

12. Knee pain exercises. Advice pamphlets developed in conjunction with Wolverhampton City clinical commissioning group (CCG)/Primary care trust (PCT) and approved by the Chartered Society of Physiotherapy. http://www.arthritisresearchuk.org/arthritis-information/conditions/osteoarthritis-of-the-knee/knee-pain-exercises.aspx
13. Thomas EH, Marianne SG (2007) Conservative management of sports In-juries. chapter 16: the knee. 2da edición, p 694
14. Liu X-J, Wang J, Zheng H-J (2006) Optimum design of the 5R symmetrical parallel manipulator with a surrounded and good-condition workspace. Robot Auton Syst 54:221–233 (Elsevier 2006)

Design and Construction of a Translational Parallel Robot for Drilling Tasks

R. Maldonado-Echegoyen and E. Castillo-Castaneda

Abstract This article describes the design, position kinematics analysis and construction of a translational parallel robot for drilling tasks based on two 5-bar mechanisms. The final design provides a large workspace and a simplified geometrical analysis compared with Delta translational parallel robot. Forward and inverse geometric models are also presented. Deformation analysis using ANSYS was performed to obtain a detailed mechanical design considering forces acting during a classical drilling process. The robot was built and validated for vertical drilling tasks on soft materials.

Keywords Translational parallel robot · Drilling task · Deformation analysis · Kinematics · Student paper

1 Introduction

For the use of high-speed machines it is essential to develop robust machines with small moving masses. A step in this direction is the use of parallel mechanisms that have become an option for manufacturing machine tools at low cost and high level of performance. Some machine tool builders like Ingersoll, Giddings and Lewis, Toyoda, and Hitachi have developed machine tools using parallel mechanisms based essentially on the Stewart-Gough platform, the main disadvantage is the limited workspace [1–3].

Briones [4] designed a translational robot based on two 5-bar mechanisms coupled to two prismatic joints for machining purpose; this prototype has a large workspace and a simplified geometry. Doukas et al. [5] proposed a finite element model to

R. Maldonado-Echegoyen (✉) · E. Castillo-Castaneda
CICATA Unidad Queretaro, Instituto Politecnico Nacional, Mexico, Mexico
e-mail: rodrigorobotics@hotmail.com

E. Castillo-Castaneda
e-mail: ecastilloca@ipn.mx

M. Ceccarelli and E.E. Hernández Martinez (eds.), *Multibody Mechatronic Systems*,
Mechanisms and Machine Science 25, DOI 10.1007/978-3-319-09858-6_48

predict the accuracy of a robot arm during a milling process for different positions under a variety of load conditions. Makhanov et al. [6] proposed tool path optimization for milling tasks to increase path accuracy. Cai et al. [7] developed a new type of parallel machine tool able to perform tasks such as grinding, milling and other machining processes. Kim and Tsai [8] presented a 3-DOF translational parallel manipulator called Cartesian Parallel Manipulator (CPM). The manipulator consists of a moving platform that is connected to a fixed base by three limbs. Siciliano [9] performed a study on the kinematics of the Tricept robot, which comprises a three-degree-of- freedom parallel structure having a radial link of variable length. Glavonjic and Zivanovic [10] presented a new 3-DOF spatial parallel mechanism for horizontal and vertical milling machines with rather regular shape of the workspace, similar to serial machines. There is a hybrid system where the x-y movements are realised by a parallel scissors architecture whereas the z-axis is located in the table, called The Specht Xperimental, as shown in [11]. The Orthoglide [12], features three fixed parallel linear joints which are mounted orthogonally and a mobile platform which moves in the cartesian x-y-z space with fixed orientation.

This paper presents a machine, for drilling tasks in soft metals, based on a translational parallel mechanism. Forward and inverse geometric models are presented and deformation analysis is performed to assess the accuracy considering the forces reported in the literature for a drilling process.

2 Robot Description

In Fig. 1, the 0 frame corresponds to fixed platform or base, while the mobile platform is indicated as P. The base is linked to the mobile platform by four identical arms G1, G2, G3 and G4 and two prismatic joints Pr_1 and Pr_2. Each arm consists of

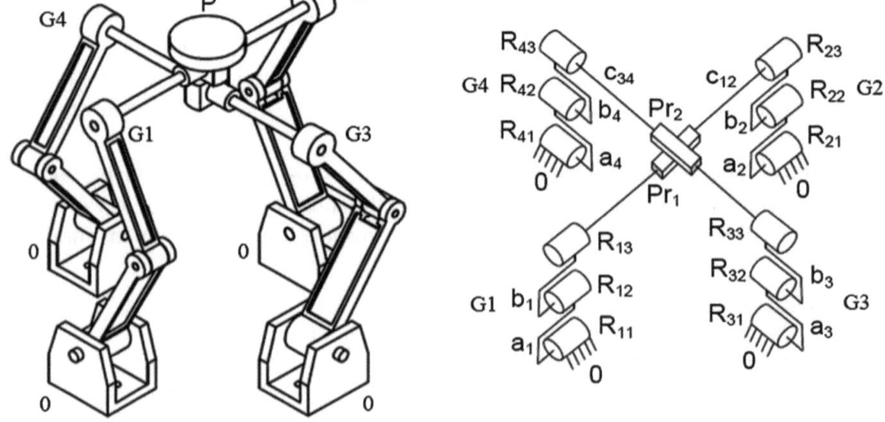

Fig. 1 Two 5-bar robot

two links, a_1, b_1; a_2, b_2; a_3, b_3; a_4, b_4, connected by rotational joints R_{12}, R_{22}, R_{32} and R_{42}. The lower end of link a_1, a_2, a_3 and a_4 is attached to the base by joints R_{11}, R_{21}, R_{31} and R_{41} respectively.

The G1 and G2 arms are joined by bar C_{12}, and two rotational joints R_{13} and R_{23}, thereby forming a 5-bar mechanism. Similarly, G3 and G4 arms, form a second 5-bar mechanism composed by the R_{33} and R_{34} joints and C_{34} bar [1].

3 Kinematic Analysis

Figure 2 shows joint variables θ_i, link lengths a_i and b_i and constant distances d_i and h_i. Points P_1 and P_2 have constants d_1 and d_2 in axis Y respectively, while P_3 has constant d_3 in axis X. The vector equation for arm-*i* is given by:

$$\overrightarrow{\boldsymbol{OA_i}} + \overrightarrow{\boldsymbol{A_iB_i}} + \overrightarrow{\boldsymbol{B_iP_i}} = \overrightarrow{\boldsymbol{OP_i}} \tag{1}$$

That can be expressed as follows:

$$\begin{bmatrix} P_{1x} \\ P_{1y} \\ P_{1z} \end{bmatrix} = \begin{bmatrix} a_1\cos(\theta_1) + b_1\cos(\gamma_1) \\ -d_1 \\ a_1\sin(\theta_1) + b_1\sin(\gamma_1) \end{bmatrix} \quad \begin{bmatrix} P_{2x} \\ P_{2y} \\ P_{2z} \end{bmatrix} = \begin{bmatrix} a_2\cos(\theta_2) - b_2\cos(\gamma_2) \\ d_2 \\ a_2\sin(\theta_2) + b_2\sin(\gamma_2) \end{bmatrix}$$
$$\begin{bmatrix} P_{3x} \\ P_{3y} \\ P_{3z} \end{bmatrix} = \begin{bmatrix} d_3 \\ a_3\cos(\theta_3) + b_3\ \cos(\gamma_3) \\ a_3\sin(\theta_3) + b_3\ \sin(\gamma_3) \end{bmatrix} \tag{2}$$

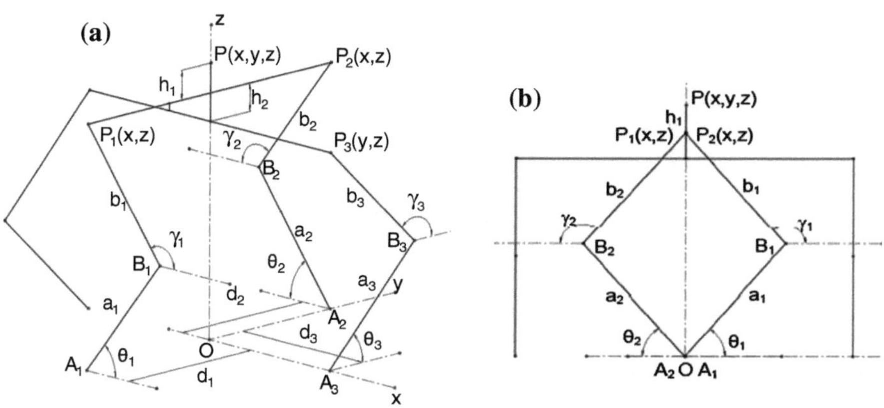

Fig. 2 **a** Geometric description of the robot, **b** lateral view, plane X–Z

P_1, P_2, and P_3 are related with the coordinates of the mobile platform P by:

$$P_x = P_{1x} = P_{2x}, \quad P_y = P_{3y}, \quad P_z = P_{1z} + h_1 = P_{2z} + h_1 = P_{3z} + h_1 + h_2 \quad (3)$$

3.1 Inverse Geometric Model

Inverse geometric model defines the relationship between known cartesian coordinates, P_x, P_y and P_z of the mobile platform and the desired joint coordinates of each actuator, θ_1, θ_2 and θ_3.

From Eq. 2, adding squares of first and third lines:

$$(P_{1x} - a_1\cos(\theta_1))^2 + (P_{1z} - a_1\sin(\theta_1))^2 = (b_1\cos(\gamma_1))^2 + (b_1\sin(\gamma_1))^2 \quad (4)$$

Simplifying, we obtain the following quadratic equation:

$$(k_{13} - k_{11})\tan^2\frac{\theta_1}{2} + 2k_{12}\tan\frac{\theta_1}{2} + k_{11} + k_{13} = 0 \quad (5)$$

where:

$$\begin{aligned} k_{11} &= -2a_1P_{1x} \\ k_{12} &= -2a_1P_{1z} \\ k_{13} &= P_{1x}^2 + P_{1z}^2 + a_1^2 - b_1^2 \end{aligned}$$

Therefore, the angle θ_1 can be calculated as follows:

$$\theta_1 = 2\tan^{-1}\left(\frac{-k_{12} \pm \sqrt{k_{12}^2 - k_{13}^2 + k_{11}^2}}{k_{13} - k_{11}}\right) \quad (6)$$

In a similar way:

$$\begin{aligned} \theta_2 &= 2\tan^{-1}\left(\frac{-k_{22} \pm \sqrt{k_{22}^2 - k_{23}^2 + k_{21}^2}}{k_{23} - k_{21}}\right), \\ \theta_3 &= 2\tan^{-1}\left(\frac{-k_{32} \pm \sqrt{k_{32}^2 - k_{33}^2 + k_{31}^2}}{k_{33} - k_{31}}\right) \end{aligned} \quad (7)$$

where:

$$
\begin{aligned}
k_{21} &= -2a_2P_{2x} \\
k_{22} &= -2a_2P_{2z} \\
k_{23} &= P_{2x}^2 + P_{2z}^2 + a_2^2 - b_2^2 \\
k_{31} &= -2a_3P_{3x} \\
k_{32} &= -2a_3P_{3z} \\
k_{33} &= P_{3x}^2 + P_{3z}^2 + a_3^2 - b_3^2
\end{aligned}
$$

3.2 Forward Geometric Model

Due to the symmetry of the robot, the forward geometric model is solved by taking only one 5-bar mechanism. For example, for the G1 and G2 arms (see Fig. 3a) and are given by

$$P_{1x} = c_1\cos\left(\frac{180 + \theta_1 - \theta_2}{2}\right) \tag{8}$$

$$P_{1z} = c_1\sin\left(\frac{180 + \theta_1 - \theta_2}{2}\right) \tag{9}$$

where:

$$c_1 = \sqrt{a_1^2 + b_1^2 - 2a_1b_1\cos(\varphi)} \tag{10}$$

P_{3y} can be determined using the 5-bar mechanism formed by G3 and G4 arms, see Fig. 3b, considering that (elbow to the right) and according to Eq. 2:

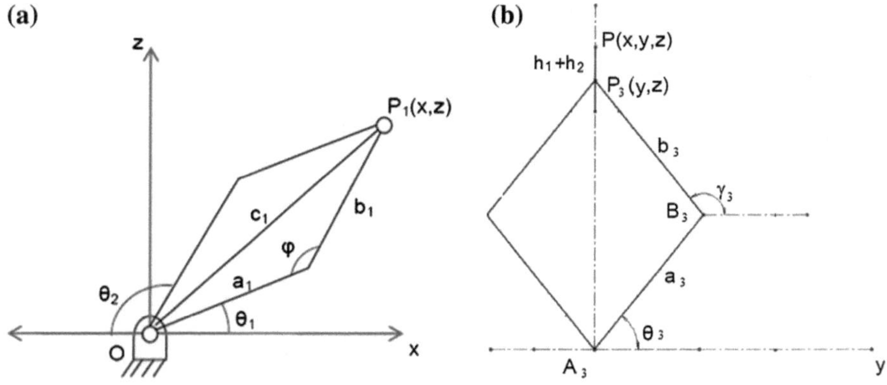

Fig. 3 **a** A 5-bar mechanism, plane Z–X, **b** lateral, plane Z–Y

$$P_{3y} = a_3 \cos(\theta_3) - b_3 \cos(\gamma_3) \tag{11}$$

where γ_3:

$$\gamma_3 = \sin^{-1}\left(\frac{P_{1z} - h_2 - a_3 \sin(\theta_3)}{b_3}\right) \tag{12}$$

Finally, mobile platform position is:

$$P_x = P_{1x} \qquad P_y = P_{3y} \qquad P_z = P_{1z} + h_1 \tag{13}$$

4 Workspace

The inverse geometric model of the robot was used to calculate the shape and volume of the workspace by selecting a volume, with uniformly distributed points, large enough to ensure that the robot workspace is contained. The points belong to the workspace if there is a valid solution of the inverse kinematics model. Figure 4 shows the resulting workspace corresponding to the length values shown in Table 1.

The robot has a larger workspace compared to the Delta translational parallel robot [13]. The robot workspace has a regular shape, which is very convenient for machining tasks.

5 Drilling Forces

A drilling force model [14] was considered to simulate the robot behavior under load conditions, Fig. 5 shows F_x, F_y, F_z components of drilling forces. For design purposes, maximum force values were: F_x = 300 N, F_y = 300 N, F_z = 1000 N. Drilling conditions were: Feed Rate 132 mm/min, Depth of cut: 2.2 mm, Spin speed 1100 rpm, Material: Al 7039, Tool diameter: 7.698 mm. It should be mentioned that drilling forces depend on the material, the tool diameter, the tool speed and the depth of cutting [15].

6 Deformation Analysis

Deformation analysis was performed using ANSYS software. The robot was considered a static structure; in this case, 3 different mobile platform positions along z-axis were analyzed: (a) 243 mm, (b) 328 mm and (c) 398 mm. We considered that

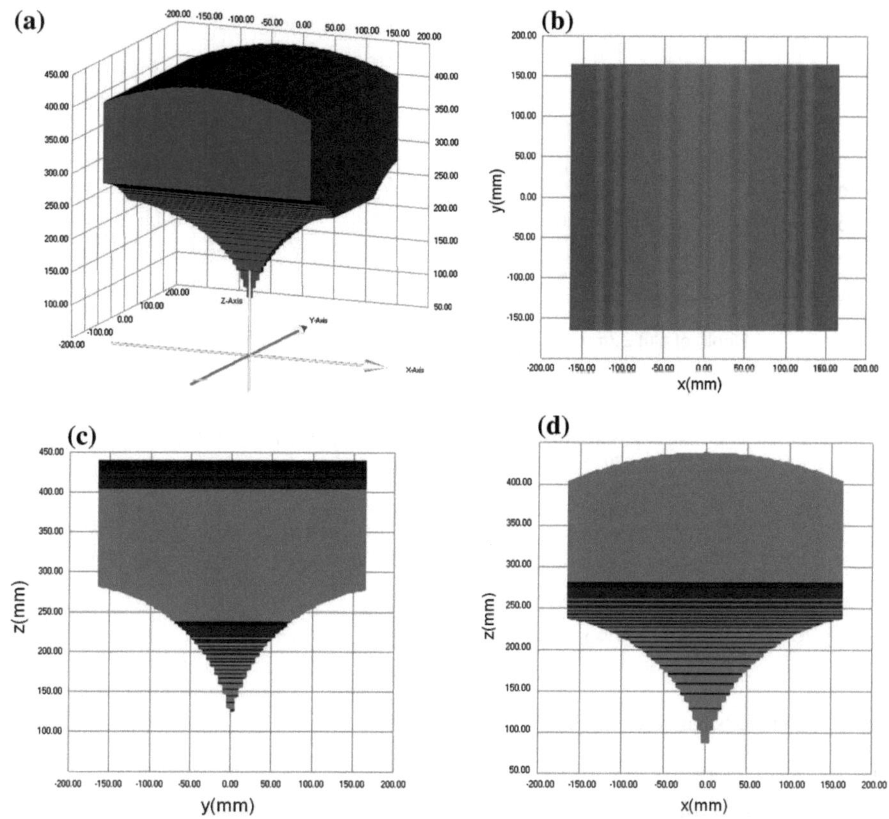

Fig. 4 Robot workspace, **a** plane X–Y–Z, **b** plane X–Y, **c** plane Y–Z, **d** plane X–Z

Table 1 Numerical values of the robot geometric model

Variables	Dimensions (mm)
$a_1 = a_2 = a_3 = a_4$	200
$b_1 = b_2 = b_3 = b_4$	200
$h_1 = h_2$	40
$d_1 = d_2 = d_3$	195

drilling forces are applied on mobile platform. The simulation results are shown in Fig. 6, deformation values are represented by a color scale.

Table 2 shows deformation analysis for the different mobile platform positions.

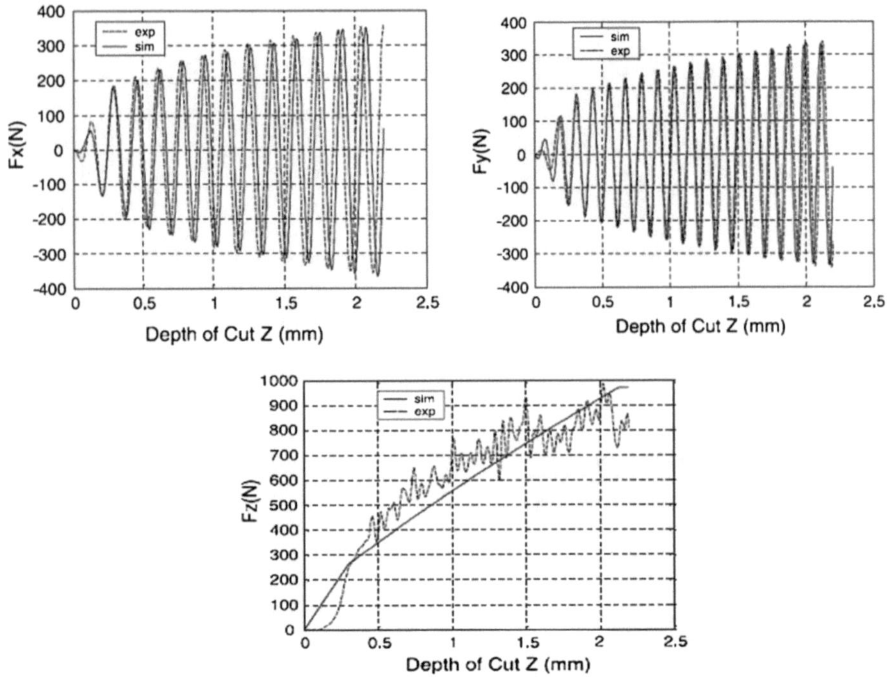

Fig. 5 Dynamic behavior of drilling forces

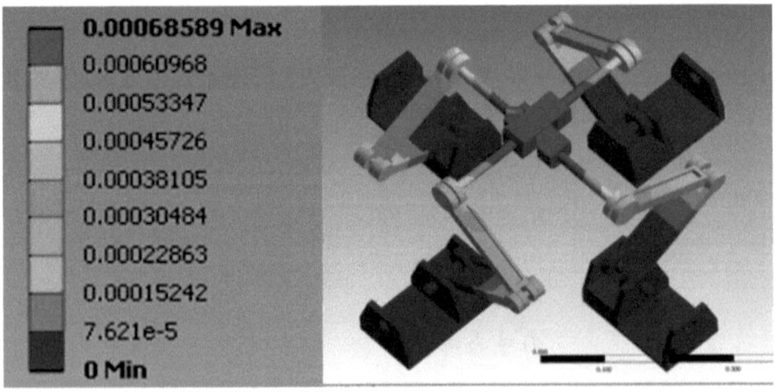

Fig. 6 Analysis of deformation at a height of 243 mm

7 Construction

The links and the mobile platform are made of aluminum. Prismatic joints are manufactured by a rectified and preloaded stainless steel bar of 14 mm diameter and slides using a linear bearing. Linear bearing are mounted on the mobile plat-form at

Table 2 Deformation at different heights

Heights (mm)	Maximum deformation (mm)
243	0.68
328	0.24
398	0.4

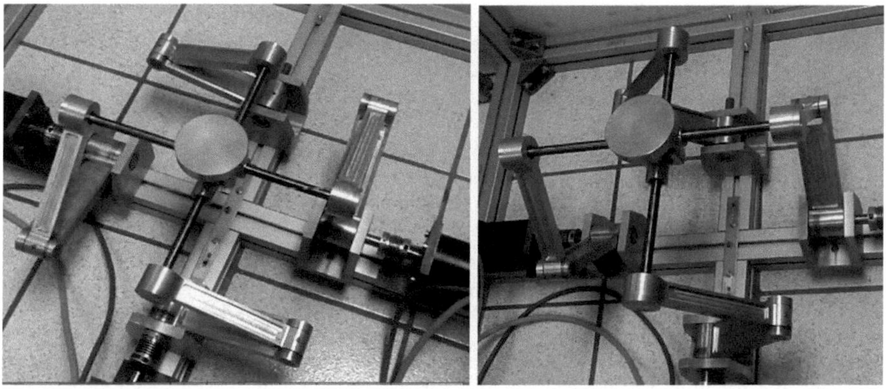

Fig. 7 Final robot prototype

90° from each other, which limit the mechanism to translational motion only. The maximum stroke of prismatic joints C_{12} and C_{34} are 346 mm (Fig. 7).

The robot is actuated by three AC servomotors, BSM AF-275-80N model, from BALDOR by means of a multiaxis control module NEXT-MOVE E100.

8 Conclusions

The robot workspace is larger compared with similar parallel robots such as Delta translational parallel robot. A static deformation analysis at 3 different heights was conducted using ANSYS, considering maximum forces generated by a drilling task that were obtained from previous work. As future work we have: machining on soft materials, determine the accuracy for trajectory tracking tasks, and complete the kinematic analysis in velocity and acceleration.

Acknowledgments The authors thank to SIP-IPN for its financial support.

References

1. Merlet JP (2006) Parallel Robots, 2nd edn. Springer, Hermes, Paris
2. Tsai L-W (1999) Robot analysis: the mechanics of serial and parallel manipulators. Wiley, New York

3. Zhang D (2010) Parallel robotic machine tools. Springer, Boston, MA
4. Briones JA (2009) Diseñó, Análisis y Construcción de un robot paralelo traslacional. Centro de investigación en ciencia aplicada y tecnología avanzada, Instituto Politécnico Nacional, Querétaro, Mexico
5. Doukas C, et al. (2012) On an empirical investigation of the structural behavior of robots. In: 45th CIRP Conference on Manufacturing Systems
6. Makhanov SS, Batanov D, Bohez E, Sonthipaumpoon K, Anotaipaiboon W, Tabucanon M (2002) On the tool-path optimization of a milling robot. Comput Ind Eng 43:455–472
7. Cai Q, et al. (2001) A study on kinematics and dynamics of a 3 DOF parallel machine tool. J Mater Process Technol 111(1–3):269–272
8. Kim HS, Tsai L-W (2003) Design optimization of a cartesian parallel manipulator. J Mech Des 125(1):43
9. Siciliano B (1999) The Tricept robot: inverse kinematics, manipulability analysis and closed-loop direct kinematics algorithm. Robotica 17:437–445
10. Milutinovic DS, Glavonjjc M, Kvrgic V, Zivanovic S (2005) A new 3-DOF spatial parallel mechanism for milling machines with long X travel. Ann CIRP 54(1):345–348
11. Weck M, Staimer D (2002) Parallel kinematic machine tools—current state and future potentials. CIRP Ann Manuf Technol 51(2):671–683
12. Chablat D, Wenger P (2000) Architecture Optimization of a 3-DOF Translational Parallel Mechanism for Machining Applications, the Orthoglide. Adv Robot Kinematics 113:305–314
13. Maya M, Castillo E, Lomeli A, Gonzalez-Galvan E, Cardenas A (2013) Workspace and payload-capacity of a new reconfigurable delta parallel robot. Int J Adv Robot Syst 10:1
14. Pirtini M, Lazoglu I (2005) Forces and hole quality in drilling. Int J Mach Tools Manuf 45 (11): 1271–1281 (Manufacturing Automation and Research Center, Department of Mechanical Engineering, Koc University, Sariyer, 34450 Istanbul, Turkey)
15. Hamadea RF, Ismailb F (2005) A case for aggressive drilling of aluminum. J Mater Process Technol 166:86–97

A Solution to the Approximate Spherical Burmester Problem

Jérémie Léger and Jorge Angeles

Abstract The synthesis of spherical linkages is of the utmost importance because (a) it poses challenges to the kinematician that are not present in the planar case and (b) its spatial counterpart depends on the synthesis of a spherical linkage. While the synthesis of spherical linkages for rigid-body guidance is a classic subject, and well documented in the literature, this has been limited to the exact-synthesis case, with four and five prescribed poses. The extension to approximate synthesis, more realistic and more appealing to the mechanism designer, has been reported in the past, but the synthesis method proposed therein is too cumbersome to be readily implementable. The approach proposed here obviates the constraints imposed by the unit vectors determining the center point and the circle point of each of the two dyads making up the four-bar linkage, thereby ending up with an unconstrained nonlinear least-squares problem. An example is included, that illustrates the procedure.

Keywords Spherical four-bar linkage · Synthesis · Rigid-body guidance · Motion generation · Approximate · Unconstrained optimization · Master's student · Student paper

1 Introduction

The problem of rigid-body guidance was first formulated for planar four-bar linkages by Burmester [1], hence the eponym associated to the problem. The *Burmester problem* is formulated as the synthesis of one dyad—a rigid link with two lower kinematic pairs [2] at two of its points—at a time, usually represented as a line

J. Léger (✉) · J. Angeles
Montreal, Canada
e-mail: jleger@cim.mcgill.ca

J. Angeles
e-mail: angeles@cim.mcgill.ca

M. Ceccarelli and E.E. Hernández Martinez (eds.), *Multibody Mechatronic Systems*, Mechanisms and Machine Science 25, DOI 10.1007/978-3-319-09858-6_49

segment. The two end points defining the segment are the center point B, which represents the center of a revolute (R) joint that couples the link to the (mechanism) frame, and the circle point A_0, representing the center of a second R joint that couples this link to the coupler link. The problem hinges in finding the position vectors $\mathbf{a}_0$ and $\mathbf{b}$ of points A_0 and B, respectively, which leads to four unknowns, i.e., two pairs of Cartesian coordinates.

Burmester showed that the problem admits exact solutions for a maximum of five prescribed poses of the coupler link—i.e., one equation per prescribed pose, without counting the *reference pose*. If four poses are prescribed, then, as Burmester showed, the problem admits infinitely many solutions. The locus of these solutions lie in a cubic termed the *circle-point curve* $\mathcal{K}$ for A_0 and in a second cubic termed the *center-point curve* $\mathcal{M}$ for B. The name of the former comes from the circle described by point A as it adopts various positions A_j, for $j = 0, 1, \ldots, m$, that correspond to the m prescribed poses. However, A_0 and B cannot be independently chosen from $\mathcal{K}$ and $\mathcal{M}$, respectively, as the position vectors of the two points are coupled by the *synthesis equations*.

The Burmester problem is amply studied in the specialized literature [3] for exact synthesis; for approximate synthesis the problem has also been studied, although to a lesser extent [4]. The spherical version of the Burmester problem is also recorded in books [3, 5]. It is known that this problem, as its planar counterpart, admits exact solutions for up to five prescribed poses of its coupler link. If four poses are prescribed, then the problem admits an infinity of solutions, lying in a spherical center-point and, correspondingly, in a spherical circle-point curve. Again, the circle point and the center point cannot be arbitrarily chosen in their corresponding curve, $\mathcal{K}$ and $\mathcal{M}$, respectively, for they are coupled by the synthesis equations.

Some attempts to solve the spherical Burmester problem for approximate synthesis have led to rather cumbersome optimization methods that are not readily implementable, e.g., [6]. Reported in this paper is an alternative approach to the solution of the spherical rigid-body guidance four-bar linkage that is based on unconstrained nonlinear least squares, which can be readily implemented with scientific software commercially available. The paper includes an example whereby 61 attitudes of the coupler link are prescribed, including the reference pose.

2 The Approximate Synthesis Problem

The problem at hand leads to m synthesis equations when m is greater than the number of design variables. In the case of the spherical Burmester problem, the synthesis equations, taken from [7], are

$$f_j = \mathbf{a}_0^T(\mathbf{Q}_j - \mathbf{1})\mathbf{b} = 0 \quad j = 1, \ldots, m \tag{1}$$

where $\mathbf{a}_0$ and $\mathbf{b}$ are the unit vectors defining a dyad at the reference pose, which, in the spherical case reduces to an attitude, and $\mathbf{Q}_j$ is the rotation matrix that carries the coupler link from the reference attitude to the jth attitude. Both dyads of the spherical four-bar mechanism are shown in Fig. 1, where * denotes the second dyad.

Notice that both $\mathbf{a}_0$ and $\mathbf{b}$ must always be of unit norm. This complicates the optimization procedure, since constrained optimization would be required in this form. To avoid the complication, the vectors are written in terms of the latitude φ and longitude ϑ of each of the points. These vectors thus become

$$\mathbf{a}_0 = \begin{bmatrix} \cos\varphi_a \cos\vartheta_a \\ \cos\varphi_a \sin\vartheta_a \\ \sin\varphi_a \end{bmatrix}, \quad \mathbf{b} = \begin{bmatrix} \cos\varphi_b \cos\vartheta_b \\ \cos\varphi_b \sin\vartheta_b \\ \sin\varphi_b \end{bmatrix} \tag{2}$$

In this form, the four design variables are evident. They are the longitude and latitude of both points. The problem of approximate synthesis in this case is formulated as

$$\min_{\mathbf{x}} \frac{1}{2}\sum_{j=1}^{m} f_j^2(\mathbf{x}), \quad \mathbf{x} = [\,\varphi_a \quad \vartheta_a \quad \varphi_b \quad \vartheta_b\,]^T \tag{3}$$

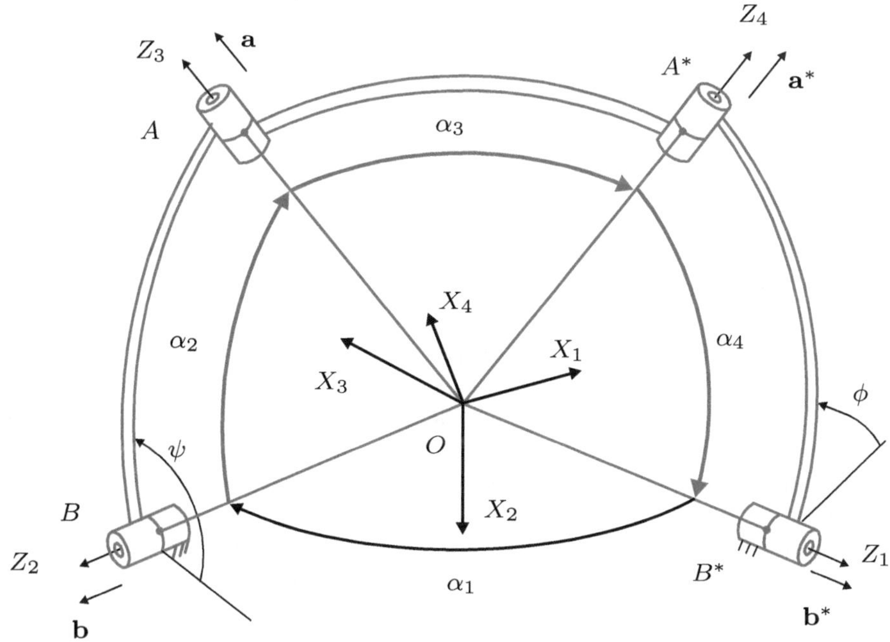

Fig. 1 A spherical four-bar linkage for function generation

The solution to this nonlinear least-squares problem can be found iteratively by means of the Newton-Gauss method. At each iteration, an overdetermined linear system of equations is solved in a least squares sense, namely,

$$\mathbf{J}_k \Delta\mathbf{x}^k = -\mathbf{f}(\mathbf{x}^k), \quad \mathbf{f}(\mathbf{x}^k) \equiv \begin{bmatrix} f_1(\mathbf{x}^k) \\ f_2(\mathbf{x}^k) \\ \vdots \\ f_m(\mathbf{x}^k) \end{bmatrix} \tag{4}$$

where $\mathbf{x}^k$ denotes $\mathbf{x}$ at the kth iteration, while $\mathbf{J}_k$ is the Jacobian $\mathbf{J}(\mathbf{x}) \equiv \partial\mathbf{f}/\partial\mathbf{x}$ evaluated at $\mathbf{x}^k$. Then, the solution is updated using a damping factor α that is introduced if the need arises[1]:

$$\mathrm{x}^{k+1} = \mathrm{x}^k + \alpha\Delta\mathrm{x}, \quad 0 < \alpha < 1 \tag{5}$$

The jth row of the Jacobian is found as

$$\mathbf{J}(j,:) = \begin{bmatrix} \frac{\partial f_j}{\partial \mathbf{a}_0}\mathbf{A} & \frac{\partial f_j}{\partial \mathbf{b}}\mathbf{B} \end{bmatrix} \tag{6}$$

where, from Eq. 1

$$\frac{\partial f_j}{\partial \mathbf{a}_0} = \mathbf{b}^T(\mathbf{Q}_j - \mathbf{1}) \tag{7}$$

$$\frac{\partial f_j}{\partial \mathbf{b}} = \mathbf{a}_0^T(\mathbf{Q}_j^T - \mathbf{1}) \tag{8}$$

and A, B are the 3×2 matrices defined below:

$$\mathbf{A} = \begin{bmatrix} \frac{\partial \mathbf{a}_0}{\partial \varphi_a} & \frac{\partial \mathbf{a}_0}{\partial \vartheta_a} \end{bmatrix} = \begin{bmatrix} -\sin\varphi_a \cos\vartheta_a & -\cos\varphi_a \sin\vartheta_a \\ -\sin\varphi_a \sin\vartheta_a & \cos\varphi_a \cos\vartheta_a \\ \cos\varphi_a & 0 \end{bmatrix} \tag{9}$$

$$\mathbf{B} = \begin{bmatrix} \frac{\partial \mathbf{b}}{\partial \varphi_b} & \frac{\partial \mathbf{b}}{\partial \vartheta_b} \end{bmatrix} = \begin{bmatrix} -\sin\varphi_b \cos\vartheta_b & -\cos\varphi_b \sin\vartheta_b \\ -\sin\varphi_b \sin\vartheta_b & \cos\varphi_b \cos\vartheta_b \\ \cos\varphi_b & 0 \end{bmatrix} \tag{10}$$

As shown in [8], for the planar case, the problem of approximate synthesis for rigid-body guidance admits multiple minima, while only two are required for the mechanism. The method introduced here gives a solution for each initial guess of the design vector. The procedure should be repeated a minimum of two times with

[1] If the procedure is found to diverge, then damping is introduced to enhance convergence.

different initial guesses. The designer would carefully choose each initial guess to place the circle points and center points where she or he thinks it is more appropriate, given the conditions of the design task at hand. The procedure is illustrated with the synthesis of a door mechanism. In this way the method is simpler than a previously reported method [6]. However, it requires good engineering insight, since two good initial guesses must be provided while attending the specifications of the design problem that motivated the synthesis task.

3 Example: The Synthesis of a Door Mechanism

In this example, a sports-car door with two landmark points C and R of the door will undergo a trajectory that takes the points from the door closed attitude to its open attitude. The longitude and latitude of each point at its closed and open attitude are given in Table 1. These data are compatible in that the two spherical triangles OC_0R_0 and OC_mR_m are congruent, with O denoting the center of the sphere.

3.1 Selecting a Trajectory

To describe the intended trajectory from pose 0 to pose m, the problem is divided in a sequence of two rotations. The first rotation progressively takes point C_0 to C_m by a rotation about a fixed axis, the one normal to the vectors $\mathbf{c}_0$ and $\mathbf{c}_m$ pointing at C_0 and C_m, respectively. The second rotation completes the trajectory by rotating the door about the vector $\mathbf{c}_i$ pointing to C_i. Each point of the trajectory can then be written as

$$\mathbf{Q}_j = \mathbf{Q}_{2j}\mathbf{Q}_{1j} \tag{11}$$

Using this scheme, only the angles of rotation of $\mathbf{Q}_{1j}$ and $\mathbf{Q}_{2j}$ must be selected to fully describe the jth attitude. This is achieved by first using cycloidal functions to generate the angles and then by discretizing these functions into m points. The form of each angle ϕ_j using a cycloidal function is

$$\phi_j = \Delta\phi \; cycl(t_j) \tag{12}$$

Table 1 Landmark points at closed and open attitudes

	Open attitude (j = 0)		Closed attitude (j = m)	
	Longitude (°)	Latitude (°)	Longitude (°)	Latitude (°)
Point C	0	0	60	60
Point R	30	0	30	60

where

$$cycl(t_j) \equiv t_j - \frac{1}{2\pi}\sin(2\pi t_j) \tag{13}$$

and t varies from 0 to 1. At $t = 0$, the door is at attitude 0 and at $t = 1$ at attitude m. The angle $\Delta\phi_1$ is simply the angle between $\mathbf{c}_0$ and $\mathbf{c}_m$. The angle $\Delta\phi_2$ is found after having found $\Delta\phi_1$ by solving the equation

$$\mathbf{r}_m = \mathbf{Q}_{2m}\mathbf{Q}_{1m}\mathbf{r}_0 \tag{14}$$

where $\Delta\phi_2$ is the angle of rotation of $\mathbf{Q}_{2m}$. A solution to this equation is

$$\Delta\phi_2 = \text{sign}\{[(\mathbf{Q}_{1m}\mathbf{n}_0) \times \mathbf{n}_m)]^T\mathbf{g}_m\}\arccos\left((\mathbf{Q}_{1m}\mathbf{n}_0)^T\mathbf{n}_m\right) \tag{15}$$

where sign$(\cdot)$ is the signum function of $(\cdot)$ and $\mathbf{n}_0$, $\mathbf{n}_m$ are the normals to vectors $\mathbf{c}_0$, $\mathbf{r}_0$ and $\mathbf{c}_m$, $\mathbf{r}_m$, respectively, defined as:

$$\mathbf{n} = \frac{\mathbf{c} \times \mathbf{r}}{||\mathbf{c} \times \mathbf{r}||_2} \tag{16}$$

The discretization is done by choosing a set of values of variable t in the interval [0, 1]. Since the open and closed attitudes are important to the trajectory, it would make sense to place more discrete points near $t = 0$ and $t = 1$ and less around $t = 0.5$. This is achieved by using the function below

$$t(x_j) = \begin{cases} 0.5(1 - \sqrt{1 - (x_j + 1)^3}) & \text{if } x_j \leq 0.5 \\ 0.5(1 + \sqrt{1 - (-x_j + 1)^3}) & \text{if } x_j > 0.5 \end{cases} \tag{17}$$

its plot being shown in Fig. 2, and

$$x_j = \frac{2j}{m} - 1 \tag{18}$$

3.2 Finding the Mechanism

Having found suitable trajectory points to describe the motion of the door using 61 attitudes, the synthesis of the mechanism is now addressed. The Newton-Gauss method, described in Sect. 2, is used, as implemented in Matlab with Householder reflections, which yielded two solutions for each dyad, with the initial guesses and damping factors in Table 2.

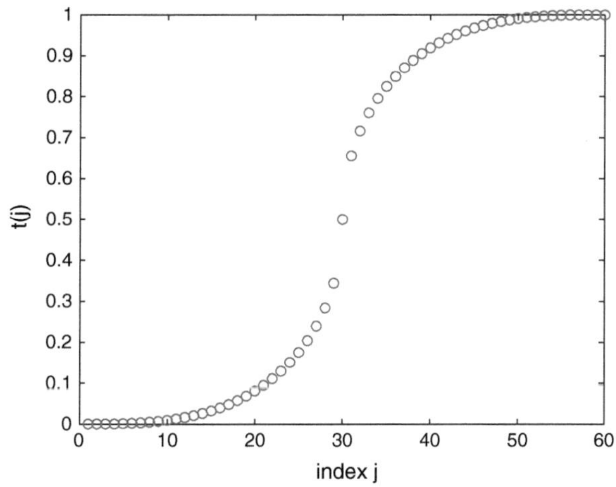

Fig. 2 Discretization of t

The initial guess for points A_0 and B in Table 2 were chosen to be points of the door for A_0 and expected corresponding center point for B by looking at the trajectory.

The solutions to the two dyads of the mechanism are shown in Table 3. The Denavit-Hartenberg (DH) parameters of the linkage are also shown in this table, while the corresponding mechanism is displayed in Fig. 3.

Notice from Table 3 that the number of iterations is high. This is caused by the low damping factors and is unavoidable since low damping is demanded by the procedure to converge. Without low damping, there is an oscillation of the objective

Table 2 Initial guesses and damping factors

	First leg	Second leg
Latitude of A_0 (°)	20	15
Longitude of A_0 (°)	20	30
Latitude of B (°)	−90	30
Longitude of B (°)	30	130
Damping factor	5×10^{-4}	1×10^{-4}

Table 3 Least-squares solution

DH parameters			First dyad	Second dyad
α_1	119.54	Latitude of a_0 (°)	29.19	15.01
α_2	120.67	Longitude of a_0 (°)	5.66	31.72
α_3	27.89	Latitude of b (°)	−91.63	29.82
α_4	89.31	Longitude of b (°)	30.02	129.73
		RMS error	3×10^{-3}	3×10^{-3}
		Number of iterations	329	85

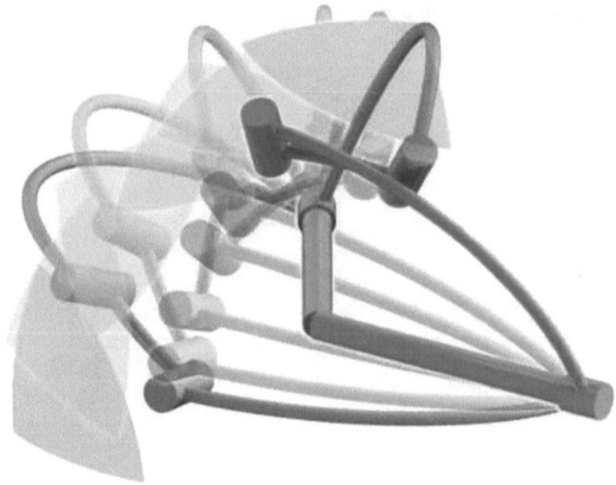

Fig. 3 Synthesised linkage

function and the solution diverges. With this high number of iterations and a modern processor, a solution to the mechanism is found under 30 s, which is reasonable for the task of designing a single mechanism.

4 Conclusions

The spherical Burmester problem for the approximate synthesis of a four-bar linkage was formulated as an unconstrained nonlinear least-squares problem, then solved using the Newton-Gauss method, as implemented in Matlab, which resorts to Householder reflections (HHR). This feature is important, for, although the bread and butter of numerical analysts and control engineers, HHR are less common in kinematics circles. The merit of HHR lies in that they preserve the condition number of the rectangular Jacobian. The unconstrained formulation was possible by resorting to the spherical coordinates of both the center point and the circle point of the dyad under synthesis. The procedure proposed here was illustrated with the synthesis of the door mechanism for a sports car, which has been proposed as a synthesis example in the past, within the context of the exact spherical Burmester problem.

References

1. Burmester L (1888) Lehrbuch der Kinematik. Arthur Felix Verlag, Leipzig
2. Hartenberg RS, Denavit J (1964) Kinematic synthesis of linkages. McGraw-Hill Book Company, New York
3. McCarthy JM, Soh GS (2011) Geometric design of linkages. Springer, New York

4. Liu A-X, Yang T-L (1999) Finding all solutions to unconstrained nonlinear optimization for approximate synthesis of planar linkages using continuation method. ASME J Mech Des 121:368–374
5. Chiang CH (1988) Kinematics of spherical mechanisms. Cambridge University Press, Cambridge
6. Liu Z, Angeles J (1992) The constrained least-square optimization of spherical four-bar linkages for rigid-body guidance. Trans Can Soc Mech Eng 16(1):47–60
7. Angeles J, Bai S (2010) A robust solution of the spherical Burmester problem. In: ASME 2010 international design engineering technical conferences
8. Yao J, Angeles J (2000) Computation of all optimum dyads in the approximate synthesis of planar linkages for rigid-body guidance. Mech Mach Theory 35(8):1065–1078

Mechatronic Design of a Mobile Robot and Non-Linear Control

J. Hernández, J. Torres and S. Salazar

Abstract In this work design of a mobile robot is presented, which is used as laboratory tool to test various control algorithms. This work also presents a variable control structure for the mobile robot. The internal control scheme uses a PID controller to control the motors angular speed, while the external loop is associated to the kinematic model of the robot and is also responsible to generate the reference velocity trajectory for each motor. The proposed control uses two types of state feedback control algorithms, which are switched from one to another at the point one encounters with a singularity.

Keywords Mobile robot · Path tracking · PID control · Input-output control · Student paper

1 Introduction

A mobile robot is an electromechanical device which is capable of moving in a workspace with different levels of autonomy. The level of autonomy is determined by the capacity of the robot to perceive working ambient through sensors for modifying its behavior accordingly. An extensive work has been done in the development of mathematical models for the mobile robots propelled by wheels [1–3]. It is common to consider that the robot wheels rotate without sliding on the contact surface. It is said that the model of a robot mobile is no-holonomic because the velocities along of coordinate axis satisfy no integrable conditions [3]. Some related papers approach similar problems (see [4, 8]). We consider the problem of

J. Hernández (✉) · J. Torres · S. Salazar
Department of Automatic Control Centro de Investigacion y de Estudios Avanzados del I.P.N (CINVESTAV) and UMI LAFMIA CNRS-CINVESTAV, Mexico, Mexico
e-mail: jhernandez@ctrl.cinvestav.mxs

J. Torres
e-mail: jtorres@ctrl.cinvestav.mx

M. Ceccarelli and E.E. Hernández Martinez (eds.), *Multibody Mechatronic Systems*, Mechanisms and Machine Science 25, DOI 10.1007/978-3-319-09858-6_50

tracking trajectory, which has been extensively treated in the literature. In this work the design of a robot mobile is presented, it consist of a number of sensors, a onboard PC, mechanical structure and electronics devices that allow certain autonomy for the mobile robot.

Several articles ans reports have been written over the past ten years, on the problem of controlling wheeled robots. Following simple control laws, based on tangential linearization or heuristic methods [3] and more general controllers have been proposed on the basis of non-linear control theory [1, 7]. This article focuses on the trajectory tracking issue. We develop an approach to control in two parts: the first responsible for defining speed profiles and the second devoted to obtain the desired response of the actuators.

1.1 Mobile Robot Dynamic Model

A large class of mechanical nonholonomic systems is described by the following form of dynamic equations based on Euler [2]:

$$M(q)\ddot{q} + C(q,\dot{q})\dot{q} + G(q) = B(q)\tau + J^T(q)\lambda \tag{1}$$

While the nonholonomics constraint is:

$$J(q)\dot{q} = 0 \tag{2}$$

where q is the n dimensional vector of configuration variables, $M(q)$ is a symmetric positive definite $n \times n$ matrix, $C(q,\dot{q})$ presents the n vector of centripetal and coriolis torques, $G(q)$ is the n vector of gravitational torques, $B(q)$ is the $n \times r$ input transformation matrix $(r<n)$, τ is the r dimensional vector of inputs and λ the Lagrange multipliers of constrained forces. A simple structure of differential drive mobile robot is shown in Fig. 1. Two independent analogous DC motors are the actuators of left and right wheels, while one or two free wheel casters are used to keep the platform stable.

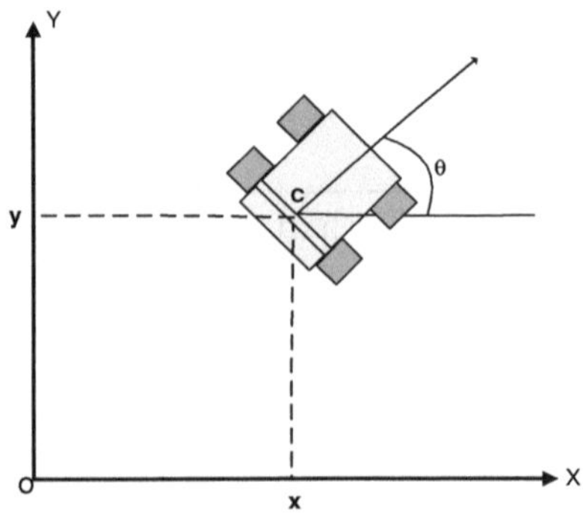

Fig. 1 Coordination of mobile robot

Pose vector of robot in the surface is defined as $q = (x, y, \theta)^T$. Where x and y are the coordinates of point c; center of axis of wheels, and θ is the orientation angle of robot in the inertial frame. From these equations we can obtain the kinematic model of a mobile robot as shown in [1], which is given as follows:

$$\begin{aligned} \dot{x} &= v\cos(\phi) \\ \dot{y} &= v\sin(\phi) \\ \dot{\theta} &= \omega \end{aligned} \tag{3}$$

These are the equations of kinematics model, where:

$$\begin{bmatrix} v \\ \omega \end{bmatrix} = T\begin{bmatrix} \omega_d \\ \omega_i \end{bmatrix} = \frac{r}{2}\begin{bmatrix} 1 & 1 \\ \frac{1}{l} & -\frac{1}{l} \end{bmatrix}\begin{bmatrix} \omega_d \\ \omega_i \end{bmatrix} \tag{4}$$

where ω_d and ω_i represent the angular velocities of the right and left wheels respectively.

2 Robot Descriptions

The mechanical structure consists of two bars joined in the middle by means of a shaft that allows them to pivot. Each side-bar is issued of a DC motor having gearboxes to amplify the power transmitted by a chain system to the opposite wheel fixed to the same bar, see Fig. 2. In such a way that just the front and rear whell are coupled to the corresponding motor. The axes of the wheels are integrated with the incremental type encoders thereby allowing in a subsequent process to estimate the relative position of the robot. The robot has integrated a laser as main sensor that gives information on the surrounding environment. Figure 2 shows the mechanical structure of the robot [6].

The robot used in this work was partially constructed in our laboratory, on hardware that enables communication with various sensors and the systems that

Fig. 2 Mechanical structure

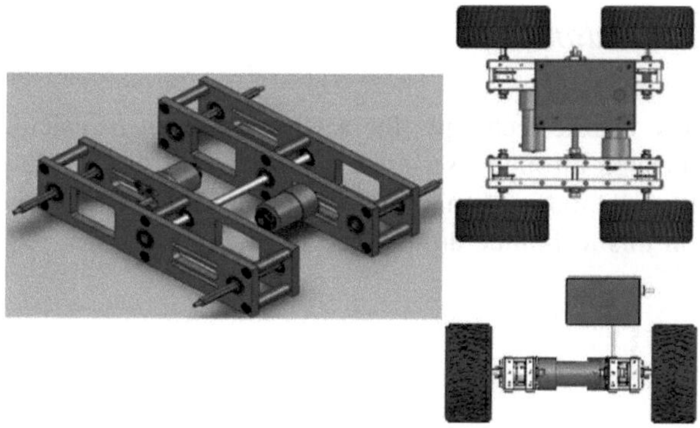

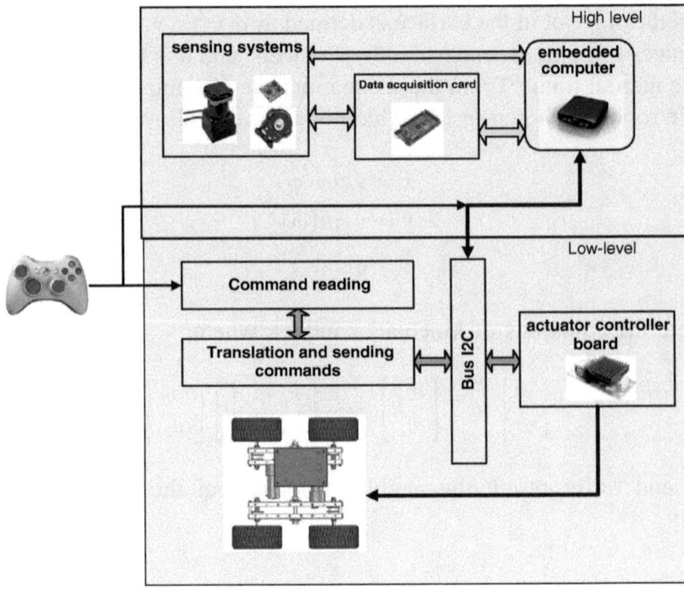

Fig. 3 Architecture for robot control

require the robot to operate, which is integrated in an embedded computer. The design was based on different modules as shown in Fig. 3.

Each module has capabilities that allow you to update the state of the robot and define the behavior of the actuators of the robot. Different tests were performed to validate the proper relative position of the robot by odometry technique and the use of inertial unit (IMU). The robot can be fully autonomous embedded computer that controls all the actions that the robot performs. The electronic design proposed keeping in mind to include additional sensors or to include more complex system that provide greater benefits to the robot. The Fig. 4 shows the embedded system of the robot.

2.1 Odometry

The classic technique for a wheeled robot to calculate its position is to track its location through a series of measurements of the rotation of the wheels of the robot, this method is often known as "Odometry". The fundamentals and ways of working on this technique are defined in [4].

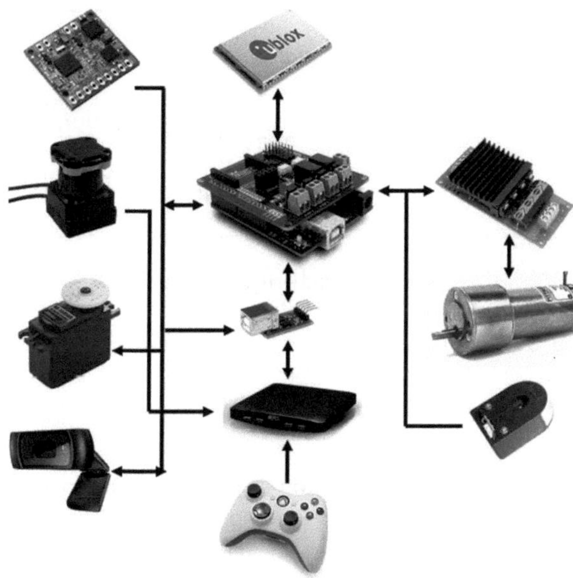

Fig. 4 Embedded system for robot

3 Control Scheme

The central idea of the control in the linealization is transform algebraically a nonlinear system to another totally or partially linear in order that linear control techniques can be implemented. This kind of control has been applied in some practical cases such as the control of helicopters, aircrafts, and industrial robots. The control scheme is presented in the Fig. 5. Where you can see two control loops; the internal loop is associated to the dynamic model of the actuators and is composed by a classic PID control and the external loop is a variable structure control and is responsible of the control of the kinematic model. The variable structure control switches between a static linearizable control and a dynamic linearizable control.

The reason for interchanging between these two control laws has by objective avoiding singularities of each scheme. Static control: presents a singularity with $\phi = \frac{k\pi}{2}$, $k = \pm 1, \pm 2, \ldots$ And we can achieve convergence in the coordinates x and ϕ to the desired coordinates x_d and ϕ_d. Dynamic control: present singularities with $\phi = \frac{k\pi}{2}$, $k = \pm 1, \pm 2, \ldots$ and $v = 0$. We can achieve convergence in the robot coordinates x, y, ϕ to the desired coordinates x_d, y_d, ϕ_d.

Considering these characteristics, the operation of the robot start with the static control due to the dynamic control has a singularity when $v = 0$. However when the linear velocity is greater than a certain threshold, the operation switches to the dynamic control.

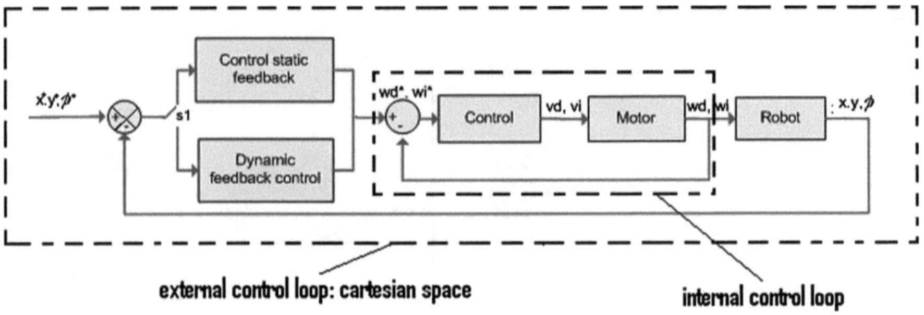

Fig. 5 Control scheme

3.1 Kinematic Model Control

The position and orientation of a mobile robot over a plane can be described by its coordinates (x, y) with respect to a fixed reference system and the angle ϕ is given by the robot with respect to X axis how is shown in Fig. 1. Kinematic model Eq. (3) can be seen as a transformation:

$$\Gamma : [v(t), \omega(t)] \rightarrow X(t) = [x(t), y(t), \phi(t)] \tag{5}$$

which maps pairs of velocity trajectories $[v(t), \omega(t)]$ to mobile robot trajectories in the cartesian space. Given a desired trajectory $X_d(t) = [x_d(t), y_d(t), \phi_d(t)]$, the inverse kinematic problem is to determine its velocity profile $[v_d(t), \omega_d(t)]$ such that:

$$\Gamma[v_d(t), \omega_d(t)] = X_d(t) \tag{6}$$

We propose as a trajectory to follow a parabola that is described by the equations:

$$\begin{aligned} y_d(t) &= x_d^2(t) \\ \phi_d(t) &= \arctan(2x_d(t)) \\ x_d(t) &= a \sin\left(\frac{2\pi}{P}t\right) \end{aligned} \tag{7}$$

where the parameters are chosen as follow: $a = 0.9$ y $P = 120$.

3.2 Static Feedback Linearization

Considering the kinematic model Eqs. (3) and (4), we desire as objective that the variables $x(t)$, $\phi(t)$ follow asymptotically determined trajectories given by x_d and $\phi_d(t)$ respectively. From the Eq. (3) we obtain:

$$\begin{bmatrix} \dot{x} \\ \dot{\phi} \end{bmatrix} = A(\phi)\begin{bmatrix} \omega_d \\ \omega_i \end{bmatrix} = \begin{bmatrix} \frac{r\cos(\phi)}{2} & \frac{r\cos(\phi)}{2} \\ \frac{r}{2l} & -\frac{r}{2l} \end{bmatrix}\begin{bmatrix} \omega_d \\ \omega_i \end{bmatrix} \tag{8}$$

Choosing the control variables ω_d, ω_i and considering u_1 and u_2 as new control variables:

$$\begin{bmatrix} \omega_d \\ \omega_i \end{bmatrix} = A^{-1}(\phi)\begin{bmatrix} u_1 \\ u_2 \end{bmatrix} = \begin{bmatrix} \frac{1}{r\cos(\phi)} & \frac{1}{r}l \\ \frac{1}{r\cos(\phi)} & -\frac{1}{r}l \end{bmatrix}\begin{bmatrix} u_1 \\ u_2 \end{bmatrix} \tag{9}$$

Therefore it is said that the control variables $x(t), \phi(t)$ will be governed by the following linear time invariant differential equation.

$$\begin{bmatrix} \dot{x} \\ \dot{\phi} \end{bmatrix} = \begin{bmatrix} u_1 \\ u_2 \end{bmatrix} \tag{10}$$

Choosing the control variables as:

$$\begin{bmatrix} u_1 \\ u_2 \end{bmatrix} = \begin{bmatrix} \dot{x}_d - \alpha e_x \\ \dot{\phi}_d - \beta e_\phi \end{bmatrix} \tag{11}$$

Therefore the tracking error is governed by the following equations:

$$\begin{aligned} \dot{e}_x + \alpha e_x &= 0 \\ \dot{e}_\phi + \beta e_\phi &= 0 \end{aligned} \tag{12}$$

Under this control scheme we have that $e_x(t) \to 0, e_\phi(t) \to 0$ when $t \to \infty$ if $\alpha, \beta > 0$, which implies at the same time $x(t) \to x_d(t)$ and $\phi(t) \to \phi_d(t)$ when $t \to \infty$.

We conclude that the velocity profiles should be governed by the following equation:

$$\begin{bmatrix} \omega_d \\ \omega_i \end{bmatrix} = \begin{bmatrix} \frac{1}{r\cos(\phi)} & \frac{1}{r}l \\ \frac{1}{r\cos(\phi)} & -\frac{1}{r}l \end{bmatrix}\begin{bmatrix} \dot{x}_d - \alpha(x - x_d) \\ \dot{\phi}_d - \beta(\phi - \phi_d) \end{bmatrix} \tag{13}$$

3.3 Dynamic Feedback Linearization

Considering the kinematic model given in Eq. (3), we propose a new control variable named λ_1 so that:

$$\dot{x} = v\cos(\phi) = \lambda_1 \tag{14}$$

where v is obtained and represents the first control law:

$$v = \frac{\lambda_1}{\cos(\phi)} \tag{15}$$

Replacing v in Eq. (3) and deriving the equation we find:

$$\ddot{y} = \dot{\lambda}_1 \tan(\phi) + \lambda_1 \omega \sec^2(\phi) \tag{16}$$

Introducing a new control variable λ_2 and making equal to the previous equation we can obtain:

$$\omega = \frac{\lambda_2}{\lambda_1}\cos^2(\phi) - \frac{\dot{\lambda}_1}{\lambda_1}\sin(\phi)\cos(\phi) \tag{17}$$

In order to ensure the convergence of the control variables $x(t), y(t)$ to the pre-established trajectories $x_d(t), y_d(t)$, we select the variables λ_1 and $\lambda_2$as follow:

$$\begin{aligned} \lambda_1 &= \dot{x}_d - \alpha_1(x - x_d) \\ \lambda_2 &= \ddot{y}_d - \beta_2(\dot{y} - \dot{y}_d) - \beta_2(y - y_d) \end{aligned} \tag{18}$$

Finally we replace the computed values in the kinematic model in Eq. (4), therefore the velocity profiles should be:

$$\begin{bmatrix} \omega_d \\ \omega_i \end{bmatrix} = \begin{bmatrix} \frac{1}{r} & \frac{l}{r} \\ \frac{1}{r} & -\frac{l}{r} \end{bmatrix} \begin{bmatrix} \frac{\lambda_1}{\cos(\phi)} \\ \frac{\lambda_2}{\lambda_1}\cos^2(\phi) - \frac{\dot{\lambda}_1}{\lambda_1}\sin(\phi)\cos(\phi) \end{bmatrix} \tag{19}$$

In summary, the trajectory tracking problem can be interpreted as a wheel regulation velocity problem according to the profile given in expressions Eqs. (14) and (19).

4 PID Control

The proportional integral derivative control is perhaps the control scheme most widely used in industry, due to its excellent performance and robustness for parametric disturbances and load; another advantage over more complex control schemes is that it only requires to measure the variable to control, hence we consider a reference system for comparing the performance of other control schemes [5].

5 Experimental Results in Real Time

For obtaining a correct performance in the proposed control scheme we assume that each motor wheel can provide the velocity profile defined by the proposed control law, in this way, we do some tests for each motor, so that they made possible to find PID control gains which allow to follow the reference velocities. The Fig. 6 shows the response of both motors to a soft and rising reference signal.

The Fig. 7 shows the simulation of velocities profiles for both wheels, which are necessary to get the trajectory, according to the proposed control scheme.

The design parameters of the static closed loop control law were given by $\alpha = \beta = 1$, in the dynamic case $\alpha_1 = \beta_1 = 3$ and $\alpha_0 = \beta_0 = 1$, the initial conditions $(x(0), y(0), \phi(0), \omega(0)) = (0, 0, 0, 0)$ were fixed. The obtained results for both wheels are showed in the Fig. 8 where we can see the form of the signals.

The Fig. 9 shows the path followed by the robot, which corresponds to a parabola.

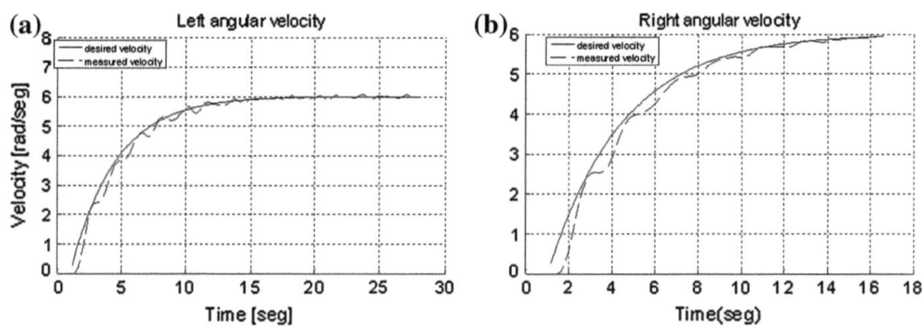

Fig. 6 Responding to a reference input: **a** left engine, **b** right engine

Fig. 7 Velocity profiles for each wheel

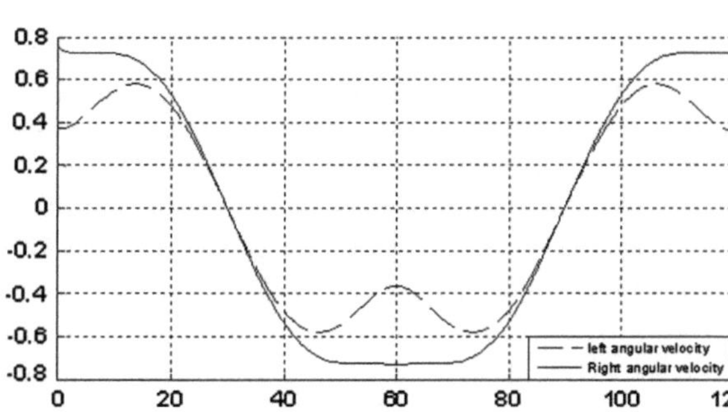

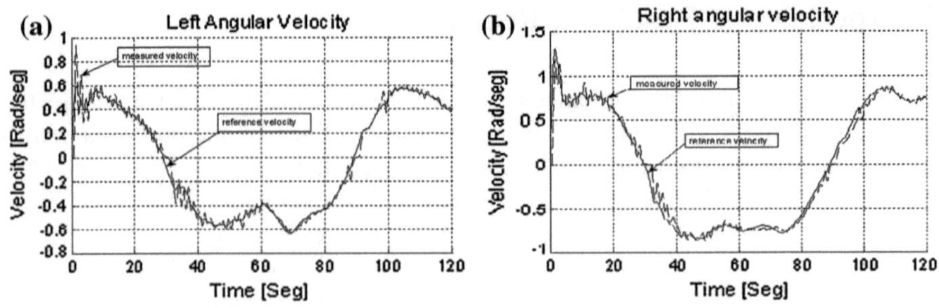

Fig. 8 Reference angular velocities and measured velocities: **a** left engine, **b** right engine

Fig. 9 Path in the plane followed by the robot

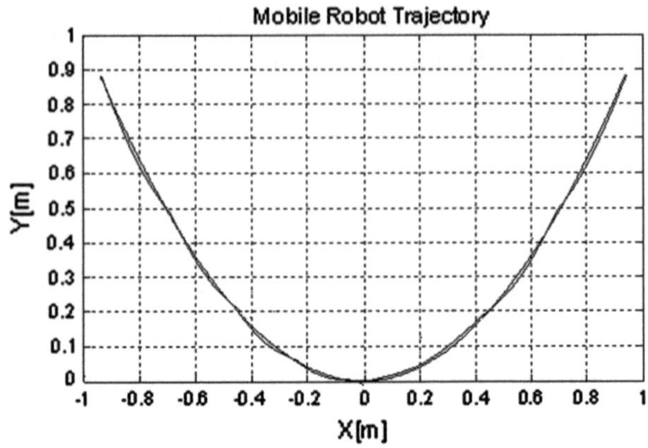

6 Conclusions

It is shown the design of a mobile robot platform allowing the test of tracking control schemes due to its computer and sensors embedded architecture. This laboratory prototype seems to be useful when dealing with the trajectory tracking problem. We validated the performance of the robot with the control of tracking proposed. Actually, the implementations of more advanced tracking techniques are under study with the aim of having outdoor autonomous navigation.

References

1. de Wit CC, Siciliano B, Bastin G (1996) Theory of robot control. Springer, New York
2. Campion G, Bastin G, d'Andrea-Novel B (1996) Structural properties and classification of kinematic and dynamic models of wheeled mobile robots. IEEE Trans Robot Autom 12:46–62
3. Isidori A (1995) Nonlinear control systems, 3rd edn. Springer, New York

4. Borenstein J, Feng L (1994) UMBmark a method for measuring, comparing and correcting dead-reckoning errors in mobile robots. University of Michigan Technical Report UM-MEAM-94-22 (December)
5. Ogata K (2010) Modern control engineering, 5th edn, Prentice Hall, New Jersey
6. Treviño JH, Muñoz JT (2013) Electromechanical design: reactive planning and control with a mobile robot, In: Proceedings of the IEEE, 11th International Conference on Electrical Engineering, Computing Science and Automatic Control (CCE), pp 13–18
7. Ozkan M, Canigur E (2012) Model reference adaptive control of a non-holonomic wheeled mobile robot for trajectory tracking. In: Proceedings of the IEEE International Symposium on Innovations in Intelligent Systems and Applications (INISTA), pp 1–5
8. Tomas S, Bricaire EA, Villa MV (2002) Non-linear discontinuous control of a mobile robot. In: Computation and systems. I.P.N. México, special volume, pp 42–49

Decentralized Supervisory Control of an AMS Based on the ISA Standards

E.G. Hernandez-Martinez, S.A. Foyo-Valdes, E.S. Puga-Velazquez and J.A. Meda-Campaña

Abstract This paper presents the modeling and supervisory control of a real Automated Manufacturing System. The plant model is obtained from the process definition and material-handling tasks, according to the ISA-88 and ISA-95 standards, including the communication of different local controllers. Then, a hierarchical-decentralized supervisory control scheme is implemented where the low-level is composed of a decentralized supervision of the task precedences and storage limitations restrictions. The high-level supervisor deals only with the product recipes using the subset of events related to the process tasks only. Therefore, the scheme provides a clear separation between the equipment operations and the product manufacturing actions. The supervisory control is implemented on a PC communicated through a network of PLC's and local controllers with industrial robots.

Keywords Discrete-event systems · Finite-state automata · Supervisory control · Automation · Student paper

1 Introduction

The coordination control layer of Automated Manufacturing Systems (AMS's) is an important topic of industrial and academic research. It must consider the equipment technologies, communication protocols and product scheduling to establish a

E.G. Hernandez-Martinez (✉)
Universidad Iberoamericana Ciudad de México, México City, México
e-mail: eduardo.gamaliel@ibero.mx

S.A. Foyo-Valdes · E.S. Puga-Velazquez · J.A. Meda-Campaña
ESIME Zacatenco, Instituto Politécnico Nacional, México City, México
e-mail: sergiofoyo@live.com.mx

E.S. Puga-Velazquez
e-mail: erika_selenep@hotmail.com

J.A. Meda-Campaña
e-mail: jmedac@ipn.mx

M. Ceccarelli and E.E. Hernández Martinez (eds.), *Multibody Mechatronic Systems*, Mechanisms and Machine Science 25, DOI 10.1007/978-3-319-09858-6_51

desired operational behavior which enables flexible and concurrent equipment tasks [1] for the manufacturing of different and concurrent products. Market conditions, customer requirements and delivery times require maximal concurrence of tasks and less reprogramming time of the local controllers for the manufacturing of new products or introducing different production orders. In this context, the formalism of Finite State Automata (FSA) have been widely used to model the logical event sequences of a class of AMS's describing all concurrence situations, tasks asynchrony, blocking and other conflicts on the AMS [2, 3]. Some advantages of FSA are the building of complex models through the mixing of individuals models of equipment. Also, it is possible to model individual specifications, for instance, security procurement, resource sharing or desired process behaviors to synthesize supervisors using Supervisory Control (SC) Theory [4]. The main drawback of the FSA and SC is the exponential increasing of the models for complex real AMS's. However, SC theory proposes alternative schemes based on decentralized and hierarchical supervision.

Some works report experiences and approaches of SC schemes applied to real AMS's [5]. For example, modular non-conflicting supervisors are designed in [6], and [7] studies hierarchical interfaces-based architectures. Resource failures disturbances are included in [8] and [9] analyzes the deadlock avoidance while [10] explores modular architectures and its translation to a PLC-based local control. Finally, [11–13] describe the SC implementation on some AMS's prototypes.

In standard SC, the manufacturing rules related to specific products are included as other specifications in the supervisor synthesis procedure. However, this often requires updating of the supervisor when the production rules change. Some works propose to separate the product supervision in hierarchical and modular SC architectures [14]. Thus, the change of product orders involves only the recalculation of the high-level product supervisors. Some important works are [15] where the supervision is based in terms of equipment tasks-status models and [16] based in terms of control events named *imperative*, i.e. some controllable events related to the start of main process tasks, including tools to prove the product feasibility. Also, the product supervisors separation is supported by the industrial standards ISA-88 [17] and ISA-95 [18] that suggest to divide the equipment and procedures models from the product manufacturing actions (product recipes). Works related to the application of these standards to discrete-event manufacturing systems can be consulted in [19, 20].

This paper describes a real implementation of a hierarchical-decentralized SC scheme with the following features:

(1) The low-level equipment operations are grouped as main tasks or generic routines according to the ISA-88 and ISA-85 standard. Every task is programmed off-line and involves the communication of different local controllers.
(2) The tasks are classified as *process tasks*, related to the workstations, and *material-handling tasks*, related to robot manipulators, storage and conveyor belts. The beginning and end of tasks represents the events of the discrete-event plant model.

(3) To avoid the state explosion problem in the SC, we obtain a two-level hierarchical scheme, where the low-level is composed of the conjunction of two decentralized supervisors related to process specifications (task precedences) and capacity limitations of the automatic storages units, respectively. The high-level supervisor deals with the product recipes and their specifications (production orders) using only the events related to the process tasks.
(4) The control architecture is implemented in WinCC® programming language communicated to PLC's using a PROFIBUS-DP network.

2 Preliminary Concepts

2.1 FSA and Supervisory Control Theory

An event is defined as an asynchronous occurrence that generates a spontaneous change of the system. For instance, in an AMS, an event could be the arrival of a workpiece at a machine, the completion of a task, or a machine fault. A *language* defined over the event set Σ is a set of finite-length strings formed from the members of Σ. A regular language can be represented by a *Finite State Automaton* or automaton [2]. It is a device that is capable of representing a language according to well-defined rules. A deterministic automaton is a six-tuple

$$G = \{X, \Sigma, \delta, \Gamma, x_0, X_m\} \tag{1}$$

where $X = \{0, \ldots, n\}$ is the set of states, $\sum = \{e_1, \ldots, e_m\}$ is the event set, $x_0 \in X$ is the initial state, $X_m \subseteq X$ is the set of marked states, $\delta : X \times \Sigma \to X$ is the (partial) transition function and $\Gamma : X \to 2^{\Sigma}$ is the active event function, i.e. $\Gamma(x)$ is the set of all feasible events at state x. Marked states serve to distinguish those event strings that have some special significance.

The *synchronous product* is an automaton operation that joins the behavior of two automatons that operate concurrently. It allows to model and mix simple automatons to synthesize complex models.

The application of SC is necessary, where a plant automaton represents the uncontrolled behavior of the system, which needs to restrict some sequences of events that violate a security or process specification or arrive at blocking states. Then, a supervisory automaton is synthesized from based on a set of specifications. Thus, the plant communicates to the supervisor the set of feasible events at the state x and the supervisor enables a subset of events $\Gamma_C(x) \subseteq \Gamma(x)$ that satisfies the system specifications at this state. The plant automaton G and specifications automaton S are obtained by the synchronous product of the system elements and specification models, respectively. Then, the supervisor automaton S/G (read S controlling G) is synthesized by the supremal controllable sublanguage algorithm *supcon* from G and S.

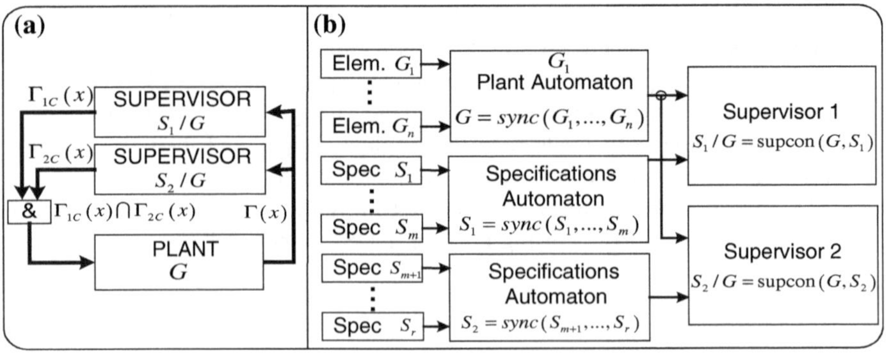

Fig. 1 Decentralized supervisory control. **a** Decentralized SC scheme, **b** decentralized SC synthesis procedure

The main disadvantage of (monolithic) SC is the possible state explosion for the case of large and complex systems and the respective computational costs. An alternative is the Decentralized SC scheme [2, 4] shown in Fig. 1 for the case of two supervisors S_1/G and S_2/G. The specification models are divided to generate two supervisors and the *conjunction* of the supervisors actions enables an event e just when it is enabled by S_1/G and S_2/G simultaneously, i.e. if $\in \Gamma_{C1}(x) \cap \Gamma_{C2}(x)$. Thus, each supervisor is responsible of local specifications and the overall specification is the intersection of all local specifications [4].

2.2 ISA-95 and ISA-88 Standards

The ISA-88 and ISA-95 standards provide the modeling framework to classify equipment and control procedures related to the manufacturing of products. Also, they establish the interface between management systems, coordination control and low-level control technologies [18]. ISA-88 is related to batch control systems [17] however, the main concepts have been extended to other continuous and discrete processes [20]. The next section describes the application of ISA-88 and ISA-95 to the modeling of case of study.

3 AMS Description and Tasks Definition

Figure 2a shows the AMS of the case of study. It is composed of an Automatic Hexagonal-shaped storage (HS) of 7-level and 6 columns with touch sensors and spindle motor, a six-DOF CRS Gantry robot (GR), automatic Lathe (LT) and Mill (MI) Boxford-CNC machines, a XY Cartesian Robot (CR) with pneumatic gripper together to a 6X7 array Matrix-shaped storage (MS), a conveyor belt (CB) with loading and unloading devices and automatic detection of pallets and finally an

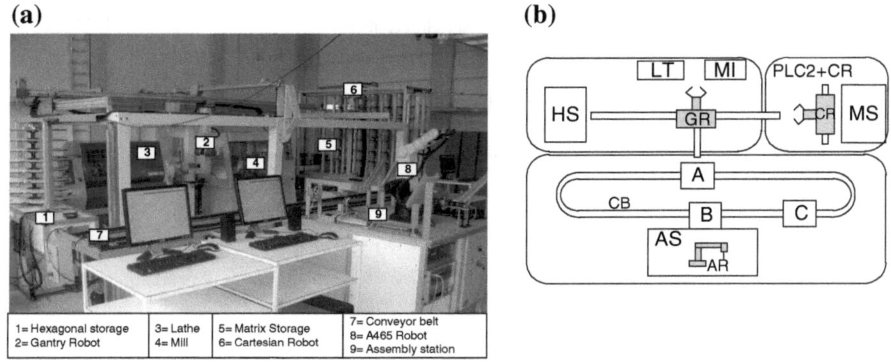

Fig. 2 **a** Photo of the AMS, **b** Scheme of the AMS

Assembly Station (AS) composed by pneumatic assembly devices, temporary storages and a six-DOF CRS-A465 Assembly Robot (AR) with interchangeable grippers and pneumatic clamping mechanisms.

The raw material is manually placed in the HS. Then, the GR transports the workpieces from HS to LT or MI and the manufacturing parts from this process to the CR that stores the parts on the some columns of the MS. When an assembly product is required by the production order, the CR and GR carry out the parts from the MS to the CB or the GR transport it directly from the CNC machines to the CB. The CB moves the parts to the AS. When the AS accumulates the necessary parts, the AR and the pneumatic units make the product. The Final Product (FP) is again transported by the CB, GR and CR and it is stored in the MS for the consumers. Note that the MS has specific columns for LT and MI parts, FP and empty pallets for every kind of storage. The AMS architecture can be decomposed in three main groups as shown Fig. 3b with their respective local controllers which are integrated in the control network. A control PC communicates with a master PLC only. A PROFIBUS-DP network carries communication from the master PLC to three slaves PLC's in an open-chain configuration. The sensors and actuators of the HS, MI, LT, MS, CB and AS are connected directly to the PLC's whilst the GR, CR and AR are connected to the PLC's through the General Purpose of Input-Output (GPIO) of their respective local motion controllers.

Based on the task decomposition for the AMS proposed in the Fig. 3. The main strategy is to separate the material-handling system from process workstations of the AMS. Due to the physical restrictions of functionality, some equipments work together always, therefore they are considered as a unique functional unit, for example GR and CR. Note that the AR appears twice because it performs as a material-handling system sometimes and participates the AS process too. Figure 3 also shows the PCM at the task level of the AMS equipment. A detailed list of these tasks is given in Fig. 4 where it is considered only a task for LT and MI and seven final assembly products for simplicity. The product notation *MLL* denotes an assembly composed by one MI piece and two LT pieces and so on.

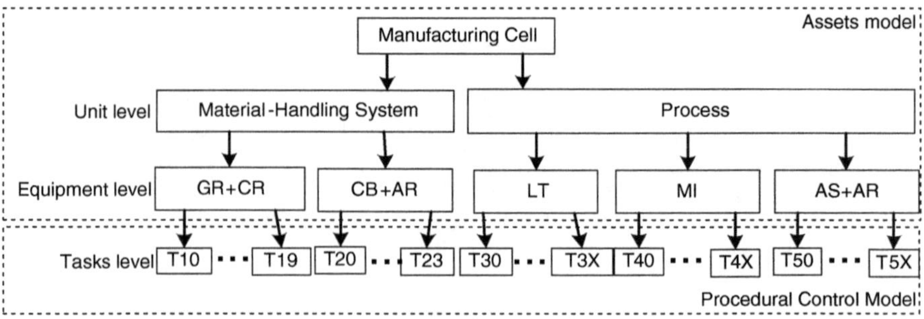

Fig. 3 AM and PCM of the AMS

Manuf. Cell		
Material-Handling System	GR+CR	T10. GR transports from HS to LT T11. GR tansports from HS to MI
		T12. GR+CR transport from LT to MS T13. GR+CR transport from MI to MS
		T14. GR+CR transport a LT part from MS to CB (point A) T15. GR+CR transport a MI part from MS to CB (point A)
		T16. GR transports from LT to CB (point A) T17. GR transports from MI to CB (point A)
		T18. GR transports a empty pallet for FP from MS to CB (point A) T19. GR transports from CB (point A) to MS (FP column)
	CB+AR	T20. CB+AR transport a LT part from CB (point A) to CB (point B) and local buffers of AS. The pallet and base get out by CB (point C)
		T21. CB+AR transport a MI part from CB (point A) to CB (point B) and local buffers of AS. The pallet and base get out by CB (point C)
		T22. CB transport empty pallet for FP from CB (point A) to CB (point B) T23. CB transport FP from CB (point B) to CB (point A)
Process	LT	T30. Machining process of LT
	MI	T40. Machining process of MI
	AS+AR	T50. Assembly of the product ML T51. Assembly of the product MLL T52. Assembly of the product MLLM T53. Assembly of the product MLLL T54. Assembly of the product MLLLM T55. Assembly of the product MLLLL T56. Assembly of the product MLLLLM

Fig. 4 Description of task

4 Supervisory Control Architecture

The proposed SC architecture is shown in Fig. 5a. The plant automaton is obtained from the synchronous product of the equipment tasks as shown Fig. 5b where sT_{ij} (controllable event) and fT_{ij} (uncontrollable event) are the beginning and end events of the task T_{ij}. The SC is divided on two levels.

The low-level SC establishes a decentralized supervision of task precedences and storage limitation, respectively. The specification automatons for the synthesis procedure of the task precedences supervisor is obtained from the tasks precedence diagram shown in Fig. 6 where every arrow establishes a End-Start logic precedence between pair of tasks. Note that the tasks Precedence represent only the correct logic dependence of the tasks during the AMS execution and it does not refer to product sequences. They obey process and security restrictions. For example, the Task T_{30} begins only if the GR put it one workpiece (finish of task T_{10}), etc.

The storage limitations supervisor is related to the capacity of the columns of the HS, MS and local buffers of the AS. It can be analyzed using queues diagram and its translation to the specification queues automatons, as shown in Fig. 7, where the last state corresponds to the maximum storage capacity.

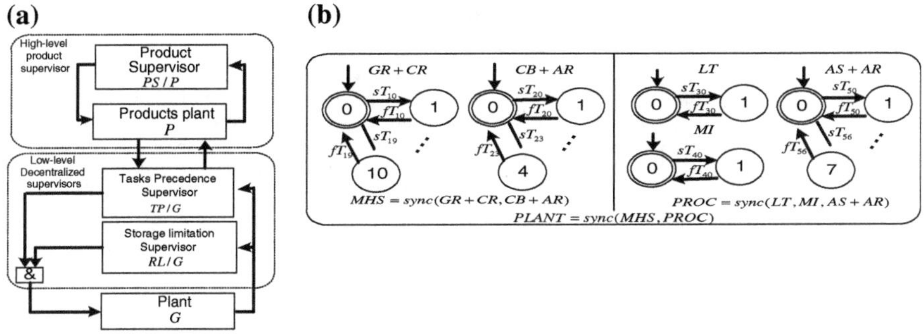

Fig. 5 **a** SC architecture, **b** plant automaton

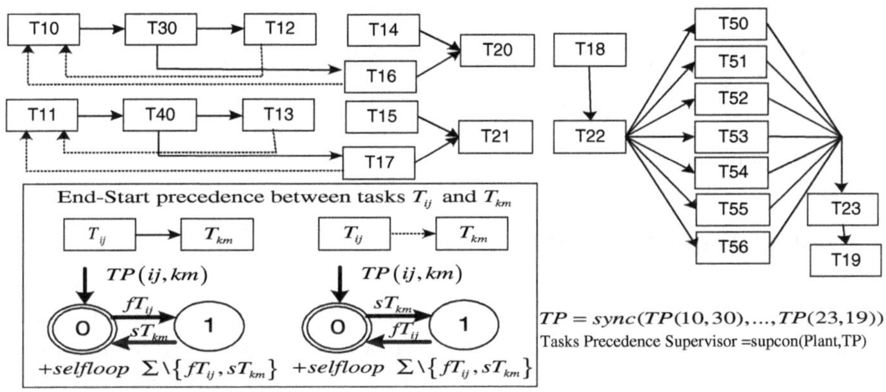

Fig. 6 Precedences of tasks

Note that the marker states establishes that it is desirable that storage of raw material and available empty pallets remain full whereas the originally empty buffers of manufacturing parts and final products must be filled during the tasks execution.

The high-level SC is related to the product definition and supervision based on the begin of process tasks only. The products are modeled and synchronized to obtain the plant automaton. Then, a set of specifications construct the product specifications automaton that generates the product supervisor. These specifications are related to product orders or logic dependence of chains of products. The start-task events of the controlled plant automaton are communicated to the low-level SC which enables these commands to the local controllers.

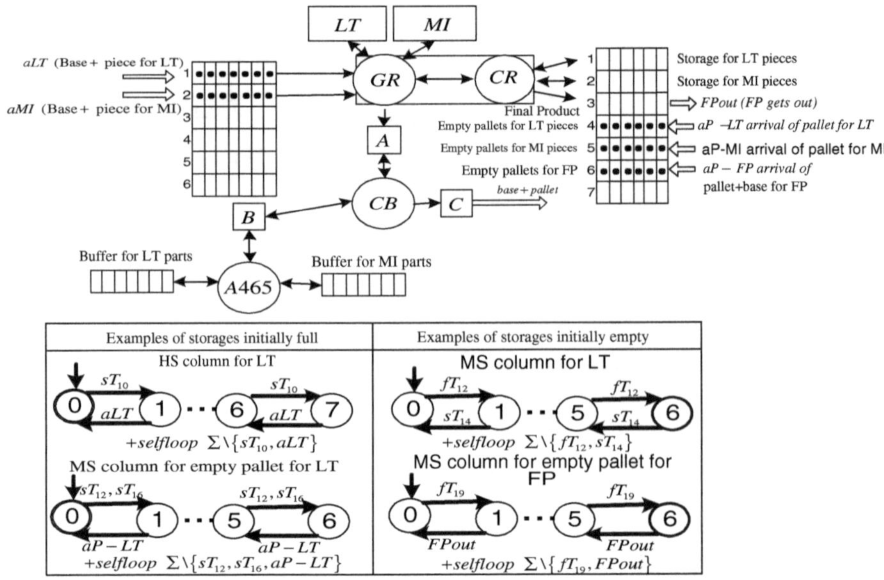

Fig. 7 Queues diagram of the AMS

5 Control Implementation

The SC implementation is illustrated in Fig. 8a. The application is divided in the Event Window (EW) and SC Window (SCW). All commands related to the communication between the software interface with the AMS are defined in the EW. Thus, the events can be operated from the EW performing a manual mode. Clearly, the begin of tasks are translated in virtual inputs for the ladder diagram of the master PLC whilst the end of tasks are virtual outputs or marks from the PLC to

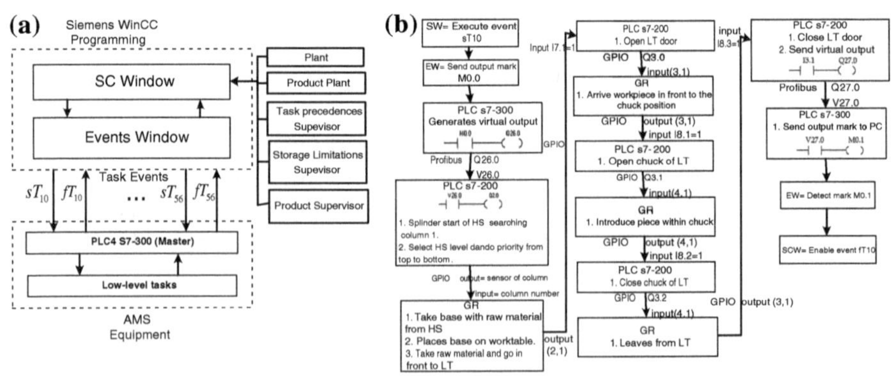

Fig. 8 **a** SC implementation, **b** Example of a low-level task programming

the PC. The EW has a bidirectional communication with the SCW where the plant and supervisor automatons are uploaded using their transition matrix representation. When the algorithm begins, the plant and supervisors automatons run concurrently. For every actual state, the feasible events are enlisted for both the Plant and Supervisors. When the supervised product plant permits a start of process tasks, it automatically enables this event to the EW that sends the respective command to the AMS. When a material-handling process is available or an end of task is detected, the algorithm selects the appropriated transition and the automatons jump to other states. This process continues indefinitely until the algorithm stops. The language in Siemens WinCC® is event-based programming and permits the code import from VisualBasic® or C++. The low-level programming of the tasks in the AMS is an important and long-time training activity. Figure 8b shows a schematic diagram of the task T_{10} where the GR transport a material raw from HS to LT. The routine begins with the start-event given by the SCW and EW. Then, a virtual output is communicated to the master PLC which sends the information to the local controllers related to the task, in this case, a Siemens PLC s7-200, the GR controller and the LT controller. The local controllers are communicated along the whole task and finally send the end of task signal to the master PLC and the PC for the execution of the control algorithm.

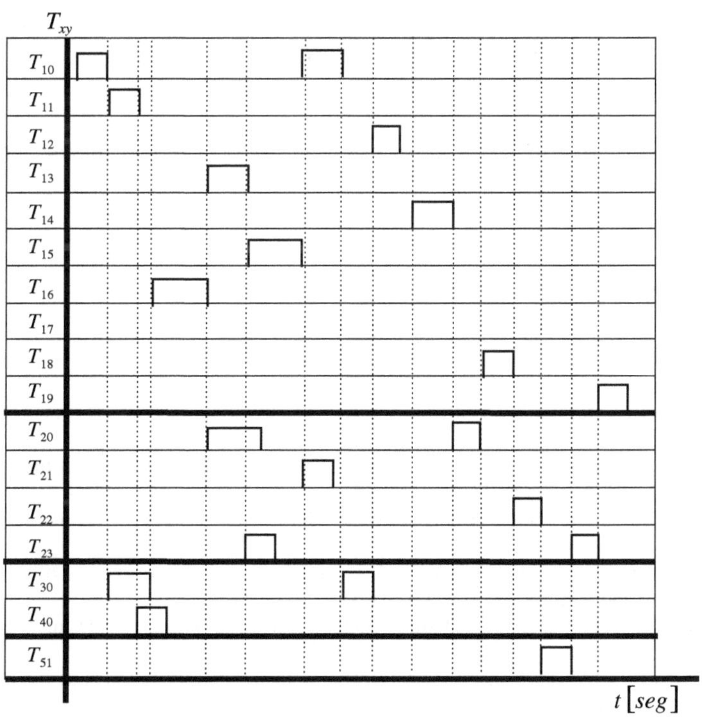

Fig. 9 Phase space diagram of the task

Figure 9 shows the execution of tasks, controlled by the supervisors when a MLL product is required. To two parts “L” machined by LT are achieved using the direct path T_{16} to CB, and the sequence T_{12}, T_{14} using the MS temporaly, respectively. The part “M” is machined by MI, stored in MS and moved to the CB trough the tasks T_{13}, T_{15}. After that, the AS produces the MLL product using the task T_{51}, and the task T_{19} puts the product in the FPout. Note the concurrence of some tasks, for instance in the case of $T_{11} - T_{30}$, $T_{13} - T_{20}$ and $T_{16} - T_{40}$. All the sequence of tasks obeys the precedence and limitations established in the decentralized supervisor.

6 Conclusion and Future Works

This paper presents the implementation experience of a SC architecture where the low-level supervision is decentralized to avoid the state explosion of the automatons and where the product manufacturing rules are carry out separately from the control equipment supervision. The modeling and control split is based on the ISA-95 and ISA-88 industrial standards where the reprogramming of the system for the manufacturing of new product orders or introducing new products is translated in the recalculation of new product supervisors in the software not requiring the reprogramming of the low-level tasks contained in the local controllers. As future works, we will explore the task execution based on Petri models and hybrid architectures using stochastic automata for production time optimization.

Acknowledgments To Universidad Iberoamericana, CONACyT (SNI scholarship) and IPN with project 20140659, and scholarships EDI, COFAA, and PIFI.

References

1. Groover MP (2008) Automation production systems and computer integrated manufacturing. Prentice Hall, Englewood Cliffs
2. Cassandras C, Lafortune S (2008) Introduction to discrete event systems. Kluwer Academic, Boston
3. Caillaud B, Darondeau P, Lavagno L, Xie X (2002) Synthesis and control of discrete event systems. Kluwer Academic, Dordrecht
4. Wonham MW (2009) Supervisory control of discrete-event systems, available on website. University of Toronto
5. Castillo I, Smith JS (2002) Formal modeling methodologies form control of manufacturing cells: survey and comparison. J Manufact Syst 21(1):40–57
6. Nourelfath M, Niel E (2004) Modular supervisory control of an experimental automated manufacturing system. Control Eng Pract 12(2):205–216
7. Leduc J, Lawford M, Pengcheng D (2006) Hierarchical interface-based supervisory control of a FMS. IEEE Trans Control Syst Technol 14(1):654–664
8. Yalcin A (2004) Supervisory control of automated manufacturing cells with resource failures. Robot Comput Integr Manufact 20(2):111–119

9. Yalcin A, Boucher TO (2000) Deadlock avoidance in flexible manufacturing systems using finite automata. IEEE Trans Robot Autom 16(4):424–429
10. Queiroz H, Cury J (2002) Synthesis of local modular supervisory control for a manufacturing cell. In: 6th international workshop on discrete event systems, pp 377–382
11. Chandra V, Huang Z, Kumar R (2003) Automated control synthesis for an assembly line using discrete event system control theory. IEEE Trans Syst Man Cybern C 33(2):284–289
12. Costa GO, Santos EAP, Busetti MA (2004) Design and implementation of a low cost control system for a manufacturing cell. In: IEEE conference on robotics, automation and mechatronics pp 281–286
13. Hernandez EG, Bricaire EA, Nava JA, Sanchez A (2005) A design strategy of discrete event controllers for automated manufacturing systems. In: 2nd international conference on electrical and electronics engineering, pp 361–364
14. Sanchez-Carmona A, Aranda-Bricaire E, Hernandez-Martinez EG (2006) Design of coordination controllers for a class of discrete-event manufacturing systems. In: IFAC symposium on information control problems in manufacturing, pp 283–288
15. Pétin JF, Gouyon D, Morel G (2007) Supervisory synthesis for product-driven automation and its application to a flexible assembly cell. Control Eng Pract 15(5):595–614
16. Sanchez A, Aranda-Bricaire E, Jaimes F, Hernandez E, Nava A (2010) Synthesis of product-driven coordination controllers for a class of discrete-event manufacturing systems. Robot Comput Integr Manufact 26(4):361–369
17. Instrument Society of America, ISA-88.01 (1995) Batch control systems, part 1. Models and terminology, ISA Standards
18. Instrument Society of America, ISA-95.1 (1999) Enterprise-control system integration, part1. Models and terminology, ISA Standards
19. Sanchez A, Parra LF, Baird R, Macchietto S (2002) Hybrid modeling and dynamic simulation of automated batch plants. ISA Trans 42(1):401–420
20. Nava JA (2005) Architecture for the control of continuous, discrete and batch systems (in Spanish). Master Thesis, Cinvestav

A Planar Cobot Modelled as a Differential Algebraic System

Omar Mendoza-Trejo and Carlos Alberto Cruz-Villar

Abstract Collaborative Robots (cobots) are devices designed for direct interaction with human operators in a shared workspace. In such devices, the human provides the necessary force for the movement of the system, while the cobot provides a virtual guiding surface. The main purpose of the virtual guiding surfaces is to direct the movement performed by the human such that a desired path can be followed. This work presents simulation results of a planar 2-DOF cobot with differential gears modelled as a differential algebraic system, that is, the well known dynamic model of a planar 2-DOF robot is subject to the algebraic constraint imposed by the velocities that exist on the differential gear train. Moreover, the force exerted by the human operator is decomposed into its cartesian components and modelled as a PD controller.

Keywords Collaborative robots · Differential algebraic system · Continuously variable transmission · Student paper

1 Introduction

In 1995 Northwestern University and General Motors Company began a project aimed to design a support device for human workers in material handling. In the automobile assembly phase, human workers performed activities (parts-picking, identifying defective parts, fitting parts) that can not be replaced by an automated system. To cope with problems presented in the assembly phase Colgate [1, 2]

O. Mendoza-Trejo (✉) · C.A. Cruz-Villar
Sección de Mecatrónica, Cinvestav, Av. Instituto Politécnico Nacional, 2508 Mexico, Mexico
e-mail: omendoza@cinvestav.mx

C.A. Cruz-Villar
e-mail: cacruz@cinvestav.mx

M. Ceccarelli and E.E. Hernández Martinez (eds.), *Multibody Mechatronic Systems*, Mechanisms and Machine Science 25, DOI 10.1007/978-3-319-09858-6_52

proposed a way to create a motion guide known as virtual surface which is completely passive and therefore intrinsically safe.

Although cobots were designed for applications in the automotive industry, the operating principle of cobots can be used in many areas, for example in different welding processes [3] and rehabilitation therapies [4, 5].

The collaborative robots can be considered as constrained mechanical systems and although the motion equations of this kind of systems can be addressed by techniques that allow to convert the whole system into an ODE system and in this way to use conventional ODE solvers, the obtained models could modify the relationships between variables involved in the constraints, implying numerical errors in the system solution (issues that are diminished by handling the system as a differential algebraic one) [6, 7].

Moreover, as far the authors know, cobotic systems reported in literature only present experimental results, nevertheless, simulation results of the overall system model (mechanical structure-human) can be of great help to get a best experimental device. This is because simulation results allow to verify the system performance with different values of parameters design (mass, inertia, gear ratios) without the need to build a prototype for each set of parameters.

Due to that described in the previous paragraphs, this paper presents simulation results of a planar 2-DOF cobot modelled as a differential algebraic system, where the exerted force by the human operator is modelled as a PD controller.

2 Operating Principle of Cobots

In a cobotic system the human operator provides the necessary force to perform the movements and the cobot provides a virtual guide surface. The cobots have two operating modes: free mode and virtual surface mode. In the free mode, the human operator can move the cobot without any constraint. In the virtual surface mode, only motion along a desired virtual surface is allowed, this mode is performed via the control action [8]. The virtual surfaces are created by using continuously variable transmissions (CVTs) [9–12]. The CVTs do not provide energy, but constrain the speed ratio among the joints. This section presents a brief description about CVTs based on differential gear trains.

2.1 CVTs Based on Differential Gears

For cobots based on CVTs with differential gears, there exist two connection methods, the serial connection and the parallel connection. In the serial connection method, the first joint is connected to the output of a differential gear train, while the second joint is connected to the differential gear by a belt. The parallel connection method (Fig. 1) has an additional differential gear train connected to the second

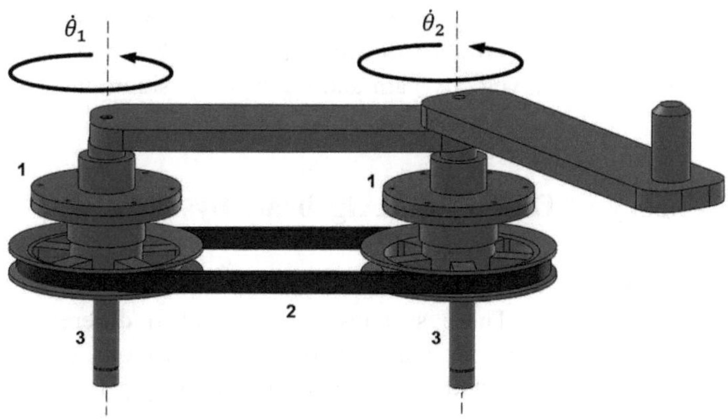

Fig. 1 Cobot of 2 degree of freedom

joint. This connection method has the ease of having several operation modes dependent on the actuators and the transmission ratio between the differential gears [12–14].

In Fig. 1 a 2-DOF cobot is shown, components 1 are the differential gear trains, component 2 is the connection system and components 3 are the motors (which are of low power) responsible for the control action. When the human operator moves the system by the handle, a movement in both second and first links is produced. Considering that the motors (control action) are not active, the transmission ratio between the first and second joints is fixed, so the whole system behaves like a one degree of freedom mechanism along a fixed path. By activating the control system (motor drives), the transmission ratio can be controlled, implying that a desired path can be forced.

2.2 Cobot Control

Cobot's control is different from control of conventional robots. In conventional robots, motors are directly connected to the joints and are synchronized to implement a movement of the endpoint. Moreover, motors installed on cobots are not used to actuate joints, that is, motors only have constraint effects on the joints. The endpoint trajectory control of cobots is performed via the synchronization of the joint velocities [10, 12, 15]. For purposes of this work, the control law established in [12] is used. Equation (1) states the control model.

$$V_s = V(I + G_V \Delta T + G_R \Delta R) \quad (1)$$

where V_s the is control velocity vector, V is the actual velocity vector of the endpoint, ΔT the is vector error of direction motion, ΔR is the vector error of position motion, G_V is the velocity gain and G_R is the position gain.

3 Basic Theory of Differential Algebraic Systems

A differential algebraic system is a set of differential equations subject to a set of algebraic constraints [16]. These systems are also called differential algebraic equations (DAEs) and are present in a big variety of applications, e.g., constrained mechanical systems, fluid dynamics and control engineering [6].

Present knowledge of nonlinear DAE systems is limited to some morphologies. An example of these morphologies is the semi-explicit DAE (ODE with constraints) and can be expressed as:

$$\begin{aligned} \dot{y} &= \mathrm{f}(t, y, z) \\ 0 &= \mathrm{g}(t, y, z) \end{aligned} \tag{2}$$

where $\dot{y} = \mathrm{f}(t, y, z)$ is a differential equation and $0 = \mathrm{g}(t, y, z)$ is an algebraic equation. More detailed information about more morphologies to represent DAE systems can be found in [7, 17–19].

3.1 DAE Solvers

There are two main approaches to deal with differential algebraic systems. The first approach is based on reducing or converting the DAE system into an ODE system with the same DAE's behaviour [7]. Once the system is reduced or converted, the DAE system can be solved by using conventional methods to solve ODE, as for example the Runge-Kutta method.

The second approach consists on establishing the equations modelling the system in such a way that an available DAE solver can be used. The main drawback that present the DAE solvers is that almost all can only deal with DAE systems of at most index 3. Some available solvers of DAE are DASSL/DASPK developed by Petzold [6], RADAU5 developed by Hairer and Wanner [19] and MEBDF developed by Abdulla and Cash [20].

4 COBOT's Modeling as a Differential Algebraic System

In this section the dynamic model of the 2-DOF kinematic chain and the kinematic modelling of the CVT based on differential gears are presented. The dynamic modelling of the kinematic chain has been addressed in many works, so the dynamic model of the kinematic chain is well known. Considering that this work is focused on a planar 2-DOF robot where a human provides the movement force F, the dynamic model of the system is represented as:

$$M(\theta)\ddot{\theta} + C(\theta, \dot{\theta})\dot{\theta} = J^T F \tag{3}$$

From the dynamic model of the kinematic chain and the kinematic model of the differential gear train, it is possible to state the differential-algebraic system, where the dynamic model represents the set of differential equations while the kinematic model represents the algebraic constraint.

4.1 Kinematic Model of the Differential Gear Train

Figure 2 shows the basic components of a differential gear train. Component 1 is the carrier gear, component 2 is the sun gear, component 3 is the planet gear and component 4 is the ring gear.

To solve the kinematic problem, it is necessary to take the carrier gear as reference. Let us consider ω_c as the angular velocity of the carrier gear, ω_s the angular velocity of the sun gear, ω_p the angular velocity of the planet gear, ω_r the angular velocity of the ring gear and r_c, r_s, r_p and r_r their respective radii. Equations 4, 5 and 6 represent the transmission ratios of the different combinations that can be obtained in the differential gear train.

$$\frac{r_p}{r_r} = \frac{\omega_r - \omega_c}{\omega_p - \omega_c} \tag{4}$$

$$\frac{-r_p}{r_s} = \frac{\omega_s - \omega_c}{\omega_p - \omega_c} \tag{5}$$

$$\frac{-r_r}{r_s} = \frac{\omega_s - \omega_c}{\omega_r - \omega_c} \tag{6}$$

Let us consider that $r_p = r_s$. Now, taking into account that $r_r = r_c + r_p$ and $r_c = r_s + r_p$, the next relationships are established:

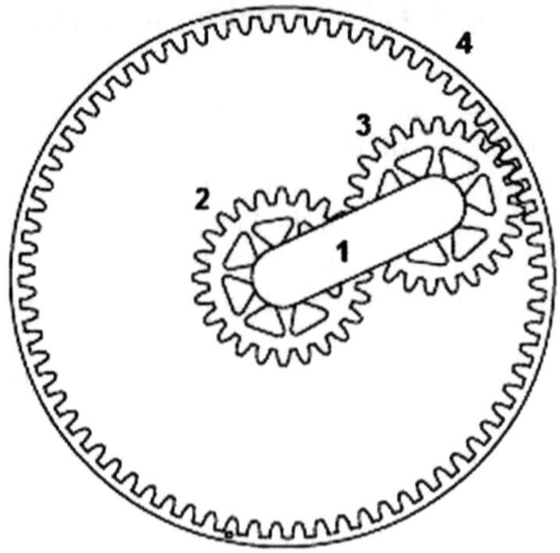

Fig. 2 Differential gear train

$$\begin{aligned} r_c &= 2r_p \\ r_p &= r_p \\ r_s &= r_p \\ r_r &= 3r_p \end{aligned} \tag{7}$$

4.1.1 Constraint for Serial Configuration

There exist two connection methods of the differential gear train in cobots. In serial configuration, it is necessary one differential gear train. Substituting the values of Eqs. (7) in (6) the following expression is obtained:

$$4\omega_c - 3\omega_r - \omega_s = 0 \tag{8}$$

where ω_c, ω_r and ω_s are the angular velocities of the first and second links and the control velocity respectively. The angular velocity of the sun gear is connected to a motor, which will have the task of changing the speed ratio in the differential gear train.

4.1.2 Constraint for Parallel Configuration

In parallel configuration, two differential gear trains are needed. Whereas the differential gear trains are identical and the ring gears are connected by a belt, the following expression can be obtained:

$$\omega_{s_1} - \omega_{s_2} = 4\omega_{c_1} - 4\omega_{c_2} \tag{9}$$

where ω_{s_1} and ω_{s_2} are the angular velocties of the motors, while ω_{c_1} and ω_{c_2} are the angular velocities of the first and second links respectively.

Therefore, given that $\omega_c = \dot{\theta}_1$ and $\omega_r = \dot{\theta}_2$ in Eq. (8), the differential algebraic system of index 2 that determines the performance of a cobot with differential gear trains in serial configuration is as follows:

$$\begin{aligned} M(\theta)\ddot{\theta} + C(\theta, \dot{\theta})\dot{\theta} = J^T F \\ 4\dot{\theta}_1 - 3\dot{\theta}_2 - \omega_s = 0 \end{aligned} \tag{10}$$

Given that $\omega_{c1} = \dot{\theta}_1$ and $\omega_{c2} = \dot{\theta}_2$ in Eq. (9), the differential algebraic system of index 2 that determines the performance of a cobot with differential gear trains in parallel configuration is the following:

$$\begin{aligned} M(\theta)\ddot{\theta} + C(\theta, \dot{\theta})\dot{\theta} = J^T F \\ 4\dot{\theta}_1 - 4\dot{\theta}_2 - \omega_{s_1} + \omega_{s_2} = 0 \end{aligned} \tag{11}$$

5 Simulation Results of the Cobotic System

In this section, simulation results of a planar 2-DOF cobot with differential gear trains, modelled as a differential algebraic system are presented. The exerted force by the human operator is modelled as a PD controller [21]. In simulations the human-cobot system should perform a desired trajectory. The human operator provides the necessary force to the system to perform desired trajectories, while the control action Eq. (1) redirects the movement of the system to minimize the error in the trajectory tracking. Figure 3 shows a schematic diagram of the whole system.

The DAE system is solved using the computer program *Scilab*® [22].

The length of the links are $L_1 = 0.3$ m and $L_2 = 0.3$ m, their centers of mass are located at half of those distances. The masses of these links are $m_1 = m_2 = 1.134$ kg.

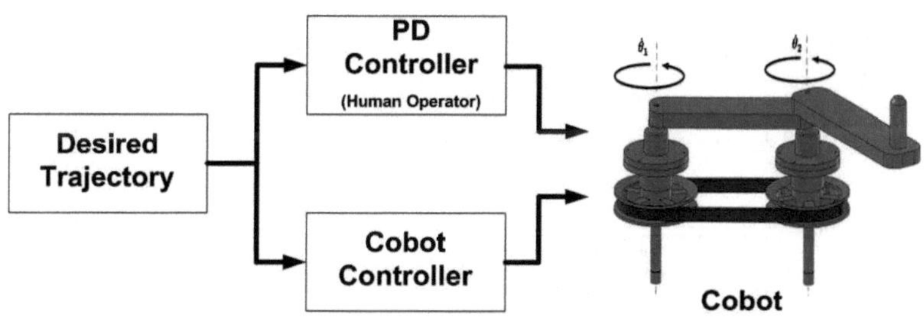

Fig. 3 Schematic diagram of the system

5.1 Cobotic System with Serial Connection Method

This section considers that the system (human-cobot) should perform two desired trajectories (a straight line and a circle).

5.1.1 Straight Line

The straight line is on the “X” axis and has a length of 25 cm, that is, the straight line must begin in the coordinates (−0.35, 0.4 m) and finish in the coordinates (−0.1, 0.4 m). The system starts from the repose, and the end effector of the kinematic chain is in the coordinates (−0.35, 0.4 m).

Figure 4 shows the trajectory tracking when the controller of the cobotic system is *off* and when the controller is *on*. If the value gains of G_R and G_V of Eq. (1) are zero it is said that the controller is *off*.

As we can see in Fig. 4, the cobot controller with gains G_R and G_V (tuned to follow the desired path), improve the trajectory tracking. Table 1 shows the errors in trajectory tracking. These errors are calculated using the following expression:

$$Se = \int_{t_o}^{t_f} \left(\mathbf{P}_{di} - \mathbf{P}_{geni}\right)^2 dt \tag{12}$$

where $\mathbf{P}_{di}$ is the desired trajectory and $\mathbf{P}_{geni}$ is the generated trajectory by the end effector.

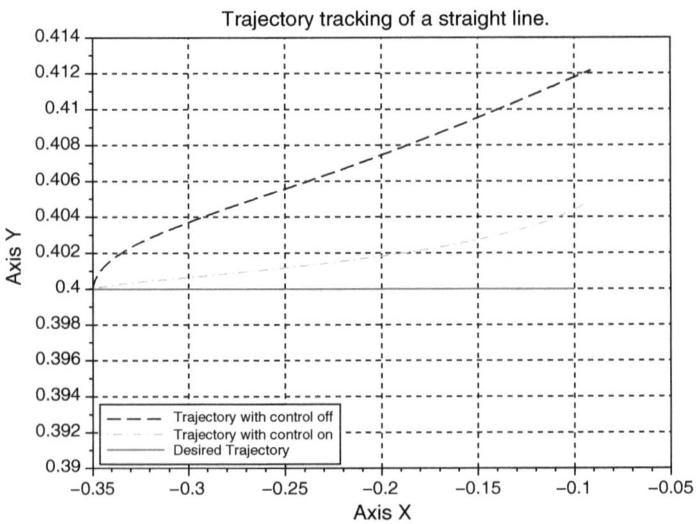

Fig. 4 Straight line

Table 1 Errors in trajectory tracking

–	Error without control (m^2)	Error with control (m^2)
Straight line	1.1936	0.0618
Circular path	0.0198	0.0167
Sinusoidal path	0.0003596	0.0002792

5.1.2 Circular Path

The circular path that is desired to be followed has a radius of 0.2 m with center in (−0.2, 0.3 m). The system starts from the repose, and the end effector the kinematic chain is in the coordinates (−0.2, 0.5 m). Figure 5 shows the trajectory tracking for the circular path when the control cobot is *off* and when the cobot control is *on*. A zoom near the coordinates (0, 0.3 m) is done to clearly see the performance of both trajectory trackings. Table 1 shows the errors in trajectory tracking. These results demonstrate that the trajectory tracking is better when the cobot control is *on*.

5.2 Cobotic System with Parallel Connection Method

This section considers that the system should perform a sinusoidal trajectory. The sinusoidal trajectory is given by the equation $y = 0.1 * \sin(8\pi x)$ $[\mathrm{m}]$ where $x \in [0.0.25]$ $[\mathrm{m}]$ and begins in the coordinates (−0.2, 0.2 m). The system starts from the repose, and the end effector of the kinematic chain is in the coordinates (−0.2, 0.2 m). Figure 6 shows the trajectory tracking for the sinusoidal path, when the cobot control is *off* and when the cobot control is on. A zoom near the coordinates (0.38, 0.29 m) is done to clearly see the performance of both trajectory trackings. Table 1 shows the errors in trajectory tracking. These results demonstrate that the trajectory tracking is better when the cobot control is *on*.

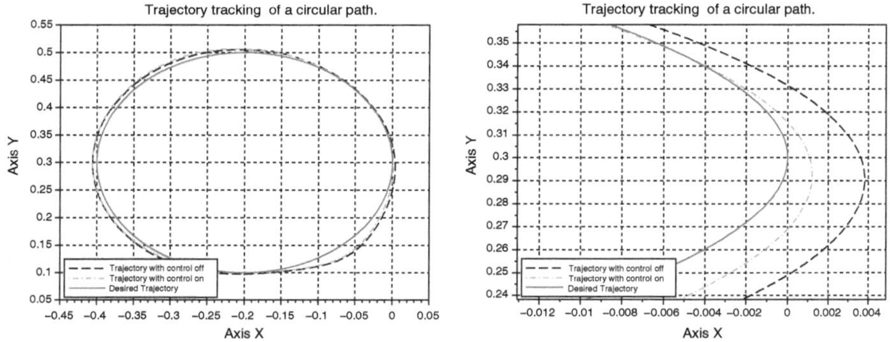

Fig. 5 Circular path

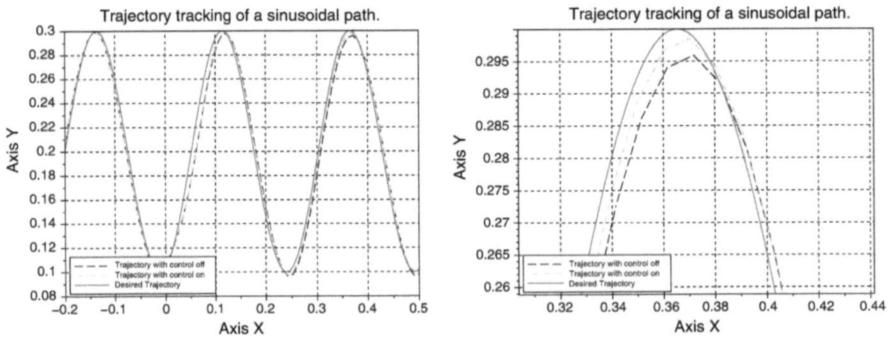

Fig. 6 Sinusoidal path with cobot control off and cobot control on

6 Conclusions

A planar 2-DOF collaborative robot modelled as a differential algebraic system is presented. This mathematical model of cobots allows to simulate the system performance, which, can be of help to get a best experimental device.

Three simulation results of trajectory tracking are presented. In simulations, the exerted force by the human operator is modelled as a PD controller, where the PD controller can represent a skilled human operator or a non-skilled human operator. When a non-skilled human operator is operating the system, a controller that redirects the movement of the cobot is needed, on the other hand, when a skilled human operator is operating the system, the cobot control is not necessary.

References

1. Colgate JE, Peshkin MA, Wannasuphoprasit W (1996) Nonholonomic haptic display. In: IEEE international conference on robotics and automation, pp 539–544
2. Peshkin MA, Wannasuphoprasit W, Akella P, Colgate JE (2000) Cobots in material handling. Northwestern University, Evanston
3. Mustafa S, Marić B (2011) Assisting manual welding with robot. Robot Comput Integr Manuf 27:818–828
4. Pan P, Lynch K, Peshkin M, Colgate E (2005) Human interaction with passive assistive robots. In: 9th International conference on rehabilitation robotics, 2005
5. Worsnopp T, Peshkin M, Lynch K, Colgate E (2006) Controlling the apparent inertia of passive human-interactive robots. J Dyn Syst Meas Control 128 (1):44–52
6. Brenan KE, Campbell SL, Petzold L (1989) Numerical solution of initial-value problems in differential-algebraic equations. Classics in applied mathematics, North-Holland, New York
7. Costa-Casteló R, Grió R, Basañez L (1998) DAE methods in constrained robotics system simulation Computación y Sistemas, vol 1, pp 145–160, Mexico
8. Moore CA, Peshkin MA, Colgate JE (1999) Design of a 3R cobot using continuously variable transmissions. In IEEE International Conference on Robotics and Automation (ICRA), 1999

9. Wannasuphoprasit W, Akella P, Peshkin MA, Colgate JE (1998) Cobots: a novel material handling technology. ASME 98-WA/MH-2
10. Gellispie R, Peshkin MA, Colgate JE (2001) A general framework for Cobot control. IEEE Trans Robot Autom 17(4):391–401
11. Peshkin M, Colgate E, Wannasuphoprasit W, Moore C, Gillespie B, Akella P (2001) Cobot architecture. IEEE Trans Robot Autom 17(4):377–390
12. Dong Y, Zhang L, Dunmin L, Bernhardt R, Surdilovic D (2004) A novel cobot and control. In: Proceedings of the 5th world congress on intelligent control and automation, China
13. Surdilovic D, Bernhardt L, Zhang L (2003) New intelligent power assist systems based on differential transmission. Robotica 21:295–302
14. Zhang L, Lu D, Wang L, Shen J, Bernhardt R (2003) Dynamics of five-bar COBOT ussing differential mechanism. J Mar Sci Appl 2(2)
15. Wannasuphoprasit W, Cillespie RB, Colgate JE, Peshkin MA (1997) Cobot control. In: Proceeding of the IEEE 1997 international conference on robotics and automation
16. Gear Charles W (1971) Simultaneous numerical solution of differential-algebraic equations. IEEE Trans Circuit Theory CT-18:89–95
17. Moore C, Peshkin M, Colgate E (2003) Cobot implementation of virtual paths and 3-D virtual surfaces. IEEE Trans Robot Autom 19(2):347–351
18. Hairer E, Lubich C, Roche M (1989) The numerical solution of differential-algebraic systems by Runge-Kutta methods. Springer, Berlin
19. Hairer E, Norsett SP, Wanner G (1991) Solving ordinary differential equations. In: Series in Computational Mathemathics, 1991. Springer, Berlin
20. Ascher UM, Petzold LR (1998) Computer methods for ordinary differential equations and differential-algebraic equations. Society for Industrial and Applied Mathematics (SIAM), Philadelphia
21. Méndez -Iglesias J, Parra-Vega V, Ruiz-Sánchez F (2005) Identification of the human behavior in virtual environment tasks as a non-linear control block. In: Proceedings of the 16th IFAC world congress, 2005
22. Gomez C, Bunks C, Chancellier J (1999) Engineering and scientific computing with Scilab. Springer Science + Business Media, New York

Design and Implementation of an Affective Computing for Recognition and Generation of Behaviors in a Robot

Rodolfo Romero Herrera, Francisco Gallegos Funes and Maria Adela Soto Alvarez del Castillo

Abstract In this paper adapting a robot to a social approach is presented in the selection of behaviors for interaction in real environments, which represents an emotionally charged, causing attention to focus on the most relevant aspects of the surroundings for the realization of the software system; which is part of a branch called affective computing, where there is a classification for emotion exhibited by a system. In this classification, the machine is included within the application for displaying and perceiving simulated emotions. Determining the emotion is by using templates based on probability theory of Markov. In this project the user interaction function is modified for vary with respect to the emotional state of the agent, which is determined according to the environment in which it is. Based in the Kinect sensor, emotional states according to body language of people and positions is detected. The system recognizes that the compatibility of the emotional state has been entered against it by users or persons with whom the project is tested. After recognizing the emotional state; a robot mimics the movements of the human pretending to have such emotions.

Keywords Markov · Computer affective · Kinect · Robot · Recognizing · Student paper

R. Romero Herrera (✉) · M.A. Soto Alvarez del Castillo
Escuela Superior de Computo, Instituto Politécnico Nacional, México, Mexico
e-mail: rromeroh@ipn.mx

M.A. Soto Alvarez del Castillo
e-mail: adela.alvarezdelcastillo@facebook.com

F. Gallegos Funes
Escuela Superior de Ingeniería Mecánica y Eléctrica, Instituto Politécnico Nacional, México, Mexico
e-mail: fgallegosf@ipn.mx

M. Ceccarelli and E.E. Hernández Martinez (eds.), *Multibody Mechatronic Systems*, Mechanisms and Machine Science 25, DOI 10.1007/978-3-319-09858-6_53

1 Introduction

The word robot was first used in 1921 when the Czech writer Karel Capek (1890–1938) premieres at the Prague National Theatre his work Rossum's Universal Robot (RUR). Its origin is the Slavic word "robota", which refers to the work done forcibly [1].This term had fallen into disuse, but it had not been for the writers of the literary genre of science fiction [2].

Robots were originally conceived as fictional entities, unemotional, which performed repetitive tasks. But they are capable of much more [3].

Social robotics has emerged as a paradigm still unresolved [4]. The Turing Test, proposed by Alan Turing in 1950, attempts to discern when a machine can be considered intelligent, and is based on a simple premise: a judge is in a room and begins to ask to a human and a computer; both located in different rooms, and if the judge is unable to distinguish with certainty what the answer machine and who the person, then you can consider that the machine is intelligent [5, 6].

The body language is the most basic form of human communication, having been studied by psychology and sociology through expressions that can reveal various feelings that words fail to say. So while maintaining a conversation we can see that our partners perform various movements while talking [7]. The language of our body is the most spontaneous expression of our thoughts. Through our gestures, gaze, posture, movements, we can show our moods or emotional states such as happiness, sadness, anger, etc. [8].

The Kinect sensor bar contains two cameras, a special infrared light source, and four microphones. It also contains a stack of signal processing hardware that is able to make sense of all the data that the cameras, infrared light, and microphones can generate. By combining the output from these sensors, a program can track and

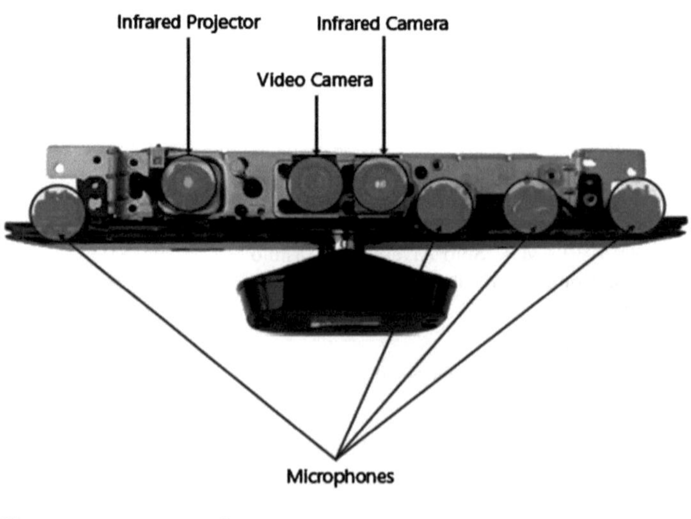

Fig. 1 A Kinect sensor unwrapped

recognize objects in front of it, determine the direction of sound signals, and isolate them from background noise. Figure 1 shows a Kinect with the cover removed. You can see the two cameras in the middle and the special light source on the left. The four microphones are arranged along the bottom of the sensor bar. Together, these devices provide the "view" the Kinect has of the world in front of it.

2 Detection and Tracking Positions with Skeletal

Skeletal means skeleton and is based on an algorithm to identify body parts of people within the field of view of the Kinect sensor and perform body tracking. This feature allows us to identify gestures and /or positions to be taken to detect emotions. See Fig. 2. To achieve joints where points (Joints) obtained are used [9].

Fig. 2 Some of the postures or poses detected by the Kinect sensor expressing emotion

Fig. 3 Skeleton of emotional expression

From “SkeletonFrame” it is possible to capture images. Once the skeleton is obtained, it must obtain the positions of the body parts. The number can be called JointID (joints), which can be identified as spots (small circles). See Fig. 3. Using the “position” property will get the position relative to the sensor. With the positions of the different body parts can use Markov processes [10].

3 Statistical Surveys

Dividing a surface either which generates Table 1. If we pass any objects on this surface, there will be different passing frequencies in the cells. Matrices frequency and probability are built considering the positions and changes from one period to another among them, generating a matrix of 5 × 5.

In this matrix we find that the cell in row A with the B column indicates how many times the speck (green dot in Fig. 2) is B since been in A, the cell located at the line D and the D column indicates how many times the continuous ball in D since it was in D previously. The transition probability matrix probabilities handled as shown in Table 2.

The above matrix is called the transition matrix of a step and gives the probability that the system moves from one state to another in a step that is in a transition. It can also be used to specify the probability that a system moves from one state to another in any number of steps or transitions, if the odds of a step do not change over time. By the following example you can see how Bayes networks relate and

Table 1 Frequency matrix

	A	B	C	D	E
A	408	62	0	0	0
B	61	406	68	0	0
C	0	63	402	60	0
D	0	0	59	301	55
E	0	0	0	55	256

Table 2 Transition probability matrix of a step or first order

	A	B	C	D	E
A	0.8680	0.1319	0	0	0
B	0.1140	0.7588	0.1271	0	0
C	0	0.12	0.7657	0.1142	0
D	0	0	0.1421	0.7253	0.1325
E	0	0	0	0.1612	0.8387

Fig. 4 Movement b to c

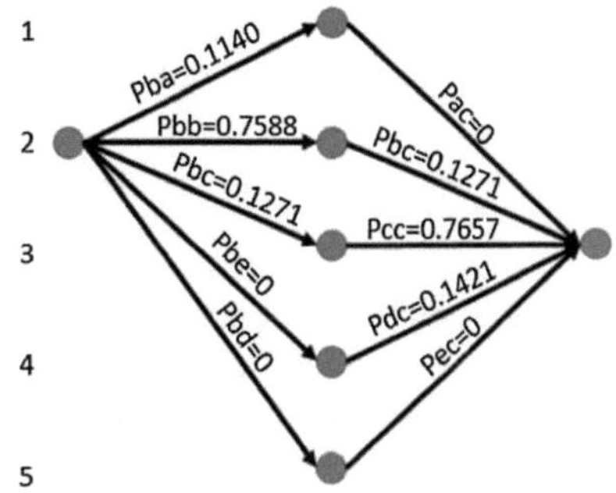

answer the question: If the system is in state B, what is the probability that the state is in C after two stages? In Fig. 4 the calculations are shown.

The number of each arrow represents the chance to take that particular step; for example, in a system of arrows going from B to B and then from B to C. The probability of the first step is 0.7588 and the second step is 0.1277. The two probabilities that the system travels that route are then multiplied. The probability of any possible routes may be calculated in the same way.

$$\begin{aligned} P^2_{B,C} &= (0.1140 \times 0) + (0.7588 \times 0.1271) + (0.1271 \times 0.7657) + (0 \times 0.1421) \\ &\quad + (0 \times 0) \\ &= 0.1937 \end{aligned}$$

Generally speaking, if $P_{i,\,j}$ is the probability that the system goes from state i to state j in one step and $P_{2i,\,k}$ is the probability that the system goes from i to k in two steps, then:

$$P^2_{i,k} = \sum_{todo\ j} (P_{i,j})(P_{j,k}) \tag{1}$$

Equation 1 can be used to calculate P^2i, k for all combinations of i and k. Values can then be presented in matrix form as shown in Table 3 and will have a transition matrix constructed in two steps [10].

Table 3 Likelihood of a matrix markov chain of second order

	A	B	C	D	E
A	0.7685	0.2146	0.0168	0.0000	0.7685
B	0.1855	0.6061	0.1938	0.0145	0.1855
C	0.0137	0.1829	0.6178	0.1703	0.0137
D	0.0000	0.0171	0.2119	0.5636	0.0000
E	0.0000	0.0000	0.0229	0.2521	0.0000

The process shown in Eq. (1) can be used to calculate the transition matrix of three steps and in general:

$$P_{ij}^{(n)} = \sum_{r \in E} (P_{ir}^{(k)})(P_{rj}^{(n-k)}) \quad (2)$$

By calculating the transition probabilities of two, three or n steps, a list of all the routes are obtained. It can give you a simple way to get the transition probabilities through matrix multiplications.

4 Robot Puppet Type

The system is used with the principal goal to control a humanoid robot type. The Fig. 5 shows a block diagram "Puppet" system. The infrared image is captured by the Kinect using digital image processing and recognition of patterns based on stochastic processes. The information obtained is sent to the servo motors to position the puppet according to captured with the Kinect sensor. For the construction of the puppet were used micro servo Power HD and HS311 standard to control movement of the robot from 0° to 90° and 0° to 180° Each of the engines corresponds to an input of the Arduino board which controls movement.

4.1 *Kinect*

As already mentioned the kinect sensor is used in the project to detect different body parts and thus obtain its position. To set and get data from the Kinect is used Processing program. Within this program you need to install the Seller Simple OpenNI containing image processing functions to use the data provided to us by the kinect, such as gesture recognition functions, hands Fig. 6.

Fig. 5 Titere

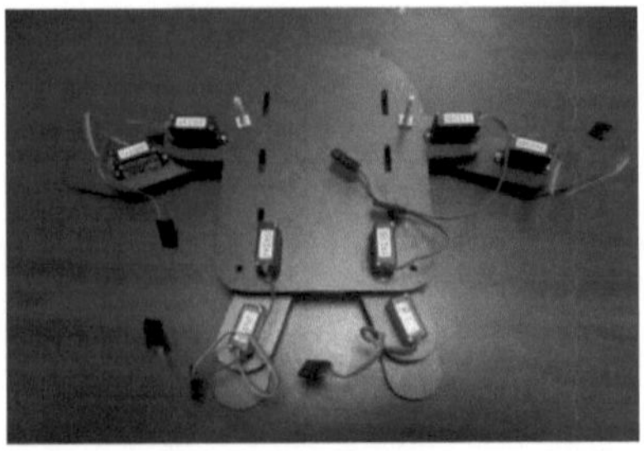

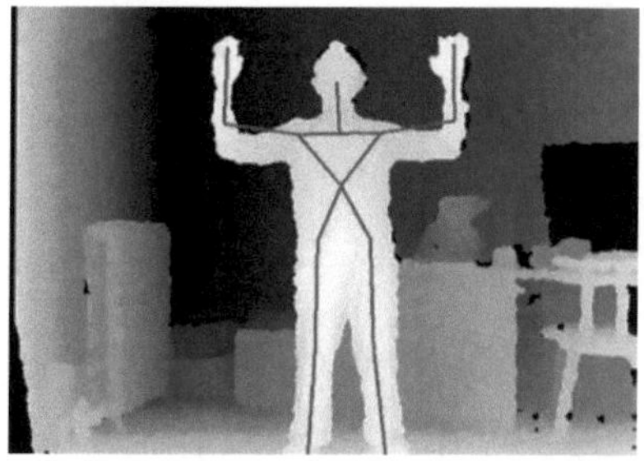

Fig. 6 Detection skeleton

4.2 Space and Transformation

Each data set is defined within geometric coordinate plane or space. Depth and Video space are measured in pixels in the X and Y position starting from the lower left position of the plane. Z corresponding to the depth dimension and is measured in millimeters. The X axis extends to the right while Y it does down. The X axis is a range from −2.2 to 2.2 (7.22″) for a total expansion of 4.2 meters; while the Y axis is from −1.6 to 1.6 (5.25″); the Z-axis ranges from 0 to 4 (13.12433″).

The functions detect the movements of a person standing in front of the camera. The individual must remain as shown in Fig. 7 for a calibration. After calibrating a position sensing of a limb to the other, so you can calculate the angle at each joint limb blocks; later matches the value with a percentage of time Pulse with Modulation (PWM) which activates.

Depending on the movements of the robot, the movements of the motors are generated. See Fig. 6.

4.3 Arduino

The control module Arduino is an open source programming tool that allows us to perform several tasks, among which are the sensor readings, output activation, reading inputs, use of timers, serial communication and modules PWM. These last

Fig. 7 Robot positioning

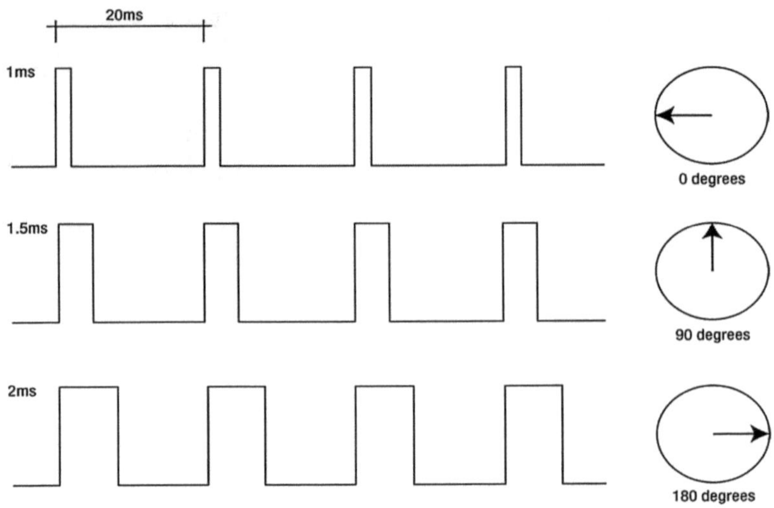

Fig. 8 Positioning servo

two tools were used in the project, because communication with the computer is performed by the serial protocol, and for positioning the servo was used a PWM.

The actuators are positioned according to the pulse being received, as shown in Fig. 8. Testing the state "Neutral" with the user ID 1,160 cm tall, the following results were obtained.

The program welcomed being able to observe the captured image of the state, which is the session 1.

Kinematics must have the engine, which it is say, how the robot moves and positions must acquire initially defined, so that the movements of the human body generate appropriate movements of actuators and consequently reproducing the motion of a human being is achieved previously trained.

Figure 9 shows a skeleton displacement angles generating angles-time PWM servomotors.

5 Results

It tests with the state "Neutral" with the user ID 1 of 160 cm tall. The program welcomed being able to observe the captured image of the state, which is the session 1.

It results by clicking on the button; a new window appears with the results of the user with ID 1, Session 1. We can see in the Fig. 10, the matrix shows the points where each body part is. Compare with the state entered Neutral, by clicking the button with the image of an eye, the matrix is shown with points more frequently. In the button with an arrow, it shows the equivalence of the state, in this case a 59.230 %.

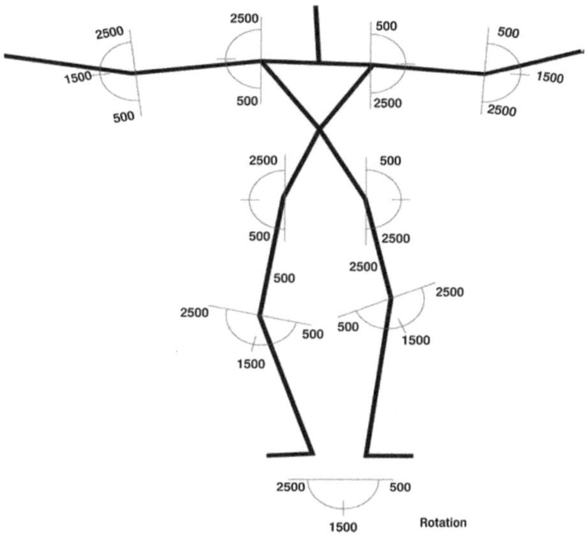

Fig. 9 Positioning servo

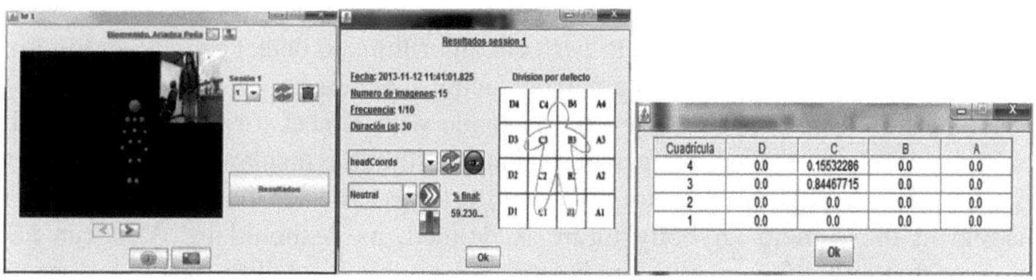

Fig. 10 Result of session 1

Fig. 11 Result of session 1

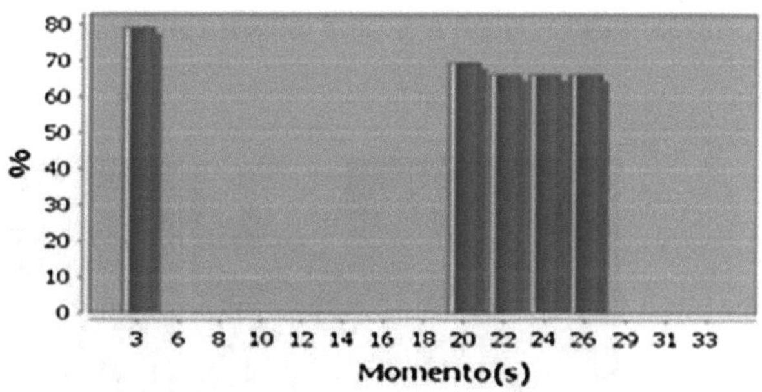

The compatibility level is shown in the graph of Fig. 11.

The tests performed to calculate the error are shown in the graph of Fig. 12, where the accuracy of the data provided by the Kinect is observed. It is observed that the Kinect operates correctly between 60 cm to 5 m.

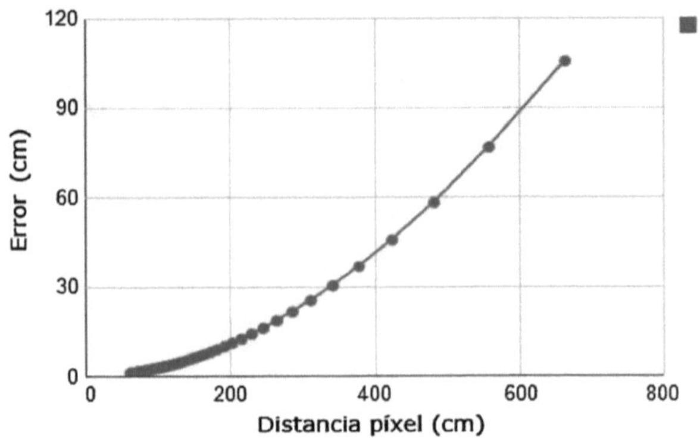

Fig. 12 Error in function of distance

For Tracking skeletonization and the ideal distance is between 2 and 2.5 m. Within this range there were no problems for detecting and tracking movements.

The human body works 24 h a day for that reason the puppet was subjected to stress tests continuously working. Only one gear of a servomotor is damaged, and as to the tracking movements generating no problem was observed.

Another critical point is that we analyzed audit performance under different lighting environments and the light intensity did not affect.

On the other hand is required to perform algorithms to detect a single individual, as confused when more than one person facing the camera.

In the Fig. 13 we can see the capture made with Kinect for a 180 cm tall. The only requirement to detect the person is that is within the display area of the sensor. Any problem at different detector positions; you can even see that when there are no people in the camera anybody figure is defined, as it should be. You can also observe that only the human figure is captured, so this is totally isolated. It does not detect tables, chairs, etc.

Figure 14 shows a woman in different positions. We must start from a position of always initialization. With this it should be clear that is compared the transition probabilities obtained from the movement of the green flecks. Even the person has

Fig. 13 Examples capture various positions

Fig. 14 Testing different scenario and height

partially left the field of view of the camera; it can still be detected, but is not considered as part of the recognition of affective states.

Once the figure detected the likelihood that a figure is related to a specific table of probabilities of transition from one affective state is detected. These probability tables were determined by Markov processes. Ergo, there is a probability table for the affective state of distress, sadness, happiness, fear, surprise, disgust and a neutron state; which means that there is no emotion, but that different emotions are nullified. Thus Table 3 and the graph of Fig. 5 are obtained.

6 Conclusions

A tool like Kinect greatly facilitates the development of projects since it has two tools that make a difference; one is having an infrared projector with a camera in the same property and the other a powerful programming tool with features such as tracking and skeletonization.

The Arduino feature allows easy adaptation to Kinect with a simple programming environment and a high level language. Due to the control with low error rates decrease which is feasible using data correction techniques. Currently remote control is real so remote interactions will soon appear art, and many branches of engineering and science opening for virtual, augmented and mixed reality.

The use of the Kinect sensor as a detector of affective states is highly efficient. Skeletonization detector and tracker as expressions of human body lead to a simple tool. The tables of transition probabilities obtained by Markov processes, allow us to have a benchmark to determine the emotional state of the person.

Although affective expressions through the body may vary depending on the region and customs; can be determined for each individual their custom tables; but they even ca be compared with some other generic.

The known expressions were implemented for the robot player played. By interacting with users, they lose track of whether you are in front of a machine, similar to the proof of Turin effect; since they take it as if you were really alive.

Acknowledgments The Local Committee wants to acknowledge the support received from IPN (Instituto Politécnico Nacional).

References

1. Origen y Desarrollo de la Robótica (2013, Julio 26). http://proton.ucting.udg.mx/materias/robotica/r166/r63/r63.htm
2. Desarrollo de robots basados en el comportamiento, Sergi Bermejo Sánchez Univ. Politéc. de Catalunya, 31/12/2004, 236 p
3. Inteligencia artificial: Modelos, técnicas y áreas de aplicación Francisco Escolano Editorial Paraninfo, 2003, 370 p
4. Desarrollo de robots basados en el comportamiento, Sergi Bermejo Sánchez Univ. Politéc. de Catalunya, 31/12/2004, 236 p
5. Avances en robótica y visión por computador José Andrés Somolinos Sánchez, Univ de Castilla La Mancha, 05/07/2002, 285 p
6. F. Prieto Pérez (2004) Niveles de Percepción, Vol 2, pp 226–235, Ars Medica. Revista de Humanidades, España
7. Dautenhahn K, Billard A (1999) Bringing up robots or the psychology of socially intelligent robots: from theory to implementation. In: Proceedings of the autonomous agents, 1999
8. García Sánchez I, Pérez Ordaz R, Calvo Lluch A (2013) Expresion corporal. Una práctica de intervención que permite encontrar un lenguaje propio mediante el estudio y la profundización del empleo del cuerpo, no. 23, pp 19–22, España
9. Kranmer J, Burrus N, Echtler F, Herrera D (2012) Hacking the kinect, apress. Springer, USA
10. Cappe O, Loulines E, Ryden T (2005) Inference in Hidden Markov models. Springer, USA

Author Index

M. Ceccarelli and E.E. Hernández Martinez (eds.), *Multibody Mechatronic Systems*,
Mechanisms and Machine Science 25, DOI 10.1007/978-3-319-09858-6

MIX
Papier aus verantwortungsvollen Quellen
Paper from responsible sources
FSC® C105338

If you have any concerns about our products,
you can contact us on
ProductSafety@springernature.com

In case Publisher is established outside the EU,
the EU authorized representative is:
Springer Nature Customer Service Center GmbH
Europaplatz 3, 69115 Heidelberg, Germany

Printed by Libri Plureos GmbH
in Hamburg, Germany